Design and Equipment for Restaurants and Foodservice

A Management View

FOURTH EDITION

Chris Thomas
Edwin J. Norman
Costas Katsigris

WILEY

Library of Congress Cataloging-in-Publication Data

Chris Thomas

Design and equipment for restaurants and foodservice: a management view / Chris Thomas, Edwin J. Norman, Costas Katsigris. — Fourth edition.

pages cm

Includes index.

ISBN 978-1-118-29774-2 (hardback)

1. Food service management. 2. Food service—Equipment and supplies. 3. Restaurants—Design and construction. I. Thomas, Chris, 1956- II. Edwin J. Norman III. Katsigris, Costas.

TX911.3.M27K395 2013

647.95068—dc23

2013019342

Printed in the United States of America

10 9 8 7 6 5 4 3 2 1

Contents

7

DESIGN AND ENVIRONMENT   180

8

SAFETY AND SANITATION   214

9

BUYING AND INSTALLING FOODSERVICE EQUIPMENT   243

10

STORAGE EQUIPMENT: DRY AND REFRIGERATED   283

11

PREPARATION EQUIPMENT: RANGES AND OVENS   313

12

PREPARATION EQUIPMENT: FRYERS AND FRY STATIONS   347

13

PREPARATION EQUIPMENT: BROILERS, GRIDDLES, AND TILTING BRAISING PANS   372

Preface

This *Fourth Edition* of *Design and Equipment for Restaurants and Foodservice: A Management View* brings exciting changes to the look and content of the book. Many are the result of adding a new co-author, Edwin "Ed" Norman, FCSI, CSI, to our team; he amplifies the expertise provided in past editions by former restaurateur and longtime college foodservice instructor, Costas "Gus" Katsigris.

One key change you'll note right way—based on consistent response from users of previous editions—is that this *Fourth Edition* has been pared down with regard to ancillary information. For example, information on items like table linens and glassware, and how to negotiate a building lease, has been removed. Instead, we have chosen to focus specifically on the two topics promised in the title—**design** and **equipment**.

Few foodservice projects can be tackled on their own. So, the new focus of the book is on selecting and working with the team members you need to bring your project to successful completion. You will find a great amount of detail in the chapters about utilities and appliances, and it is all aimed at giving you a basic, working knowledge of these topics. As a result, you should be able to ask pertinent questions, make better decisions, and even troubleshoot as your foodservice operation takes shape.

ORGANIZATION AND NEW TO THIS EDITION

The book begins by examining these key questions in most new foodservice businesses: Do you need to hire a design firm? If so, exactly what can designers do for you, and how do you evaluate their services and prices?

In **Chapter 1: Getting Your Project Started**, you will find information about the types of consultants and designers that work on foodservice projects, along with guidelines for selecting the right one(s) for you and determining the fairness of their fees. We also cover business plan basics and how to do market research to fine-tune your concept before hiring anyone to design it. And we introduce the Request for Proposal and Scope of Work; these are the documents that outline responsibilities and deadlines for any design project.

In **Chapter 2: Principles of Kitchen Design**, we swing open the kitchen doors for an in-depth look at where to locate the heart of any foodservice business—and it isn't always at the back of the house. It may be advantageous to have the kitchen staff on display, working in full view of guests on customized equipment that is as attractive as it is useful. There is also a short section on the special space and equipment requirements of banquet kitchens.

Chapter 2 includes multiple new illustrations to augment explanations of flow patterns in kitchen design. The goal: Don't just try to fit equipment into a space; make the space work for you and for the equipment you need to produce a particular menu. Advice for doing this is key to this chapter. We also mention the USDA's Food Establishment Plan Review, a health-related look at your design. It might be a requirement of the local health department and should be seen as a chance to do things right before construction gets underway.

In this edition, we've grouped all the front-of-house areas into **Chapter 3: Front-of-House Atmosphere and Design**. From outdoor dining and entryways to restrooms and banquet spaces, here you will find solid advice for making this square footage work well for those types of foodservice business that include public spaces. Creating profitable multipurpose space and building flexibility into your space plans are among the topics. We have retained the discussion of bar design, with its unique space requirements.

We're just as thorough in **Chapter 4: Planning Back-of-House Support Areas**. Here you will find extensive information about planning a dish room. One of the points made succinctly in this chapter is that every task performed in an area affects the way it is designed. Good dish room design, for instance, requires you to think about how dishes are bused, scraped, stacked, washed, and stored. There is similar information about the space requirements and necessary equipment for receiving and storage areas. Throughout this chapter, the emphasis is on staff comfort and safety.

Chapter 5: Electricity and Energy Management and **Chapter 6: Gas, Steam, and Water** introduce the utilities used in foodservice: electricity, gas, water, and steam. We've kept all the basic information intact and sharpened the focus on managing your utility use instead of feeling as though you're at the mercy of it. This includes discussions and examples of renewable energy use and guidelines for green certification, plus a section on energy conservation. There is new information on the topics of bottled water and chloramines.

It's back to design in **Chapter 7: Design and Environment**, with a discussion of the critical environmental components of any building that contribute to its overall atmosphere and the comfort of guests and staff: lighting, color, sound, and air quality. We have added information about energy-efficient advances in lighting technology and heating, air conditioning, and ventilation (HVAC).

Chapter 8: Safety and Sanitation contains practical information about safety and sanitation. We have added a new section on environmentally friendly cleaning supplies as well as details about the benefits and challenges of composting foodservice waste.

In **Chapter 9: Buying and Installing Foodservice Equipment**, we have kept and updated all the basic information a person needs to purchase equipment: how to calculate the total cost of ownership, whether it's OK to buy used equipment, and how to write specifications for the purchase. We also provide plenty of details about the next logical steps: safe installation and training staff to use and maintain the equipment.

A new priority in many types of foodservice is using fresh, local ingredients. At the same time, all operators must pay close attention to food costs. **Chapter 10: Storage Equipment: Dry and Refrigerated** contains the information necessary to plan for receiving and storage of ingredients, ensuring you get exactly what you ordered, and storing it to minimize waste.

The back of the book—**Chapters 11 through 17**—is devoted to specific equipment types or categories. The text includes tidbits about new technology, updated and improved features, and updated photos.

Finally, we have added to the popular features located between some chapters with new examples of **Facility Designs** and three new **Conversations** with people who are design experts in their fields.

FEATURES

- An **Introduction** and **Learning Objectives** appear at the beginning of each chapter, to give students a preview of its focus.
- We've scoured the leading industry publications for interesting highlights, lists, and short features to augment the points made in the text, and these are in boxes located throughout the chapters. They're organized into five categories:
  - **The Dining Experience.** Information that relates directly to front-of-house and how guests perceive the concept and comfort level, the delivery of food and service, and so on.
  - **Budgeting and Planning.** The bottom-line financial aspects of running a business, from market research to taxes, to leasing equipment.
  - **In the Kitchen.** Design solutions for kitchen and other back-of-house areas: storage, dish rooms, receiving areas, and more.

- ■ **Building and Grounds.** Maintenance and engineering recommendations, as well as external concerns, such as landscaping and lighting parking lots.
  - ■ **Foodservice Equipment.** How to buy, install, and maintain specific pieces of equipment.
- ■ Chapters end with a short **Summary** of major points and **Study Questions**. The latter can serve as pretest reviews or discussion topics in the classroom and/or online. The **Glossary** at the end of the book has been updated wth all the new terminology introduced in this edition.

One of the things that we believe sets this book apart from other books on this topic is our passion for giving students real-world advice. That's why we include, between some of the chapters, interviews with professionals from all aspects of the foodservice world, from chefs and restaurant managers, to equipment design and repair experts. There are 9 **Facility Design** and **A Conversation With** features throughout the book. There are three new "**Conversations**" in this edition.

SUPPLEMENTAL OFFERINGS

A comprehensive online **Instructor's Manual** with **Test Bank** accompanies this book and is available to instructors to help them effectively manage their time and enhance student learning opportunities.

The **Test Bank** is specifically formatted for **Respondus**, an easy-to-use software program for creating and managing exams that can be printed to paper or published directly to Blackboard, WebCT, Desire2Learn, eCollege, ANGEL, and other eLearning systems. Instructors who adopt this book can download the test bank for free.

A password-protected Wiley Instructor Book Companion website devoted entirely to this book (www.wiley.com/college/katsigris) provides access to the online **Instructor's Manual** and the text-specific teaching resources. The **Respondus Test Bank** and the **PowerPoint** lecture slides are also available on the website for download.

ACKNOWLEDGMENTS

Chris Thomas says: Ed Norman has been interested in foodservice management since college—in fact, it was his major at Iowa State University. Today, given his more than 40 years of not only designing foodservice facilities but also operating them, I could not have been paired with a more experienced coauthor. In addition, he is incredibly good-natured and knows how to meet a deadline! So, thank you, Ed—it's been great working with you.

Ed Norman says: I would like to thank Chris Thomas, my coauthor and a remarkable writer—for without her assistance, coaching, words of wisdom, and gentle push I would never have known how to get started on such a daunting task as a book-length project. It's a lot like designing a foodservice facility—with many parts and pieces that must be coordinated for a great end result. I would also like to thank my son, Eric Norman, for allowing me to move more work to his desk, thus giving me the valuable time to revise this book.

We would also like to thank new friends at Wiley. Mary Cassells is the person who reached out to Ed to ask about the possibility of his involvement in this revision. She kept us both on task and promptly answered questions as they arose. We very much appreciate her time and patience. Our thanks also to Gus Katsigris for pioneering this book for three editions and please know that we will take exceptional care of it going forward.

Special thanks to David Davis and Anne Alenskis at Idaho Power Company and Byron Defenbach at Intermountain Gas Company for reviewing and making utility-related recommendations, which helped update the highly technical Chapters 5 and 6.

From its first edition, this book has included many photos and illustrations. We have updated them as needed to reflect the latest design trends and equipment technology. Many of the folks listed here are Ed's good friends and leading experts in their facets of the industry.

Special thanks for streamlining the process of obtaining photos and permission to use them in this edition. Listed alphabetically by company, they include:

Todd Griffith, Bill Groleau, Julie Hang, and John Muldowney
Alto Shaam

Mark Klosterman
Apex Marketing

Andrew Padden and Wayne Sirmons
BSI, LLC

Mark Brenner and Beth McHale
Eagle Group

Jason Emanuel and Steve Follett
Follett Corp.

Andrea Brown and Lee Walter
Franke Foodservice Systems, Inc.

Robin Alfano and Dave Rolston
Hatco Corp.

Graham Keavney
Isabel Cruz Restaurants

Rob Geile and James Piliero
ITW/Traulsen

Selim Bassoul and Darcy Bretz
Middleby Companies

Kellie Wood
Unified Brands/Groen

Bill Dolan
Viking Range

Tamra Nelson and Jackie Van Zwol
Woodstone

In addition, we owe a debt of gratitude to the foodservice instructors who agreed to give us feedback that guided the improvements to this and previous editions of the book through their thoughtful reviews and survey responses. They include: Andrew O. Coggins, Pomplin College of Business and Virginia Polytechnic Institute and State University; Jeffrey Elsworth of Michigan State University; Ken Engle of The Art Institute of Washington; Pamela Graff of Minneapolis Community and Technical College; Nancy Graves of University of Houston; William Jarvie of Johnson & Wales University; Lisa Kennon of University of North Texas; Diane Miller of Liberty University; Stephani Robson of Cornell University; Donald Rose of Oklahoma State University; Kathleen Schiffman of St. Louis Community College; Andrew Tolbert of Texas Tech University; Carol Wohlleben of Baltimore International College; and Jessie Yearwood of El Centro Community College.

We know how valuable your time is, and we appreciate your willingness to help.

Now, read on for everything you ever wanted to know—and a few things you probably never expected to learn—about foodservice design and equipment.

Chris Thomas
Edwin J. Norman, FCSI, CSI

1 GETTING YOUR PROJECT STARTED

In this book, you will learn about designing and equipping restaurants and other food facilities. The many types of foodservice and facilities differ greatly in their design and equipment needs, but many basic elements apply to all of them. No matter which type of facility you have in mind, you will soon realize the incredible amount of planning and the intensive thought that must go into good facility design as well as how the equipment you select contributes to a successful operation.

Our goal is to provide you with this basic knowledge. By reading the book, you will gain valuable insight into designing and equipping your facility.

After reading this chapter, you will be able to:

■ Define preliminary planning steps for foodservice businesses.

■ Create a business plan and feasibility studies.

■ Calculate prime cost estimates.

■ Identify types of designers and consultants, and understand how to work with them.

■ Determine the scope of work and fees for your project.

■ Explain professional and ethical considerations.

If you feel confident that you can design your kitchen and equip it properly, then you may be able to spare yourself the cost of hiring a design consultant. However, most people become frustrated with the design process because it is quite complex. You must provide a lot of information to an architect or contractor and be able to fully answer all the many questions that arise.

For a workable foodservice design project, it is also necessary that you understand the prevailing building codes, life safety codes, health code requirements—both for the building design and the equipment—and the Americans with Disabilities Act (ADA), and that you have at least a general working knowledge of building systems. Less experienced individuals will have difficulty fully understanding the process and the information required to execute a new or remodel foodservice project. The solution is to hire consultants to tell them what they don't know and guide them through the process. A good consultant makes it a priority to stay on top of code requirements, best practices, and technology—and brings to the attention of their clients issues they are unaware of or unfamiliar with. The goal of the consultant is to deliver the best possible outcome for the client—you—and the project.

1-1 INITIAL PLANNING AND PROJECT CONSIDERATIONS

Before you hire professionals of any kind, it makes sense to know exactly what you're hiring them for—and the better you can describe your wants, wishes, and requirements, the better they will be able to act on them. So, it's important to define your project as best you can before selecting your project team. If it's a restaurant, consider the components in (Illustration 1-1).

What else does this entail? First, you must create a basic business plan that describes your concept and explains the necessary enhancements to the new or remodeled space that will make the *concept* work effectively and efficiently. A preliminary menu is part of this plan, outlining the types of food and beverages you intend to serve to customers. A realistic projected budget—with a forecast of all start-up costs, operating income, and expenses—is another important part of the plan. This is sometimes called a *financial feasibility study*.

Regardless of whether you are embarking on your own project or a project for a company, putting the concept and business plan on paper is necessary. These vital documents are critical in obtaining bank financing or the financial assistance of investors. If you are a franchisee, or employed by a company, much of this preliminary information may already be provided to you—but some of it will be uniquely yours.

Your business plan also serves as a basis for discussion with many of the individuals you will need on your team (see Illustration 1-2). It is not only a means for you to convey your thoughts on the project and the outcomes you desire but also a launching pad for dialogue with each person who reviews it. Professionals will each have their own questions to help them grasp your vision and help you fine-tune it. The answers will enable them to identify ways in which they can help direct you to successful completion.

Having a written plan also helps you stay true to your vision in the face of the many twists and turns your project might take as it progresses. At the construction phase, things don't always go as scheduled—weather causes delays, inspectors don't show up when they are supposed to, a key item is out of stock just when it's needed, and so on.

If you have a clear view of the big picture—of where and what your food business will be in the future—it has a much better chance for survival. You know where you are going and have thoroughly researched the potential advantages and pitfalls.

In determining the potential success of your concept, you need to find out if it will:

- Work in the location you have chosen.
- Generate sufficient sales to realize a profit.
- Have a certain amount of staying power no matter what the economy does.

This is why we suggest doing additional research to include with your business plan. Potential investors will most definitely want to see the proof that you have thought through these items thoroughly and put them in writing. The written document is your *feasibility study*, the research you have done to justify the implementation of your concept.

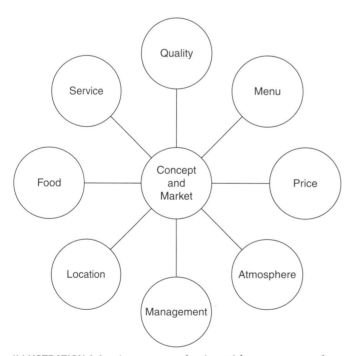

ILLUSTRATION 1-1 A restaurant begins with a concept and a potential market. All other components revolve around these two important considerations.

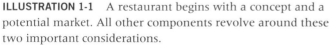

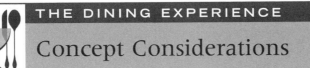

THE DINING EXPERIENCE

Concept Considerations

What are the variables in creating a concept? Here are the four big ones:

1. *Food.* What type of food will be offered? What is the style of preparation? How extensive will the menu be? If the food is to be sold to customers, what will the price range be?

2. *Service.* How will the food be made available to the customers: self-service? counter service? or full service, where the guest is seated and a waiter takes orders? In each of these situations, the overall aim is for all guests to feel reasonably well cared for by the employees who serve them.

3. *Design/Décor.* There are as many options here as there are food businesses. In general, however, the building's exterior should be inviting. Its interior should be comfortable and clean. The furnishings, lighting, and even the ambient noise level should reflect the style of eatery.

4. *Uniqueness.* In marketing, you'll hear the term *unique selling proposition*, or USP. A USP is like a signature. Everyone's is a little bit different, and the difference makes it special in some way.

In restaurants and employee cafeterias, for example, a good concept has USPs that enable it to attract and retain patrons. Eating establishments with the best USPs provide instantly noticeable differences that distinguish them from their competitors, the other places in the area where people could choose to dine.

There are two basic types of feasibility studies and, depending on the type of facility you are planning, you should do them both.

FINANCIAL FEASIBILITY STUDY

We've mentioned the financial feasibility study, which covers the money matters—income versus outgo—plus the costs of getting started.

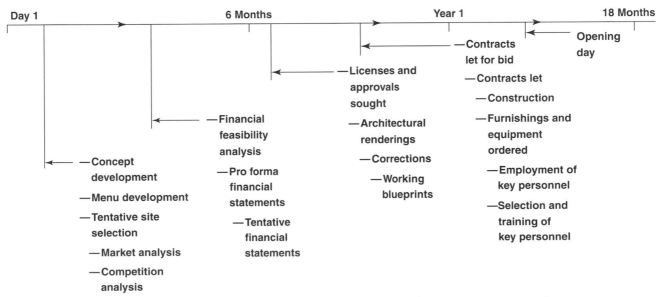

ILLUSTRATION 1-2 This timeline shows the major points in the foodservice planning process. It may be two years or more before the restaurant's doors open to the public.

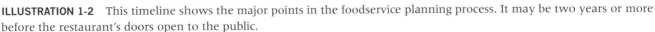
Source: John R. Walker, *The Restaurant: From Concept to Operation, 5th Edition.* This material is licensed by permission of John Wiley & Sons, Inc., 2011.

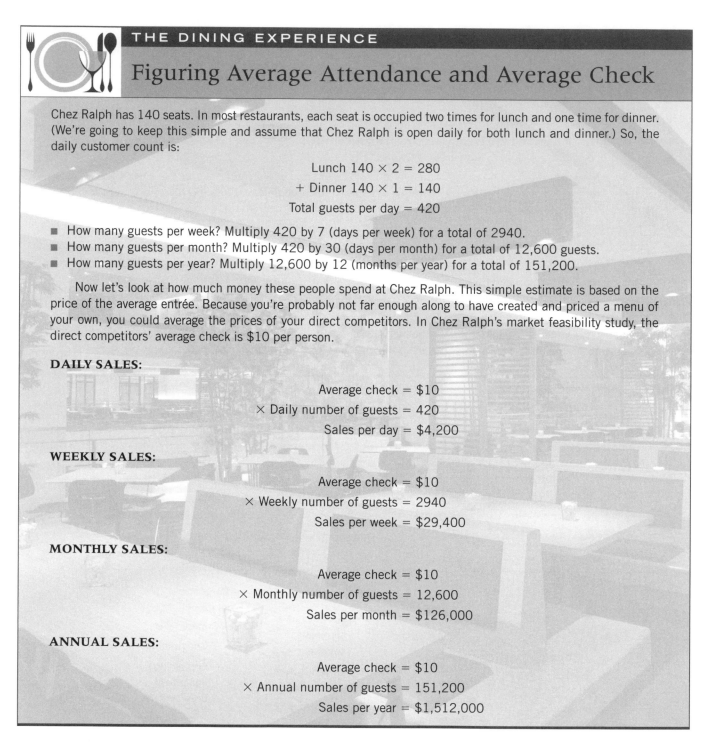

THE DINING EXPERIENCE

Figuring Average Attendance and Average Check

Chez Ralph has 140 seats. In most restaurants, each seat is occupied two times for lunch and one time for dinner. (We're going to keep this simple and assume that Chez Ralph is open daily for both lunch and dinner.) So, the daily customer count is:

Lunch 140 × 2 = 280

+ Dinner 140 × 1 = 140

Total guests per day = 420

- How many guests per week? Multiply 420 by 7 (days per week) for a total of 2940.
- How many guests per month? Multiply 420 by 30 (days per month) for a total of 12,600 guests.
- How many guests per year? Multiply 12,600 by 12 (months per year) for a total of 151,200.

Now let's look at how much money these people spend at Chez Ralph. This simple estimate is based on the price of the average entrée. Because you're probably not far enough along to have created and priced a menu of your own, you could average the prices of your direct competitors. In Chez Ralph's market feasibility study, the direct competitors' average check is $10 per person.

DAILY SALES:

Average check = $10

× Daily number of guests = 420

Sales per day = $4,200

WEEKLY SALES:

Average check = $10

× Weekly number of guests = 2940

Sales per week = $29,400

MONTHLY SALES:

Average check = $10

× Monthly number of guests = 12,600

Sales per month = $126,000

ANNUAL SALES:

Average check = $10

× Annual number of guests = 151,200

Sales per year = $1,512,000

Anyone who borrows money—or seeks to become a franchisee, for that matter—must possess two types of credibility: professional and financial. Professionally, the bank wants to see a résumé loaded with career highlights and pertinent experience. If you have had little professional experience in foodservice, be prepared to explain exactly how your previous jobs and other skills have prepared you for success in this venture. In terms of financial stability, you will be expected to submit two to three years of income tax returns and a full slate of current personal debts and savings. If there are blemishes on your credit report, be prepared to explain them.

The financial portion of your business plan must include several key calculations.

PROJECTED SALES

If you will be selling your food, the first requirement is to project sales levels for each day, each week, and the entire year. However, if you haven't even opened the doors yet, how in the world do you calculate *that*? You must first determine:

1. The number of guests you plan to serve each day
2. The average amount that each guest will spend (known as the **average check** or *check average*)

Just remember, you can't make accurate projections unless you have realistic numbers to start with. Also, in putting your final plan together, it is always smart to slightly underestimate sales. After all, every seat will probably not be filled for every meal, every day. Leave yourself a buffer.

PROJECTED EXPENSES

The three major (and often, for business owners, the most depressing) costs are food, beverage, and labor, which collectively are known as **prime costs**.

The National Restaurant Association (NRA) conducts extensive annual studies to compile the expenses and percentages for many different types of restaurants. Because the organization uses national averages, the figures are only estimates. Taking a look at them, however, will be helpful in making your own calculations.

Let's say, after studying the most recent NRA Restaurant Industry Operations Report, you decided your food costs will total no more than 30% of your business's income and beverages will cost 10% to 12% of its income. (Decide how much food people buy versus how much beverage—let's estimate about 85% food and 15% beverage.) We'll add labor costs to that in a moment. But, first, using the numbers we have, let's calculate the daily food and beverage costs based on these figures. (See Calculating Food and Beverage Costs on p.6.)

Determining these major expenses may seem a bit overwhelming at first, but it is really a simple, logical process based on your own estimates of how many people will walk into your eatery and how much money they will spend there.

LABOR COSTS. Every food-related business has labor costs, even if it is not open to the general public—and you cannot decide how many people to hire or what to pay them until you have calculated labor costs. For restaurants, the NRA has survey data that can help; other types of businesses have similar resources. Remember that labor costs are always higher the first few months of business, as it takes extra time to train the staff and you may hire more people than you need at first.

Labor costs are approximately 33% of the total sales of a restaurant, depending on the type of operation. For example, McDonald's reports labor costs as 25% of sales, but other businesses report higher percentages based on the services they provide to their customers. Another factor is employee benefits, which can swing from as little as 2% to 4% on average. It is important to include these measures in your business plan. Have an accountant or other professional help you determine a reasonable estimate of both labor and benefits costs based on your location.

If a restaurant owner estimates its labor costs will be 33% of total sales, using the calculations we've already made for daily sales, the numbers look like this:

■ Daily Labor Cost

$3,360 (daily total sales) × 33% (0.33) = $1,108.80

■ Monthly Labor Cost

$1,108.80 (daily labor cost) × 30 (days per month) = $33,264

■ Annual Labor Cost

$33,264 (monthly labor cost) × 12 (months per year) = $399,168

THE DINING EXPERIENCE

Calculating Food and Beverage Costs

FOOD SALES:

$3,360 (daily total sales) × 85% (0.85) = $2,856

FOOD COSTS:

$2,856 (daily food sales) × 30% (0.30) = $856.80

BEVERAGE SALES:

$3,360 (daily total sales) × 15% (0.15) = $504

BEVERAGE COSTS:

$504 (daily beverage sales) × 12% (0.12) = $60.48

Multiply these totals by 7 to obtain your weekly cost estimates:

FOOD:

$856.80 (daily cost) × 7 (days per week) = $5,997.60

BEVERAGE:

$60.48 (daily cost) × 7 (days per week) = $423.36

Multiply these totals by 30 to obtain your monthly cost estimates:

FOOD:

$856.80 (daily cost) × 30 (days per month) = $25,704

BEVERAGE:

$60.48 (daily cost) × 30 (days per month) = $1,814.40

Multiply the monthly totals by 12 to obtain your annual estimates:

FOOD:

$25,704 (monthly cost) × 12 (months per year) = $308,448

BEVERAGE:

$1,814.40 (monthly cost) × 12 (months per year) = $21,772.80

MARKET FEASIBILITY STUDY

If yours will be a competitive business open to the public, such as a restaurant, a *market feasibility study* defines the target customer, analyzes the competition, and looks closely at the trade area around the restaurant.

How you go about this research depends on whether you have a site in mind already, or whether you have a concept and are searching for the best place to locate it. Site-specific research focuses on data for the immediate and nearby neighborhoods; research to pick a location probably includes data for an entire city, to be narrowed down as your site choices narrow.

CUSTOMER RESEARCH. If you have decided on your concept, this is the time to define it so well that you can convince investors it is worth financing. The study's goal is to pinpoint the average, most frequent guest at your business. To do this, you need *demographics* on these folks: their age, gender, income per household, level of education, number of kids, ethnic group, religious affiliation, and so on. Categorize them by lifestyle and see how much you can find out about them.

Also consider the life cycle of the potential population. Singles marry and have children, or not. Traditional families may end up as two single-parent households after divorce. Empty-nesters eventually retire and become affluent, middle-, or low-income senior citizens, active or inactive. Gauging these life-cycle trends can help you fine-tune a concept that won't lose its appeal because its primary group of customers is dwindling.

This type of research can be time- and labor-intensive, and it may be easier to purchase data on some topics, so build in a modest budget for it. Most investors understand market research is an investment that will not pay off for many months. However, they also understand it is absolutely necessary.

Where do you find demographic information? Think about which government agencies in your city or county would want to know these same things, and ask if they'll share their statistics with you. You will find information not only about the customers themselves but also the area in which you want to locate.

The public library is another promising place to start. If nothing else, look for a reference book there called *The Insider's Guide to Demographic Know-How*, which lists more than 600 sources of demographic information and tells you how to analyze the data when you get it. Nielsen Site Reports is one place to purchase market reports by city, ZIP code, and more, further broken down by consumer spending, traffic volume, senior/generational information, and many other categories. The U.S. Census Department tracks statistics in about 20,000 categories and says marketers and researchers request only about 10% of them. You can find plenty of numbers to crunch.

SITE RESEARCH. Research into the trade area and location begins after site selection is well underway. Can the location support your concept? Here you evaluate the strength (or weakness) of the local economy. How much industrial, office, or retail development is going on? In what shape are nearby houses and apartments, and what is the vacancy rate? How much is property worth? What's the crime rate? How often do businesses and homes change hands? In short, learn about the overall stability of the area.

This part of the study should also include details about a specific site: its visibility from the street, public accessibility to the driveway or parking lot, availability of parking, city parking ordinances or restrictions, and proximity to bus or subway lines.

Don't forget to add details about proximity to a museum, park, hotel, sports facility, college, military base—anything that would serve as a regular crowd generator for you. With the potential location of your business as the center, the 5-mile radius around the site is your prime market for customers. You'll want to get to know this area almost as well as you know your own home.

Site selection information specific to the foodservice industry is available through the National Restaurant Association, including calculations known as the Restaurant Activity Index (RAI) and Restaurant Growth Index (RGI). The RAI is an indication of a population's willingness to spend money eating out. The RGI is a statistical prediction of cities in which a new restaurant has the best chance of succeeding—as-yet-undiscovered or underutilized markets. In both surveys, the number 100 is the benchmark; scores above 100 indicate better-than-average prospects, while scores below 100 indicate poorer prospects.

You might be dining out a lot to do the research for this section. Within this 5-mile radius you must find out, in great detail, what other types of businesses are your direct competitors. If you're planning to open a restaurant, zero in on any others that have concepts even remotely similar to yours. Take notes as you observe their seating capacity, menu offerings, prices, hours of operation, service style, uniforms, table sizes, décor—even the brand of dishes they use is valuable information. Will your concept stand up to their challenge? In your market feasibility study, you might classify the competition in one of two ways:

1. The existing direct competitors seem to have more business than they can handle, so there's room for you.

2. Even though they are direct competitors, they have distinct weaknesses (outdated décor, overpriced menu, limited parking) that give you viable reasons to enter the market. If their concept is poorly executed, test your own skills by figuring out why.

IN THE KITCHEN

How Much Space Is Necessary?

In most foodservice projects, the most expensive square footage is the kitchen and food-related spaces. The primary reason is that many mechanical, electrical, and plumbing connections are required. If you have been in commercial kitchens, stop and think about all the lights, equipment, and exhaust systems you have seen. Consider all the items that must be plugged in or wired to an electrical circuit, plumbed to a drain, connected to refrigeration condensers, supplied with hot and cold water, and/or located to ventilate the cooking equipment under the exhaust hoods.

Therefore, it is not unusual for the kitchen space to represent a total construction and equipment cost of $450 per square foot or more. By comparison, other common building uses—offices, mailrooms, break rooms, and so on—cost about $125–$150 per square foot.

As a result of the high cost of this space, it is vitally important that the kitchen be neither too big nor too small for the tasks that will take place there. This requires careful thought about the operation, its operational style, menus, and production requirements, as well as the number of meals to be served. The "correct" size for each department should be dictated by a good understanding of the processes that will occur in each area. In Chapters 2 and 3, you will learn more about the space requirements for particular areas.

Architects have a general idea of how much space a typical foodservice operation may require—but unless they specialize in foodservice facilities, their estimates are generally based on industry averages or past experience.

A qualified designer/consultant can help determine if your specific facility can function properly in the space available and can point out areas where special consideration for certain functions has not been addressed.

If you need assistance with any of these steps—developing a concept, doing market research, or preparing a business plan specifically for foodservice—you may wish to employ a professional Management Advisory Service (MAS) consultant early in the process. You will learn more about MAS consultants in this chapter.

1-2 FINDING A QUALIFIED DESIGNER OR CONSULTANT

When you have a plan and sufficient background information to share with a team, it's time to recruit the team! Like a good sports team, it will comprise a number of skilled individuals with different but complementary areas of expertise.

An architect may be necessary if you want a building designed and built from the ground up, or if demolition and reconstruction are among your options.

You will almost surely need an engineering team, which might be composed of civil, structural, mechanical, and electrical engineers. In some cases, even more specialized engineering may be required; an architect or engineer can explain those specific disciplines if necessary.

You may need an interior designer to bring your *theme* and atmosphere to life in the space. Many architectural firms have interior designers on staff—however, it is important to be sure they have the experience necessary for your type of project. Designers familiar with the needs of the hospitality industry will understand the importance of details like surfaces that are easy to sanitize and overall durability in a demanding environment. They also are aware of specialized needs for front-of-the-house and back-of-the-house design (many of which are detailed in Chapters 2 and 3).

Table 1-1 lists many skills and specialties provided by design and management consultants. As you can see, there is truly a wealth of knowledge to be tapped, depending on your needs and your budget.

The professional team contributes creativity, expertise, knowledge of industry trends, and experienced workers who are familiar with construction, utilities, and other considerations.

TABLE 1-1

Design and Management Consultant Services

Accounting and finance	Architectural design
Beverage system design	Business strategy
Capital budgeting	Code compliance
Compliance certification	Concept development
Contract management	Culinary development
Dietary and nutrition	Distribution and procurement
Due diligence	Energy and environment
Equipment surveys	Executive coaching
Facility assessments	Feasibility studies
Finance raising	Food production systems design
Food safety and hygiene	Franchising
Human resources	Imagineering
Interior design	IT systems, sourcing, and management
Kitchen design	Laundry design
LEED planning and design	Legal advice and litigation support
Management recruitment and development	Marketing and promotion
Master planning	Menu development and engineering
Operating procedures and systems	Operations review and reengineering
Operator RFP selection and monitoring	Quality management
Revenue generation	Space planning
Strategic financial analysis	Sustainability
Training	Waste management design
Workshops and education	Workstation ergonomics and design

These companies can save time and often get better deals on equipment, furnishings, fabrics, and other items from suppliers. However, the team's most important contribution may be objectivity. Don't be surprised if a consultant discourages you about some aspect of design or equipment you think you truly want—but that, in the consultant's experience, did not work well in other situations. After all, the consultant's job is to apply experience and expertise to look out for you and your best interests.

Because most projects are intensive and very much a team effort, your selection of a designer/consultant should be carefully thought out. The Internet is a good place to begin your search for professionals in your area who possess the qualifications you seek. The Foodservice Consultants Society International (www.fcsi.org) has a search engine to help you find consultants all over the world.

It is prudent to interview several candidates to find the one you feel has the best knowledge to handle your project. Here's a list of typical questions you might ask during an initial interview:

■ What is your foodservice operational and design background?

■ Have you worked in or had exposure to a foodservice facility of this type?

■ Describe your approach to a new project.

■ How do you ensure you'll be reasonably accurate in providing what is needed?

■ How will you make my facility function according to my needs?

■ How will you select and size the equipment?

■ Explain some ways you try to reduce labor and operational costs in the design process or with equipment selection.

■ How do you charge for your services? Is anything specifically not included in your fees? Anything that requires additional fees?

- Do you (personally or as a company) carry liability insurance for the projects you work on? What other types of insurance do you have that might be pertinent to this project?
- What are the payment terms for your services?
- What value do you deliver that your competitors may not?

The goal of this interview is to gain a comfort level. Do you feel at ease with the designer? Do you believe you can trust this individual or team with your project? Ask to see examples of their work and, of course, be sure to check their professional references.

Once you have selected a short list of designers, you can then start dialogue that will go into the finer aspects of your project. Negotiate fees as necessary.

1-3 TYPES OF DESIGNERS AND CONSULTANTS

The terms *designer, consultant,* and *design consultant* are used almost interchangeably. All refer to the professionals hired to help clients plan the layout and equipment necessary for a foodservice operation.

It is not our intention to sway your opinions about the type of designer or design firm you should partner with for your facility. Rather, our aim is to make you aware of some of the differences in how these designers typically operate, and how each approach may or may not work for you. Your own knowledge, experience, and opinions about how a good operation is designed will dictate the types of research you must do to select a qualified consultant.

In our experience, most of these individuals or firms fall into several basic categories. A brief description of each follows.

DESIGN CONSULTANTS

More people than ever before refer to themselves as "consultants" as they offer their services to the foodservice industry. The primary reason is that, as the industry matured, many former operations people opted to start or work for consulting practices. The term *consultant* has a professional ring to it. Today, there are companies that focus on design, others that focus on operational specialties, and some that offer a blend of both. You will save time up front by interviewing only those consultants that offer the specific services you need. However, it also helps to learn about the additional services they can provide that you might not have considered but that just might benefit you.

FCSI-AFFILIATED DESIGNERS

There is only one professional organization that represents true, fee-based consultants and designers in this facet of the design industry. The Foodservice Consultants Society International (FCSI) is a trade group with approximately 1400 members at various membership levels. Those members who have demonstrated sufficiently high-caliber work and background carry a "professional" consultant status within the society and are required to earn yearly education credits, just as engineers and architects are, to retain their professional status. The society also has a strict code of ethics to which its members adhere, which includes keeping the client's best interest first. Professional members are the only ones who can place the "FCSI" designation behind their name on business cards, documents, and the like. (Other members can state they are members but can't use the initials.) Among FSCI members worldwide are approximately 450 with professional status.

OTHER FEE-BASED DESIGNERS

Other fee-based consultants and designers do not belong to FCSI. They run private practices and elect not to affiliate with the organization. These consultants may do excellent work, but you must qualify them as part of your evaluation process. Anyone can call themselves a consultant—the proof is in the work product and the satisfaction of their clients. You will want to ask questions of any potential consultant until you feel comfortable with their knowledge base and view their previous work to see how they address certain elements of design. You may also want to ask them questions about new trends in design or equipment. That should help you determine whether they'll be able to deliver the "latest and greatest" outcomes you're looking for.

EQUIPMENT DEALER-DESIGNERS

Many foodservice equipment dealers offer design services at greatly reduced fees, as long as you purchase your heavy equipment from their dealership. This can be a disadvantage if the designer is selecting equipment based on the company's profit margin derived from sales of a particular brand rather than on the needs of your facility. The designer's overall experience and knowledge should also be factored into the equation, and you should ask the same questions of the dealer-designer that you would a fee-based consultant. Often, a dealer-designer is not able to attend as many industry seminars and training sessions as a fee-based consultant and therefore may have limited knowledge of new trends.

Caution is also warranted because you are generally required to purchase equipment as part of your agreement. It's smart to ask what the fee will be if for any reason you decide not to buy your equipment, or to buy only some of it, from this dealer. The designer's credentials should be carefully examined for past experience. An equipment dealer-designer may show examples of projects the dealership installed but that someone else designed.

INTERIOR DESIGNERS

An interior designer may be employed, either directly by you or as part of your architectural team. These professionals offer design ideas that may include furniture and accessories, finish selections, draperies, and so on. Some interior designers call themselves restaurant designers, although they may not offer kitchen design services or may work in partnership with a kitchen designer. In short, it is important to qualify the services each designer can offer and then select those that fit your needs.

If an interior designer is part of the project team, it is especially important to schedule regular coordination meetings that include all team members—interior designer, food facilities designer, management advisory consultant (a job title to be explained momentarily), and architect—so everyone is clear about what is expected for the look, feel, and theme or concept of the entire facility.

MANAGEMENT ADVISORY SERVICES (MAS) CONSULTANTS

MAS consultants, whether individuals or firms, offer design services plus a wider range of other services. For example, one MAS consultant may offer *concept development* services for restaurants; another may be more qualified to help with such financial matters as raising capital or operational profitability.

MAS consultants have the same ethical standards and commitment to continued education as FCSI's professional design members, and can be located on the FCSI website.

Again, the incredible variety of services you'll have to choose from makes it critical to identify exactly what you need and then ask the questions to determine whether a consultant can provide that type of expertise for you. A targeted approach on your part is essential to getting the results you desire.

1-4 WHO IS THE CONSULTANT'S CUSTOMER?

Once you have selected your MAS and/or design consultant(s), you must determine whether they will answer directly to you or to someone else, such as the architect or general contractor for the project. In many cases, the design consultant answers first to the architect and only then to the owner. This isn't necessarily bad, except you as the owner are paying the bills—and the architect can dictate specific parameters that affect the project cost and then hold the consultant to those parameters. A good architect serves as the conduit between consultant and owner.

However, you may want the design consultant answering directly to you because of the complexities of your operation. If that is the case, the coordination of consultant and architect then becomes your responsibility. If you add the services of an MAS consultant, you may wish to control the activities of both consultants so that all aspects of the project are coordinated by you, and you then share the information with the architect.

Here's one reason this clear chain of command is important: Much of the pre-design programming of the foodservice facility should take place before the architect blocks out the square footage. In many cases, this does not happen, and the consultant must then justify the need for more or less square footage. This can add costs if the entire process is not well orchestrated before the more detailed design work begins. If you are dealing with a new concept that requires additional square footage beyond the norms, the architect needs to know the concept is not typical and should therefore not be based on industry averages or previous experience. If you think about restaurants such as those of the Rainforest Café chain, you will see that certain design elements require additional space within the facility. A careful evaluation of how the concept might affect total project cost is important in advance of the design process.

By clearly determining who answers to whom, and doing so very early in the process, you will help head off problems during design and construction. All team members should know the protocols you have established and who makes the final call regarding specific elements in the design process.

1-5 ESTABLISHING A SCOPE OF WORK AND FAIR FEES

Once you have selected the consultants you would like to work with, it is time to obtain a fee proposal for their work. Your *request for proposal (RFP)* must concisely spell out what the consultant will provide as a work product, with a basic timeline and due date. Typically, projects are due in stages, and these intervals and multiple deadlines must be spelled out.

To protect yourself from cost overruns, it also is necessary to spell out, in great detail, the services you want the consultant(s) to provide. This part of the RFP is called the ***Scope of Work***. Its goal is to keep your project moving on an acceptable timeline and get you what you need from the consultant(s). Be explicit about the number of meetings, number of site visits, specific documents, and work product the consultant must plan for and deliver in his/her proposal. Consultants base their fees on the information and specifics contained in the Scope of Work. Everything you discussed or thought about in previous meetings or discussions must be reduced to writing in this document and included with the RFP.

BUILDING AND GROUNDS

A Sample Scope of Work

A TYPICAL SCOPE OF WORK FOR A PROJECT MIGHT READ LIKE THIS:

Consultant shall prepare a preliminary program of the space and provide it to the owner on or before 1/10/14. Immediately following approval of the program from the owner, the consultant shall begin schematic design along with other members of the design team until such a time that a final design is agreed upon by all parties (no later than XX date).

Upon approval of the final design, the consultant shall provide all design drawings, MEP (mechanical, electrical, plumbing) schedules, equipment cut sheets, specifications, and other required documentation to the architects/engineers and owner as well as the final contract documents that will be used for bidding the project.

Upon substantial completion of construction, the consultant shall visit the facility and provide a final inspection of the facility and equipment and submit a punch list for any missing or unsettled issues to the owner. The consultant shall plan for, in their fee proposal, XX number of face-to-face meetings and XX number of site visits during the planning, design, and construction of the facility.

You can ask consultants, architects, engineers, and others to provide you with sample Scope of Work documents for their services, which you can then modify to meet your needs. The objective is for you to get all the services you want for fair and competitive fees.

The Scope of Work also protects the consultants by clearly outlining their responsibilities and deadlines. Here it is wise to acknowledge a phenomenon commonly known as *scope creep*. As the name suggests, scope creep is what happens when an owner expects more for his money than what was identified in the Scope of Work, even though all parties used this document to determine their fees. If you happen to have missed a small detail or item that should have been part of the Scope of Work, it likely won't be a problem. However, if you push for or insist on more work than the fee was based on, your consultant may demand a higher fee for this additional and heretofore unknown work. Typically, this will fall under a clause for Additional Services that may be referenced in the vendor's proposal, with an hourly rate to be applied for the extra work.

No one wants to leave your project hanging, but everyone wants and deserves to be paid for the services they provide. With this in mind, you must carefully plan and consider every element of your needs when defining the Scope of Work to avoid the added costs and unnecessary stress of scope creep once your project is underway.

When you have all of your fee proposals in hand, add them up to determine your costs. These are called *soft costs*, and other items such as permits and licenses will be added to them as you go through the process. *Hard costs* are the actual costs of construction, materials, labor, and equipment to build the project. You must address both hard and soft costs in your project's financial planning.

As you peruse these proposals, how do you know if a fee is fair and reasonable? The bids for certain facets of the work may be close to one another on cost, or there may be a significant spread between high and low cost. Generally speaking, a major difference between high and low bids is a signal that the Scope of Work was misunderstood or is not complete enough to provide the bidders with sufficient information. On the other hand, if the bid amounts are fairly close, the Scope of Work was most likely conveyed in a manner that allowed all bidders to understand the specific requirements of the project. There are other possibilities, of course, but those are good starting points for comparison.

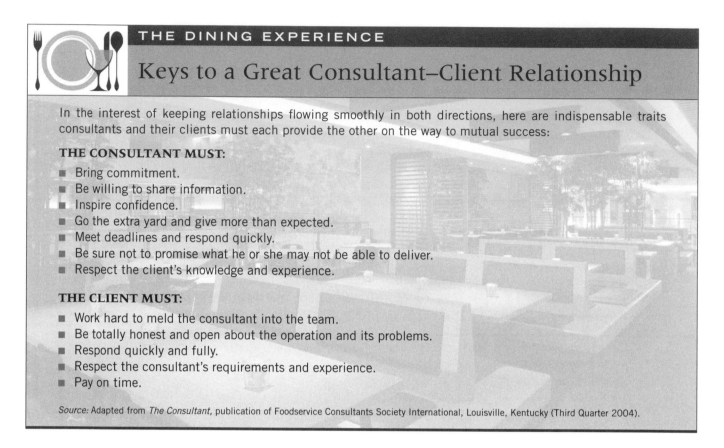

THE DINING EXPERIENCE

Keys to a Great Consultant–Client Relationship

In the interest of keeping relationships flowing smoothly in both directions, here are indispensable traits consultants and their clients must each provide the other on the way to mutual success:

THE CONSULTANT MUST:

■ Bring commitment.
■ Be willing to share information.
■ Inspire confidence.
■ Go the extra yard and give more than expected.
■ Meet deadlines and respond quickly.
■ Be sure not to promise what he or she may not be able to deliver.
■ Respect the client's knowledge and experience.

THE CLIENT MUST:

■ Work hard to meld the consultant into the team.
■ Be totally honest and open about the operation and its problems.
■ Respond quickly and fully.
■ Respect the consultant's requirements and experience.
■ Pay on time.

Source: Adapted from *The Consultant*, publication of Foodservice Consultants Society International, Louisville, Kentucky (Third Quarter 2004).

A true comparison of consulting fees must be correlated to the quality and experience of the bidder. A seasoned consultant or firm may demand higher fees than one less experienced or knowledgeable. As long as you can understand the possible differences in outcomes, you can move in the direction with which you will be most comfortable. For example, let's say the high bidder is a knowledgeable and experienced consultant who really brings a lot to the table. Is it worth paying the higher fee because you believe you'll have a better overall comfort level with this person? If the less experienced designer is selected, can you get by with a little less help or lower quality? In many cases, your choice will depend on the size of the project. A small to midsize project will almost surely not require the same degree of experience as a large, complicated project.

There is no single rule for determining a fair consulting cost—except that, as your consultant's client, you want the best for the lowest cost. If you are spending $1.5 million to build a restaurant, would it seem proper to spend only $5,000 for the kitchen design—which is, in fact, the heart of the operation? Would a $30,000 fee at 2% of the construction cost be reasonable if you knew the outcome would substantially enhance the total operation? The objective is to keep your soft costs in control while building a team capable of delivering the outcomes necessary for a well-designed and properly equipped food facility.

Most consultants provide a fee rationale for your review, and they can help you more fully realize their value if you question the amount of their fee. If you have your heart set on a particular consultant over the others you have shortlisted, then by all means discuss your concerns about the fee with him or her prior to making a final decision. You may find he was confused about the Scope of Work, or she incorporated items that were discussed previously but omitted from the Scope of Work. Most often, consultants consider your best interests and want to be fair just as much as you do. They will base their fees on the services they feel you need.

1-6 BUSINESS PROFESSIONALISM AND ETHICS

One consideration during the entire process of building your project team is the ethical and professional attributes of each person or firm. You want people who are going to play by the rules and treat you with fairness and respect throughout the process—even when, as often happens, situations get tense mid-project. If you get the nagging feeling that an individual is somehow questionable, seems difficult to get along with, or doesn't present himself professionally, it is generally best to bypass him. You want and need people you can trust with your money and your reputation. No one wants to suffer monetary or operational losses due to unethical or unprofessional acts or behavior.

Your consultant(s) and designer(s) can provide you with invaluable information on how to lay out and equip your operation. Their experience and knowledge can save you a lot of money over the life of your restaurant or other foodservice facility. It is your job to help them understand what you perceive as important and what your budget will allow. It is their job to give you the best and most for your money. Together you can achieve great outcomes, within budget, and ensure yourself the greatest chance for business success.

The relationship between you and your consultant shouldn't stop once the work is done and the doors are open for business. Use this valuable professional relationship to fine-tune your operation once you have had the chance to work in your new facility.

Reach out to the consultant with issues that still need attention. You may find that refinements and redefining of space may be beneficial, especially if you are presented with outcomes different than those you planned for. A high-quality consultant/designer can identify the problems and determine appropriate fixes. After all, she is still on your team—and if you've built a solid relationship based on trust, fairness, and mutual respect, she has every reason to want you to succeed.

■ SUMMARY

If you're reading this, you've made it through Chapter 1—and now surely realize what a complex undertaking it is to start or revitalize a foodservice business. The chapter begins early in the process with the steps required to put a preliminary concept and budget on paper, both to attract financing and to explain what you want prospective designers and consultants to do to make your vision a reality.

Steps are detailed for creating a concept and a business plan. Basic calculations are provided for a financial feasibility study, as are research tips and techniques for a market feasibility study. The latter should include customer and competitor research as well as site research.

In terms of financial data, you have now learned how to estimate customer counts (average attendance), how much customers spend on a meal (average check), and, based on those figures and some industry percentages, how to calculate food, beverage, and labor costs.

You are now also aware of the various categories of designers and consultants who work in the restaurant industry and the types of skills they can provide for foodservice projects. Sample questions are included for the initial interviews with these professionals.

The chapter contains brief summaries of what is contained in a Request for Proposal and a Scope of Work, two written documents necessary for getting bids for consulting work, as well as a discussion of "who's the boss of whom?" You may want everyone on your team to report to you, or you might be more comfortable having team members report to a general contractor or the architect. The importance of a clear chain of command cannot be emphasized strongly enough.

A discussion of fees offers guidelines for determining what's fair in terms of consulting costs. The chapter ends with basic tenets of ethics and professional behavior to smooth the way for a positive and productive consultant–client relationship.

■ STUDY QUESTIONS

1. Select an existing foodservice business in your area and describe what the site research would include for a *market feasibility study*.

2. Do online research to locate three design consultants or consulting firms. From the information you find, pick the one you think would be the best fit for a foodservice project. Explain your reasoning.

3. Add two more questions to the list on pages 9–10 that you would ask prospective designers and consultants before hiring them.

4. What is the most common cause of scope creep?

5. Do you think your attorney should be part of the project team? Why or why not?

2

PRINCIPLES OF KITCHEN DESIGN

■ INTRODUCTION AND LEARNING OBJECTIVES

The kitchen is the heart of any foodservice business. Like a human heart, its job is to pump and circulate life, in the form of food, through the rest of the operation. Therefore, kitchen placement affects the quality of the food, the number of people who can dine at any particular time of day, the roles and workload of the kitchen employees and servers, utility costs, and even the atmosphere of the front-of-house space. Remember, each of these elements also figures in the overall profitability of the business.

In this chapter, we explore what constitutes good kitchen design. If you pose the question to 10 foodservice professionals, you could get 10 distinctly different answers. However, if you dig a little deeper, you will realize that the design of any commercial kitchen must be based on the menu and the style of cooking that will be employed to produce the variety of products to be offered. The fact that the menu, in almost every case, is the driving force behind the design actually makes it easier to identify the specific needs of the workers as well as their specific workstation and production needs.

Design refers to overall space planning; it defines the size, shape, style, and decoration of space and equipment in the kitchen. The *layout* is the detailed arrangement of the kitchen floor and counter space—where each piece of equipment will be located and where each workstation will be. A *work center* is an area in which workers perform a specific task, such as tossing salads or garnishing plates. When several work centers are grouped together by the nature of the work being done, the whole area is referred to as a *work section*: cooking section, baking section, and so on.

After reading this chapter, you will be able to:

■ Describe the trends in modern kitchen design.

■ Explain how to budget for the kitchen you want.

■ Identify where to put your kitchen within your facility.

■ Explain how to create flow patterns that make the service system and work centers run smoothly.

■ Describe the food safety considerations when designing a kitchen.

■ Describe the guidelines for equipment placement.

■ Describe the unique design needs of service areas and each part of the kitchen.

A poorly designed kitchen makes food preparation and service more difficult than it should be, and it can even undermine staff morale. So, if a new foodservice operator has little cash to spend on professional designers, that cash is probably best spent planning the location and design of the kitchen—the one area where equipment, ventilation, plumbing, and general construction costs combine for a major investment.

Saying that the menu is the primary driver of the design makes it seem simple, but kitchen design is often a difficult and complex task. Once you know the menu and the style of production, the next task is formidable: the skillful laying out of the workstations and work centers within the space you have. And to create this layout, you must know the types and sizes of equipment you will need.

Most buildings, whether new or existing, have their own unique sets of constraints that must be dealt with. Columns, beams, doorways, and other construction-related details may leave you with interesting and challenging obstacles. The architects and builders who originally worked on the space don't always provide for what a foodservice operation requires in terms of adjacencies and clearances. Even the most carefully planned new construction project can yield challenges that must be dealt with during construction.

The natural tendency in kitchen design is to try to fit the equipment into the space available rather than make the space work to fit the specific needs of the operation. If you can be involved in the building design, then it is in your best interest to spend the time necessary to work with the architect and contractor.

If you will occupy space in an already constructed building, it is important to size up the space, looking specifically for obstacles that could cause problems. On the preliminary drawings of a space, home in on its possible shortcomings: insufficient electrical capacity, not enough room for adequate waste disposal, and the like. Instead of living with these problems, why not tell the architects what you require and where you want it? If it is going to be *your space*, you must be aggressive about whether, and how, it can be made to work for you. A professional food facilities designer will be able to quickly point out the obstacles and requirements to the design team, or to you personally, if you are considering an existing building.

It's smart to design the maximum amount of flexibility into any foodservice setting. Here are types of flexibility to consider: multiple uses for equipment, and how that may affect the design of the work sections; mobility of the equipment within the kitchen space; operational flexibility; and labor flexibility. Shortsighted decisions early in the planning process impede flexibility later.

An important early consideration may be optimistic but is necessary nonetheless—and that is the eventual need for expansion. In addition to your own square footage, look at the space next door. Would it suit you if you needed to use it? Would it be expensive to modify? Would expansion be possible into that space without too much disruption of business? Even the locations of your building's exterior walls are important. Expanding into an existing parking lot is possible. If your building is adjacent to a garden area, you might be able to design an expansion that works with it to create outdoor dining space. But if a street is on the other side of the wall, expansion is simply not possible in that direction.

Owners and managers of restaurants and other foodservice venues are constantly bombarded with new ideas, concepts, and plans for using their kitchen space wisely by reducing operating costs or increasing productivity. Appliances, gadgets, and space-savers of all sorts may be tempting, but if you don't have a kitchen that is well planned in the first place, you may be throwing money away by purchasing them.

The three basic kitchen-related costs (and, in parentheses, the ways to reduce them) are:

1. Labor (increased productivity)
2. Utilities (increased energy efficiency)
3. Food (menu flexibility and planning)

What does kitchen design have to do with these costs? Careful space planning gives the business owner or manager the best possible environment and tools with which to control these three critical cost areas.

2-1 TRENDS IN KITCHEN DESIGN

Industry professionals say that now, more than ever, kitchen design is driven by consumer demands and economic factors. In many cases, we are seeing smaller and more efficient kitchens. Experts see this trend as the result of three things: a shortage of qualified labor; an ever-increasing battle for space in general for business uses; and budget constraints, including the demand for an increased return on investment.

At the same time, however, demand for more fresh and healthy food is rising. You may have heard the phrase "farm to table" used to describe this recent movement, which encourages people to limit their exposure to processed foods and the many additives that are part of our food products. This trend changes the dynamics of the kitchen, as it generally requires more pre-preparation of fresh product.

For many years, designers have scaled back the footprints of commercial kitchens, ostensibly saving money by saving space. Now, we must carefully address how to intelligently add back some of what was taken out—a shift that, in our view, is more likely a movement than a trend. That is, it may be here to stay for some time, rather than just a year or two.

Here are a few more examples of what is trending in today's commercial kitchen universe.

DISPLAY KITCHENS

A *display kitchen* is one where much of the food preparation is done in full view of customers. Watching a busy kitchen staff at work is interesting to most of us. It whets the appetite and gives guests the feeling they are being catered to with a meal that is freshly prepared as they look on. For today's more sophisticated diner, perceptions of quality, freshness, and presentation are just as important as how the food tastes. A well-functioning display kitchen also accentuates the sense of showmanship that certainly is part of the culinary arts. It enhances the total dining experience by being part of the atmosphere and the evening's entertainment value. It presents opportunities for the culinary staff to interact with guests. Of course, this may influence the type of person you hire as a staff member. Not everyone is good at, or comfortable with, conveying such a public image. But for chefs who enjoy the limelight, something magical happens when they can see the patrons, and vice versa.

In a display kitchen—and for that matter, anywhere food is prepared in full view of guests—it is important that designers create work centers that are well organized and can be kept looking clean. Take advantage of undercounter storage space to minimize messiness. Provide a hand sink for handwashing and a utility sink to rinse kitchen tools and utensils right at the work center. Select surfaces that are easy to clean, from wall paints to countertop materials. Think carefully about where to dispose of waste. Ventilation should receive professional attention to minimize the grease buildup that naturally comes with cooking and to avoid the possibility of smoke wafting into the dining room.

One nice design detail is to install half-walls to create what is sometimes called a *semi-open kitchen*. The staff can be seen from the waist up, preparing food, but the inevitably messy and unsightly aspects of cooking—soiled pots and pans, stacks of plates, the dirty floor, and the like—cannot. Nothing should go on in a display area that indicates any type of volume cooking. Instead, the emphasis is on individually prepared dishes.

Food preparation in view of the guests addresses another modern-day concern: food safety. Most guests believe that when food is prepared in full view, the staff is more conscious of safe food-handling practices than they would be closeted away at a prep station in the back of the house.

Today, as the cost of foodservice space keeps climbing, there is some financial urgency behind such a multitask environment. The modern operator must maximize profit per square foot of space and risks failure by using space extravagantly and having to pay the higher costs of heating, cooling, and insuring it. Combining at least part of the kitchen with the dining area is one way to conserve space. Yet the display kitchen is generally more expensive: up to $550 per square foot or more, compared to $250–$350 for a more prudent back-of-house commercial kitchen. When it is in public view, everything from equipment to walls to preparation surfaces has to look good.

A final word about display or semi-open kitchens: They should be considered as an option only when the menu and food preparation techniques actually lend themselves to display—pizza dough being twirled overhead, steaks being flame-broiled over an open grill, sushi receiving attention to intricate detail, or the sauté station producing a variety of dishes.

APPLIANCES ON DISPLAY. In terms of equipment, we've noticed massive, brick wood-burning ovens (or gas-fired counterparts) as display kitchen staples. They're not exactly portable, weighing up to 3000 pounds, but they're attractive, energy-efficient, and quite functional—they can turn out a pizza in three to five minutes.

Induction rangetops have found their way into display kitchens because they are sleek-looking, easy to clean, speedy, and energy-efficient. Induction cooking works by creating an electromagnetic field that causes the molecules (in this case, of a pan) to move so rapidly that the pan—not the rangetop—heats up, in turn cooking the food inside. In other words, the magnetic field prompts other magnetic items (i.e., metal cookware) to heat, while its ceramic surface stays cool to the touch. Not every metal pan is suited to induction cooking, but specific, multiple metal pans are made for this purpose. Cleanup is as simple as wiping off the cooktop surface; there are no spills seeping into burners and no baked-on messes. A 2.5-kilowatt induction burner puts out the equivalent cooking power of a 20,000 BTU burner on a typical sauté range.

Yet another display kitchen requirement is the rotisserie oven or grill. We usually think of whole chickens browning perfectly in a glass-fronted rotisserie cabinet, but there are now attachments that allow you to bake pasta, casseroles, fish, vegetables, and more. From countertop units no more than 30 inches wide to floor models 6 feet in width, rotisseries may be purchased as gas-, electric-, or wood-fired. Ease of cleaning should be a consideration when choosing a rotisserie unit that will be placed in view of patrons.

Finally, the *cooking suite* is a real boon to today's hardworking chef on display. A cooking suite (or *cooking island*) is a freestanding, custom-built unit into which just about any combination of kitchen equipment can be installed. Instead of a battery of heavy-duty appliances against a wall, a cooking suite allows workers to man both sides of the island (see Illustration 2-1). This arrangement concentrates cooking activities, improving communication because appliances and personnel are centrally located. Often a cooking suite requires less floor space than a conventional hot line, with shorter electrical and plumbing connection lines. Menus that include mostly sautéed, grilled, or charbroiled items would be well served by a cooking suite.

THE MARCHÉ KITCHEN

You are in a *Marché kitchen* (pronounced mar-SHAY) if you can walk up to a standalone counter, place an order, and get fresh food, cooked to order, as you wait. This display-style retail concept with European origins is different from a display kitchen in that the diners stand and watch the action instead of being seated and waited on. Its novelty and excitement come from the combination of freshness and the commercial equipment moved from back of the house to right up front. It's almost as if the wall between dining room and kitchen has been peeled away and you're practically standing inside a clean, attractive, and well-lit kitchen full of griddles, fryers, broilers, refrigerators, steam-jacketed kettles, and all the trappings of big-deal cooking. Food is prepped, cooked, and served in one place, which saves steps for the workers and doesn't require the heavy hauling of ingredients from kitchen to serving line.

Most Marché kitchens are designed with some flair—blonde wood, tiled pedestals and warmer trays, faux finishes on ventilation hoods and equipment—to make them more attractive. In some, customers pay for each course separately as they receive it on the serving line.

THE SMALL, TECHNOLOGICALLY SUPERIOR KITCHEN. Another idea taking hold today is simply a small kitchen with carefully orchestrated work centers, all designed with ergonomics and efficient labor utilization in mind. It is created to use the fewest possible number of people to perform

ILLUSTRATION 2-1   A cooking suite or cooking island.

Courtesy of Viking Commercial, Fullerton, California.

each task and outfitted with the highest-quality equipment to minimize downtime. Greater use of technology (touch screens, programmable appliances) improves customer service and saves time. Commonsense touches—trash receptacles built into the counters, compact storage for individual areas to minimize steps wasted in fetching supplies—mean the crew is able to work smarter, not harder. Plug-and-use (portable) equipment combinations (steamer/ovens; refrigerated space beneath rangetops) that are easy to service, with surfaces that are easy to clean, also are popular. Fast cooking equipment with small footprints has played an important role in getting the kitchen space to a minimum—but again, this is only workable when your space fits the tasks to be performed therein.

2-2   MAKING THE NUMBERS WORK

Deciding how much money to spend on design, construction, and equipment is a crucial first step—not just a total figure, but an estimate of how funds are to be allocated and how much time each step will take. First, does the design budget realistically match the concept in size and scope? The budget should reflect the type of market in which the finished restaurant will operate. There is a tendency to overspend at the front end of a project, so much so that the finished facility can't be profitable for many months while construction and design costs are being paid off. So early on, set your ego aside and, likewise, refuse to accept grandiose plans

IN THE KITCHEN

The Chef-Designed Kitchen

As professional chefs assume near-celebrity status, their input as part of the design team is indispensable. What you really have with a highly qualified chef is a person who understands food and the important details of a functional kitchen. Combining that know-how with the technical expertise of engineers or designers means a greater range of workable possibilities.

Such is the case with Tru, a Chicago restaurant that is a joint effort of chef Rick Tramonto, pastry chef Gail Gand, and Richard Melman, a longtime restaurateur and chairman of Lettuce Entertain You Enterprises. Their 200-square-foot kitchen produces 800 to 1000 meals nightly, with a staff of no more than 23 at peak periods (see Illustration 2-2).

A cooking suite solves the traffic problem, with a central cooking area surrounded by garde manger, fish and meat prep, and saucier stations along the walls. A special station handles the Tru house specialty, amuse bouches—bite-sized pre-dinner palate-teasers. A separate pastry room includes granite counters, mixers, convection ovens, and cold storage.

To ensure freshness, Tru accepts deliveries twice daily—with separate walk-in storage areas for meat, fish, and produce. The kitchen preps and vacuum-packages these items in single-serve portions for quick later use. Refrigeration beneath the cooking surfaces keeps food at optimum temperatures until it is needed.

ILLUSTRATION 2-2   Executive Chef Rick Tramonto and Executive Pastry Chef Gale Gand designed the kitchens at Tru Restaurant in Chicago.

Courtesy of Tim Turner Photography.

from consultants or designers who may have good ideas but don't seem to be thinking about your cost parameters. If you have done your due diligence in selecting a consultant/designer, that individual or company should understand your situation and your budget and should not present ideas you cannot afford.

In fact, the lack of early budgeting and planning is the undoing of many a promising concept. It's true that it is difficult to decide months in advance what to spend on design, construction, décor, furniture, fixtures, equipment, china, flatware, glassware, table linens, smallwares, publicity, and other pre-opening expenses, but a good budget should include all of these and more, plus a contingency amount of about 20% to cover any cost overruns. For renovations or **adaptive reuse** projects in older structures, the contingency amount may be even higher than 20%, as problems often crop up due to the age or condition of buildings, utilities, and fixtures as well as the special care and additional permits often required to refurbish them. It is also more difficult to budget for a facility that is part of a larger project—say, a hotel restaurant—than a freestanding restaurant, because often you are bound by the design decisions of others.

Underestimating your budget will not impress your banker or your investors. It is more likely to compromise your concept and misrepresent your ideas to the contractors you hire to work with you.

All your costs will vary widely and will depend on many factors in your market: economic conditions, location, the type of restaurant you want, and so on. The key to controlling them is planning. Let's take an equipment purchase as an example. Spend time with the salespeople or manufacturers' representatives. Pay attention during their equipment demonstrations. Don't order something that is more, or bigger, than you need. Check Chapter 9 of this text for guidelines on equipment specifications and what to look for in a warranty or guarantee.

Price new equipment before you look at used; that way you have an idea of what is truly a good deal. Look for equipment that can perform more than one function: A combi oven, for instance, can hold, bake, or steam. If you order it custom-made, give the manufacturer enough time to do it right. Then, well before it arrives, be sure the preliminary utility connections (called **rough-ins**) are in the correct places. Check widths of doors and the intended space where the equipment will be installed to be certain they are wide enough. These are all small steps, and many of them don't specifically have to do with budgeting. But they all add up to one huge factor: making a smart investment you can live with. A simple mandate when buying used equipment is "buyer beware." Find out as much as you can about the equipment, and ask for a guarantee or warranty if at all possible.

In creating a budget, it is reasonable to expect cost estimates up front from the designer, architect, or consultant you hire after she has received enough information from you to prepare her own fee estimates. And remember, "costs" and "fees" are different animals.

One factor that has major cost implications is *time*. If work crews don't show up on schedule or materials you've ordered don't arrive promptly, the delays can become costly. Contractors may charge expensive overtime, and there'll be no time to dispute any details you may not be satisfied with. In some cases, you may have to delay your opening. So, along with your budget, you must develop a timeline to help you pinpoint when you have to spend certain amounts. If your concept is fully developed and you've already obtained the site for your foodservice facility, it is reasonable to assume:

- The design phase should take no more than 16 to 18 weeks. This includes the time to get necessary permits from authorities and the bidding process to hire contractors.
- The construction and/or remodel process should not take more than 16 to 24 weeks.

Change orders—requests to alter your requirements after the initial agreements with contractors—can increase costs. However, in some situations modifications are needed that were not anticipated. In this case, you should present the details (in writing) to the contractor. This document is called a **bulletin**, and it usually includes a drawing and/or specifications for the exact requirements.

BACK OF HOUSE

Electrical ($ per square foot of kitchen space) _____
Plumbing ($ per square foot of kitchen space) _____
Mechanical ($ per square foot of kitchen space) _____
Fire protection/sprinklers _____
Kitchen walls, ceilings (ceramic, tile, etc.) _____
Kitchen equipment _____
Kitchen smallware _____

SUBTOTAL _____

Note: To determine the total cost PER SQUARE FOOT, divide the total number of dollars on this part of the list by the number of square feet for the back of the house.

FRONT OF HOUSE

Millwork ($ per square foot) _____
Finishes (e.g., drywall, paint, tile, carpet, window
    treatments) ($ per square foot) _____
Specialties (e.g., door glazing) ($ per square foot) _____
Lighting (ceiling and wall fixtures, dimmers) _____
Sound (systems and treatment) _____
Furniture (tables, booths, chairs, stools) _____
Accessories (e.g., artwork) _____
Plateware _____
Glassware _____
Flatware _____
Linens _____

SUBTOTAL _____

Note: To determine the total cost PER SQUARE FOOT, divide the total number of dollars on this part of the list by the number of square feet for the front of the house.

FEES

Architect _____
Engineering firm _____
Mechanical contractor _____
Designer and/or consultant _____
Legal fees (e.g., permits, licenses) _____

SUBTOTAL _____

FEES

Façade treatment _____
Signage _____
Parking lot, curbs, sidewalk construction _____
Exterior lighting (e.g., parking lot) _____
Landscaping (design and installation) _____
Sprinkler system for landscaping _____

SUBTOTAL _____

**ADD ALL FOUR SUBTOTALS TO ARRIVE
AT YOUR TOTAL PRELIMINARY BUDGET.**

ILLUSTRATION 2-3 Budget Checklist.

Illustration 2-3 is a checklist for a budget to construct a theoretical restaurant. It is a freestanding 6900-square-foot eatery that will have 175 seats and 40 barstools.

2-3 BASIC KITCHEN DESIGN GUIDELINES

In Chapter 1, we discussed your options for getting assistance with kitchen design. Whether you rely on the advice of consultants or equipment manufacturers, heed the following basic guidelines:

- In addition to defining the concept itself, define the goals you expect to achieve with it. Decide on your menu before the design process begins.
- Be clear about how much money you intend to spend. Trust the professionals who pay attention to your budget, not the ones who consistently try to talk you into spending "just a little bit extra."
- Make a commitment to yourself to choose the most energy-efficient equipment in your price range.
- Consider future growth that may require additional space and utilities.
- Consider equipment and workspaces (tables, shelves) that are movable. Think about maximum flexibility.
- Separate the stages of food production so raw materials can be prepared well in advance.
- Look for equipment that has a good reputation and a proven performance record. As much as possible, each piece should be multifunctional, simple to operate, easy to maintain, and energy efficient.
- Control costs in places where customers won't notice and the kitchen crew won't be affected.

Some experts suggest the overall cost of your kitchen equipment should not exceed one-third of the total investment for the entire facility.

STAFF COMFORT AND SAFETY

Today's kitchen designers also consider the comfort and safety of the people who work in them. They realize the science known as human engineering will have a positive effect on workers' productivity and morale. Each of the next topics must be addressed to make the kitchen the healthy heart of the facility.

- *Sufficient space to perform the required tasks.* Elsewhere in this chapter we discuss the production and preparation areas in a commercial kitchen. Each of these requires a different amount and configuration of space. A baker shaping dinner rolls, for example, has different space needs than a waiter filling iced tea glasses. Space requirements are generally influenced by:
  - The number of persons working in a specific area.
  - The amount and types of equipment required in the work area.
  - The amount of storage required for immediately accessible supplies.
  - The types of products being produced in the area.
  - The clearance required for moving equipment, opening appliance doors, and so on.
- *Adequate aisle space.* If an aisle is not wide enough, employees will struggle to work comfortably in the space, and it may even be insufficient for compliance with the federal

Americans with Disabilities Act (ADA). If the aisle is too wide, employees add steps and fatigue to their workdays. In addition to foot traffic, rolling carts and equipment are always around. Where space is at a minimum, some narrow aisles can be declared one-way. Overall, however, kitchen aisles should be at least 36 inches wide—wider if they carry two-way traffic or mobile cart traffic.

■ *Intelligent design to minimize injury risks.* Most people have to reach, twist, lift, and bend to perform their jobs, but if these motions are repetitive or excessive, they become unnecessarily strenuous and can prompt injuries ranging from back problems to carpal tunnel syndrome. For instance, most culinary tasks require surface space. It must be the correct height for the task and located within easy reach of the employees who will use it. In Chapter 8, you will learn more about proper work-surface heights. Putting heavier equipment on rolling carts or ordering it with **casters** also prevents back injuries.

■ *Properly designed equipment, in good working condition.* Sharp knives, red-hot rangetops, and motorized equipment are part of any kitchen, but they need not invite disaster. Look for safety features as you purchase. On heated equipment, for instance, check the amount of insulation and order insulated handles and door-locking safety mechanisms.

■ *Comfortable temperatures and humidity control.* Many foodservice operators are concerned about making their guests comfortable—but what about the employees who spend entire workdays there? The ideal balance of fresh air, humidity, and air movement is a technical topic best left to ventilation experts. What we have noticed is that many commercial kitchens pay attention to grease control because it is part of their fire code requirements, but overall the kitchen space is not properly air-conditioned. Current design best practices suggest the ambient temperature in the kitchen should be no higher than 78°F to 82°F for maximum worker comfort. In Chapter 7, you will learn more about kitchen ventilation requirements and recommendations.

■ *Adequate lighting for the required tasks.* For kitchens, adequate lighting includes attention to glare and shadows as well as light levels. Fatigue sets in and errors multiply when lighting is insufficient. Good lighting also is necessary to monitor the sanitation of food, surfaces, and utensils. (See Illustration 2-4.) Workplace lighting is also part of the discussion in Chapter 7.

■ *Noise control and abatement.* Kitchens can be noisy places, from chefs barking orders to the whirring of appliances and the clatter of dish rooms. It is no longer sufficient to keep the din from the kitchen from spilling into the dining area. Chapter 7 offers ways to control noise as part of the space design.

We can also learn to plan wisely by thinking of the commercial kitchen as a manufacturing facility: With a combination of labor and raw materials, it turns out product. The unique aspect of foodservice is that the finished product is sold in an outlet attached to the factory.

Very often, in fact, serving food is a secondary operation or amenity of a corporation where the primary business is something other than foodservice—a school, a prison, a hospital, and so forth. Those who run the parent corporation must understand that operating a foodservice facility is a business in and of itself, and it operates differently than most other types of businesses because it deals with a highly perishable product. Health and life safety are of critical importance in foodservice, which of course affects the design of the facility.

As in any type of manufacturing plant, foodservice productivity is highest when the assembly lines and machinery are arranged in a logical, sequential order. In kitchens, this includes everything from tossing a salad to turning in orders so no guest is left waiting too long for a meal.

SERVICE SYSTEMS AND FLOW PATTERNS

A major issue that must be addressed before deciding on a kitchen design is the way in which food will be delivered to guests. This is known as the **service system**. A large operation, such as

ILLUSTRATION 2-4 A buffet installed with a breath guard for hygiene purposes.

Courtesy of BSI, LLC, Denver, Colorado.

a hotel, can have more than one service system at work simultaneously: elegant tableside service, room service, and casual bar service. At the other end of the spectrum, quick-service restaurants employ service systems that emphasize speed and convenience, including takeout service and the fast-food option of standing at the same counter to order, pay for, and wait for a meal served within minutes.

Each service system has subsystems; together, they encompass every aspect of the progression of food from kitchen to table and back to the dishwashing area. This progression is called *flow*, much like the traffic flow of a busy street grid. The two types of flow to consider when planning your kitchen design are product flow and traffic flow. ***Product flow*** is the movement of all food items from their arrival at the receiving area through the kitchen to the guests. ***Traffic flow*** is the movement of employees through the building as they go about their duties. The ideal, in both types of flow systems, is to minimize backtracking and crossovers—again, to make sure the "streets" don't get clogged.

Every foodservice operation has several basic ***flow patterns***.

The raw materials to create each dish have a back-to-front-to-back flow pattern. They arrive at the back of the operation, in the kitchen, where they are prepared. Then they travel to the front of the operation, to be served in the dining area. Finally, they return to the back again, as waste or, in the case of a buffet type of operation, as leftovers. The movements are so predictable they can be charted, as shown in the functional flow diagram of Illustration 2-5.

IN THE KITCHEN

Human Engineering Checklist

FLOORS

■ Adequate number of floor drains to keep floors dry

■ Carborundum chips in quarry tile in slippery or wet areas

■ Slip-resistant wax on vinyl floors

■ Ramps and handrails in receiving area, and storage space for carts and hand trucks

■ Floor mats for comfort of workers who must stand in one place for long periods

■ Kitchen floor level with walk-in refrigerator floor

■ Heavy slope of floor or floor troughs around steam-jacketed kettles to encourage quick drain-off of hot liquids to floor drains

■ Coved corners of floors where they meet the wall, for ease of cleaning

MATERIALS HANDLING

■ Hand trucks and carts for moving all foods

■ Strong, easy-to-clean shelving

■ Portable shelving

■ Ladders for reaching stored goods on high shelves

■ Carts for the movement of processed foods from production area to refrigeration and then to service area

UTENSIL HANDLING

■ Knife racks

■ Easy-to-clean utensil drawers with removable inserts

■ Utensil drawers at every workstation and table

■ Overhead utensil racks

FOOD PRODUCTION EQUIPMENT

■ Compliance with National Sanitation Foundation International (NSFI) standards

■ Compliance with ADA requirements

■ Portable equipment, if needed in more than one department

■ Portable bins for flour, sugar, and salt

■ Wall-hung or mounted on legs for ease of cleaning

■ Free of burrs, sharp edges, and difficult-to-reach areas

■ Safety equipment and guards on equipment, such as shields for mixing machine

■ Disposals in all production areas (if permitted by local codes)

■ Open rail-type undershelving that permits crumbs and small particles of food to fall to the floor

■ Marine edge on all tables with sinks (to prevent water from spilling on floor)

■ Adequate space for parking equipment from other departments (bread racks, raw ingredients from stores, etc.)

WAREWASHING EQUIPMENT

■ Pot storage racks beside potwashing station and in or near each work area

■ Storage containers for soiled linen

■ Box, glass, and metal can container in each major work area

■ Utensil sorting table

- Paper and bone container at dishwashing station
- Pre-rinse, power or hand
- Cleaning supply storage
- Hose reel
- Cart wash-down area

SERVICE AND DINING

- Condiments and support service equipment available near the point of service
- Convenient dish drop-off
- Easy-to-clean chairs without cracks that accumulate crumbs
- Minimum number of steps from food pickup to point of service

Source: John C. Birchfield, *Design and Layout of Foodservice Facilities,* 3rd ed. (Hoboken, NJ: John Wiley & Sons, Inc., 2008).

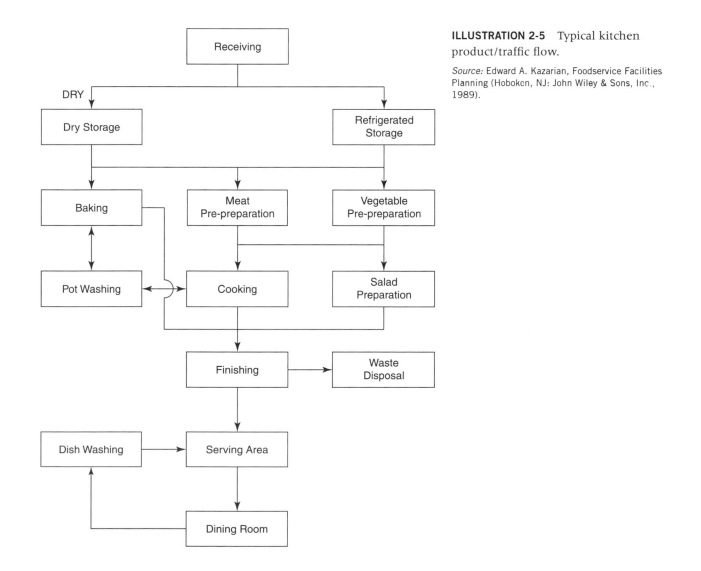

ILLUSTRATION 2-5 Typical kitchen product/traffic flow.

Source: Edward A. Kazarian, Foodservice Facilities Planning (Hoboken, NJ: John Wiley & Sons, Inc., 1989).

Another type of traffic pattern to consider is the flow of the service staff as waiters pick up food, deliver it to guests, and clear tables. On a busy night, the whole system really does resemble a freeway. As you might imagine, there is always the possibility of disaster if someone makes a wrong turn. The key to managing these three types of flow is that each should not interfere with the others.

Within the kitchen, each cooking section has its own distinct flow pattern. It could be a pattern of steps the chefs follow to put each dish together or the methodical way the dishwashers scrape, sort, and wash dishes and dispose of waste.

The service systems and flow patterns of your business should guide your kitchen design. An operation with huge numbers to feed in short periods will differ from one that also feeds large numbers but over a longer period. Can you see how?

The distance from the kitchen to the dining area is one important consideration, and kitchen designers have devised numerous strategies to cope with it. You may have noticed that, at some restaurants, the waiters are expected to do quite a few food-related tasks outside the kitchen, at waitstations closer to the guests. They might slice and serve bread, ladle soup, arrange and dress salads, or pour beverages. The idea is to speed service and preserve the (sometimes inadequate) kitchen space for actual cooking tasks.

Another critical decision to be made early in the design process: Should waiters come into the kitchen to pick up food, or should it be handed to them through a **pass window** between the kitchen and dining area? Although the pass window is considered informal, it can be used in a fancier restaurant, perhaps masked from public view by a wall or partition.

Each of these items—distance and kitchen access—helps determine your flow patterns. In a perfect world, flow patterns would all be straight lines that do not intersect. However, this ideal is rarely achieved. One simple rule is that the faster you want your service to be, the more important it is that your flow patterns do not cross. In a fast-service scenario, the flow lines must be short and straight. The next time you are standing at a fast-food counter, notice how few steps most of the workers have to take to pour your soft drink, pick up your burger, and bag your fries. Speed is the desired outcome.

The reverse is true in a fine-dining establishment, where the work may all be done in the kitchen in order to enhance the feeling of a leisurely dining experience. No clattering plates, no bustling waitstations here.

Now that we've looked at the flow of people as they perform their duties, let's follow the *food flow line*: the path of raw materials from the time they enter the building to the time they become leftovers.

The **receiving area** is where the food is unloaded from delivery trucks and brought into the building. Most restaurants locate their receiving areas close to the back door. The next stop is storage—dry storage, refrigerated storage, or freezer storage—where large quantities of food are held at the proper temperature until needed.

Food that emerges from storage goes to one of several preparation, or prep, areas for vegetables, meats, or salad items. Slicing and dicing take place here to prepare the food for its next stop: the production area. The size and function of prep areas varies widely, depending mostly on the style of service and type of kitchen.

When most people think of a commercial kitchen, what they imagine is the **production line**. Here the food is given its final form prior to serving; boiling, sautéing, frying, baking, broiling, and steaming are the major activities of this area. The food is plated and garnished before it heads out the door on a serving tray. And that's the end of the typical food flow line.

Several kitchen work centers are not included in the typical food flow sequence but are closely tied to it. For instance, storage areas should be in close proximity to the preparation area to minimize employees' walking back and forth. Some kitchens have a separate **ingredient room**, where everything needed for each recipe is organized, to be picked up or delivered to a specific workstation. Storage is much more useful when it is placed near the prep area rather than near the receiving area, saving steps for busy workers. The bakery is usually placed between the dry storage and cooking areas because mixers and ovens can be shared with the cooking area. A meat-cutting area is also essential. It should be close to both refrigerators and sinks for safety and sanitation reasons as well as for ease of cleanup. Remember, however, that some kitchens simply are not big enough to accommodate separate, specialized work centers. Kitchen space planning is a matter of juggling priorities, and it involves continuous compromise.

As you juggle your priorities, think about each task being done in each work center. How important is it to the overall mission of the kitchen? Might some duties be altered, rearranged, or eliminated to save time or space? Some of the ideas to discuss here are frequency of movement between pieces of equipment, the distance between pieces of equipment, temporary landing areas where raw materials or finished plates can sit until needed, wheeled equipment that can be rolled from one site to another, and parking space for equipment when it's not being used.

Simply stated, if work centers are adjacent without being cramped, you save time and energy—and if people who work in more than one area have handy, unobstructed paths between those areas, they can work more efficiently.

One work center that is often misplaced is the pot sink, which always seems to be relegated to the most obscure back corner of the kitchen. True, it isn't the most attractive area, but think of the many other work centers that depend on it. The typical kitchen generates an overflow of pots and pans. Why isn't the pot sink placed closer to the production line to deal with the mess?

And, speaking of pots, think carefully about where to store them. Both clean and dirty, they take up a lot of space and require creative storage solutions. Often pot/pan racks can hang directly above the sink area, giving dishwashers a handy place to store clean pots directly from the drainboard. (Remember that anything stored near the floor has to be at least 6 inches above it for health reasons.)

2-4 FLOW AND KITCHEN DESIGN

Let's take a look at some common flow plans for food preparation. The most basic, and most desirable, flow plan is the **straight line**, also called **assembly-line flow**. Materials move steadily from one process to another in a straight line (see Illustration 2-6). This type of design minimizes backtracking; it saves preparation time and confusion about what's going out of the kitchen and what's coming back in. The straight-line arrangement works well for small installations because it can be placed against a wall and adapted to the cooks' duties.

Where space is inadequate to arrange food preparation in a straight line, a popular and efficient choice is the **parallel flow**. There are four variations of the parallel design:

1. *Back to back.* Equipment is arranged in a long, central counter or island in two straight parallel lines (see Illustration 2-7). Sometimes a 4- or 5-foot room divider or low wall is placed between the two lines. It's primarily a safety precaution, meant to keep noise and clutter to a minimum and to prevent liquids spilled on one side from spreading onto the other. However, such a wall also makes cleaning and sanitation more difficult.

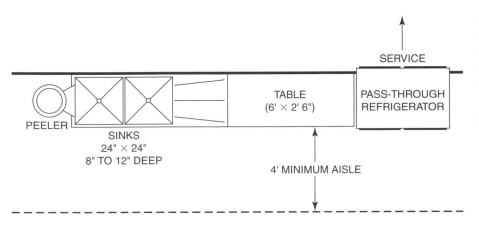

ILLUSTRATION 2-6 Straight-line arrangement. In a straight-line or assembly-line kitchen, food and materials are passed from one work center to another in a straight line.

Source: Robert A. Modlin, ed., *Commercial Kitchens*, 7th ed. (Arlington, VA: American Gas Association, 1989).

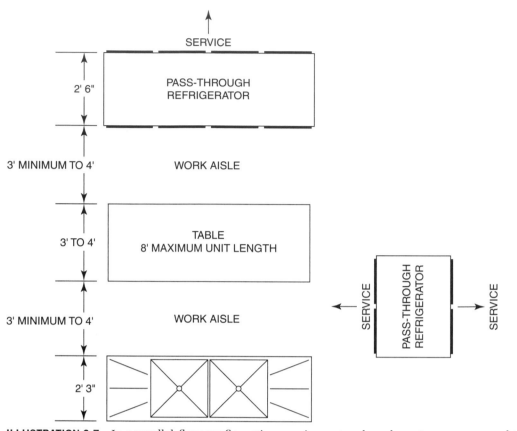

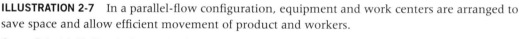

ILLUSTRATION 2-7 In a parallel-flow configuration, equipment and work centers are arranged to save space and allow efficient movement of product and workers.

Source: Robert A. Modlin, ed., *Commercial Kitchens*, 7th ed. (Arlington, VA: American Gas Association, 1989).

The back-to-back arrangement centralizes plumbing and utilities; you may not have to install as many drains, sinks, or outlets, as both sides of the counter can share them.

A back-to-back arrangement in which the pass window is parallel to (and behind one of) the production areas is sometimes known as a *California-style kitchen*.

When the pass window is located perpendicular to the production line, it may be referred to as a European-style kitchen design. The advantage of the European design is that each cook on the line can see the progression of multiple dishes that make up one table's order.

2. *Face to face.* In this kitchen configuration, a central aisle separates two straight lines of equipment on either side of the room. Sometimes the aisle is wide enough to add a straight line of worktables between the two rows of equipment. This setup works well for high-volume feeding facilities like schools and hospitals, but it does not take advantage of single-source utilities. Although it is a good layout for supervision of workers, it forces people to work with their backs to one another, in effect separating the cooking of the food from the rest of the distribution process. Therefore, it is probably not the best design for a restaurant.

3. *L-shape.* Where space is not sufficient for a straight-line or parallel arrangement, the L-shape kitchen design is well suited to access several groups of equipment and is adaptable for table-service restaurants (see Illustration 2-8). It allows the placement of more equipment in a smaller space. You'll often find an L-shaped design in dishwashing areas, with the dish machine placed at the center corner of the L.

4. *U-shape.* This arrangement is seldom used, but it is ideal for a small space with one or two employees, such as a salad preparation or pantry area. An island bar, such as the ones in T.G.I. Friday's restaurants, is another example of the U-shape at work.

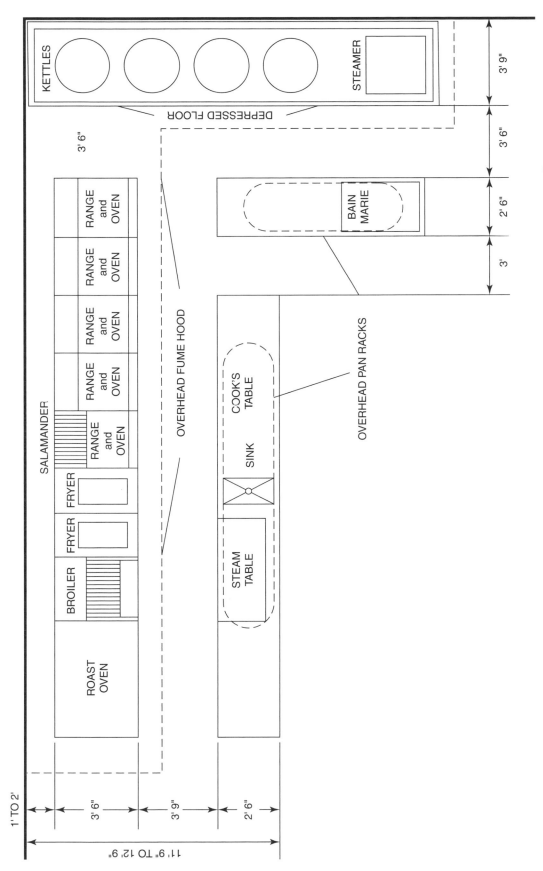

ILLUSTRATION 2-8 The L-shape kitchen layout works well in long, rectangular spaces and allows more equipment to be placed in a smaller area.

Source: Robert A. Modlin, ed., *Commercial Kitchens*, 7th ed. (Arlington, VA: American Gas Association, 1989).

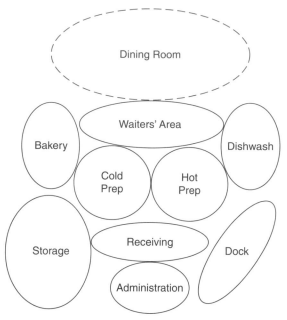

ILLUSTRATION 2-9 Bubble diagrams are rough sketches that show the relationship among areas of a restaurant or among parts of the kitchen or dining area.

Source: Lodging: The Magazine of the American Hotel and Lodging Association, Yardley, Pennsylvania.

There are also circular and square kitchen designs, but their limited flow patterns make them impractical. Avoid wasted space if you can by making your kitchen rectangular, with its entrance on one of the long walls to save steps.

The more foodservice establishments you visit, the more you will realize that the back of the house is a separate and distinct entity from the rest of the business, with its own peculiar problems and solutions. Correct flow planning sometimes means breaking down each kitchen function into a department of sorts and then deciding how those departments should interact. They must also interact with the external departments of the facility: dining room, bar, cashier, and so on.

A good way to begin the design process—both for the overall business and for the kitchen—is to create a *bubble diagram*. Each area (or workstation) is represented as a circle, or bubble, drawn in pencil in the location you've decided is the most logical for that function. If two workstations will share some equipment, you might let the sides of their circles intersect slightly to indicate where the shared equipment might best be located. The finished diagram will seem abstract, but the exercise allows you to visualize each work center and think about its needs in relation to the other centers. Illustration 2-9 shows this type of bubble diagram, and there are other ways to use it. You can also lay out a kitchen using a diamond configuration, situating the cooking area at one point of the diamond shape and other crucial areas in relation to it at other points, as in Illustration 2-10. Notice that this layout

ILLUSTRATION 2-10 The diamond shape can also be used to create a diagram that shows the four primary work areas (prep, cooking, dishwashing, and storage) in relation to each other, plus unobstructed paths for kitchen and waitstaff.

Source: Adapted from Restaurant Hospitality magazine (October 1998).

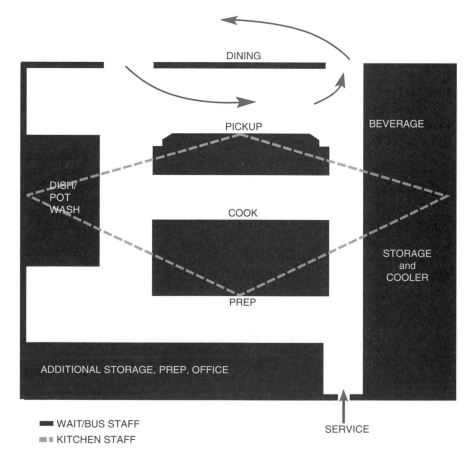

minimizes confusion (and accidents) with a separate kitchen entrance and exit. This allows the people who bus the tables to deliver soiled dishes to the dishwashing area without walking through the entire kitchen to do so.

An alternative to drawing diagrams is to list each work center and then list any other work center that should be placed adjacent to it. Conversely, list any work center that should *not* be next to it. For example, it's probably not a good idea to have the ice maker and ice storage bin adjacent to the frying and broiling center.

2-5 SPACE ANALYSIS

We now begin to analyze the kitchen section by section, to help you approximate how much space you will need for each. Of course, not all kitchens will need all areas we mention here. In general, however, they are:

- Preparation areas
- Production areas
- Bakery area

In Chapter 4, you will learn more about the other back-of-the-house functions that support the kitchen:

- Warewashing (dishwashing) area
- Storage areas: dry and refrigerated
- Receiving area
- Office
- Employee locker rooms, staff restrooms

For now, let's briefly discuss the serving and cooking functions of the operation and the space requirements of each.

PREPARATION AREAS

We'll use the term *preparation areas* to refer to food preparation. After all, a lot more activities than cooking are crammed into most kitchens. Here are some of the major work sections you may find.

FABRICATION. The fabrication area is where raw (or processed) foods begin their journey to their final destination: the guest's plate. Sometimes referred to as *pre-prep*, it is here the kitchen staff break down prime cuts of beef, clean and fillet fish, cut up chickens, open crates of fresh produce, and decide what gets stored and what gets sent on to the other parts of the preparation area. In planning for each area, start with a flowchart to determine which functions should be included. Illustrations 2-11 and 2-12 show the flowchart and the resulting layout for a butcher shop, one of the most common types of pre-prep sites for kitchens that don't buy meats custom-cut.

If the facility plans to handle its own meat-cutting duties (and many do to save money), you'll need room for a sink, a heavy cutting board, portion scales, meat saws, grinders, and slicers. Some of these items can be placed on mobile carts and shared with other areas of the kitchen.

PREPARATION. In the preparation area, foods are sorted further into individual or batch servings. The loin that was trimmed in the fabrication area is cut into steaks, lettuce and tomatoes are diced for salad assembly, shrimp is battered or peeled. Ingredients are also mixed: meatloaves, salad dressings, casseroles. Salad and vegetable prep areas are found in almost every foodservice setting. They are busy places, and their focus must be on efficiency. When designing the layout,

ILLUSTRATION 2-11 A flowchart is the next step after a bubble diagram in planning the movement of materials through your kitchen.

Source: Robert A. Modlin, ed., *Commercial Kitchens*, 7th ed. (Arlington, VA: American Gas Association, 1989).

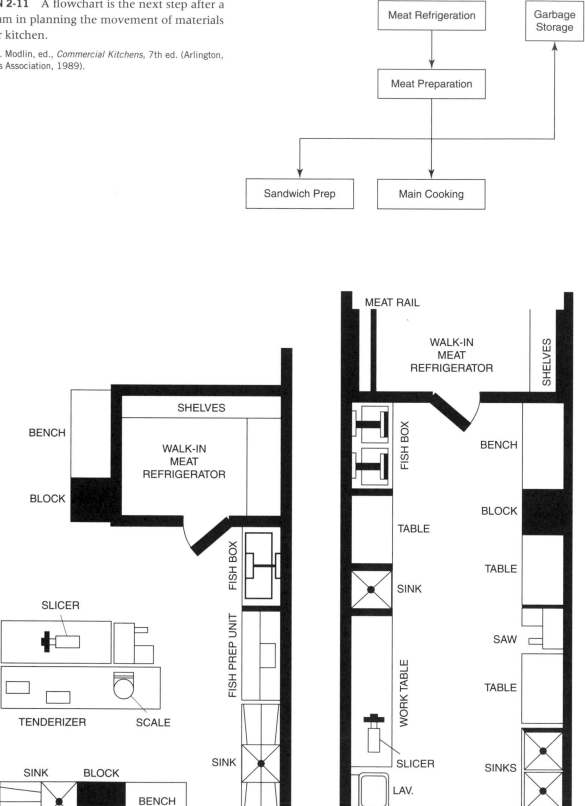

ILLUSTRATION 2-12 After you've completed a flowchart, it is easier to draw the actual layout of a work center.

Source: Robert A. Modlin, ed., *Commercial Kitchens*, 7th ed. (Arlington, VA: American Gas Association, 1989).

remember the need for worktables, compartment sinks, refrigerators, and mechanical equipment. Order some worktables with food and condiment wells that are cooled from beneath with ice or refrigeration systems, allowing easy accessibility.

A prep area with distinct requirements is the *garde manger*, which encompasses both food preparation and decoration or garnish. The garde manger area is the source of cold foods: chilled appetizers, hors d'oeuvres, salads, pâtés, sandwiches, and so on. If ice carvings are part of the repertoire, this is where they are created. Obviously, refrigeration is of paramount importance here, as are knife storage and room for carving tools, plus hand-held and small appliances: ricers, salad spinners, graters, portable mixers, blenders, juicers. Color-coded bowls, cutting boards, knives, scrub brushes, and even kitchen towels all help avoid cross-contamination among types of raw foods.

Some kitchens refer to their cold-food prep area as the **pantry**, although traditionally, a pantry is a station that supports and reports to the *garde manger* chef.

PRODUCTION. Yes, it's finally time to do some cooking, in the production area. This area is divided into hot-food preparation, usually known as the **hot line**, and cold-food preparation, called the *garde manger* or *pantry*. Production is the heart of the kitchen, and all the other areas are meant to support it.

HOLDING. As its name suggests, the holding area is where either hot or cold foods are held until needed. The holding area takes on different degrees of importance in different types of foodservice operations. Basically, the larger the quantity of meals produced, the more critical the need for holding space. For banquet service and in cafeterias and hospitals, food must be prepared well in advance and stored at proper temperatures. In fast-food restaurants, the need is not as great, but it still exists.

Many hospitals are now moving toward room-service menus in order to avoid long holding periods and deliver fresher foods. This is viewed as an amenity by patients, as they can order food from a menu at their leisure rather than sticking to a preset mealtime as in the past.

ASSEMBLY. The final activity of the preparation area is the assembly of each item in an order. At a fast-food place, the worktable is where hamburgers are dressed and wrapped and where fries are bagged. At an à carte restaurant, it is the cook's side of the pass window where steak and baked potato are put on the same plate and garnished. Again, in large-scale foodservice operations, assembly takes up more room.

The menu and type of cooking you do will determine the makeup of your production area. Will you need a fabrication area at all if yours is a fast-food franchise that uses mostly prepackaged convenience foods? Conversely, cooking from scratch will probably require a lot of space for preparation, baking, and storage. **Batch cooking**, or preparing several servings at a time, will also affect your space allocation. Finally, the number of meals served in a given period should be a factor in planning your space. Your kitchen must be able to operate at peak capacity with plenty of room for everyone to work efficiently. For a hotel with banquet facilities and for an intimate 75-seat bistro, this looks very different.

PRODUCTION AREAS

Now let's delve into greater detail about the hot line and the pantry areas and what goes on in each. Much of today's line layout can be traced to the classic French brigade system, in which the cooks or chefs stood in a row, and each was responsible for preparing certain food items—one fish, another vegetables, and so on.

On the hot line, you must determine placement of equipment based on the cooking methods you will use. There are two major ways to cook food: dry-heat methods (sauté, broil, roast, fry, bake) and moist-heat methods (braise, boil, steam). The difference is, of course, the use of liquid in the cooking process.

A kitchen that is well laid out may position its volume cooking at the back and its to-order cooking at the front. In large-volume food production, combi-ovens, rotating deck ovens, steam-jacketed kettles, and tilting braising pans will be found together. In smaller (batch or à la carte) kitchens, fryers, broilers, open-burner ranges, and steam equipment are grouped together.

In kitchens that require both large- and small-batch cooking, the large-volume, slower-cooking equipment is placed behind a half-wall under an exhaust hood (see Illustration 2-13). On the other side, or in front of the wall, is the equipment used for smaller-volume or quick-cooking dishes. Because the smaller dishes require more attention, putting this equipment at the front of the wall allows them to be watched more carefully.

In most communities, it is a law that all heat- and/or moisture-producing equipment must be located under ventilating hoods. Also, all surface-cooking equipment (rangetops, broilers) must be located so the automatic fire-extinguishing (sprinkler) system can reach them. Ventilating hoods are expensive, so the tendency is to crowd a lot of cooking equipment into the small space beneath them.

Here are the stations or sections on the typical hot line.

BROILER STATION. A single or double-deck broiler is found here.

GRIDDLE STATION. The width of the griddle determines its capacity. It can be either a floor-standing model or a tabletop griddle that sits atop a counter. The tabletop model is a smart use of the space below it, where the cook can store supplies. This is a good place to put refrigerated drawers, so meats can be kept chilled until they are ready to be cooked.

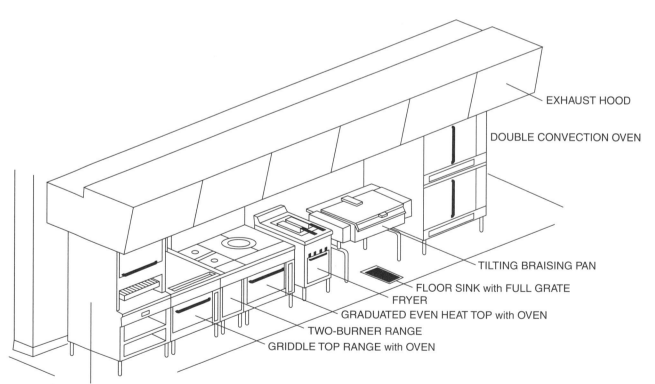

EXHAUST HOOD

DOUBLE CONVECTION OVEN

TILTING BRAISING PAN

FLOOR SINK with FULL GRATE

FRYER

GRADUATED EVEN HEAT TOP with OVEN

TWO-BURNER RANGE

GRIDDLE TOP RANGE with OVEN

UPRIGHT BROILER with CABINET BASE

ILLUSTRATION 2-13 The hot line, or battery of cooking appliances, in a typical foodservice kitchen. They are arranged together under an exhaust hood system.

Source: Carl R. Scriven and James W. Stevens, *Manual of Equipment and Design for the Foodservice Industry* (Albany, NY: Thomas Learning, 1989).

SAUTÉ STATION. Here we find three types of ranges: the flat-top sectional range, with individual heat controls for each section; the ring-top range, with rings of various sizes that can be removed to bring flames into direct contact with the sauté pan; and the open-top range, also sectional, with two burners per section. Ranges are now being built with step-up burners that allow the sauté cook to work on lower burners near him and higher burners to the rear of the range. Ovens can be placed tabletop in this area, or they can be installed below the range. Above the range is often a mini-broiler known as a ***salamander***, used for quick duties like melting butter or as a holding area.

SAUCE STATION. Not every kitchen has room for this station, where soups, sauces, and casseroles are prepared on rangetops or in steam-jacketed kettles.

HOLDING STATION. Once again, this area is designated for holding finished food before it is assembled on plates or put onto trays. Individual stations can have their own holding areas, or one central area may be used. Either way, holding areas often include a dry or wet steam table or hot-water baths for keeping sauces warm.

UTILITY DISTRIBUTION CENTER. In high-volume production kitchens, such as banquet and catering operations, the installation of a ***utility distribution system*** is recommended. A utility distribution system is designed to provide all the necessary services (gas, electricity, hot and cold water, and steam) to the cooking equipment placed under the exhaust canopy (see Illustration 2-14). Connected directly to the canopy, it's like having a huge fuse box that controls all the utilities, with a single connection for each of them, plus emergency shutoffs and inspection panels to allow quick access to (and easy cleaning of) all components. As cooking equipment is added or moved, the system can be adapted, rearranged, or expanded.

We'll discuss utilities more in Chapters 5 and 6, but here are a couple of terms you should be familiar with in discussing a utility distribution system. Most systems are shaped like a big H.

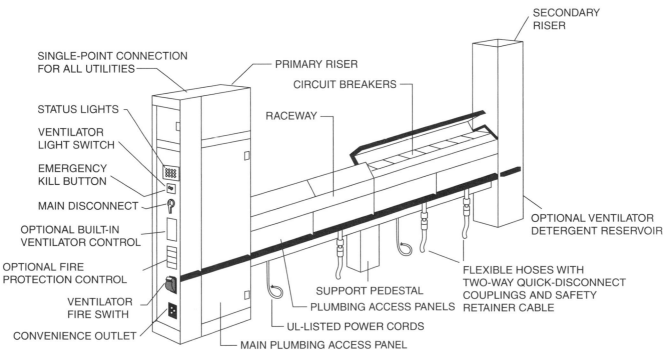

ILLUSTRATION 2-14 A utility distribution system is the central location for all utilities needed for cooking equipment.

The two vertical, upright posts are called *risers*. The larger one, which houses the main utility connections, is called the *primary riser*; the smaller one is the *secondary riser*, which serves mostly for balance. The horizontal section between them is known as the *raceway*. Here you will find the outlets that connect the system to the cooking equipment.

PANTRY. The pantry is where cold foods are prepared for serving. Preparation responsibilities here may include salads, sandwiches, cold appetizers and entrées, and desserts. Sometimes it is referred to as the *cold kitchen*. Illustration 2-15 shows the components of this complete workstation.

A two-compartment sink is a must for this area, as salad greens are washed and drained here. Worktables with cutting boards should be adjacent to the sink, and it is helpful if the sink itself has corrugated drainboards.

Refrigeration is also necessary, as cold storage is required for many ingredients and a holding area for cold prepared foods. In restaurants, this could be a pass-through setup, with sliding glass doors on both sides so pantry employees can slide finished items into the refrigerator and servers can remove them as needed from the other side. In banquet or other high-volume foodservice operations, a walk-in refrigerator may be more appropriate. Inside the walk-in, mobile carts can store finished trays of plated salads for easy access. If reach-in carts are used, make sure they have tray slides on them that will hold large sheet pans.

For sandwich-making, the pantry needs slicers and other types of cutting machines as well as mixers for making dressings. Because often these items are shared with the fabrication area, the equipment should be placed on rolling carts.

Another requirement is a way to keep breads enclosed so they do not dry out. If recipes require toasted bread, at least one commercial toaster is needed.

When the pantry is also the site of dessert preparation, additional refrigeration and counter space are needed. You must find room for a freezer and an ice cream chest, with its special water well to store scoops. The latter must meet local health ordinance requirements. Soft-serve machines may also be needed. Again, mobile carts equipped with shelf slides can be helpful, both to store and to transport products.

ILLUSTRATION 2-15   A pantry area layout.

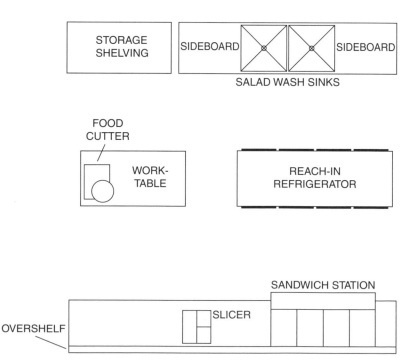

BAKERY AREA

We use the term *baking* in referring to foods that start as batter or dough. Breads, desserts, and pastries are the output of the typical restaurant bakery area, which has its own characteristics and needs.

Before designing a bakery area, ask yourself:

■ What baked goods will we prepare here, and what baked goods will we purchase from outside sources?

■ Will we bake from scratch or use premixed items and/or frozen doughs?

Often, to bake or not to bake on site is strictly a business decision, as baking on site requires additional room, equipment, and skilled personnel. Dozens of hybrid arrangements are used in foodservice: A restaurant may decide to purchase all breads but make desserts in house, or vice versa. Pizza operations must do baking on site, but other than that, variations abound.

Within the kitchen, a bakery generally takes on one of two basic forms. It is either a compact area located next to the hot line, like the sample layout shown in Illustration 2-16, or it is a distinct area of its own, like the larger bake shop shown in Illustration 2-17.

When the bakery is a compact area, the setup may be as simple as a baker's table with storage bins beneath, which ovens and mixers shared with other areas. A larger, standalone baking area has other distinct attributes, some of which are discussed next.

MIXING STATION. Here you will find an array of large floor or tabletop mixers and their accessories. This station, of course, requires electrical power and a generous amount of storage space. The station also contains a table with scales, where ingredients are weighed, measured, and mixed.

PROOFING STATION. This is where mixed dough is held to rise, with controlled temperature and humidity. Proofing boxes (enclosed cabinets for this purpose) require electricity, a water source, and drainage capability. Mobile racks also roll around in this section, storing ***sheet pans*** and dough.

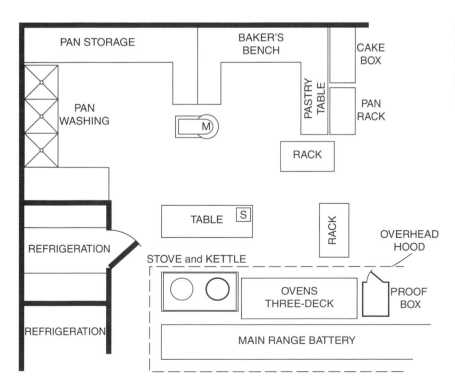

ILLUSTRATION 2-16 The bakery area is an integral part of many kitchens. Here is a layout for a small corner site.

Source: Robert A. Modlin, ed., *Commercial Kitchens*, 7th ed. (Arlington, VA: American Gas Association, 1989).

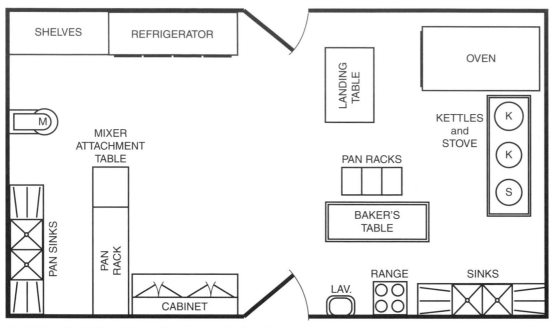

ILLUSTRATION 2-17 A layout for a large, high-volume bakery.

Source: Robert A. Modlin, ed., *Commercial Kitchens*, 7th ed. (Arlington, VA: American Gas Association, 1989).

FORMING AREA. This is where dough is shaped into rolls, pies and cakes are put in pans, and the like. In small kitchens, the mixing station sometimes must double as the forming area. The tasks performed here call for bakers' tables with tops approved by local health authorities and for mechanical dough dividers and shapers. The mobile racks from the proofing station can stop here on their way to the oven.

BAKING STATION. Ovens and exhaust hoods take up most of the space here. Oven types depend entirely on what is being baked. If there is only one type of product, in large quantities, a carousel or reel oven is preferable. The combination steamer/oven can add moisture to the baking process to make harder crusts for some types of bread. Deck ovens are stacked, each with individual temperature controls, for baking two types of products at once. Convection ovens circulate hot air inside the ovens during the baking process—good for breads but not so great for some types of pastries and cheesecakes. There are many options, depending on your menu needs.

FINISHING STATION. A baker's worktable is a requirement here, where pastries and breads are given their final form prior to serving. This includes slicing, decorating, glazing, and the like. Room is needed for mobile racks so product can be rolled in and out of the station.

In creating a bakery area, remember that baked goods require a variety of storage environments ranging from room temperature with good air circulation to constant refrigeration. If this is also the general dessert-making area, leave room for the ice cream chest, freezer, and soft-serve machine.

What can a bakery area share with other parts of the kitchen? Refrigerator space might be shared with the pantry or the service area. Mixers can be shared with the preparation area, especially if they are tabletop models on rolling carts.

The bakery should be located near the pot sinks, as baking generates a lot of dirty dishes. Some bakery areas have their own pot sinks as well as hand sinks. It's also handy to be close to the dry storage and walk-in refrigerator areas, for easy access to raw materials.

Worktable space should be no less than 6 feet per employee, and adequate lighting is important because of the delicate work done here, such as cake decorating. Landing space for baked goods is needed, especially in front of the ovens. Allow plenty of room for sheet pans, which measure 18 × 26 inches. Proofing boxes should also be placed near the ovens.

For preparing fillings and syrups, a portable tilting steam-jacketed kettle may be needed near the ovens. Another special requirement may be a rangetop burner for melting sugar or heating sauces.

OTHER CONSIDERATIONS

What are the desirable characteristics of a foodservice kitchen door? First, it should be light-weight. It should open easily, as it will be opened often—and by people who have their hands full with trays of food, bus tubs, and other heavy items. Prolong every door's useful life by making sure it has a metal **kickplate** or **scuff plate** over the bottom area on both sides—the spot most likely to be kicked open by hurrying feet. Finally, kitchen doors should always contain eye-level windows, so employees don't barrel through and hit each other accidentally.

These kinds of doors come in standard sizes, with a variety of styles depending on the size of the kickplate, the color, and whether you want guards for the doorjambs.

Every kitchen needs work surfaces, which as you know are often referred to as **landing spaces**. These surfaces may be mobile or permanent. They are usually made from stainless steel (see Illustration 2-18) and can be ordered with shelves, sinks, slots for sheet pans, drawers, and even overhead pot racks for additional storage. As you study your options, think about safety. Can anyone catch a pocket or button on drawers? Make sure they have recessed handles. If you'll be hosting cooking classes or demonstrations, worktables can be purchased with removable, adjustable overhead mirrors or chalkboards. There are many to choose from! Just be sure you have allocated sufficient room for them in your kitchen layout.

Another decision to make is what type of edges your worktables will have. Edges can be rounded (called a *rolled edge*), curved upward (a *marine edge*), raised, straight, and so on. If your employees must slide heavy containers from one table to the other, you will need different kinds of table edges than if your goal is simply to prevent spills.

In Chapter 8, you will learn more about the correct heights for work surfaces, to ensure the comfort of your kitchen staff. Worktables also come in standard widths (usually 30 inches) and lengths (from 24 inches up, in 1-foot increments). Tables longer than 72 inches require under-bracing for additional support. As you design your kitchen space, remember that long tables are

ILLUSTRATION 2-18 Stainless steel worktables are important in any kitchen.

Courtesy of InterMetro Industries, Wilkes-Barre, Pennsylvania.

often inefficient because of all the steps it takes to walk around them. If you can do the same job with two shorter tables and a bit of space between them, do it. Better yet, use tables with casters that allow them to roll. Locking casters act as brakes, so you can keep them from rolling if you wish. Casters are capable of carrying heavy loads without squeaking, rusting, or corroding in the intense heat and humidity of a kitchen environment. The heavier the load, the larger the casters you will need. The bearings in each wheel allow them to roll with ease under weight; roller bearings carry more weight than ball bearings, but ball bearings roll more easily.

Placement of sinks, water supply, and electrical outlets should be priorities in designing any kitchen. We'll focus more on specific utility needs in Chapters 5 and 6, and on specific types of equipment in Chapters 10 to 16.

Floor and wall materials should be damage-resistant, with easy-to-clean surfaces. Ceramic glazed tile on walls will withstand both heat and grease. The latest innovation, fire-resistant panels (FRP), is now being used extensively in commercial kitchens; these panels are easy to clean and less expensive than tile. For floors, quarry tile that contains carborundum chips is an excellent option that provides natural slip resistance. There should be a floor drain in front of every sink in the preparation area and a floor drain for every 6 linear feet of your hot line. One smart alternative to individual drains is to cut a 4-inch-deep trough along the hot line floor containing several drains, covered by a metal grate.

Finally, you may not think of it as equipment, but you'll need plenty of trash receptacles—and places to put them—throughout the kitchen. Trash cans should be lightweight plastic, covered, and on dollies so they can roll around. Always use trashcan liners to make them easier to empty.

2-6 SERVICE OR BANQUET KITCHENS

If your business plans include space for private dining—meeting rooms or separate catering areas—you may also need a **banquet kitchen** to service these areas properly. This kitchen probably will not see daily use, but when it is needed, it is a labor-intensive, production-oriented place that requires powerful, reliable, and multifunctional equipment. If your banquet and special events business becomes successful, you will stretch your main kitchen resources awfully thin without an extra banquet kitchen, and you'll run your waitstaff ragged if the meeting rooms are located far from the main kitchen. Banquet kitchens are sometimes called *service kitchens*.

Think of your service kitchen as an extension of your main kitchen. The purpose of the banquet kitchen is to make only the final food preparation before serving to the banquet or meeting crowd. Only modest, one-day storage is needed here, and because most foods will be delivered partly or completely prepared, directly from the main kitchen, separate prep areas are not needed in the banquet kitchen. All cleaning, peeling, slicing, and butchering can take place in the main kitchen. You do not need a separate dishwashing area either.

It is true that a service kitchen makes money only when it is in operation. However, don't think of this investment as a waste if it is not constantly busy. Remember, when the lights are out and nobody is using it, it is still depreciating as an asset, not actively costing operating dollars or shrinking your bottom line.

How much space will you need for these additional kitchen facilities? It depends mostly on how many people you can seat in your banquet rooms. In the Sofitel Hotel in midtown Manhattan, an 800-square-foot production kitchen in the basement also functions as the prep kitchen for the hotel's on-site restaurant, and a second, 700-square-foot kitchen adjacent to the hotel's banquet rooms is where the meals are heated, finished, and assembled.

A service kitchen of 75 square feet can accommodate seating of 50 to 100; for 1000 seats, you'll need as much as 500 square feet of kitchen space. The rule of thumb is 50 square feet of kitchen space for every 75 to 100 seats. In addition, you will need about ½ square foot per seat for storage of banquet tables and chairs when they are not in use.

IN THE KITCHEN

Designing a Service Kitchen

TOP PRIORITY FOR CATERING/SERVICE KITCHENS

1. Stainless-steel tables for plating food
2. Combi oven/steamer
3. Cook-and-hold oven
4. Hot food holding boxes
5. Steam table
6. Mixer
7. Tilting kettle
8. Braising oven or tilting braiser
9. Salamander
10. Rangetop
11. Reach-in and walk-in refrigeration
12. Sink, with hot and cold water
13. Beverage containers
14. Ice bin or (better) ice machine
15. Electrical outlets for all portable equipment
16. Storage for linens, plateware, flatware, glassware
17. Storage for tables and chairs

NICE TO HAVE BUT NOT ABSOLUTELY NECESSARY

1. Portable steam table
2. Portable salad bar
3. Fryer, broiler, griddle
4. Three-compartment sink
5. Dishwashing machine

In designing a layout for your service kitchen, look at portable equipment you can use to hold and serve food; it will be the most adaptable. If the banquet rooms are close to the main kitchen, you may be able to assign a certain amount of main kitchen space as your banquet area. At its simplest, it could be a few stainless-steel tables on which banquet food is plated for waitstaff to pick up and deliver. This area may also include a separate beverage dispenser and ice machine so waiters from banquet and main dining areas won't trip over each other using the same facilities. Think carefully about traffic flow for busy times of day.

Depending on how busy your banquet/catering business becomes, additional cooking equipment may be needed on the main hot line, or perhaps other stations can be expanded. Although this approach does enable you to keep the food preparation in one space, the downside is that the additional equipment is needed only when there is a banquet. At other times, it either sits idle or is used by the staff simply because it's sitting there, which is inefficient. The real solution is to separate the service kitchen from the main kitchen from the start.

Illustration 2-19 shows a large banquet department located adjacent to a main kitchen.

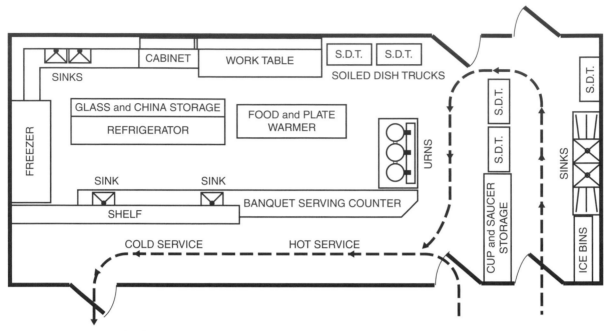

ILLUSTRATION 2-19 The banquet kitchen requires a separate layout and flowchart.

Source: Robert A. Modlin, ed., *Commercial Kitchens*, 7th ed. (Arlington, VA: American Gas Association, 1989).

2-7 FOOD SAFETY AND KITCHEN DESIGN

This part of the kitchen design process is enough to tarnish the stainless-steel dreams of any chef: It concerns food safety, a topic no one likes to talk about but that makes headlines nonetheless any time patrons become ill when dining out. The U.S. Centers for Disease Control and Prevention (CDC) estimates there are 48 million cases of food-borne illness annually in the United States. The causes of about 95% of them are never pinpointed; however, if a pattern is established that points to a certain product, supplier, or restaurant, the chance for legal liability—or at least, very bad publicity—is major.

The U.S. Food and Drug Administration (FDA) developed a preliminary food safety system known as a Food Establishment Plan Review Guide. The idea is to complete the plan review and submit it to the local health department before a food-related business opens its doors. This way, work does not have to be redone to comply with the health regulations—they are followed from the beginning of construction.

A plan review is now a requirement in many cities and states, and it certainly is a good idea for new businesses in the design phase. The checklist and guidelines cover many food-buying, handling, and preparation concerns. The initial questionnaire is quite extensive, but it is an excellent tool for pinning down the particulars of the new operation: numbers of seats and meals to be served, type of service, staffing, and so on. Documents to be submitted with the questionnaire are proposed menus; a site and floor plan showing aisles, storage spaces, restrooms, basements, workflow, and areas where trash may be stored; and manufacturers' specifications and cut sheets for the equipment that has been ordered.

A flow plan is another requirement, including the flow patterns for food, dishes, utensils, and waste. Also necessary is a list of foods that will be prepared more than 12 hours in advance and a safety plan (based on the HACCP system; see the next section) for each of these food categories. This includes the methods by which they will be thawed, cooked, cooled, chilled, kept hot until service, and the like, including temperatures. For cold storage, the plan must list how foods stored in the same refrigerated space will be protected from cross-contamination.

The Plan Review Guide also covers equipment: everything from color-coded cutting boards that prevent cross-contamination to water temperature requirements for dish machines to certification requirements that equipment meets safety sanitation standards. There must be plenty of handwashing sinks, conveniently located for use by all employees; and the FDA suggests that at least one person in the kitchen be certified in food safety by a recognized authority.

The FDA's Plan Review Guide can easily be obtained online by typing the term into any search engine.

DESIGN AND HACCP COMPLIANCE

As food moves from farm to processing plant, then to the distributor and on to the commercial kitchen—and finally finding its way to the customer—there are many critical points at which storage and holding temperatures could easily slip into the danger zone of 41°F to 140°F. This is the temperature range that promotes rapid growth of most bacteria in food.

You'll learn more about food safety and inspection compliance in Chapter 8, but you should already be well aware of the *Hazard Analysis of Critical Control Points*, or HACCP system, in the design phase of your business. It is a seven-step process to identify critical control points, meaning points at which food must be handled in a specific manner to keep it safe for consumption; determining limits for safe and unsafe handling at each point; then documenting and complying with your controls.

What does this have to do with design? Today's kitchen can use a combination of advanced technology and an intelligent layout to minimize many contamination risks. Examples:

- Install reach-in coolers in every prep area so employees don't have to trek to the walk-ins all the time. This keeps the walk-in temperature steadier and also keeps batch ingredients at their optimum temperatures until use.
- Make ovens, fryers, ranges, and even storage racks mobile so they can be moved away from walls more easily for cleaning around them.
- Place handwashing sinks closest to the stations that will need them most. Allow separate handwashing facilities for people who handle produce and people who handle raw meats. Electronic sensors on the sinks can turn water on and off and dispense hand soap so employees don't touch the faucets.
- Store raw and finished foods in separate refrigerators.

In an HACCP-compliant kitchen, everything flows forward. Ingredients move in a single direction, without backtracking, from receiving to storage to pre-prep and prep to holding and/or serving. What is not served or cannot be safely held is discarded. Equipment also moves without backtracking; it arrives in an area clean and sanitized, remains there until its job is done, and is cleaned and sanitized again before being used in another area.

■ SUMMARY

For any commercial foodservice business to function well, its kitchen must function well. Kitchen design is a complicated combination of making every square foot useful while figuring out how best to work around the possible limitations of the site. As discussed at length in this chapter, kitchen design is most successful when the business owners have determined their menu and their service system—what will be cooked, and how it will be served to their guests.

Kitchens are organized into work sections, and each work section is composed of work centers where certain tasks are performed. Layout and equipment placement are determined by the duties of each section and the appliances needed to perform the tasks. Kitchen design should always allow for flexibility in case the menu or concept changes.

Decide how workstations relate to each other by creating a bubble diagram, with each bubble (circle) representing a task or area. Let the circles intersect to indicate shared functions or equipment needs.

Flow patterns are critical in designing a kitchen. Distance and kitchen accessibility are the keys to creating workable flow patterns. In this chapter, examples of flow patterns are shown for cafeteria dining, food preparation and storage areas, and dish rooms.

Even aisle space is important in space planning, because it affects guest comfort (in the dining area), ease of equipment operation (in the kitchen), and safety (in both areas).

The biggest modern-day concern of health departments is the ability of commercial kitchens to comply with food safety and sanitation requirements. Food safety has many design implications—everything from where food is stored (individual refrigerators for each work section; separate storage for separate products, to minimize cross-contamination), to where trash is discarded, to where handwashing stations are placed and how many are needed, to mobility of equipment. It is easier to make these decisions on paper in the design phase than to figure them out—and deal with health inspectors' concerns—after the business is open.

■ STUDY QUESTIONS

1. Briefly explain how each of these factors—labor, utilities, and food—is affected by kitchen design.

2. Briefly describe the three basic flow patterns that foodservice facility designers must take into account.

3. What is the difference between a fabrication area, a preparation area, and a production area? In your view, could any of these be combined if space is at a premium?

4. Name three types of space that are often overlooked in kitchen design.

5. What equipment would you install in a separate service kitchen for banquets? What equipment would you share with your main kitchen?

6. What factors do you consider when deciding how much space to allocate for your kitchen production area?

7. City ordinances require that all cooking equipment on the hot line be located near two things. Explain which two, and why.

8. What special functions must be included in designing a pantry area if desserts will be made there?

9. Which workstation should be located closest to the pot sinks, and why?

10. Why do we say that, in an HACCP-compliant kitchen, "everything flows forward?" Explain the meaning of, and reasons for, this concept.

Kendall College

Chicago, Illinois

SITUATION

Kendall College opened in Chicago in 1934. It had a public restaurant for hands-on foodservice experience for students of the hospitality management and culinary arts programs, but the set-up was not ideal—for instance, servers had to pass through about 200 feet of public space to deliver food from the kitchen to the dining area. The school wanted an updated kitchen and dining space to showcase its fine dining quality as well as to improve learning opportunities for students with the latest equipment and technology. Budget for the project: $1.6 million.

CHALLENGES

- Finding sufficient space for up to 18 student chefs and 2 instructors in the fine dining kitchen and up to 8 student chefs and 1 instructor in an adjacent deli kitchen.
- Working with a 100-year-old building: floors not level, spaces tight, columns placed inconveniently.
- A mandate from the college that every aspect of the design be as "green" as possible.
- Getting approval from the city of Chicago for the kitchen ventilation system. City codes were in the process of being updated but did not allow for some of the newer technology.

COMMENTS FROM THE DESIGN TEAM

- "The project budget was conservative, and the owner [school] wanted to reuse as much equipment as possible while still bringing forward a new-looking facility that would make a strong statement of Kendall's professionalism."—Ed Norman, FCSI, MVP Services Group, Inc.
- "The goal inside the fine dining kitchen and the deli kitchen was to create an elegant but understated backdrop for all the cooking activities through use of white glazed ceramic tiles on the walls and medium-tone gray monolithic flooring."—Agnieszka Chapman, interior project manager, Perkins + Will
- "The flow of the restaurant and deli kitchen combine classic and traditional, which I think is wonderful."—Benjamin Browning, dining room chef and instructor
- "We would have liked a larger shared cooler (for the two kitchens), but we had to compromise its size to fit other elements. But all projects require a process of considering absolute and not-so-necessary desires and requirements."—David Kipley, Kendall College board member

HIGHLIGHTS OF THE DESIGN

- There are now two kitchen spaces: one for fine dining, the other a deli kitchen to serve as a separate banquet area for student and staff dining, and special events.
- Large glass windows between the kitchen and dining area allow guests to watch every step of their meal preparation and culinary students to see diners enjoying what they have prepared.
- Extensive use of time-saving and energy-efficient cooking equipment, such as induction ranges and pressurized steam jacketed kettles.
- A pastry station, which the old kitchen did not have.
- A state-of-the-art exhaust system that tracks all HVAC functions and displays energy use information on a large-screen monitor in the kitchen.

RESULTS

The new dining room opened in 2010 with a focus on contemporary American cuisine and locally sourced ingredients. It has warranted mentions in the prestigious *Michelin Guide* and *Zagat Survey*, which praise the overall experience as well as the ambience of the space. From the *Michelin Guide Chicago 2011*: "Imagine that dining out could be a sneak peek at a future Michelin-starred chef. That's a real possibility in this dining room and kitchen."

TEAM MEMBERS

- *College president:* Dr. Karen Gersten
- *Dean, School of Culinary Arts:* Renee Zonka, RD, CEC, CHE, MBA
- *Executive director, School of Culinary Arts:* Chris Koetke, MBA, CEC, CCE
- *Interior design:* Agnieszka Chapman, LEED AP, senior associate, Perkins + Will, Chicago, Illinois
- *Architect:* Mark Jolicoeur, principal, Perkins + Will, Chicago, Illinois
- *Foodservice consultant:* Ed Norman, FCSI, CSI, principal, MVP Services Group, Inc., Chicago, Illinois, and Dubuque, Iowa
- *Equipment dealer:* Schweppe Equipment, Lombard, Illinois
- *Owners' representative:* Construction Methods, Inc., Northbrook, Illinois
- *Construction:* Pepper Construction, Chicago, Illinois
- *Engineering:* Mechanical Services Associates, Crystal Lake, Illinois
- *Structural engineers:* C.E. Anderson and Associates, Chicago, Illinois

Source: The full article about this project first appeared in the April 2011 issue of *Foodservice Equipment and Supplies* magazine. Donna Boss wrote the original article; our encapsulation and the layout appear with permission.

3

FRONT-OF-HOUSE ATMOSPHERE AND DESIGN

■ INTRODUCTION AND LEARNING OBJECTIVES

All the physical surroundings and decorative details of a foodservice establishment combine to create its *atmosphere*, the overall mood that also may be referred to as the *ambience* or energy of the space. In this very competitive industry, most merchants understand that good design is a way to set a restaurant apart from its competition, bringing the theme and concept to life. Today, good design also includes energy efficiency, building in ways to conserve natural resources while making the most effective use of available space.

Our focus in this chapter is on the public spaces that are part of many foodservice businesses. Without necessarily being able to pinpoint specifics, any person will tell you that attractive surroundings seem to make a meal better. Even in large, industrial cafeteria settings, small but significant touches contribute to a warm and inviting feeling: greenery, fresh produce displays, the use of artwork, inventive bulletin boards. The idea, no matter what your theme or price range, is to make people feel welcome, safe, and cared for.

After reading this chapter, you will be able to:

- Identify the design details that contribute to atmosphere.
- Explain how the front-of-the-house space is planned and subdivided.
- Describe the guidelines for specific types of public space: entryways, restrooms, bar areas.
- Identify special-use spaces that can increase profitability.
- Identify the factors involved in selecting chairs and tables.
- Describe the guidelines for deciding whether or not to include a bar in your establishment, and basic bar design components.

In short, so many good choices are available in today's market that good food is not the only element that will build a steady clientele. From an inviting entryway to clean restrooms, there is much planning to be done.

3-1 CREATING AN ATMOSPHERE

The development of a successful design begins with a firm concept for the business, which was discussed in Chapter 1. The concept will help determine the atmosphere of the space.

From the corporate cafeteria to the most upscale restaurant, any public dining space can be designed to welcome its customers. Efficiency, value, and convenience are the hallmarks of modern eateries, and all of these attributes can be reflected in their design. In many places, kitchen activity is "out front"—with display kitchens, wood-burning ovens, sauté stations that showcase the chef and staff members for diners to watch. Restaurants with celebrity chefs have learned to show them off. Fresh ingredients are displayed, and sensory details are not overlooked. The message? "Our food is fresh, and we care about this!" It is no surprise that a display kitchen generates higher dollar volume than does a traditional kitchen hidden behind swinging doors, although we realize this is simply not possible in some situations. When it is, however, the guest experience is enhanced, the food quality generally improves, and profitability increases.

One good rule for developing atmosphere is to provide a change of pace. When we go to the movies, for instance, we shift gears mentally as we enter the lobby, then the theater, selecting a seat and preparing to watch the show. You should already have researched the needs and interests of potential customers in your site evaluation. Now ask yourself: What would be a welcome change of pace for *them*?

Some examples: Lunch in a bright, casual café provides a respite from the everyday office cubicle. Outdoor dining on an umbrella-covered patio beats the heck out of a windowless skyscraper. Kids deserve a chance to relax and take a break from their classroom routines when they eat in the school cafeteria. And whether it's an upscale dining room or a neighborhood brewpub, a glowing fireplace is incredibly inviting when the weather is cold. Can you think of others?

Knowing a bit about **environmental psychology** will help. This is the study of the deep, even primal reasons people feel certain ways about seating, lighting, music, and other design elements. Two basic human needs are the root of guests' behavior: security and stimulation. Humans like to have their own special space, sufficient for comfort and protected from intrusion. They also like their environment to be interesting and engaging, within limits.

Thinking about your own dining experiences can help you grasp this concept more fully. Do you notice that when given the choice, most people would rather sit at a booth on the perimeter of a dining area than out in the middle of the room at a table? When too much empty space is around us, we feel exposed, and that makes us uncomfortable. If the dining area is too stimulating—too busy, too loud—we are likely to turn and leave. If it is exactly the opposite—we've got the place to ourselves—we also are uncomfortable. Most folks are social creatures who appreciate that they have chosen a popular place—and feel a little self-conscious when they're the only customers.

The key to good dining-space design is to find the right balance between comfort, security, and the guests' tolerance for stimulation, and the target market is key. Younger customers appreciate vivid contrasts in light and color, and they don't mind it if you crank up the music, while the over-50 crowd generally feels out of place in a poorly lit or noisy atmosphere. Will yours be a gathering place for singles or the ultimate power lunch spot for the business crowd? a haven for families? an escape for empty-nesters? Examples of market-driven design details: The choice of large, round tables always leaves room for a latecomer to join his or her dining party. A wide aisle, stretching from the entryway to the bar, spotlights people as they enter and creates an impression of bustling movement. Elevating the bar or raising the tables at the perimeter of the room provides guests with a better view of other tables and activity.

Two distinct paths lead to successful dining design. One is to create the latest hot spot, a trendy statement that becomes a must-visit for sophisticated clientele. Nothing is unimaginable—and everything doable—if the budget allows it! The second path is to create a distinctive dining experience with staying power. Design in this case should include simple, classic details that will still look good in 5 years.

Regardless of the path, *simplicity* seems to be the new watchword in restaurant design. Instead of cluttering walls and shelves with memorabilia, today's destination restaurants are more like public buildings, clean and unobtrusive. Furnishings are attractive, comfortable, and functional. Use of light, interesting fabrics, and stylish shapes have replaced clutter, putting the emphasis on the food rather than the décor.

In Chapter 7, we discuss the specifics of lighting, color, sound and noise control, heating, and air conditioning.

DESIGNING FOR COMFORT

There is some debate about exactly how much impact the look and feel of a foodservice business has on its success as a place to eat. After all, the most important elements are food and service. However, no one can argue that it is the look and feel of the space that invites people to go in—or influences them to pass it up.

In a table-service restaurant, customers spend up to 25% of their time waiting—waiting for tables, waiting to order, waiting for their food to arrive, and so on. They don't feel comfortable if the space is too crowded, and they don't feel comfortable if it seems too empty.

Privacy is a key factor to consider. In your foodservice business, will it be important to your customers to see and be seen? If so, then seating them in a dark corner booth will make them feel slighted. If you want people to choose your place for intimate, romantic getaways, then a brightly lit table at a banquette, with other diners at each elbow, won't work. Perhaps you see the need for both options in the same space? Without answers to crucial questions about what your customers expect from their dining experience, it will be impossible to manipulate your space to meet those needs.

Your location will be one key to diners' needs and expectations. In major metropolitan areas, it is not uncommon to cram as many tables as possible into a public dining space. New Yorkers, for instance, are accustomed to tiny apartments and crowded subways, so perhaps they tolerate the jostling and closeness better than people in suburban or rural areas. It is true that a packed restaurant indicates success and excitement—and designers, who realize this, sometimes exploit it in their layouts.

It may be exciting...but is it safe? The New York City fire codes neither specify nor limit the number of seats based on total square footage. Instead, they require that restaurants maintain clear corridors, 3 feet wide, leading to fire exits. Restaurants with 74 seats or more are required to have two fire exits; those with fewer seats must still have unobstructed aisles, but they can lead to a single exit.

Another consideration in your public-space design is changing demographics. For instance, by the year 2030, about 20% of the U.S. population will be over 65 years of age. For these diners, setting and food quality will replace novelty in order of importance. Changes in patrons' eyesight and hearing should be reflected in design—some of the most common complaints about restaurants in general, from older diners, is that they are "too noisy" or the menu is too difficult to read in low light. Buffets are popular with the older generation; taking leftovers home is perceived as a good value. Nutritional choices (and information about them) will be requested. Senior diners prefer the privacy of a booth to a table and, on a busy evening, would rather wait in a spacious lobby or waiting area than a bar. The population is also becoming more ethnically diverse, with more African-American and Latino customers in most markets.

Don't underestimate the theatrical nature of the foodservice business. Putting food and beverages on display leads to impulse buying, thus increasing sales. Whether you sell wine, appetizers, or decadent desserts, if you are serving the public, think of ways your design will allow you to show off your wares. Speaking of impulses, one of the strongest is vanity— which, in some types of food business, also has an impact on dining room design. Many diners like to see and be seen, particularly in upscale eateries. Having a chef's table right in the kitchen can be fun and exciting for them. Offering a communal table for solo diners to meet and mingle is another option.

The bottom line for any of these factors is how they affect the guests. When you take them all into account, you see that the square footage of a room is meaningless as a measure of space. There is so much more to it!

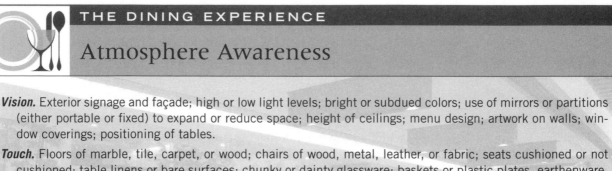

THE DINING EXPERIENCE

Atmosphere Awareness

Vision. Exterior signage and façade; high or low light levels; bright or subdued colors; use of mirrors or partitions (either portable or fixed) to expand or reduce space; height of ceilings; menu design; artwork on walls; window coverings; positioning of tables.

Touch. Floors of marble, tile, carpet, or wood; chairs of wood, metal, leather, or fabric; seats cushioned or not cushioned; table linens or bare surfaces; chunky or dainty glassware; baskets or plastic plates, earthenware, or fine china; plastic, stainless steel, or silver flatware; paper on which the menu is printed.

Sound. Loud or subdued conversation; type and volume of music, live or on the sound system; dishes being cleared; kitchen or bar noise; hum of central heat or air-conditioning system; street noise.

Smell. Aromas of baking or spices; rancid odor of fryer oil that needs changing; colognes of guests and staff; wood in fireplace; smell of new carpet, fabrics, or linens; restroom air fresheners.

Taste. A cool drink; a fluffy soufflé; a crisp onion ring; a perfectly cooked steak; a hot curried dish.

Temperature. The ambient temperature of a room; body heat of guests and staff; heat from the kitchen or coffee station; breeze created by ceiling fans; airflow when seated directly above (or below) a vent or open window; direct sunlight or use of window coverings; hot food served hot; cold food served cold.

Motion. The effort it takes to get to a table or chair; the way servers negotiate the dining room with trays; the setup of waiting lines for buffets (to people waiting, as well as those already dining); activity within the dining space as viewed through windows; outside activity as viewed by diners.

3-2 PLANNING FRONT-OF-HOUSE SPACES

For most foodservice businesses, exterior factors help generate a guest's first impression. These may include:

- *Convenience of location.* Are guests frustrated when they arrive because it was so difficult to navigate the traffic or negotiate the turn into the parking area? Was it almost impossible to find a parking space? Any of these will generate an almost automatic growl, "This better be good...."

- *Outside signage.* Does the business have signage that indicates what type of place it is or what type of food it serves? Is the signage well lit for nighttime hours? Are there indications of specific products (neon beer signs, etc.) that may hint at the atmosphere within? Is there a place to post your menu?

- *Architecture.* Does the building itself suggest the type of eatery it is, or whether it has a theme? Do the grounds look clean and inviting? Is the landscaping well cared for? Even the front door conveys a message. Is it heavy? ornate? sparkling-clean glass, or cluttered with decals?

- *The outside, as seen from the inside.* What will people see when they look out the windows or sit on the patio? Are the tables closest to the windows the best seats in the house because of a breathtaking view? Is there anything you can do to enhance an ordinary view—or will the design team need to minimize it, or block it altogether?

Of course, these are not factors in all locations. However, they must be taken into account when selecting an existing space. Consider traffic patterns, city ordinances that limit the type or size of signs that can be used, and further ordinances that restrict the brightness of the outdoor lighting.

Both quick-service and table-service chain restaurants rely heavily on exterior appearance as a marketing tool. McDonald's golden arches provide instant recognition. The Cheesecake Factory and T.G.I. Friday's restaurants, although not all are alike, have similar exteriors that make it impossible to confuse them with any other chain.

For independent restaurants, bakeries, or catering companies, the only warning is that it is often difficult to inherit a once well-known location and mask its former identity without costly reconstruction. Simply repainting the distinctive pointed roof of an International House of Pancakes, for instance, and opening a Thai restaurant there instead is not a wise decision.

Let's examine each public area of a prospective foodservice establishment. Again, not all locations or business types will require each of these spaces.

ENTRYWAYS AND WAITING AREAS

For some proprietors, space planning comes down to a single consideration: Does every square foot do something to make money for the business owner? However, it isn't always that simple. In public dining, guest comfort factors *are*, in fact, rather simple. Do you want waiting guests to stand outdoors, no matter the weather, because you don't have any room for them indoors? If your concept includes ordering at a counter, where do people line up during busy times without interfering with those already eating?

In terms of environmental psychology, an entryway should be designed to show guests exactly what they are in for when they arrive. Remember, not everyone who enters the space has already committed to eating there, so the entryway is where they will decide whether they are comfortable enough to stay. If your plans include a lively bar or an open kitchen, make sure these are visible from the waiting area or host/hostess stand. Offer a view of some of the dining tables too, and seat those tables first so the restaurant looks busy.

Although some foodservice businesses eliminate reception areas and entryways to save space and money, others use them for a variety of purposes:

- The POS (point-of-service) terminal where people pay is located here.
- The host/hostess stand is located here.
- For quick-service concepts, the front counter is here.
- Menus and daily specials are displayed here.
- Newspaper stands and pay phones (yes, a few still exist!) are placed here.
- Waiting customers sit or stand here.
- Wines or prepared desserts are displayed here.
- Raw foods are displayed, perhaps in glass cases: whole salmon on ice, fresh meats or pasta, lobsters in a bubbling tank.

Whether it is the prelude to an elegant dining experience or the start of a fast-food or cafeteria line, the waiting area should be clean, properly lit, and temperature-controlled for the guests' comfort. (In recent years, we have noticed a trend toward installing thick velveteen or other heavy-fabric curtains to separate the outside temperature from the interior space.) If a waiting or serving line is the norm, make it clear where the line begins and ends, using signs, railings, and partitions.

DINING AREA LAYOUT

There are as many layouts as there are spaces for public dining spots. When designing a dining area, a well-planned scheme carefully shapes the guest's perception with these components:

- Table shapes, sizes, and positions
- Number of seats at each table

- Types of seating (barstools, upholstered armchairs, banquettes, long benches)
- Multiple floors, steps, or elevated areas of seating
- Paintings, posters, murals, bulletin boards
- Type and intensity of lighting
- Planters (and what's in them), partitions, and screens
- Attention to sightlines, to block any undesirable view (restroom, kitchen, service areas) or, in schools or high-security facilities, to allow for unrestricted view of all diners
- Muffling of distracting noises (clattering dishes, outside traffic, nearby construction)
- Placement of service areas (coffee stations, dirty dish bins, etc.)

Each of the preceding considerations plays a role in creating the *flow pattern* of the dining area, the process of delivering food and beverages to customers (or making it easy for them to serve themselves). These are the logistics, the methods and routes used to transfer items from the kitchen to serving stations or dining tables and, finally, to the dishwasher. How well can waiters manage full trays of food? How far is it from the kitchen to the dining area, or from the buffet line to the dining area? Where are the guests' checks prepared, and are they delivered to the tables? Where do customers pay for their food? Does anyone have to hike up and down stairs? When seating guests, does the hostess sometimes seem more like a traffic cop?

Even though other facets of design can be out of the ordinary, when it comes to flow patterns, the simpler the better. Customer and employee safety should be paramount. Remember that most of us tend to walk to the right of oncoming persons. Think of the room as a neighborhood and the flow as the major streets in that neighborhood. Avoid traffic congestion, and everybody likes living there.

A good starting point is to confirm with your local fire department the correct maximum occupancy for your space and plan from there. Then, consider the maximum number of seats allowed in the space as well as the average time you want customers to spend at a table. Generally, the faster the turnover, the greater the need for clear flow patterns that do not cross. Conversely, if dining is to be leisurely, flow should be designed mostly to make the waitstaff seem as unobtrusive as possible.

In terms of environmental psychology, most people would rather sit with some type of architectural fixture on at least one side of them—a wall or window or column—that helps them define their space. Another design rule is that tables of different shapes and sizes should be mixed to create visual harmony. A lineup that is too orderly evokes more of a military mess hall image, although in some settings, that is exactly what is appropriate.

How close is too close when determining table position? A good general guideline is to allow 15 square feet per seat. This figure includes aisles and waiters' service stations but not entryways or restrooms. The figure might change with the shape or size of the room and the sizes of the tables and chairs.

Various types of facilities have industry-accepted standards for dining space allocation. For instance, the more upscale a restaurant is, the more elbow room you allocate for each customer. A fine-dining establishment should allow 15 to 18 square feet per guest; a moderately priced restaurant, 12 square feet per guest; in banquet settings, a minimum of 10 square feet per person. Generally, the lowest space-per-seat allocation is in a school cafeteria, with 8 to 10 square feet.

One factor to take into account is the *seat turnover*, or *seat turn*, which is the number of times a seat is occupied during a mealtime. How much turnover you have depends on the method of service, the time of day, the type of customer, the type of menu and atmosphere, and even the availability of alcoholic beverages at the restaurant.

Another reality of seat turnover is that, even when your dining room is "full," all the seats may not be. A party of two may occupy a table that could seat four, and so on. This partial vacancy rate can be as high as 20% in table-service restaurants or 10% to 12% at cafeterias or coffee shop counters. Vacancy rates don't apply to facilities in which meals are all eaten at the same time, such as prisons and military mess halls. For most eateries, however, table sizes can help control the vacancy rate. Arrange tables for two (*deuces*, or *two-tops*) so they can be easily pushed together to create larger tables if necessary. Quick-service establishments can also try stools and countertops or classroom-style seats with *tablet arms* to accommodate people eating alone.

The variables in table arrangement are endless (see Illustration 3-1). Notice the diagrams don't take into account items like columns, doorways, and architectural features (unusual wall placement) that often exist.

Finally—no surprise here—people usually eat at mealtimes. This means most foodservice businesses are dealing with guests only 20% to 25% of the hours they are open. You will soon learn the patterns to your own customers' dining habits, which will influence the efficiency of your table sizes and groupings. At lunch, for example, you may find at a busy downtown café that half of the customers arrive as twosomes, another 30% are singles or parties of three, and the remaining 20% are parties of four or more.

CAFETERIA DINING. For large-group feeding facilities—cafeterias, hospitals, prisons—the service area takes on a complexity rarely seen in a table service or fast-food eatery. An institutional kitchen may need as much as 2000 to 3000 square feet of service area, as this is where serving lines are set up in a multitude of combinations:

> Straight serving line (see Illustration 3-2)
>
> Shopping center system
>
> Scramble (or free-flow) system

The **straight line** is exactly what its name implies. In terms of speeding customers through the food line, it is the slowest-moving arrangement, as most guests are reluctant to pass slower ones in front of them. However, single or double straight lines are still the most common design in commercial cafeterias because they take up the least space and the average guest is comfortable with the arrangement. As customers must walk by all the food choices, they are also more likely to make an impulse purchase (see Illustration 3-3).

The *shopping center* (also called a **bypass line**) is a variation of the straight line. Instead of being perfectly straight, sections of the line are indented, separating salads from hot foods and so on. This makes it easier for guests to bypass one section. In serving lines where burgers, omelets, or sandwiches are prepared to individual order, the bypass arrangement keeps things moving.

The *free-flow* or **scramble system** is designed so that each guest can go directly to the areas he or she is interested in. (Once in a while, you'll hear it referred to as a **hollow-square system**.) Food stations may be laid out in a giant U-shape, a square with islands in the middle, or just about any shape the room size will permit (see Illustration 3-4). This design can be attractive but is often confusing for first-time customers. You are most likely to find this layout in an industrial cafeteria, where employees eat every day and soon become familiar with it. Scramble systems offer fast service and minimal waiting. They also allow for some types of exhibition cooking, including items grilled, stir-fried, or sliced to order.

SELECTING CHAIRS AND TABLES. A major component of dining space design is the type of seating that will be offered. Different types of tables and chairs take up very different amounts of space, and the styles you choose will make a statement about the atmosphere of the establishment.

Popular types of seating include chairs, stools, booths, and banquettes. Booths offer a certain feeling of privacy or intimacy, but tables and chairs are more adaptable because they can be moved around as needed. A **banquette** is an upholstered couch fixed to the wall, with a table placed in front of it. Banquettes are a hybrid of booth and table—more adaptable than a booth, but they still must hug the wall. Banquettes happen to be fashionable at the moment. Not only can they be upholstered in any number of stylish fabrics but also they maximize seating by filling up corners and allowing more guests to be seated than would fit at individual chairs. Barstools, either at bars or taller cocktail-style tables, are the most casual seating option.

TYPICAL SEATING LAYOUTS

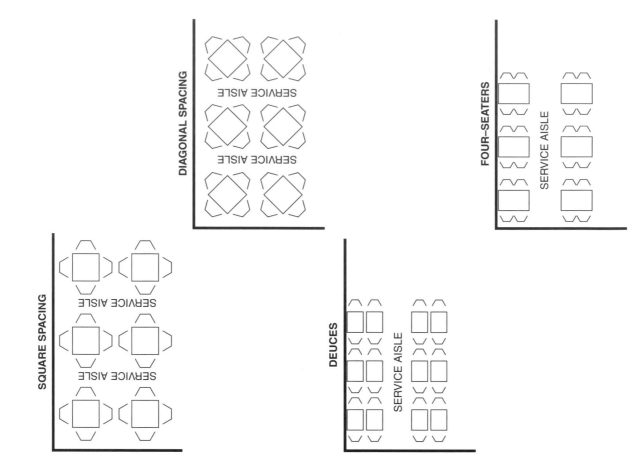

SQUARE SPACING

SERVICE AISLE

SERVICE AISLE

DIAGONAL SPACING

SERVICE AISLE

SERVICE AISLE

DEUCES

SERVICE AISLE

FOUR-SEATERS

SERVICE AISLE

FLOOR PLAN GUIDE

All suggestions given are approximate and minimum. It should be pointed out that no rule of square feet per person can be exact, because too many variables exist. The space consumed by entry and kitchen door aisles, for example, is almost equal in a room of 800 square feet and a room of 1600 square feet, but the percentage of the room used is less in the latter. Seating capacities can be determined only by a final layout, but the approximate capacity of a room can be determined by this rough guide:

Banquet or institutional seating: 10–12 square feet per person
Cafeteria or lunchroom seating: 12–14 square feet per person
Fine dining: 14–16 square feet per person

SUGGESTED TABLE SIZES:

	Banquet Institutional	Lunchroom Cafeteria	Fine Dining
2 persons	24" × 24"	24" × 30"	24/30" × 30/36"
4 persons	30" × 30"	30" × 30"	36" × 36" or 42" × 42"
4 persons	24" × 42"	24/30" × 48"	30" × 48"
6 persons	30" × 72"	30" × 72"	48" diameter
8 persons	30" × 96" or 60" diameter	30" × 96"	60/72" diameter
10 persons	72" diameter	30" × 120"	96" diameter

Note: In self-bussed tray service cafeterias, tables should be of adequate size to accommodate the trays.

SUGGESTED MINIMUM AISLE DIMENSIONS:

	Customer Access Aisles	Service Aisles	Main Aisles
Institutional Banquet	18"	24/30"	48"
Lunchroom Cafeteria	18"	30"	48"
Fine Dining	18"	36"	54"

Note: Allow 18" from edge of table to back of chair in use. For diagonally spaced tables, allow 9" more between corners of tables than needed for the type of aisle needed (e.g., for 30" service aisle, allow 39").

As rough rules of thumb, remember that tables laid out diagonally will increase seating capacity, and a smaller quantity of tables with greater seating per table increases seating capacity but reduces flexibility.

ILLUSTRATION 3-1 Floor plan guidelines.

Source: Carl R. Scriven and James W. Stevens, Manual of Equipment and Design for the Foodservice Industry (Albany, NY: Thomson Learning, 1989).

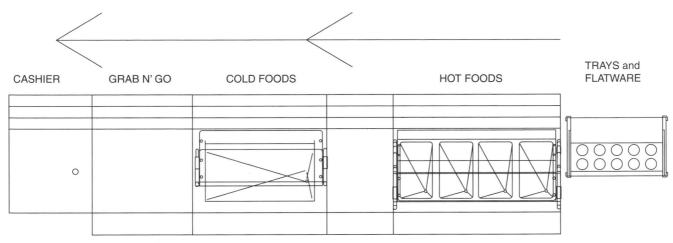

ILLUSTRATION 3-2 Serving line.

Courtesy of MVP.

In general, it is best to design a dining area with a mixture of seat and table styles for maximum flexibility. Think about the ability to accommodate both large and small groups at a moment's notice. Seating also must be arranged with clearly defined aisles and tables not too far from the kitchen, as these factors affect safety as well as speed of service.

Because they are a major investment that will probably be in use for a long time, great care should be taken to choose seats that are comfortable, durable, adaptable, and appropriate to the type of dining you will offer. The typical restaurant chair has a life of 5 years, but the best ones may last 10 or more. Chairs are part of the overall design of the room, so the style you select should be in line with the image and ambience of the room. You may hear the term *scale* used in seat selection. The scale of an object is your visual perception of its size. For instance, look at the two chairs in Illustration 3-5. When you compare the captain chair to the Windsor chair, they're actually about the same size, but the Windsor chair appears lighter and more delicate. This illustrates the difference in scale between the two.

Once you have selected a chair style, home in on the technical aspects of its construction. Less expensive chairs may be glued or even stapled together—not the optimum for durability. Upkeep and maintenance are important, including whether the manufacturer will keep spare parts available over the years. Frames can be made of metal, wood, or plastic. They can be stained, dyed, painted, or lacquered; stained and dyed frames are the easiest to maintain, and dyeing allows for an endless choice of colors. Seats may or may not be upholstered. The chairs can have arms, but only if your tables are roomy enough to accommodate them. Ask about protective laminate finishes for wooden chairs, which would otherwise chip and dent easily.

Check for design flaws that would be troublesome in a public setting: Does clothing catch or snag on edges? Are any of the edges sharp enough for people to scrape or cut themselves accidentally (guests or the waitstaff)? Are the legs wobbly? How will the chair hold up when

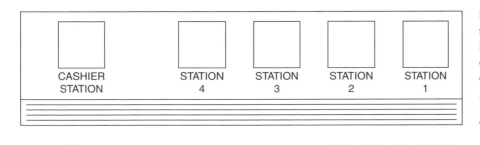

PATH OF PATRON—NO PASSING

IILLUSTRATION 3-3 Most cafeteria serving lines are set up as straight-line patterns, although on a busy day this is often the slowest-moving option.

Source: Robert A. Modlin, ed., *Commercial Kitchens*, 7th ed. (Arlington, VA: American Gas Association, 1989).

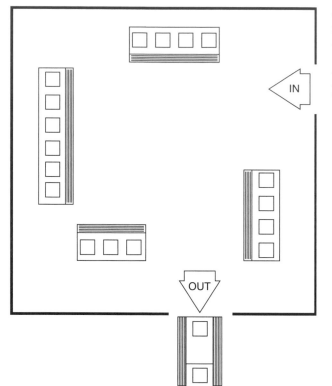

IILLUSTRATION 3-4 The hollow-square serving arrangement lets cafeteria diners wander between stations.

Source: Robert A. Modlin, ed., *Commercial Kitchens*, 7th ed. (Arlington, VA: American Gas Association, 1989).

a guest who is quite overweight sits in it? If you have many female guests, can a handbag or jacket be slung over the chair back without sliding to the floor?

It is wise to order samples of several chairs and test them for a week or two. Here are some specifics that may help in chair selection:

■ A 15-degree angle for the chair back is recommended.

■ The depth of the seat, from edge to chair back, should be 16 inches.

■ The height of the chair, from the floor to the top of the chair back, should be no more than 34 inches. Anything higher impedes servers.

■ The standard distance from the seat to the floor is 18 inches.

■ The distance between the seat and the tabletop should be 12 inches.

■ Allow 24 to 26 inches of space for each chair at a table—28 inches for armchairs.

■ For bar or countertop seating, allow 24 to 26 inches per stool.

Consider how the chairs or stools work when empty, too. Do they fit under the tables or the bar armrest? Can they be pushed in easily when not in use, to make the aisles roomier? Are they stackable? Are they easy to clean and easy to move when the floor needs cleaning?

Booth seating is another common choice. You used to see booths only in bar lounges and casual restaurants, but they also look smart in upscale eateries, where they afford a sense of privacy and even romance. Booths can save space, taking up as little as 8 square feet per person. However, booths are more labor-intensive than tables because it is harder to clean beneath and around them. Also, they can't be moved to accommodate various sizes of dining parties.

Deciding on your overall dining atmosphere will help greatly in table selection. Space usually dictates how many tables you will need, and, in most cases, you can get more square tables than round ones in the same square footage. Research shows square tables also seem to produce faster turnover, while round ones prompt guests to linger a bit longer. Attention to aesthetics may require that you blend both square and round tables in your dining area, arranging them at different angles to avoid that military mess hall look mentioned earlier.

Tables should be chosen in tandem with chairs, as they will be used together. For example, tables that are 26 inches tall work best with chairs that measure 16 inches from seat to floor; 30-inch-tall tables work best with chairs that measure 18 inches from seat to floor.

The latest trend in more casual dining situations is the counter-height table, popular enough to have migrated into home dining rooms as well. Also sometimes called *pub tables*, these are 36 inches tall rather than the standard-table height of 28 to 30 inches and, of course, require taller chairs or barstools.

When purchasing tables, check for sturdy construction. You want long use and solid service from them. Self-leveling legs or bases allow you to adjust for wobbles and also permit the table to glide easily along the floor if it must be moved. A recent invention, brand-named Table Shox, is a self-adjusting hydraulic glide that works like a tiny shock absorber to adjust to uneven floor surfaces and prevent wobbling.

Think about whether you will cover the tables with linens, butcher paper, or nothing at all. You must make an early decision about the type of finish you want on your tables, especially if you won't be using tablecloths. Choices range from marble, wood, and ceramic to durable plastic laminates such as Formica and Corian, which are stain-resistant and easy to maintain. Nowadays they come in many patterns, including faux marble, which would work even in an upscale restaurant setting.

No matter what you decide, a table should have a waterproof top, and its **base** should be placed to give customers a comfortable amount of leg room beneath. If lighting will be low and your tables draped, a simple pedestal-style base is appropriate. However, if the dining space is airy and open and the tables will not have cloths, the style of the table base can be part of your design. Table bases don't just come in chrome, brass, or black enamel anymore. Today's trends range from fire engine red and deep evergreen to plated finishes of copper, pewter, or bronze. Those little feet at the bottom of the base that hold the table steady also come in different styles. You will probably choose between the so-called four-pronged spider base and the cylindrical mushroom base.

If your business will be the site of receptions or other special events that require buffet setups, you might consider stocking some banquet tables. A few exotic shapes in the banquet table universe are designed to be modular: arc-shaped *serpentine* tables, which can be placed end to end to create an S shape; graceful, rounded *ovals;* *trapezoids* that create a solid hexagonal (six-sided) table when two are placed together; *half-rounds* and *quarter-rounds*, which can be placed at the ends of rectangular tables to round them out for buffets or extra seating space. Take a closer look at Illustration 3-6 and note the creative ways table shapes can be arranged.

Banquet tables are probably the hardest-working pieces of furniture in the business. They should be made with rugged steel underpinnings and reinforced with angled iron rails that run the length of the table. Each leg should have its own brace to lock it in place, as well as a self-leveling device sometimes known as a **glide**.

The lightweight plywood tabletop may seem flimsy in comparison to its underpinnings, but this makes the banquet table easier to transport, handle, and store. If metal edging rims the table, it should be firmly riveted to the wood as a safety feature and for ease of storage.

Almost everyone knows that seating comfort at a banquet-style event is not exactly top priority. Indeed, the most popular large group seating choices are folding chairs and stackable chairs, selected primarily for ease of storage and handling. If you're in the market to buy them, use the same guidelines you would for other chairs: comfort, durability, and adaptability. Folding chairs should fold and unfold easily, have strong locking mechanisms, and be sturdily constructed. Stack chairs, as they are often called, should be easy to stack as high as a person can reach without marring other chairs beneath them. Also ask for **wall-saver legs**, which extend beyond the chair back and prevent the chairs from rubbing or marring the wall they are stacked against.

Captain

Windsor

IILLUSTRATION 3-5 These two chairs are similar in size, but their styles give them different looks.

SERVICE AREAS AND WAITSTATIONS

Generally, whether it is called a service area or a waitstation, many foodservice facility designers view this area with some apprehension. And little wonder, as it is actually an extension of the back of the house that happens to be located in the front of the house. This is the busy zone between the kitchen and the guests, where the food production staff and waitstaff use the most efficient means to get food out of the kitchen and into the hands of their customers.

The teamwork on both sides of the pass window evolved into today's service area. At a small table-service restaurant, it may be an area no larger than 4 to 6 feet long that houses the pass window. Many eateries hire people called **expediters** to stand on either side of the window

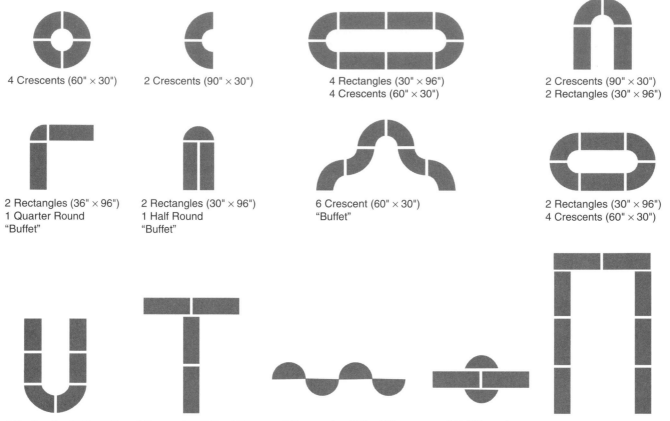

4 Crescents (60" × 30") 2 Crescents (90" × 30") 4 Rectangles (30" × 96") 2 Crescents (90" × 30")
4 Crescents (60" × 30") 2 Rectangles (30" × 96")

2 Rectangles (36" × 96") 2 Rectangles (30" × 96") 6 Crescent (60" × 30") 2 Rectangles (30" × 96")
1 Quarter Round 1 Half Round "Buffet" 4 Crescents (60" × 30")
"Buffet" "Buffet"

4 Rectangles (30" × 72") 4 Rectangles (36" × 96") 2 Rectangles (30" × 96") 4 Half Rounds 8 Rectangles (36" × 96")
2 Crescents (90" × 30") 2 Half Rounds "Buffet"
"Buffet"

IILLUSTRATION 3-6 When arranging buffet tables for large groups, the only limitation is your imagination.

THE DINING EXPERIENCE

Table Size Options

The standard table, from floor to tabletop, should be 30 inches in height; highboy cocktail tables are 42 inches in height. Here are basic table sizes and uses:

- For one or two guests: a 24 × 30-inch square table, also known as a two-top or **deuce**.

- For three or four guests: a 36 × 36-inch square table; a 30 × 48-inch rectangular table, commonly called a *four-top*; or a 42-inch-diameter round table. (There are 36-inch-diameter round tables, but they are a tight fit for four persons.)

- For five or six guests: a two-top and four-top can be joined to create seating for up to six, or use a 48-inch-diameter or 54-inch-diameter round table.

- For seven or eight guests: two four-top tables can be joined, or use a 72-inch-diameter round table.

- For bars and cocktail lounges: a 20 × 20-inch square table or a 20-inch-diameter round table.

- For bars and reception-style events: highboy tables in 24-inch, 30-inch, or 36-inch rounds. So named because they are tall (42 inches), these can also be ordered with adjustable-height bases for maximum versatility.

- Banquet spaces often use folding tables. These are typically 30 inches wide and come in two lengths: 72 inches (seats six) and 96 inches (seats eight). Narrower tables, from 15 to 18 inches wide, are designed to seat people on one side only. This is known, for obvious reasons, as **classroom-style seating**.

and facilitate ordering and order delivery. An expediter may organize incoming and outgoing orders for speed, check each tray for completeness and accuracy of the order before it is delivered, and even do a bit of last-minute garnishing as plates are finished. When standing in the kitchen to do this, the person is sometimes known as a *wheel person* or *ticket person*.

In quick-service establishments, the service areas are the counters, clearly visible to incoming customers. A recent innovation is the separation of beverages and condiments from prepared foods. Nowadays, customers fill their own drink cups, dispense their own mustard or ketchup, and so on.

Generally, you should provide a small waitstation (20 to 36 inches, square or rectangular in shape) for every 20 to 30 seats, or a large one (as large as 8 to 10 feet long and 24 to 30 inches wide) for every 50 to 75 seats.

The need for a large waitstation depends mostly on how far away it is from the kitchen and food pickup area. Placement of waitstations also depends on the availability of utilities, as the stations need electricity and water. Also, design them with counter space, work surfaces, and overhead shelving in mind. This, of course, requires separate, detailed drawings of the station itself, as shown in Illustration 3-7.

The waitstation typically has no inherent eye appeal, yet it is an absolutely necessary component of an efficient service system and must be stocked with everything the waitstaff uses regularly. A likely list will include the POS terminals for placing orders, calculators, bread and butter, bread baskets, all coffee-making and -serving paraphernalia, assorted garnishes, salad dressings and condiments, dishes and flatware, water pitchers and glasses, ice, napkins and tablecloths, bus tubs, bins for soiled linens, tray holders, and so forth. Sufficient counter and storage space will help busy servers stay organized and save time.

There are two distinct schools of thought about whether waitstations should be the repositories for some prepared foods: soups, salads, precut pies and cakes, and so on. While one group says it makes service of these items faster, another insists it also increases pilferage by staff members, therefore increasing costs.

Where should your waitstation(s) be located? The shorter the distance between waitstation and kitchen, the better off you will be. Long hikes between the two actually increase labor costs because waiters spend their time going back and forth. It is also harder to keep items at proper temperatures if they must be carried for longer distances.

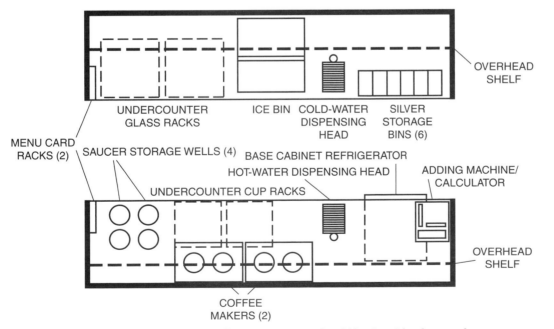

IILLUSTRATION 3-7 Even an area as small as a waitstation should begin with a layout plan, on paper or on computer.

If your service area is immediately adjacent to your kitchen, a few simple design precautions will make it safer for employees and provide a better flow pattern for service traffic. Many of these areas have swinging doors that link them to the kitchen. If possible, choose double doors, 42 inches in width, with clear, unbreakable plastic windows in each door that should be no smaller than 18 inches high and 24 inches wide. Each door should be installed to swing only one way and clearly marked IN or OUT. The doors should be separated by at least 2 feet. When refrigerators are located in the service area, glass doors will help save energy by allowing servers to see the contents without opening the door. Also, consider sliding doors instead of doors that swing open.

The floor is another important consideration. Make it nonskid, just like the kitchen floor. (Options for flooring materials are detailed in Chapter 8.) And, because there is always the possibility of a spill, be sure there are adequate floor drains. Finally, because of the intense use of this space and the interaction of your servers and production line people, this area should be well lit, like the kitchen.

OUTDOOR DINING

IILLUSTRATION 3-8 To keep outdoor dining spaces warm on chilly nights, you have lots of patio heaters to choose from. Even if such heaters are used strictly outdoors, always check with your local fire department in case of specific rules about their use, minimum distance from guests, and the like.

Courtesy of Teeco Products, Inc., Irvine, California.

Nothing is more pleasant than dining outdoors on a lovely day, and adding a patio or sidewalk tables has a number of benefits. First, this move can expand seating capacity (at least during part of the year), and it can be a good way to use otherwise wasted space. If it can be seen from the street, a patio adds liveliness and action to an otherwise sedate storefront. In addition, it can attract a different type of customer, the more casual drinks-and-snacks and after-work crowd.

On the other hand, weather-related data has been collected on many cities that can tell you just how many days a year an outdoor space could effectively be used. Be honest with yourself about whether it will pay off for you. Weather is the wild card, and some foodservice operators have learned to be creative about it. Large tabletop umbrellas can shelter guests both from rain and too much sun, but if it's a real downpour, be prepared to reseat indoors. On days when the skies appear threatening, keep a few indoor tables open just in case.

To adapt to overly warm days, put up the big umbrellas once again, install an awning (retractable or not) over the space, or outfit the patio covering with misters that spray a fine mist of water on guests. This can be refreshing and fun on a hot day—or it can make mascara run and guests feel like fresh produce on a supermarket shelf, depending on how well the misters are adjusted.

In cold weather, a number of attractive patio heaters (see Illustration 3-8) are available that use natural gas or LPG. The Grape Escape, a popular wine bar in Boise, Idaho, puts monogrammed wool blankets at its outdoor tables on chilly spring and fall evenings for guests to wrap around their shoulders or legs.

One design idea we really like is the advent of garage-door–style walls, as shown in Illustration 3-9—they're windows in cold or inclement weather but can be retracted in good weather to make indoor space into outdoor space. Isabel, a casual restaurant in the trendy Pearl District in downtown Portland, Oregon, does not allow an infamously rainy climate to hamper the possibilities for open-air dining.

Don't plan to add outdoor seating without checking with your city planning and zoning department. Most cities have rules and require special permits for outdoor seating, especially if you plan to use the public sidewalk. Tables and chairs should be chosen for ease of maintenance and durability; decide before you purchase whether you'll bring

IILLUSTRATION 3-9 Isabel's exterior, with its retractable windows, allows an easy transition to open-air dining as weather permits.

Courtesy of Isabel Cruz Restaurants.

them indoors or chain them up at night to prevent theft. Either way, collapsible or stackable furniture simplifies storage.

Patios bring other challenges, too, including traffic noise and insects. When it comes to pest control, outdoor diners have to learn to live with nature, to a certain extent. Bees, flies, mosquitoes, and birds can not only be bothersome but actually pose a sanitation risk. Outdoor pest spraying is only marginally successful, so it is important to clean up spills and remove dishes and glassware from tables as soon as possible after customers have finished using them. Keeping food covered as long as possible and using decorative citronella candles on tables are two ways to cope. Employees should be on the lookout for bird droppings and keep them cleaned up—and never with dishcloths or napkins!

Pests of the human variety also are inevitable. In Seattle, we witnessed a heated exchange between a vagrant (who had decided to seat himself on a newspaper vending box to view the outdoor diners a few feet away at a nice Italian café) and the restaurant manager. Their shouting match made everyone in a two-block radius uncomfortable. Be prepared for the uninvited "guest"—and prepare a policy that will guide you in dealing with these situations and help you keep your cool.

RESTROOMS

The single most critical public perception of a restaurant is that if the restroom is clean, so is the kitchen— and, of course, the reverse is also considered true. All too often, especially during busy times, the restrooms are neglected because no one on staff has specific responsibility for checking on them.

THE DINING EXPERIENCE

Top 10 Design Mistakes

1. *Inconsistent ambience.* Theme, space, layout, and flow patterns. Colors, materials, lighting, graphics, and artwork. Uniforms, linens, menus, dishes, and signage. All should be complement each other. When they do, it implies great attention to detail. The big restaurant chains seem to have grasped this centrality of design continuity, but small restaurants, probably in their quest to cut costs, seem to treat visual aspects of the dining experience as an unnecessary expense.

2. *Too many people involved in the decision-making process.* It is good to have candid input and a variety of opinions in the design process, but a single person must drive the design program. Otherwise, you end up with a mishmash of ideas and skyrocketing costs as you experiment with them. Be sure to include the chef in decision making.

3. *The target market is forgotten.* This happens when the owner or manager defines the tone, menu, and concept based on what he or she wants instead of what truly fits the location. This mistake is made often in suburban areas, where the customer base can afford an upscale concept but the restaurateur opts for a middle-of-the-road concept.

4. *Inadequate space between tables.* It is not unusual to underestimate the level of intimacy in a given concept, but you should leave at least 3½ feet between tables so servers can do their jobs comfortably. Casual dining can cut the space a bit, to a range of 24 to 36 inches, but fine dining calls for a full 4 feet between tables.

5. *Traffic patterns are overlooked.* This refers to the movement of people within the space, both guests and employees. Major sins include placing the kitchen too far from the dining room and restricting movement into and out of the kitchen. Good service and comfort are the result of uncluttered circulation patterns, or *flow,* which is discussed in greater detail in Chapter 3.

6. *Unrealistic budgets.* There are no fixed budget guidelines for restaurants, which results in too much variation. It can cost anywhere from $50 to $60 per square foot to design a casual eatery, while a formal dining room may run $400 per square foot. Unrealistic budgets come back to haunt a project. Most major expenses are covered, but what about the so-called soft costs—insurance, permit fees, advertising? Some consultants suggest a contingency budget of 10% to 20%, just to cover possible cost overruns.

7. *Cutbacks in non-revenue-producing space.* It is important to balance the idea of more seats in the dining room with sufficient space in the kitchen to produce the foods the menu promises. Municipal fire and building codes and requirements of the ADA (see Chapter 4) must be met.

8. *Poor lighting.* Lighting is more than the diminishment of darkness. A lighting mistake can actually be offensive to guests if the results are too harsh, too dim, or at angles that make the atmosphere uncomfortable. Pick a lighting style and stick with it rather than using four or five types of fixtures to achieve everything from walkway lighting to enhancing food presentation to making diners look good. Never light directly at eye level.

9. *Offensive colors.* The use of color affects the way the food looks as well as the mood of the dining room. Blue feels frozen in space; warm hues, such as reds or browns, are generally more inviting. Consider colors that either invigorate or relax the dining space.

10. *Forgetting the future.* Restaurants need a plan for growth so they can respond to it without major expense. You can do things like expand the dining area and add a banquet room or private dining room only if the original design was created to allow for this at a reasonable cost.

Source: Adapted from information by Howard Riell in *The Consultant*, publication of Foodservice Consultants Society International, Louisville, Kentucky (Third Quarter 1998).

Research published over the years in *Restaurant Hospitality* magazine indicates clean toilets, ample amounts of toilet paper, and soap in the dispensers are three items most definitely noticed and appreciated by customers—who are wary if these details are not attended to. A majority of people surveyed also said they get fed up with wet surfaces and overflowing trash cans, which they see as an indication that no one on the staff is checking the restroom and tidying up regularly.

Although customers don't spend more than a few minutes in the restroom, this short time affects the rest of their dining experience. So why does the concept of atmosphere seem to be left outside the restroom doors of so many foodservice businesses? Your guests' expectations are simple: cleanliness, privacy, and comfort. Why not pay the same attention to décor here as you do in the dining area, with plants or attractive wallpaper or artwork reflecting your theme or mood? Temperature control and lighting are critical here too; select warm, incandescent lighting or color-corrected fluorescents that are bright without being harsh.

Luckily, a few cosmetic touches add to the perception of cleanliness in a restroom. The use of slate or easy-to-clean tile (smaller tiles on the floor, larger ones on the walls) is one option. Stark white is not a good color for walls because it's hard to keep clean; conversely, dark colors are not perceived as clean, even if they are, so it is best to pick a lighter shade. The stall partitions don't have to be dull gray metal when there are several sturdy options in attractive, solid colors that discourage scratched-on graffiti. If money is no object, beautiful hand sinks and touch-free fixtures are sold in many styles and colors.

In design circles, there are two schools of thought about restroom placement. One group thinks they should be located near the entrance so guests can freshen up before dining or departing; the other thinks they should be located discreetly at the back of the dining area. In reality, restrooms are most often sited based on the location of the water lines—which often means near the kitchen. However, guests should never have to walk through the kitchen to use the restroom.

Another pet peeve of some diners is waiting outside a locked restroom door in a narrow hallway or in view of other guests. If at all possible, divide your restroom into a small waiting area with sink and mirror and at least one stall with a locking door. In foodservice, there generally is not enough space for restroom facilities to be lavish, but they should at least be roomy enough to be practical. Any woman will tell you how difficult it can be to wedge herself into a tiny restroom stall in order to get the door closed, but this happens more often as businesses reduce space to cut costs. A minor annoyance—but an annoyance indeed.

Amenities to consider: Exhaust fans should operate whenever the toilets are occupied, and soft incandescent lighting is preferable to harsh fluorescent lighting. Make sure the stall doors properly close and latch. Provide hooks on stall doors to hang coats or handbags. Install high-quality, undistorted mirrors, including full-length mirrors. Provide privacy screens between men's urinals. Complimentary mouthwash, hand lotion, and hairspray can sometimes be found in the women's restrooms of fine-dining establishments.

In family-oriented eateries, how about diaper-changing tables in restrooms of both sexes? (The String Bean, a family-style restaurant complete with playroom in Richardson, Texas, provides spare diapers and baby wipes in its restrooms.)

RESTROOM SIZE REQUIREMENTS. Minimum restroom space requirements are spelled out in city health ordinances and typically depend on square footage or total seating capacity. They include the number of water closets (the common legal name for toilets in stalls), urinals for men's rooms, and lavatories (washbasins) for handwashing. In the western United States, most local or state codes are based on the Uniform Plumbing Code (UPC); in the eastern United States, they're more likely to be based on the International Plumbing Code (IPC); and there's also a National Standard Plumbing Code. A current list of codes adopted by state can be found on the website of the American Restroom Association.

For the small business, with up to 50 seats, a 35- to 40-square-foot area is the absolute minimum for a toilet and washbasin, although a check with your local health department will confirm the requirements in your area. Table 3-1 is an example of fixtures suggested by author Fred Lawson in the book *Restaurants, Clubs, and Bars*. Lawson believes a facility with up to 70 seats should allow for 75 to 80 square feet of restroom space. He also believes the fancier the restaurant (the higher the check average), the roomier the restrooms should be—one urinal for every 15 guests, for instance.

The ongoing debate in many jurisdictions has a whimsical name that belies a very real concern. "Potty parity" is an answer to women's frequent complaints that the number of toilets for female guests is almost never sufficient, especially in busy, high-volume venues. While it's true that women spend more time in the restroom than men do and expect enough space for a modicum of privacy, proponents of the IPC—which requires fewer fixtures for some situations than the UPC—call the potty parity idea "faddish" and say their code is based on research.

TABLE 3-1

Ratio of Restroom Facilities per Guest

FIXTURE	MALE GUESTS	FEMALE GUESTS
Toilet	1 for every 100	2 for every 100
Urinal	1 for every 25	—
Washbasin	1 for every toilet or 1 for wevery 5 urinals	1 for every toilet

Source: Fred Lawson, *Restaurants, Clubs, and Bars*, 2nd ed. (Oxford, UK: Architectural Press, 1995).

Two additional legal requirements govern restroom space. One is that, in most cities, places that serve alcoholic beverages *must* provide separate restroom facilities for men and women; typically, no unisex toilets are permitted where alcohol sales exceed 30% of total sales. The other, which we cover later in more detail, is the ADA, which mandates accessibility measures and space requirements to accommodate guests with physical limitations.

It is advisable to have separate restroom facilities for staff and customers; however, this is not always possible. We have noticed that most restaurants with separate staff restrooms offer a 30- to 40-square-foot unisex facility with one toilet and one washbasin.

BUILDING AND GROUNDS

A Flush of Pride: The American Restroom Association

Yes, someone is tracking trends and best practices in the design, comfort, and safety of restrooms. The nonprofit American Restroom Association was formed in 2004. It hosts an annual World Toilet Summit and Expo and a World Toilet Day (in November) to call attention to those facilities that everyone seems to take for granted: their restrooms. From the group's website (www.americanrestroom.org), here are its priorities:

SCOPE OF INTEREST

- Restroom design and technology
- Restroom availability and accessibility
- Pertinent legislation and regulations
- Documenting the problems faced by people who hesitate to travel or who avoid activities that put them out of range of proper toilet facilities

GOALS

- Generate public relations campaigns that result in positive media coverage.
- Address regulatory and legislative weakness.
- Act as a health impact clearinghouse.
- Survey and develop Municipal Friendliness Ratings.
- Communicate with similar associations around the world.
- Develop suggestions, brochures, and suggested designs.
- Serve as a clearinghouse for companies and individuals promoting these products and designs.
- Develop lines of communication with mall and building managers, architects, builders, and other groups that can change restroom design.

MULTIPURPOSE SPACES

A trend in front-of-house design is to stretch the usefulness of every square foot of space beyond the rather limited scope of one or two day-parts (i.e., lunch or dinner, or lunch *and* dinner), making the facility inviting in ways that prompt guests to visit anytime, day or night. Design experts break mealtimes down in terms of guests' needs: relative solitude in the morning; speed and convenience for lunch; comfort in mid-afternoon; value in the evening; and a "scene" to participate in for late night. The Globe Restaurant in New York City's Gramercy Park is a fine example of maximum functionality:

■ The day begins with self-service breakfast items, such as muffins and coffee, ordered at a circular steel counter.

■ At lunch, the same counter becomes a buffet bar where cooks dish up soup, panini sandwiches, and pastas.

■ At dinner, the counter is transformed once again into a raw bar, where shrimp, clams, and oysters are displayed with bottles of chilled white wine.

■ Late at night, the lights are turned down, barstools emerge from on-site storage, and the counter area becomes a coat-check stand. The expediting kitchen and shelves that held soft drinks and prepared salads earlier in the day are hidden from view behind vertical blinds and shades.

One way to make dining space more profitable is to add takeout capability to a seated-dining or cafeteria concept. People who stop in to pick up food can be a lucrative target, as long as you don't take it personally that they aren't going to linger long enough to enjoy the atmosphere you worked so hard to create. Take-out counters are growing in popularity, although planning this type of space includes special considerations:

■ *Speed.* The faster you can serve to-go customers, the better. Drive-through has become the gold standard, and not just for quick-service chains. This adds entirely different design components.

■ *Space.* How many people will line up? Is there room at the counter to deal with more than one customer at one time? room for one to order while another pays?

■ *Menu.* A to-go menu can be smaller, featuring easily packaged items. It can work well for both upscale and homestyle cuisine.

Ideally, the take-out counter should be located within 30 to 50 feet of the entrance, attractively displayed to take the guesswork out of decision making and allow easy viewing of multiple choices. There should be no cross-traffic between the take-out counter and the regular sit-down dining area, as their purposes and moods are mostly different.

Of course, the school cafeteria is perhaps the most flexible multipurpose space, which often must also be used for evening meetings, indoor physical education activities in inclement weather—and as the neighborhood's polling place for local elections! Take a hint from school facilities and build adaptability right into the design.

BANQUET AND MEETING SPACE. The ability to host social events and business meetings can help with your facility's bottom line and, in fact, some types of foodservice establishments are required to have this type of flexibility, including corporate and school cafeterias, convention centers, and state capitol buildings. In terms of design and equipment, a few requirements are called for:

■ *Technical savvy.* Most events make use of some sort of audiovisual (A/V) equipment, and many require wireless Internet access, videoconferencing capability, large video screens, and more. Incorporating these into the design of the space makes a lot of sense. While you don't necessarily have to purchase and store all the A/V equipment, you must have a good relationship with at least two professional A/V companies that can subcontract for these services.

IILLUSTRATION 3-10 The Party Pleaser™ Bar is a good example of a portable bar, on 5-inch casters. It's made of stainless steel and accommodates pre- or postmix dispensing systems.

Courtesy of Lakeside Manufacturing, Inc., Milwaukee, Wisconsin.

- *Noise abatement.* Depending on the time of day and the location of the room, the design must take into account what else is going on in the building. If it's a restaurant, how can a private party be corralled sufficiently so as not to disturb other diners? If it's a hotel, sleeping guests might be a factor.
- *Banquet-related equipment.* Round banquet tables, lecterns, easels, risers or platforms, room dividers, large coffee urns, a portable dance floor—any of these might be useful for events. In each case, you can rent the items as needed, or purchase them. Another option: a portable bar, shown in Illustration 3-10.

Pre-function space is almost as important as the meeting room itself. In many situations, events begin in this area: meeting registration, coffee service, pertinent displays, and so on. Is your space conducive to this, without disturbing other guests?

Another decision you must make early on is whether to use round or rectangular tables for banquet seating. It is especially important when planning this space to allow enough room for aisles, as the waitstaff will truly be bustling (with full trays) in this environment. With the right tables and good plans on paper, the same room can take on different personalities for every occasion, as shown in Illustration 3-11.

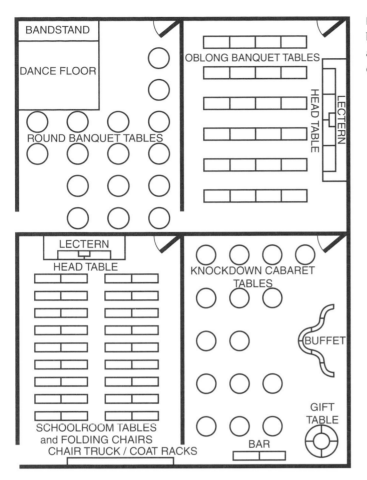

IILLUSTRATION 3-11 When choosing chairs, tables, and banquet accessories, the goal is to make the room as versatile as possible for different occasions. This figure shows four different layouts in the same space.

Here is a handy formula for calculating banquet seating: If you are using standard rectangular tables, divide the square footage of the room by 8 to find out how many seats the area will accommodate. For instance, a 500-square-foot area, divided by 8, will seat 62 or 63 persons.

When using round tables (of any standard circumference), divide the square footage of the room by 10. A 500-square-foot area with round tables will seat 50. This formula allows room for chairs as well as space for aisles. Its use is limited only by columns, entrances, or service doors that would restrict people sitting in those areas.

PRIVATE DINING SPACE. For restaurants and corporate foodservice, a private dining room can serve as meeting space, but it should not be overlooked as a profit center for special events from holiday parties to milestone anniversaries. Renting a private dining room is an option for customers who don't have the space or desire to entertain at home. Hosts can enjoy the pleasure of choosing a menu and inviting guests with none of the mess or stress of doing the cleaning and cooking—or the dishes!

Unlike commandeering a large table in the main dining room, a private room gives the host control of virtually every aspect of the event: a personalized menu, selection of table settings and linens, music, decorations, wine choices, and so on.

The host probably doesn't think of it this way, but the private dining area is also a benefit to the other people who happen to be in the main dining area. Anyone who has been seated next to a large, boisterous party knows how uncomfortable it can be. It is better to corral them in their own space.

Private dining rooms are profitable too. When the menu is planned, guests are served a particular entrée, or perhaps have a choice of two or three. The food is preordered and prepaid, so the kitchen knows how much to prepare and the manager knows how much labor to schedule. If a party is too small to make the private room profitable, consider adding a flat fee for the room rental.

IILLUSTRATION 3-12 The private dining room at Eleven Madison Park in New York City.

Courtesy of Eleven Madison Park, photo by Brian Canlis.

Experts recommend the private room be a space in the dining area that can be walled off separately or, at least, adjacent to the main dining room. They say rooms on other floors—a basement wine cellar or a second-floor balcony—don't do as well because people can't see them, so they require more marketing and word of mouth to stay booked. A few restaurants can afford to buy adjacent buildings to house their special-event rooms.

You must also consider the feasibility of operating a separate room. If a separate kitchen is not available, how can the main kitchen handle the private party's dinner order in addition to preparing meals for the rest of the dining room? Will service suffer in one room or the other? Is parking adequate for private events? Can the building accommodate the noise level?

One terrific example of a successful private dining room is at Eleven Madison Park in New York City, seen in Illustration 3-12. It features a separate space—not tucked away, but on a second floor with a full-length glass wall. It seats up to 55 guests, who can see both the dining room and the outdoor park beyond. The private dining room has its own kitchen, bar, restroom, and elevator, making it a virtual restaurant within a restaurant. It also features heavy sliding wooden panels that can split the room into two halves to hold separate, smaller functions.

3-3 TO BAR OR NOT TO BAR?

To many adults, a bar is their preferred gathering place—after work, on weekends, to mingle with friends or watch a major sporting event. Even if a bar is considered chic or upscale, it is still expected to be a friendly and comfortable place, with a style that promotes interaction among patrons, servers, and bartenders. Today's customer doesn't go to a bar to *drink*. Instead, the combination of atmosphere, entertainment, service, food, and beverages makes it a pleasant place to *be*.

In some restaurants, the bar also serves as a waiting area for diners. Most restaurants with bars place them near the entrance to avoid bar patrons having to maneuver through the dining area to get there. Chili's, the Dallas-based casual dining chain, has its bar areas front and center, with frozen margaritas in stout, frosted mugs to beckon thirsty guests.

You'll also have to think about where to put those guests who don't drink or who don't wish to sit in a bar while they wait for tables. You certainly don't want to offend them or make them feel pressured to buy something, but a menu of nonalcoholic drinks is a good idea.

In quick-service and cafeteria settings, having a bar is not usually a consideration. However, the growing trend is to provide a beverage area, away from the food counter, where customers can fill their own drink cups. This process speeds service by allowing the counter personnel to concentrate on food sales and preparation. How big will your beverage area be, and where will it be located?

As for a full-fledged cocktail lounge, make the decision—to bar or not to bar—based on how much profit you expect liquor sales to generate compared to the rest of your business and how much space you have to achieve the atmosphere you want to provide (live bands, a disc jockey, big-screen televisions, or other types of entertainment).

At T.G.I. Friday's, the cocktail area is a casual gathering place, attracting a crowd around its central, island-style bar. Even on a slow dining-out night, the bar area may be hopping. In a hotel, there might be different expectations of the bar—or more than one bar for more than one type of patron.

Today's upscale bars are much like upscale restaurants; their theatrical touches set them apart from competitors and attract the see-and-be-seen crowd. For instance, anyone who's been to Las Vegas has probably heard of Charlie Palmer's restaurant and bar Aureole, located in the Mandalay Bay Hotel and Casino. The showpiece at Aureole is the wine tower (see Illustration 3-13), a glass-encased, temperature- and humidity-controlled wine storage area

ILLUSTRATION 3-13 The wine tower at Aureole, Las Vegas.

Courtesy of Aureole, Mandalay Bay Resort and Casino, Las Vegas, Nevada.

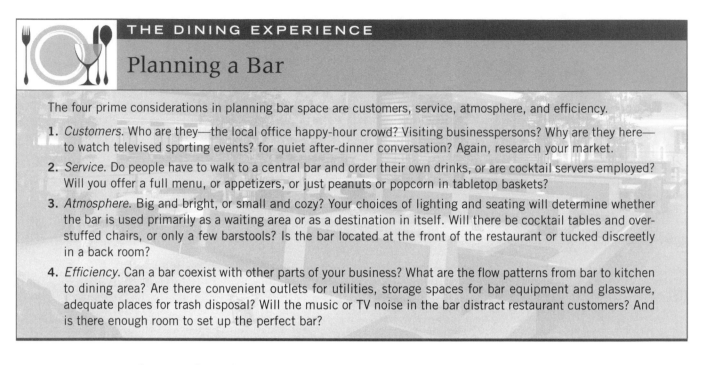

THE DINING EXPERIENCE

Planning a Bar

The four prime considerations in planning bar space are customers, service, atmosphere, and efficiency.

1. *Customers.* Who are they—the local office happy-hour crowd? Visiting businesspersons? Why are they here—to watch televised sporting events? for quiet after-dinner conversation? Again, research your market.

2. *Service.* Do people have to walk to a central bar and order their own drinks, or are cocktail servers employed? Will you offer a full menu, or appetizers, or just peanuts or popcorn in tabletop baskets?

3. *Atmosphere.* Big and bright, or small and cozy? Your choices of lighting and seating will determine whether the bar is used primarily as a waiting area or as a destination in itself. Will there be cocktail tables and over-stuffed chairs, or only a few barstools? Is the bar located at the front of the restaurant or tucked discreetly in a back room?

4. *Efficiency.* Can a bar coexist with other parts of your business? What are the flow patterns from bar to kitchen to dining area? Are there convenient outlets for utilities, storage spaces for bar equipment and glassware, adequate places for trash disposal? Will the music or TV noise in the bar distract restaurant customers? And is there enough room to set up the perfect bar?

that rises from the restaurant's bar to a height of 50 feet. Circular staircases allow guests to meander around the perimeter of the tower as they make their way from the ground-floor bar area to the restaurant below. Wines are retrieved in the tower by "wine angels," female wine stewards who are literally attached by cables to four pulley systems. They are propelled up and down the tower's seven levels using computerized joysticks, like video game players or jet fighter pilots. The tower holds 9000 bottles of wine, with more housed in a separate cellar and in off-premise storage. The entire operation is controlled by a two-way radio system, not in view of guests.

THE PERFECT BAR

Bars often are designed by someone who never has tended bar, and bar owners tend to make their choices based on the appearance of these bars without much thought to the workspace behind and beneath them. Bartenders are left to deal with the workflow issues, banging their arms and knees, jamming their fingers, and wrenching their lower backs because the tools of their trade are placed too high, too low, or with too little space for safe and efficient performance of their duties. When designing the bar area, consult professional bartenders who have high-volume operational experience. Ask, "How well will this work?"

Every bar, no matter where it is located, how big it is, or how it is shaped, has three inter-related parts:

1. Front bar
2. Back bar
3. Under bar

Here's a brief description of each, shown with basic dimensions in Illustration 3-14.

THE FRONT BAR. The front bar is where customers' drinks are served. The space is 16 to 18 inches wide and topped by a waterproof surface of treated wood, stone (such as marble or granite), or laminated plastic. Some bars have a 6- to 8-inch padded armrest along their front edge. The recessed area, nearest the bartender, is known as the ***rail*** (glass rail, drip rail, or spill trough). That's where the bartender mixes the drinks. For space, portable bar manufacturers may have a slightly recessed ***mixing shelf*** instead.

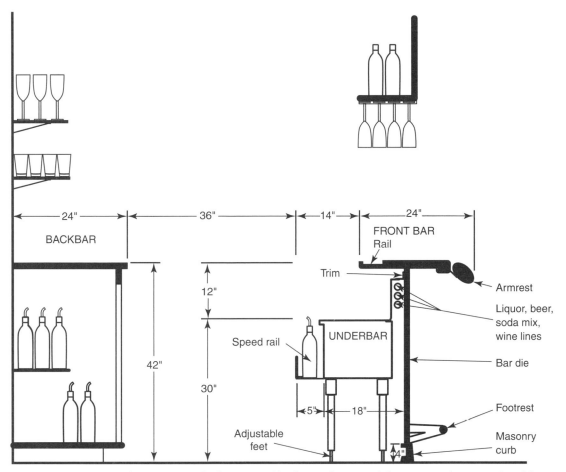

ILLUSTRATION 3-14   This diagram of a bar in profile shows the typical dimensions for a comfortable workspace.

Source: Costas Katsigris and Chris Thomas, *The Bar and Beverage Book*, 5th ed., (Hoboken, NJ: John Wiley & Sons, Inc., 2012).

The typical bar is 42 to 48 inches tall—the optimum height for the bartender's work as well as for leaning against. All bar-related equipment is designed to fit under a 42-inch bar.

The ***bar die*** is the vertical front panel that supports the front of the bar. It shields the under bar from public view. If you're sitting at the bar and kick it, what your foot hits is the bar die. On the guest side, there is usually a metal footrest running the length of the bar die, about 12 inches off the floor. It's sometimes called a ***brass rail***, a term that dates back to Old West saloons. The front bar or bar top is, of course, the horizontal surface on which drinks are served.

When selecting a bar, avoid the straight-line, rectangular model in favor of one with corners or angles, which prompts guests to sit opposite each other and visit instead of staring straight ahead at the wall. Illustration 3-15 should give you one idea. Of course, the ultimate conversation bar is the island, an oblong bar in which the bartender occupies the center. People can sit all the way around it and see each other easily. However, island bars take up a lot of space. They are also the most expensive to build.

A word about barstools: Choose them for height and comfort, and allow 2 feet of linear space per stool. Because they're high off the ground, additional rungs for footrests are more comfortable. As with any type of chair, you've got plenty of choices.

THE BACK BAR. The ***back bar*** is the wall area behind the bar structure. It serves a dual function, providing both decorative display and storage space. Sparkling glassware and all the brands of spirit are neatly arranged, usually highlighted by a mirror. This display is certainly a form of merchandising, but it's also the handiest place to store the bartender's most frequently used

ILLUSTRATION 3-15 Think about what your customers have to look at as they sit at a bar. Designs with corners or angles prompt more conversation and allow people to see and be seen.

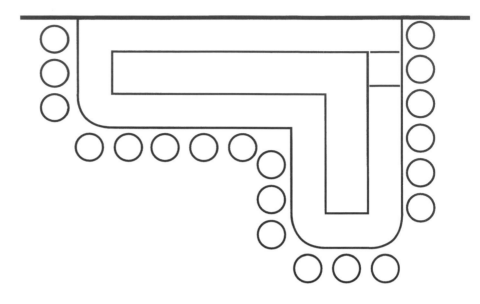

supplies. The mirror adds depth to the room while allowing guests to view others at the bar as well as the action behind them. Smart bartenders learn to use the mirror to observe the scene without being noticed.

The base of the back bar is an excellent place for refrigerated storage space. The cash register or POS system is usually found at the back bar, sometimes built in. Only 3 feet separate the back bar from the under bar; when designing this space, choose doors (preferably dual doors, made of glass) that don't impede the bartender's movements when open. Make sure there's an opening wide enough to accommodate the largest piece of movable equipment in case it must be repaired or replaced. Otherwise, you may have to lift it out.

Back-bar design requires specific plumbing and electrical considerations. You need outlets where you'll be placing appliances and plumbing for water supply, ice making, and drainage—not only drainage of melting ice runoff but also drainage that allows for proper drying of clean glassware.

THE UNDER BAR. The *under bar* is the heart of the beverage operation and deserves careful attention to design. Too often, a bar is selected because it looks good from the customer's point of view, without regard to the bartender's practical considerations. Is storage ample? Will all supplies and fountain valves be easy to access and easy to restock? Will the bar area be easy to clean? Equipment and supplies must be arranged in a compact space, but being able to use them speedily is the primary concern.

The central point of the under bar is the *pouring station*, where you'll find the automatic dispensing system for carbonated beverages and juices. The system has lines running from individual tanks through a *cold plate* at the bottom of the *ice bin* to the push-button dispensing head (commonly referred to as a *gun*).

Also in the pouring station are *bottle wells* and a *speed rail*—both places to store the most frequently used liquors and mixers. Blenders and other small bar tools are located in the under bar too. Every bartender sets it up a little differently for speed and ease of use.

The bar top should not overhang most of the under bar, and the distance from bar top to under bar should allow for comfortable stacking and store-and-pour operations. Bartenders should not have to bash a hand against the underside of the bar top to pull out a bottle of mixer.

One approach is to select glassware before designing the bar and then create the perfect places for glass rotation and storage. Other storage needs include chilling space for some wines and refrigeration for beer, both in bottles and in kegs. Ice supply is another major concern. An ice maker is usually installed in the under bar, near a double sink where ice cubes can be stored to allow the ice maker to refill itself. Another option is to put the ice maker elsewhere and keep the ice bins behind the bar refilled. This is more labor intensive, but it frees the space to install a small sink or dishwasher behind the bar. The depth and size of the ice bin should be gauged to match the expected level of business.

You'll also require a couple of other sinks. The typical *bar sink* is 7 to 10 feet long, with four compartments, two drainboards, two faucets (which each swivel to cover two compartments), and bar drain overflow pipes. Finally, a separate *hand sink* is another plumbing need to consider. If you're washing glasses behind the bar, take care not to put the glass-washing station directly in the middle of the drink-making action.

Clean glasses may be stored upside down on a glass rail, on drain boards near the ice bin, on the back bar, or in overhead racks, grouped according to type and size.

■ SUMMARY

Customers of foodservice businesses may not be able to describe it, but they know when the atmosphere of an eatery makes them feel comfortable. Increasingly, today's guests use the experience of dining out as part of their everyday life. It is convenient, reliable, and, in some cases, it is a place to entertain so they don't have to do the cooking and cleaning up themselves. Even in commercial settings such as hospitals and schools, it is beneficial to design and manage these operations with a focus on pleasing the guests and respecting their needs, from cleanliness to the ability to relax a bit and enjoy their meal.

Use what you know about your target market and the environmental psychology of how they view "comfort" to meet their needs. In this chapter, we've introduced you to many of the components of comfort and atmosphere and presented warnings about common design pitfalls. Customer perceptions are affected in every part of your establishment: from a lobby or waiting area to dining areas—indoors and outdoors—to the waitstations and restrooms. You've also become acquainted in this chapter with guidelines for deciding whether or not to have a bar, what sizes of tables you can choose from, and what types of chairs to buy, all to help create the right atmosphere for the business and the right comfort level for your customers.

One of the most important considerations in designing your space is the flow pattern used to deliver foods and beverages from kitchen to customer and back to the dish room. Like a traffic pattern, flow patterns vary depending on the type and size of the facility. They should always be designed to minimize backtracking, for efficiency and safety.

The trend of maximizing every square foot of space means some types of food businesses can offer additional services to improve profitability. A few of these options were discussed, including banquet and meeting spaces, private dining areas, and take-out or drive-through capability, and the unique design requirements of each.

■ STUDY QUESTIONS

1. Describe two examples of how environmental psychology plays into foodservice design.

2. For foodservice businesses that cannot depend on a freestanding exterior to help coax customers into the business, how can they ensure their first impression is a good one?

3. What is the single most important rule about restrooms?

4. What is a flow pattern? How does it affect the interaction between kitchen and dining areas?

5. What is the rule for determining how many waitstations to place in a dining area?

6. You have 1240 square feet of dining area, with 10 deuces and 16 four-top square tables. Arrange them to seat 86 guests, and explain your strategy.

7. What are the four primary considerations in designing a bar area? Write a short description of why each is important.

8. List three considerations for selecting chairs for a dining area.

9. If you were running a school cafeteria, which basic serving line configuration would you use? Why? Would that also be your choice for a corporate cafeteria?

10. List and describe the additional equipment needs for a banquet or meeting space.

Tom Galvin

FOODSERVICE FACILITIES DESIGNER,
GALVIN DESIGN GROUP

Tom Galvin has been a foodservice designer for more than 30 years. He says he had always been interested in architecture but was told by a high school counselor that he wasn't "creative enough" or good enough at math to pursue it. So, he majored in psychology at Bowling Green State University. Still fascinated by drawing and architectural rendering, he took classes in construction technology and interior design. After graduation, he enrolled in a masters program in architecture at Kent State University.

He was also a sound and lighting technician for some famous rock bands in the 1970s, including the Doobie Brothers, Chicago, and Foreigner, before going to work for Darden Restaurants, parent company of such well-known chain eateries as Olive Garden, Red Lobster, and Bahama Breeze. He was the manager of purchasing and food facilities design for Hard Rock Café International.

In 1999, Galvin began his own firm, the Galvin Design Group, based in Winter Garden, Florida. In all, he has been part of more than 900 design projects in 22 countries.

He also teaches courses at Boston University's School of Hospitality Administration.

Q: Years ago, that school counselor didn't exactly give you great advice. Did anyone in your college career support your interests?
A: I had a professor from Iraq, Dr. Owen—I was the only guy in the interior design class, and he thought that was great! He gave me a little office space adjacent to his because he noticed I spent a lot of time sketching and developing. And he said, "I don't know what profession you're going to go into, but pick one that no one knows much about—and then, be the best." I tell my students that today.

At Kent State, I was one of the older students. I had a job in the foodservice department, cutting meat and slicing cheese for the deli. I had a T-square in my hand one night as I was leaving, and the foodservice director asked about it. I explained it was for drawing in my architecture class.

"You know how to draw?" he asked. "Would you be interested in drawing up the kitchens and dining areas, and creating some spaces for us?"

No more cutting deli meats. That's where I got my start, and I still have the original sketch. What we ended up doing was pioneering the food court dining concept on campus. It had never really been done before, a central kitchen and themed areas to serve people.

I came up with the idea on a long walk on the campus. At Kent State, there were black squirrels that had been brought from Canada. As I watched them, I noticed how they group together—one or two squirrels, up to eight of them, and different nests of different sizes. And it hit me—I need seating areas for one, two, up to eight, where any size of group will feel comfortable, or a single diner will also feel comfortable.

The lesson from that is a designer pulls ideas from all kinds of sources. Designers work anywhere they want—they just need to be inspired. Some of my best ideas come to me between 2:00 A.M. and 6:00 A.M., when the phone's not ringing and I can think uninterrupted.

Q: How did you end up in Florida?
A: One week we had a major snowstorm in Cleveland, and I'd just had it with the snow. I sent out 350 résumés and said to myself, "Within three weeks, I'm going down to Florida for some job interviews." I lined up six interviews, got in my car, and off I went.

I took a job in West Palm Beach, went back to Ohio just long enough to pack up or give away everything I owned, and went to work for a design firm that had shown me beautiful photos of the work they did. I found out later the photos had been pulled from magazines and they'd never done

any restaurant design work! But I was stuck—I'd spent all my money moving down there. So I made the best of a bad situation for 16 months. And then I looked at the newspaper and saw a job posting for Red Lobster in Orlando. I applied and got the job. Within the first week, I discovered that everyone had their opinions about carpet color, wall color, types of booths—but no one knew much about stainless steel. Dr. Owens' advice came back to me, and when they said I had to choose between being the interior designer or the foodservice designer, I knew what I wanted to specialize in.

Q: You've said you weren't thrilled about working in the corporate world.
A: My supervisor was a tough ex-Marine. He didn't like my hair length, so I cut it a little bit. I worked long hours and did good work, but he was never really happy with me because I wasn't his choice to begin with. I tell my classes today that working in a big corporation, when your number is up, it's up, no matter what you've done or what you intend to do, and it's time to move on. You can't be down on yourself about it—there's always the risk that the bean counters in the company will decide they don't need you anymore, and they're always saying, "Cut costs, cut costs." It's a big, continuous, frustrating cycle.

Q: Can designing back-of-house space be just as cool as designing front-of-house space?
A: We integrate the back to the front because of my interior design

knowledge. Here's what a chain does: They bring you in, you design a kitchen, and then they back into a menu—which is wrong. Give me a menu and let me design the space around it, including extra areas that can change if the menu is going to change. A lot of consultants don't think to build in that flexibility for the future. And let's say you're designing kitchen space in a hotel—you learn to put most of the equipment on casters because when the current chef leaves in a few years, a new one comes in who will want to organize things completely differently. And hopefully the hood has been designed with some extra nozzles for the fire system, just in case.

Q: Other than the look and usefulness of equipment, what do you think about when you're designing?
A: Of course, you're talking with the owner about what they want, but you're also looking at energy usage, water consumption, operational hours. The things that will hit them every month are gas, water, and electric—and waste. Why shouldn't you hire a designer who wants to help you control those?

Designing back-of-house is difficult because so many things have to happen there. I take the menu, write down all the menu items, and for each item decide what equipment can generate that menu. Then, I look at that list for those pieces of equipment the menu items have in common—it could be that a six-burner with a convection oven base and a salamander top could handle 60% or even 80% of the menu.

Then I ask, "How many of these do I need—one or two? And does it go in the middle, on the left, or on the right?" That decision is based on flow—how things come into the kitchen, get dropped off in the dishwashing area, and come back out. In the perfect restaurant kitchen, you can see the dishwashing area. You can call back for your monkey dishes or platters or anything else. When the dishroom is secluded from the kitchen, you're going to increase the amount of dishes you need by 25%. You can't call for what you need when you need it, and the plates won't be hot right out of the dishroom when you need them, which creates problems with cold food.

Q: And breakage might increase too, if you have to transport the dishes farther.
A: Yes. We talked about having some rough-in space for future needs. In any kind of kitchen, let's say you have a stainless-steel table by the fryer. Instead of a 12-inch fry-dump area, or maybe it becomes a 24-inch area, or 30 or 36 inches. Below that you could have 120 volt, 208 volt, all single-phase; you could have a ¾-inch gas and a ½-inch water line, and have it all capped. So, when the seasonal menu changes, you can pull that table out and slide in a piece of equipment that creates the new menu items. When the season changes again, take it out and put in another piece of equipment, or put the table back in. You have to have that kind of flexibility.

Q: Are students surprised about how much they have to know about things like electrical requirements?
A: Yes, they are. But I tell them, when you get involved in a project like redoing a bakery or a banquet area, or the front of the house, you need to know at least enough to know whether or not a consultant knows his or her stuff. When something doesn't seem right, you need to contact another consultant.

Q: Any front-of-the-house design advice?
A: One piece of advice I can offer that we used in designing the dining areas for the Harry Potter exhibit at Universal Studios is that every dining area needs to start with a story. Think of it as a story line. You have to define exactly what you want to do—who your customers are going to be, how long you want them to stay, and so on—and be willing to hear others' input. There are lots of consultants in Orlando because of all the theme parks and theme hotels, and for large projects, you might be working on a team of 30 design consultants to create that space. Again, that's where your specialized skills will pay off.

Q: One of your design suggestions is "Create intensity." Explain what you mean and why you want to do that?
A: Well, in many cases you want people to walk into a place and say "Wow." For instance, for the Wyndham Grand Hotel, I took a European cooking island and redesigned it so you can change out

equipment on it, which is usually not possible. And I had the metal dipped in a charcoal gray finish to fit the mood of the Spanish-themed hotel. We made a brass belly-bar, and then I had the hoods wrapped with hand-pounded copper, with straps of the same wood that was used on our chef's table, which we also customized. And people say "Wow." They're not expecting to see anything like this. Think about creative ways to customize off-the-shelf pieces of equipment.

If someone calls and says, "I just want you to do a basic school cafeteria," I say, "I'm not your guy." I have to do something to create a buzz, to create interest. In a school design I did, we have a real mosaic-tile front counter illuminated by the tray shelf with LED lights. And on the back walls, we used high-resolution graphics to represent the food. I worked with a local artist to do the mosaic tile panels. There's always something you can do to make the space interesting and impressive, and it doesn't have to break the budget.

Q: What's your experience with "going green" in foodservice design? Is it something you encourage, or maybe discourage?
A: The problem with going green in the beginning is it costs 10% to 20%—maybe more—on equipment, and that's a big issue. But costs have come down. In my mind, consultants have always been green. We've always been concerned about building efficiency into our projects—it's part

of our DNA as kitchen designers. Now that the U.S. government has come up with the term, I think they're trying to use it as a way to make money on things like certifications. I know how important it is to save energy, but when someone asks me to fill out a bunch of forms to prove I'm doing it, I'm not happy with that.

Q: Where is foodservice design going as a profession, in your view?
A: I think we're going to be at a crossroads pretty soon. Equipment dealers are buying up design firms, trying to corner the market and tell clients, "We can do everything for you." The problem I see with that is the equipment dealer is pushing more of that work onto its reps. If I had a crystal ball, I could almost see where the reps and the consultants will be working together on the purchasing, procurement, and installation of the equipment, and the equipment dealer may be going away, which has been the case already in Europe. I don't want to be negative about them, but an equipment dealer today is most often only a middleman. The consultant is the one that's on-site for the project.

Right now, the Foodservice Consultants Society International (FCSI) is a firm believer that design consultants cannot procure equipment, but I think we may be forced into revisiting that bylaw because we're a global economy and the way we do things is not the way they are done elsewhere. Economics will dictate these decisions, to some extent.

4

PLANNING BACK-OF-HOUSE SUPPORT AREAS

■ INTRODUCTION AND LEARNING OBJECTIVES

We began the design process for back-of-the-house space in Chapter 2 with a discussion of the commercial kitchen and its various workstations and flow patterns. Now, we continue with a closer look at some of the support or auxiliary areas that also are part of foodservice businesses.

After reading this chapter, you will be able to:

- ■ Identify the components and flow patterns of the warewashing (dish room) area.
- ■ Identify the components and flow patterns of the receiving area.
- ■ Identify the features considered when designing storage areas.
- ■ Identify the features considered when designing employee areas.
- ■ Identify the features considered when designing office space.
- ■ Identify some of the major design requirements of the Americans with Disabilities Act.

Not every foodservice operation needs space for each of these areas, and the significance of each depends entirely on the type of operation. Using a combination of square footage and meals served, you can chart the amount of space needed, as shown on the graphs in Illustration 4-1. These graphs indicate individual space requirements for storing, preparing, cooking, and serving food, plus dishwashing and staff facilities.

Of course, your particular project may not fit into the standard guidelines, which is something for your foodservice consultant and kitchen designer to take into account. They can assist you in determining your specific needs and factor other considerations into the floor plan—but this is best accomplished early in the planning process.

You have already read in this text about how each part of a foodservice business is connected with the others. Use these relationships as a starting point to plan your back-of-the-house layout. Which areas make sense located nearest to which others? Using the method introduced in Chapter 2, you can start with a diagram (see Illustration 4-2), and work toward a high-level or overhead view of the space itself, as shown in Illustration 4-3. This is known as your *preliminary space plan.*

No great amount of detail is needed for the preliminary space plan. You will continue to improve it as you drill down to individual areas and the equipment each requires. Like the bubble diagram first mentioned in Chapter 2, this is an exercise on paper meant to help you discover how food and supplies can move logically, and with the least effort, through your operation. The idea is to minimize wasteful backtracking. All this sketching may be time-consuming—but wouldn't you prefer to make mistakes, changes, and improvements now, on paper, rather than later, with expensive, detailed equipment plans and architectural renderings in hand?

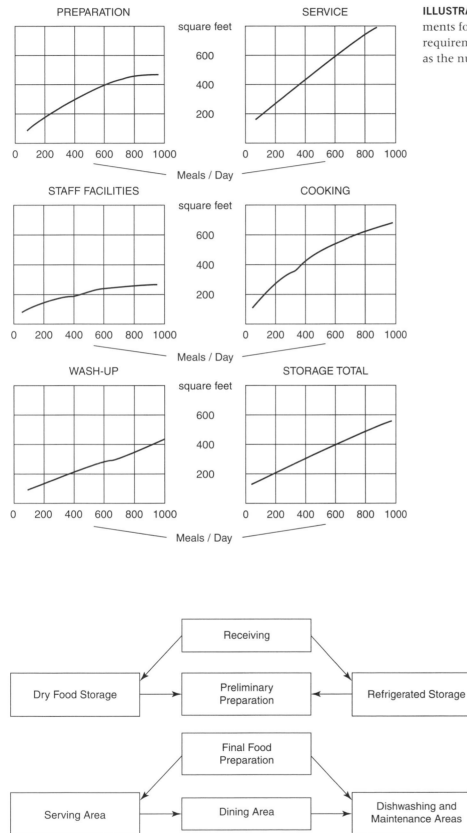

ILLUSTRATION 4-1   Common space requirements for a foodservice facility. Note that space requirements do not increase in the same ratio as the number of guests served.

ILLUSTRATION 4-2   This figure of major work centers in a restaurant and how they relate to each other is similar to a bubble diagram.

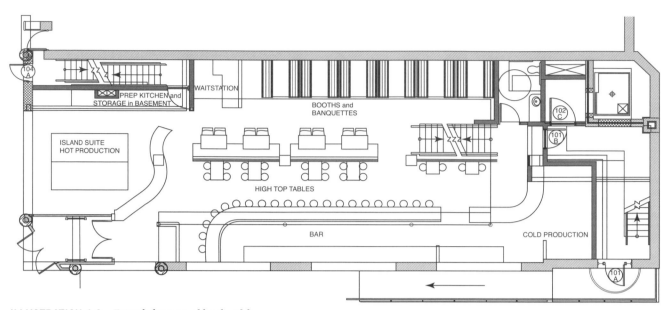

ILLUSTRATION 4-3 Rough layout of back-of-house.

Courtesy of MVP Foodservice Design, Sparks, Nevada.

As you work, remember that for some types of commercial foodservice businesses, like restaurants, the market is so volatile that chances are good you will change something—the menu, concept, or size—in 3 to 5 years. Try to give yourself room and flexibility to grow.

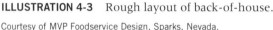 4-1 ALLOCATING WORK SPACES

Many of the considerations for kitchen space allocation are detailed in Chapter 2. Briefly, design your kitchen based on the type of menu you will serve, the number of guests to be fed at one time (or one meal), the type of cooking you will do (individual dishes versus large batches), and the service system you will use to deliver the food.

Most experts will tell you that even though all kitchens perform the basic function of meal preparation, the type of foodservice venue in which the kitchen is located will greatly affect the design and delivery systems necessary to do a proper job. Be mindful that what works well in one venue may not do well in another.

In a restaurant, for example, most experts suggest that a conventional kitchen takes up 40% to 50% of the total space. Those that use more prepared foods or have more frequent deliveries can minimize storage space and perhaps squeak by with 25% to 35% of the total restaurant space. And in most cases, some trade-offs are necessary. However, the steadily rising costs of construction have dramatically changed space requirements. For example, in a restaurant setting, the trend is to consolidate the back-of-the-house functions so more dollars can be spent on the front of the house, where guests are served and money is made. Keep costs in mind, of course, but don't allow this to cause you to compromise the amount of space needed for your production requirements. Often, cutting space ends up hampering the operation.

Beyond the restaurant kitchen, in other facilities that have kitchens, we begin to add variables to the design and complexity of the space. Hospitals, correctional facilities, schools, and office buildings each have their respective needs. Some design criteria are actually mandated by authorities, and these are the types of restrictions the designer must consider. This is especially true in corrections, where inmate and employee safety are of utmost importance.

In the long run, nothing will reflect worse on you than a facility that is unable to do what it was intended to do. When you are faced with questions you cannot adequately answer, it may be time to bring in someone who has proven knowledge and experience in design and equipment selection for your foodservice operation.

WAREWASHING

The warewashing area or dish room of a commercial kitchen is the perfect example of just how many considerations must be factored into design decisions. In this section, we describe the dishwashing process in great detail so you can see the need for certain features or precautions when designing the space.

Warewashing is the term for collecting soiled dishes, glasses, flatware, pots, and pans and scraping, rinsing, sanitizing, and drying them. It is by far the most necessary and least exciting part of the foodservice business. Whether achieved by hand or by machine, or a combination of both, warewashing is a messy job with high temperatures, high humidity, and slippery floors that require constant caution. For the restaurant owner, this is also one of the most costly areas to operate. Labor (and high turnover), utilities, and equipment are all expensive, and breakage can account for 10% to 15% of dish room expenses.

Cost is not the only reason warewashing is of major importance. Quite simply, it has a direct impact on public health. Done poorly, you jeopardize your business.

Many an operator has experienced the frustrating effects of an inadequate or poorly designed dish room. This area must be correctly sized and equipped to handle the volume of dishes, utensils, pots, and pans generated by the type of facility. If you cannot scrap, wash, rinse, sanitize, and return these items clean and ready to use within a reasonable time period, the entire operation runs slowly.

In dishwashing and scullery areas, the key is to use right sizes of equipment and tables to handle the volume of ware being delivered. The flow pattern of the area is equally important. This means allowing adequate space for employees and all necessary dish carts, plus sufficient landing space for all the items that pass through the area, both soiled and clean.

For health and sanitation reasons, it is vital to separate the dishwashing area into soiled and clean sections. Make certain that employees who move from the soiled dish area to the

---

IN THE KITCHEN

Space and Sizing Guidelines

1. The restaurant kitchen is approximately half the size of the dining room.

2. Sizing by seat count:

DINING ROOM

Deluxe: 15 to 20 square feet per seat

Medium: 12 to 18 square feet per seat

Banquet: 10 to 15 square feet per seat

BACK OF THE HOUSE (KITCHEN)

Deluxe: 7 to 10 square feet per seat

Medium: 5 to 9 square feet per seat

Banquet: 3 to 5 square feet per seat

(Add the banquet requirement to the kitchen.)

3. Food prep is approximately 50% of the back of the house.

4. Storage is approximately 20% of the back of the house.

5. Warewashing is approximately 15% of the back of the house.

6. Waitstaff circulation is approximately 15% of the back of the house.

Source: Lodging, the magazine of the American Hotel and Lodging Association, Yardley, Pennsylvania.

clean dish area have adequate and properly positioned handwashing sinks and are required to use them whenever they move between the sections.

You'll learn much more about mechanical dishwashing requirements in Chapter 16, "Dishwashing and Waste Disposal." For now, let's cover the way busing and washing affects the size and location of the dish room. First, the dirty dishes have to get there. Some types of foodservice business have bus persons or waitstaff bring the dishes directly to the warewashing area; others try to save steps by stacking the dishes in bus tubs on carts in the dining room and bringing them to be washed only when the cart is full. (The biggest mobile carts, which can carry at least a dozen of the large, oval waiters' trays, are nicknamed "Queen Marys.") In cafeterias and quick-service restaurants, guests often bus their own tables, depositing waste and trays at receptacles near the doors as they leave or placing their dirty dishes on a conveyor belt to the dish room. Emphasis should be on designing a system in which every plate, cup, and fork has its own flow pattern so it gets in and out of service quickly and efficiently.

A smooth, short flow pattern from eating areas to dish room will minimize breakage and lower labor costs. Identify the shortest route that both minimizes the natural noise and clatter of dishes and keeps them out of the guests' sight as much as possible. In institutional settings, the dish room should be placed in the flow pattern of departing guests so as not to interfere with incoming ones. Use of a conveyor belt, or placement of walls or partitions, can minimize contact with dish room sights and sounds.

The simplest space allocation method is to bring the soiled dishes from the kitchen or dining area into a straight-line configuration whereby they arrive at soiled dish tables; are scraped, stacked, washed, and sanitized; and come out sparkling clean at the other end of the line. Just scraping and stacking the items waiting to be washed takes up 60% of dish room counter space; clean dishes take up the remaining 40%. Space should be allocated for a three-compartment sink for efficient scraping and rinsing.

Sometimes the scraping and stacking is the responsibility of the bus persons or waiters who bring in the dishes. This method is known as *decoying*, and it allows the dishwashers to concentrate on rinsing the dishes and loading them into machines. No matter who scrapes, you'd better figure out where the scrapings are going to end up. A trash can with plastic liner? a waste disposal?

Glasses and flatware have special sanitation needs. As used flatware is taken off a table, it should be placed immediately in a presoaking solution, either in a sink or bus tub. When heading into the dishwasher, flatware should be placed, facing up, in perforated round containers, much like those you're familiar with in home dishwashers. Glasses and cups are arranged face down on special racks, not placed on regular dish racks with the plates. If *stemware* is used, such as tall wineglasses or water glasses, purchase special racks with individual, high compartments to prevent chipping the glass bases. Allow adequate room for storage of these specialized containers, both full and empty. In some fine-dining establishments, a separate dishwasher is utilized strictly for glassware and flatware. Separating the glassware and flatware process from that used for dishware allows these items to be cleaned in a machine that does not have large amounts of soil particulate floating in the tanks and potentially sticking to the glassware or utensils.

Neither flatware nor glassware should be towel-dried on removal from the dishwasher. Instead, place flatware on large, absorbent towels and sort it as it dries; allow glasses to air-dry before removing them from racks. Again, this requires space. Most restaurants store their glasses in the racks until they are used again, which saves storage space and time.

DISH ROOM SIZE AND CONFIGURATION. The dish room can be organized in several ways. In most cases, you will set your equipment in a straight line, an L-shape, or a hollow square, with the equipment making the sides of the square and employees standing inside. Illustration 4-4 shows sample dish rooms in each of these three common configurations.

In determining the size of your dish room, consider the size of the dining area and the number of meals served during peak periods. Unless your eatery depends on disposable or recyclable ware, each person served will generate six to eight dirty dish items. Calculate how many full dish racks your dish machine can process in one hour. Finally, realize that the best dish rooms are, on average, working at 70% efficiency.

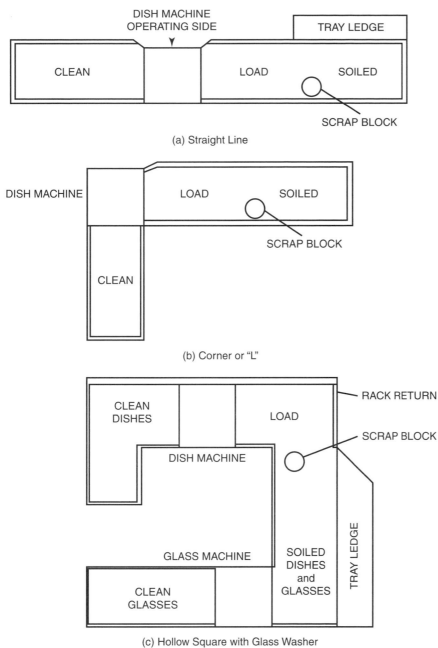

(a) Straight Line

(b) Corner or "L"

(c) Hollow Square with Glass Washer

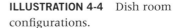

ILLUSTRATION 4-4 Dish room configurations.

Source: Robert A. Modlin, ed., *Commercial Kitchens*, 7th ed. (Arlington, Virginia: American Gas Association, 1989).

To get from dirty to clean, you'll need a dishwashing machine, which will be discussed in detail in Chapter 16. The smallest space for a single-tank dishwasher is 250 square feet, in an area approximately 20 × 12 feet. Add to that the space needed for dish carts full of bus trays, empty racks for loading dishes, and room to wash those unwieldy pots and pans. All of this requires a minimum of 40 square feet, including a 4-foot aisle to separate the pot sink from other equipment.

You'll probably need to decide on the room configuration before you order your dishwasher. Sizes vary from 30 to 36 inches wide for single-tank machines to 30-foot-wide commercial flight-type machines, which include a long conveyor belt to carry dishes into the washer. Table 4-1 contains suggestions for dish machine capacity and space requirements.

OTHER DISH ROOM CONSIDERATIONS. Inside the dish room, scraping, spraying, lifting, and carrying mean greater labor costs and the potential for disaster unless careful attention is paid to safety. Nonslip floors and adequate drainage are necessities. Usually city ordinances also specify lighting

TABLE 4-1

Typical Space Dimensions for Dishwashing Equipment

TYPE OF DISH SYSTEM	DISHES PER HOUR	SPACE REQUIRED
Single-tank dishwasher	1,500	250 square feet
Single-tank conveyor system	4,000	400 square feet
Two-tank conveyor system	6,000	500 square feet
Flight-type conveyor system	12,000	700 square feet

Source: Carl Scriven and James Stevens, *Food Equipment Facts* (New York: John Wiley & Sons, Inc., 1989).

requirements (70 to 100 foot-candles of brightness) and govern the ventilation system. Dish machines generate so much heat and humidity that proper ventilation is necessary for employee comfort; as a bonus, it also helps clean dishes dry faster. Acoustical tile for walls and ceilings in this area keep noise from spilling into the dining room.

Another item that must be carefully chosen is the type of door that links the warewashing area to the rest of the facility. Swinging double doors are not a good idea here for the same reason we discourage them between the kitchen and service areas, unless each door is installed to swing only one way, space between the doors is at least 2 feet wide, and each door is clearly labeled IN and OUT.

A final important element of a well-designed dish room is the method of handling waste, both food and other types. Waste must be handled expediently and must not pass through the clean area of the dish room during warewashing. The use of waste containers, disposers, and pulpers can help with the waste processing. You will learn more about these dish room workhorses in Chapter 16.

As you can see, good planning and understanding of proper dish room function will serve your facility well.

RECEIVING AREA

Vigilant cost control begins in the receiving area, the space set aside to receive food and beverage shipments and other deliveries. It is in the receiving area that you count and weigh items, check orders for accuracy, and refuse incorrect orders or those that don't meet your quality standards.

In creating this space, consider these three factors:

1. Volume of goods to be received
2. Frequency of delivery
3. Distance between receiving and storage areas

Some floor and counter space should be allocated for temporary storage, where things can be stacked until they are properly checked in. Weight scales, dollies, or carts will be located here, and sometimes a sink and drainboard for a quick rinse and inspection of incoming produce. Remember to plan space for waste disposal here too, which should be a separate room for trash and cleaning of soiled waste cans.

The absolute minimum receiving area space is about an 8-foot square, which allows room for a receiving table, a scale, a dolly or cart, and a trashcan. In large operations, gravity rollers or conveyor belts move materials from receiving into storage.

Most product delivery is done by truck, which means you need a loading dock outside the receiving area if at all possible. Unfortunately, your situation might require a driver to unload onto the ground and cart the product into the building by way of stairs, hallway, or elevator.

TABLE 4-2

Space Dimensions for Receiving Areas

RESTAURANTS

Meals Served per Day	Receiving Area Square Footage
200–300	50–60
300–500	60–90
500–1000	90–130

HEALTHCARE FACILITIES

Number of Beds	Receiving Area Square Footage
Up to 50	50
50–100	50–80
100–200	80–130
200–400	130–175

SCHOOLS

Meals Served per Day	Receiving Area Square Footage
200–300	30–40
300–500	40–60
500–700	60–75

Source: Carl Scriven and James Stevens, *Food Equipment Facts* (New York: John Wiley & Sons, Inc., 1989).

Be sure to build safety into these tasks, with features like handrails for delivery drivers. Delivery drivers have tough jobs—a cart loaded with food items can weigh 400 pounds or more—so they will appreciate whatever you do to make their lives (and yours) a bit easier and safer.

If you are fortunate enough to have a dock, keep in mind that without adequate backing room and proper ramps, the driver still may struggle to get products off the truck. The minimum loading dock space should be 8 feet wide and 10 to 15 feet long. Space for two trucks—from 80 to 120 square feet—is preferable and prevents congestion if you require that all deliveries be made in a short period. If you are sending finished food products off-premise (as in catering operations), then you should have two separate doors and rooms: one for waste, the other a clean area for prepared food products.

Scriven and Stevens, in their *Food Equipment Facts* handbook, suggest the receiving area space allocations for various types of facilities, as shown in Table 4-2.

Note that in hospitals and other group feeding situations, dock space is shared with other departments that also have daily deliveries. In these cases, consider adding efficient transport equipment to move supplies quickly to their storage destinations.

We've seen businesses at both ends of the receiving area spectrum. Not having any receiving area at all seems to invite cost control problems. Without a place to check materials in and out logically, the tendency is to be lax about keeping track of shipments overall. This is when inferior products, incomplete orders, and pilferage start taking their tolls. A receiving area that is too large, however, ends up the dumping ground for empty boxes, broken equipment, and anything that doesn't fit anywhere else.

GENERAL STORAGE

A commercial kitchen may "grow" in certain seasons or from year to year. This may be accomplished through the use of mobile carts and tables that can be wheeled about to create new workstations or supplement those that were originally designed. Finding a few extra square feet of hallway or aisle space when planning storage needs may serve you well in the future for unforeseen times when something must be added.

The tendency is to think of storage space as unproductive and to downplay its importance. However, more and more cities are mandating specific ratios of storage space to kitchen space.

How much storage you need depends on how much stuff you want to store. Simple? Actually, it isn't. During the planning phase, you should design storage areas that can increase in capacity without increasing in size through the use of additional shelf space. A system of grids, movable shelves, and accessories can attach directly to your wall studs. Some shelves can be hinged to swing up when needed and down when empty.

Other storage systems are rolling shelf units (referred to as *high-density storage systems*) attached to an overhead track installed on the racks or on the ceiling. As shown in Illustration 4-5, systems like Top-Track have drop-in baskets, mats, and dividers to accommodate different container sizes. A track-style system with only one aisle can give you up to 40% more storage space.

In foodservice, you must determine not only the number of products you will be storing but also the number of days they'll be there before you use them. This is known as your *inventory turnover rate*. Experts suggest a restaurant have at least enough space to store 1 to 2 weeks' worth of supplies. Believe it or not, they have calculated exactly how much storage space is needed per meal served per day when deliveries are made every 2 weeks: 4 to 6 square inches. (Get out your tape measure and start measuring those shelves!) Of course, if your location is more remote—a hunting lodge, a small island resort, a rural café with infrequent product deliveries—you will need more storage room for extended periods.

In some foodservice establishments, disaster preparedness plans may require storage of significant amounts of product as a precaution in case of emergency. Prisons, hospitals, schools, and nursing homes have contingency plans that require, in most cases, enough food for a week, a month, or more.

From this, it is clear that storage space depends on how management handles ordering and receiving and on whatever regulations apply to the type of business.

Storage considerations include overshelves, undershelves, drawers, and cabinets as well as those all-important parking or landing spaces where shared equipment is stored when not in use.

Common fixtures you must consider in setting up storage areas include shelving, racks, bins, pallets, carts, and dollies. Remember to plan for aisle space of at least 3 feet in all storage situations. Make sure racks and carts have wheels (called *casters*) to enable you to roll them when they're fully loaded.

A *pallet* is a (usually wooden) platform that sits 4 or 5 inches above the floor and measures between 36 and 48 inches square. Boxes and cases can be stacked on pallets, and full pallets can be stacked atop each other. Their portability and stackability make pallets good, flexible storage options, but make certain your local health code allows for this type of storage. A dunnage rack

ILLUSTRATION 4-5 Top-Track is a high-density storage system. Its components are attached to an overhead track installed on the ceiling.

Courtesy of InterMetro Industries Corporation, Wilkes-Barre, Pennsylvania.

(which is basically a heavy plastic or metal platform that holds food farther off the floor) may be required.

DRY STORAGE

The standard restaurant storeroom is at least 8 feet wide, and its depth is determined by the needs of the operation. If the storeroom must be wider than 8 feet, increase it in multiples of 7; this allows for two rows of shelves, each a standard 21 inches wide, plus the 3-foot aisle. Some restaurants purchase movable shelving, which is usually 27 inches wide and about 5 feet long.

The storeroom door should open out, into an aisle, to maximize storage space within the room if possible. Chemicals used for cleaning, dishwashing, and so on, may not be stored with food products, and a separate area or room should be included in your planning exercise.

The many sizes of cans and bottles that must be stored make dry storage a real challenge. Minimize this problem by storing same-size items together on shelves of adjustable heights. For storage space requirements, we refer again to *Food Equipment Facts* by Scrivner and Stevens (see Table 4-3).

REFRIGERATED STORAGE

There are three types of refrigerated storage space: the reach-in refrigerator or freezer, the walk-in cooler, and the walk-in freezer. Storage space for these appliances is calculated in cubic feet. The expert guideline for an average restaurant open for three meals a day is 1 to 1½ cubic feet of refrigerated space per meal served. For fine dining, this increases to 2 to 5 cubic feet per meal served.

TABLE 4-3

Space Dimensions for Dry Storage

RESTAURANTS

Meals Served per Day	Dry Storage Square Footage
100–200	120–200
200–350	200–250
350–500	250–400

HEALTHCARE FACILITIES

Number of Beds	Dry Storage Square Footage
50	150–225
100	250–375
400	700–900

SCHOOLS

Meals Served per Day	Dry Storage Square Footage
200	150–250
400	250–300
600	350–450
800	450–550

Source: Carl Scriven and James Stevens, *Food Equipment Facts* (New York: John Wiley & Sons, Inc., 1989).

TABLE 4-4

Full-Door Reach-In Freezers

NUMBER OF DOORS	HEIGHT (INCHES)	WIDTH (INCHES)	DEPTH (INCHES)	CUBIC FEET
1	78	28	32	22
2	78	56	32	50
3	78	84	32	70–80

Source: Carl Scriven and James Stevens, *Food Equipment Facts* (New York: John Wiley & Sons, Inc., 1989).

TABLE 4-5

Walk-In Freezers (All 7′6″ Height)

SIZE OF UNIT	SQUARE FOOTAGE	CUBIC FEET
5′9″ × 7′8″	35.7	259.9
6′8″ × 8′7″	47.4	331.8
7′8″ × 7′8″	49.0	340.2
8′7″ × 11′6″	86.4	604.8

Source: Carl Scriven and James Stevens, *Food Equipment Facts* (New York: John Wiley & Sons, Inc., 1989).

As an example, let's examine the refrigeration needs of an average restaurant (140 seats) that serves three meals daily. Using our rule of thumb, the place will need between 420 and 630 cubic feet of refrigeration.

Management has been considering a combination of a 392-cubic-foot cooler and a 245-cubic-foot freezer, and is surprised to learn that this will not be sufficient. Why? Well, only half of the total 637 cubic feet is *usable* storage space—not insulation, aisles, motor, evaporator, or fans that are part of every refrigerated unit. Divide 637 by 2, and you get only 318.5 cubic feet of space—which is inadequate for this restaurant's needs.

If you have information about the purchase history of a similar operation, you can calculate your storage requirements based on that history with minor changes as you compare the operations. As a general rule, the average case of dry, refrigerated or frozen food is roughly 1 cubic foot in size.

This is another example of why your menu is such an important consideration. Restaurants or other foodservice operations that wish to use greater amounts of fresh foods will, of course, require more refrigerated than frozen storage space. It is important to consider the current and potential future needs of the operation and plan accordingly.

Tables 4-4 and 4-5 summarize the storage capacities for standard sizes of reach-ins and walk-ins.

In determining the useful space of a walk-in unit, remember that from one-third to one-half of it will be taken up by the aisle(s), and still more by the evaporator and fans. It is no wonder that, in our opinion, most foodservice facilities allocate too little refrigeration space and then have to find ways to live with it.

EMPLOYEE AREAS

Unfortunately, providing even the most basic amenities for employees seems to be an afterthought for most foodservice businesses. It is our view that employee-only areas deserve a higher priority in the planning process because they have a direct impact on sanitation, morale,

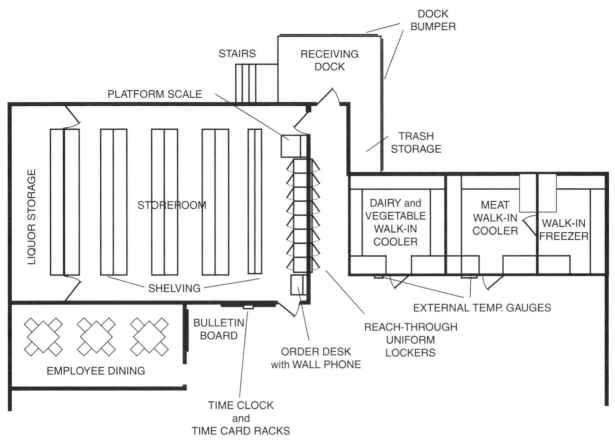

ILLUSTRATION 4-6 A layout for a receiving and storage area that includes a separate employee break room.

productivity, and security. Once again, this area should be sketched from the beginning, along with storage and receiving areas (see Illustration 4-6).

A locker room/break room area, well lit and properly ventilated, provides secure storage space for personal belongings and a place to unwind at break times. Tall, single lockers or smaller, stacked ones should have perforations to allow for air circulation. Benches in front of the lockers are a good idea. If a table or two and some chairs fit in this area, workers need not sit around in the dining area during their breaks—a practice that, though understandable, still looks unprofessional.

In larger facilities, the locker room is the site of any uniform exchange system, with a supply of clean uniforms and laundry hampers for soiled ones. The size of the employee locker room/restroom depends on the size of your staff; a general guideline is 150 feet for 10 to 20 employees and 250 feet for more than 20.

Toilets and hand sinks for employee use only are another smart idea; include enough shelf or counter space for employees to set personal toiletry items. Install a full-length mirror in the restroom or locker room too. The rule of thumb for restroom fixtures is one hand sink and one toilet for every eight employees. The ratio moves up to two of each for every 20 to 25 employees and three of each for every 25 to 35 employees. The minimum size for a single toilet and stall is 4½ square feet. Your city health department regulates the size and sanitation of employee restrooms just as stringently as the public ones. The same accessibility guidelines of the Americans with Disabilities Act also apply.

OFFICE AREA

The typical foodservice business office is tucked into the back of the house and is generally pretty messy and dingy. At country clubs, hotels, or catering companies, however, there may be offices where the guests interact with the staff to plan functions, get price quotes, and so on.

If a public office space is necessary for your operation, you should plan a minimum of 150 square feet furnished with an attractive desk and several armchairs. Space for filing cabinets, computer equipment, bookshelves, and display space for photos or albums of previous events will all be needed—especially the latter, because potential clients will want to see examples of your work.

A back-of-the-house office is more likely used for accounting, paperwork storage, and one-on-one employee meetings. It can be as small as 60 square feet or run to a relatively roomy 120 square feet. If accounting is done on site, we recommend it be housed in its own separate office to make it much easier to keep track of specific accounting paperwork.

Purchasing supplies and planning menus takes time and concentration. The office should be designed (and its filing system kept tidy enough) to allow the owner, manager, or chef to perform these tasks efficiently.

Occasionally we've seen restaurants with bars locate the office directly behind the back bar with a two-way mirror that allows management to watch the bar area. We've even seen a chef's office built almost like an airport tower—raised and glass-enclosed to allow easy observation of all activity in the busy kitchen!

Overall, restaurant office space does not have to be big or fancy. As for so many other areas, you must decide in advance exactly what you want to use the office for, and then plan the space around those needs.

4-2 SPACE PLANNING AND THE AMERICANS WITH DISABILITIES ACT

When planning a dining area or any type of public space, you must become familiar with the Americans with Disabilities Act (ADA) of 1990. Although it is a U.S. law, and this textbook will presumably be read in other countries, the ADA guidelines can be considered in any nation as a call for sensitivity to the needs of persons with physical limitations. In 1992, the ADA was revised to include a requirement that companies with 15 or more employees cannot fire or refuse to hire people with disabilities unless the impairment prevents the individual from performing the job.

Much grumbling resulted when the ADA was enacted, mostly by business owners who felt the law was ambiguous, that it was being too broadly interpreted, and that it was costing them money by requiring expensive modifications to their facilities. Lawsuits and complaints under the ADA range from customers with impaired mobility asking for wheelchair ramps to employees asking for prevention programs and workers' compensation for on-the-job back injuries.

When first signed into law, the ADA did not require restaurateurs to retrofit existing facilities immediately for accessibility. This mistakenly led some to believe their businesses were "grandfathered" and the law did not apply to them. However, the ADA is a civil rights act, and no entity is exempt from compliance. ADA relies on owner- and operator-planned alterations and barrier removal to improve accessibility in older buildings over a specified period. As businesses alter existing facilities in any way, especially in ways that affect accessibility, the areas or elements being altered must comply with ADA guidelines.

Even without alterations, businesses are not free of accessibility requirements; a plan to remove any and all barriers must be prepared, then changes accomplished over several years. For new construction (anything built after January 25, 1993), the ADA Accessibility Guidelines are mandatory. The U.S. Department of Justice, which enforces the guidelines, lists five steps every new hospitality property should take to ensure ADA compliance:

1. Obtain copies of the ADA Accessibility Guidelines and give them to the architects and building contractors. While architects and building contractors generally know the requirements of state building and fire codes, they may not be familiar with ADA requirements, which are often different.

2. Specify to your architect and building contractor that you expect your new facility to comply with ADA standards. Emphasize that ADA compliance is a top priority.

TABLE 4-6

Sample Toilet Facility Guidelines for Restaurants and Foodservice

BUILDING CLARIFICATION	USE GROUP	TOILETS FEMALES	TOILETS MALES	URINALS MALES	LAVATORIES EACH SEX	DRINKING FOUNTAIN	BATH/ SHOWER	OTHER FIXTURES	PERTINENT REGULATIONS 248 CMR 10.10(19)
Nightclubs, Pubs	A-2	1 per 30	1 per 50	50%	1 per 75				(b), (m), (n), (p)
Restaurants	A-3	1 per 30	1 per 60	50%	1 per 200				(b), (m), (n), (p)
(Hotel/Motel)	R	One Bathroom Group per Unit							(m), (q)

These are part of the plumbing guidelines for the State of Massachusetts. They may differ in other states. Courtesy of the International Association of Plumbing and Mechanical Officials.

---

3. Before construction begins, check building plans for common ADA-related mistakes. Consider having them reviewed by someone with ADA expertise.

4. Be sure the facility is being built according to the ADA requirements as shown in the building plans.

5. Inspect the facility at the completion of construction to identify ADA mistakes, if any, and have them corrected promptly.

By definition, ADA considers a "primary function area" as one in which people carry out the major activities for which the facility is used. Dining areas within a restaurant, meeting rooms in a conference center, and customer service areas in a retail shop are examples of primary function areas. Areas such as mechanical rooms, janitorial closets, employee lounges, and storage areas are not considered primary function areas under the ADA. The primary function areas must be readily accessible to all. This includes sunken or raised areas and outdoor seating areas, unless the same décor and services are provided in accessible space usable by the public and not restricted to use by people with disabilities.

The ADA gets a lot of credit for forcing a national reevaluation of attitudes toward people with physical limitations. An estimated 54 million Americans have some type of disability, and their patronage is as valuable as anyone else's, so it will benefit you to learn more about accommodating their needs, both physical and emotional. Today, hospitality employees are much more likely to take the initiative with a genuine effort to make customers feel comfortable, no matter their limitations and beyond what is required by law.

It isn't enough to know that for a wheelchair to make a 180-degree turn, 60 inches of unobstructed space is required. From the parking lot to the restrooms, the ADA most definitely affects your space planning (see Table 4-6). The IBC/ADAAG Comparison is a document of almost 400 pages, a handy reference from the International Code Council that combines the top accessibility resources for the construction trade, including the International Building Code (IBC) and the Americans with Disabilities Act Accessibility Guidelines (ADAAG). Just how technical are the requirements? Read on.

PARKING

The ADA mandates, among other specifications, a certain number of accessible parking spaces. (Although we dislike the term, these seem to be more commonly known as *handicapped spaces*.) The numbers range from a single parking space for a lot with only 25 total spaces to nine spaces for lots with more than 400 spaces; Table 4-7 shows the guidelines, and all three of the most common accessibility codes are in agreement about the quantities.

Illustration 4-7 provides design standards for accessible parking spaces. Accessible spaces must be those closest to the public entrance, and they must be clearly marked (usually with

TABLE 4-7

Numbers of Accessible Parking Spaces Required

TOTAL # OF PARKING SPACES PROVIDED IN PARKING FACILITY	MINIMUM # OF REQUIRED ACCESSIBLE PARKING SPACES
1–25	1
26–50	2
51–75	3
76–100	4
101–150	5
151–200	6
201–300	7
301–400	8
401–500	9
501–1500	2% of total

Source: 2006 IBC/ADAAG Comparison.

the universal symbol of accessibility) so the markings cannot be obscured by other parked vehicles. Most of the figures were developed to ensure wheelchair access. Each of these special spaces must be at least 8 feet (96 inches) wide, with an adjacent aisle of at least 5 feet (60 inches). Van parking spaces should be 11 feet (132 inches) wide and require an adjacent aisle of 6 feet (72 inches). A sign displaying the symbol of accessibility should identify these parking spaces as well as the additional term *VAN ACCESSIBLE* in bold letters, just below the accessibility symbol. Parking spaces should be level, with the surface slope not exceeding 1:50 ratio in all directions. This ratio means the surface should not slope more than 1 inch for every 50 inches of pavement. In addition, for every six parking spaces for disabled persons, one should be the larger-size van space.

From the vehicle to the building entrance, the path of travel must be at least 36 inches wide. For stairs, handrails between 34 and 38 inches above the stairs themselves are required. Wheelchair ramps must slope gently, with a height ratio of 1:20. If the ramp is longer than 6 feet, it must be equipped with handrails at the same height as those for stairs.

ENTRANCES

At least 50% of the entrances to a foodservice facility must be accessible to people with disabilities, including emergency exits. At nonconforming doors, signs must be posted indicating the location of the accessible entrances. If a door does not open electronically, it must be 32 inches wide. Handles, pulls, latches, locks, and other operating devices on accessible doors must be shaped to be easily grasped with one hand—no tight grip or twisting of the wrist required to open or close the door—and placed from 34 to 38 inches above the floor. Loop or lever handles are preferable to doorknobs. Eighteen inches of clear wall space is required next to the handle of a pull-open door. If there are double doors, the requirement is a 30- to 40-inch clear floor space, not counting the space the door would normally require to swing open. If there are revolving doors, an adjacent handicapped-accessible door is also required. Ramps with a rise of 6 inches should have handrails.

PUBLIC AREAS

Once inside the building, the guest in a wheelchair requires an aisle width of at least 36 inches. If counter service is offered, a 5-foot portion of the counter must be as low as 28 to 34 inches from the ground to facilitate ordering from a seated position. Tabletops must meet the same height requirements as counters. In banquet situations, if people will be sitting at a raised head table, for instance, a ramp or platform lift must be provided. Food serving lines require a minimum clearance width of 36 inches, but preferably 42 inches, to allow passage around a person using a wheelchair. Tray slides should be mounted no higher than 34 inches above the floor, and self-serve items must be positioned so they can be reached by someone in a seated position—cups must be stored horizontally, for example, instead of stacked vertically. On salad bars, cold pans may be tilted so all products are visible and may be reached easily; reach-in cooling units may have air screens instead of doors. Sneeze guard heights may have to be adjusted.

Aisles that lead to restrooms and to individual toilet fixtures also must be 36 inches wide. At restrooms that don't meet the rules, signs must be posted with directions to accessible ones. Doorways to both restrooms and individual stalls must be at least 32 inches wide. The size of accessible toilet stalls is also regulated; they must be at least 5 feet square.

Lavatories (washbasins) require clear floor space around them to accommodate the wheelchair-bound patron. This means a 30 × 48-inch space, with the rim of the basin no more than 34 inches from the floor. The bottom edges of mirrors must hang no higher than 40 inches from the floor. Soap, towel, and toilet paper dispensers should be no higher than 54 inches from the floor. Finally, whether they are placed in restrooms, lobbies, or elsewhere in the restaurant, telephones should not be mounted higher than 54 inches from the floor.

KITCHEN AREA

Two pieces of equipment that require special ADA consideration are the hand sink and the worktable. Neither can have obstacles underneath that would prevent a wheelchair-bound employee from getting close enough to safely use them, and both should be of wheelchair-friendly height.

GENERAL GUIDELINES

In the past, foodservice designers and consultants considered universal accessibility a brick-and-mortar issue. While door widths, aisles, and heights of tray slides are important today, the service aspects are equally important. Every foodservice business has its own policies, procedures, and routines that help the business operate as smoothly as possible. Sometimes these routines inadvertently make it difficult or impossible for people with disabilities to purchase services and products. ADA requires restaurants to make "reasonable modifications" in their routines when it is necessary to accommodate guests who have disabilities. Most accommodations involve making minor adjustments in procedures or providing extra assistance. ADA guidelines do not spell out how or what must be done to accomplish these "reasonable modifications," but the idea is to not exclude a customer by being unwilling to make accommodations that are fairly simple. Barriers can be more than the width of an aisle.

For instance, how can you plan to accommodate the guest with a visual handicap? A helpful server who can read the menu aloud and answer questions about it is much easer and less expensive than having Braille menus printed, especially if the menu changes frequently. It also should be noted that staff are not expected to abandon their duties to provide assistance to a person with a disability when doing so would jeopardize the safe operation of a restaurant. Each operation may adopt a means, costly or inexpensive, to assist guests in using services and buying products. A simple pad of paper and a pencil enable communication with people with hearing disabilities, as does a TTY (text telephone) connection or hiring a person who knows sign language. All comply with the dictates of ADA.

Congress has provided two types of incentives for restaurants and other businesses to assist in offsetting the cost of complying with the law. A Disabled Access Credit is available to small businesses that have 30 or fewer employees or total revenue of less than $1 million per year. The maximum allowable credit each year is $5,000, or 50% of the first $10,000 past $250, to offset the costs of removing barriers, hiring interpreters, producing documents in alternative formats (such as Braille or large print), and so on. This provision is found in Section 44 of the

Design Standards for Accessible Parking Spaces

Each accessible parking space must be at least 96 inches (8 feet) wide and shall have an adjacent access aisle. Van-accessible parking spaces must have a 96-inch (8 feet) wide access aisle adjacent to the parking space, and standard vehicle accessible parking spaces must have a 60-inch (5 feet) wide access aisle adjacent to the parking space. The accessible path of travel from the parking space to the building must start from the access aisle (note wide arrows). Access aisles are to be kept clear of all obstructions at all times.

Guidance documents:
U.S. Access Board Parking Technical Bulletin http://www.access-board.gov/adaag/about/bulletins/parking.htm
U.S. Department of Justice ADA Business Brief Re-striping Parking Lots http://www.ada.gov/restribr.htm

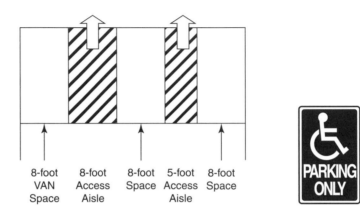

| 8-foot VAN Space | 8-foot Access Aisle | 8-foot Space | 5-foot Access Aisle | 8-foot Space |

Each accessible parking space must be designated by a sign per **NH RSA 265: 73-a.** Codes require that the sign display of the Universal Symbol of Accessibility and be mounted on a post or on a building wall so that the bottom of the sign is at least 60 inches above the surface of the parking space. Signs are to be located at the head of <u>each</u> accessible parking space. *Please do not use signs that have text with any form of the word "handicapped."*

Access aisles should be marked with diagonal stripes, preferably yellow and are to be part of the accessible route to the building. If possible, a "NO PARKING" sign should be installed.

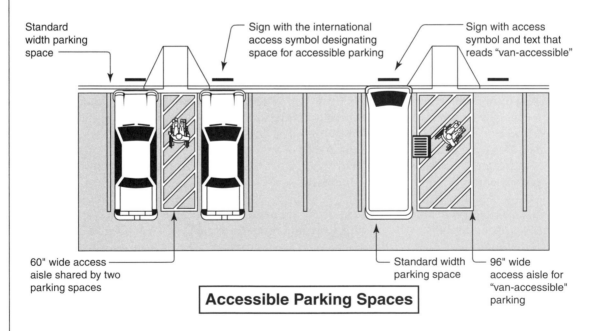

Standard width parking space

Sign with the international access symbol designating space for accessible parking

Sign with access symbol and text that reads "van-accessible"

60" wide access aisle shared by two parking spaces

Standard width parking space

96" wide access aisle for "van-accessible" parking

Accessible Parking Spaces

Reproduction attributed to: The Center for Universal Design, College of Design, NC State University, Raleigh, North Carolina.

Technical Requirements: Refer to IBC 2006 Section 1106, ANSI A117.1 Sections 502 and 503, and the ADA Standards for Accessible Design, Sections 4.1.2 and 4.6.

The amount of accessible parking spaces that must be provided is based on the total number of spaces in each parking lot.

At least one parking space must be van-accessible, and for every 6 (six) accessible parking spaces, there must be one van-accessible space.

Total Parking Spaces Provided	Required Minimum Number of Accessible Spaces
1 to 25	1
26 to 50	2
51 to 75	3
76 to 100	4
101 to 150	5
151 to 200	6
201 to 300	7
301 to 400	8
401 to 500	9
501 to 1,000	2% of total
More than 1000	20 plus one for each 100 over 1000

Accessible parking spaces must be located on the shortest accessible route of travel from the parking to an accessible building entrance.

In parking facilities that do not serve a particular building, accessible parking spaces must be located on the shortest route to an accessible pedestrian entrance to the parking facility.

Where buildings have multiple accessible entrances with adjacent parking, accessible parking spaces shall be dispersed and located near the accessible entrances. This usually means that the minimum number of accessible parking spaces must be exceeded in order to provide equal access to multiple entrances.

In multilevel parking structures, van-accessible parking spaces may be located on one level.

Committee on Architectural Barrier-Free Design
Governor's Commission on Disability
57 Regional Drive – Suite 5
Concord, NH 03301-8518
Telephone (Voice) (603) 271-4177 or (TTY) (603) 271-2774
NH Toll-Free (Voice/TTY) (800) 852-3405
FAX (603) 271-2837

ILLUSTRATION 4-7 Design standards for accessible parking spaces.
Courtesy of the United States Access Board, Washington, DC.

IRS Code. Section 190 of the Tax Code permits businesses of any size to deduct up to $15,000 each year for the cost of removing barriers in facilities or vehicles, and "barriers" in this case includes anything that would hamper effective communication. It does not include the costs of new construction. Both of these incentives are available to existing businesses, not new ones. A lot of information is available online about these tax credits, although your tax adviser is probably the best source of additional information.

Overall, the hospitality industry has been exemplary in recognizing the need for accessibility; we are, of course, in a business that focuses intently on customer service. Most of us recognize that making these adjustments is a commonsense way to cater to an important part of our clientele. The ADA helped formalize many policies and practices that, before 1990, existed informally in many locations.

■ SUMMARY

This chapter includes information about planning areas that most customers and guests never see in a foodservice business. Some of these areas are critically important to the daily operation of the business; others are simply nice to have if sufficient space is available.

When deciding how much back-of-the-house space you need, the types and quantities of foods you plan to serve—and whether the ingredients are fresh, frozen, or pre-packaged—all will help you determine receiving and storage needs. Cost of space certainly is a consideration, although when back-of-the-house functions are consolidated to save space (and therefore money), you must be able to ensure this decision doesn't hamper the efficiency of the entire operation. In other words, keep costs in mind, but don't allow them to cause you to compromise on the amount of space your production requires.

The chapter contains detailed information about how dishes are bused, scraped, stacked, washed, and stored—to illustrate the point that every step of the process figures into the overall design of the dish room.

Designing an efficient receiving area that suits the needs of your operation means determining how much food will be delivered (and whether it's fresh, frozen, or pre-packaged); how often deliveries occur; and the distance between the delivery and storage areas. The safety of your staff and delivery drivers is a chief concern in the planning process for this area.

In storage areas, it is important to factor flexibility into the design. We describe the use of modular systems where shelf space can be added or reconfigured as necessary. Determining storage needs involves knowing not only what (and how much of it) you will be storing but also for what length of time—making it necessary to determine inventory turnover rates.

A brief discussion of office space and employee break rooms and locker rooms is included in the chapter. The advice here is to first determine exactly what needs to happen in that space—then, design it to fit those needs.

Finally, it is critical that you and your designer are familiar with the Americans with Disabilities Act, which will affect many aspects of your space planning. Work with experts, if necessary, to ensure your facility is truly accessible to the public as well as to prospective employees with disabilities. This includes special attention to creating accessible parking, entrance ramps, aisle widths, and restroom facilities, to name a few.

■ STUDY QUESTIONS

1. What are the three main parts of a warewashing area?
2. Describe the sanitation needs of flatware and glassware, and explain whether and how they affect dish room design.
3. Why does a dish room require a separate flow pattern?
4. What are the two most important functions of a receiving area?
5. What are the basic items that should be part of a receiving area, even if space is tight?
6. Briefly describe safety considerations for receiving areas, both with and without loading docks.
7. Briefly summarize what you learned about selecting doors for a foodservice kitchen based on the information in this chapter.
8. List and explain three ways to make good use of storage space, both dry and refrigerated.
9. What special attention must be paid to the following items in order to meet ADA requirements? (a) hand sinks; (b) self-serve items on a buffet; (c) worktables
10. What are the five steps recommended by the U.S. Department of Justice for determining whether a space plan meets ADA guidelines?

5

ELECTRICITY AND ENERGY MANAGEMENT

Until the early 1970s, when the United States experienced its first energy crisis since World War II, most businesses were simply not focused on cutting energy use. When budgets had to be trimmed, the emphasis was on curtailing labor costs, insurance costs, and—as always—food costs. It seemed downright miserly to fret over when to turn on an oven or whether to adjust the air conditioner a few degrees warmer. Today, however, it is smart to conserve resources by operating more efficiently. At times, this requires making financial investments in equipment, mechanical or electrical systems, and the building itself. For the typical foodservice business, these investments generally make sense only if the initial expense can be recouped within 5 years.

Today, customers expect companies of all types and sizes to use limited natural resources wisely while minimizing the negative impact on the environment. This is a constant balancing act, but it can pay dividends in the community in terms of good public relations and responsible citizenship as well as save money.

After reading this chapter, you will be able to:

- Track energy usage and determine annual electricity costs.
- Identify the proper terms used to select appliances and decipher utility bills.
- Describe basic electrical principles: how power flows into your building and equipment.
- Identify tips for choosing energy-efficient equipment.
- Explain how foodservice operations are using renewable energy.
- Identify tips for saving energy in all phases of a foodservice operation.

This chapter focuses primarily on electricity. Other utilities—gas, steam, and water—and plumbing issues are covered in Chapter 6.

5-1 UNDERSTANDING ENERGY USE

Energy use per square foot in a foodservice facility is greater than in any other type of commercial building—more than triple the amount a hospital uses per square foot and at least six times what an office building uses per square foot. Using restaurants as an example, Illustration 5-1 summarizes how power is used in a survey of seven locations.

ILLUSTRATION 5-1 Where energy is used: a survey of seven restaurants.

Source: Electric Foodservice Council, "The Economics of Gas and Electric Cooking," Fayetteville, Georgia.

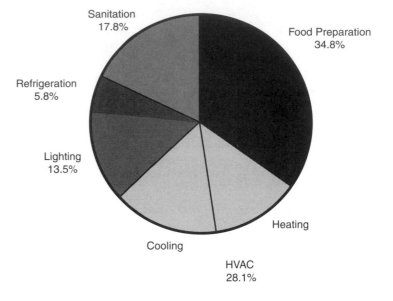

On your profit-and-loss statement, utility costs will be only 4% to 7% of your total operating expenses. However, National Restaurant Association studies suggest these expenses can be cut by as much as 20% with smarter energy consumption. Utility companies claim you can realize savings as high as 30%. These potential savings, of course, have a major and direct impact on your bottom-line profit. Table 5-1 charts the potential savings, which vary (of course) with your electric costs and amount of savings. Remember, because its energy use is so high, this type of business is especially vulnerable to fluctuations in energy costs.

The best way to save both energy and money is to plan and implement an Energy Management System, which consists of six components:

1. *Energy accounting.* A monthly tabulation of energy use and costs allows the owners of a business to track this data season to season and year to year. Put it on a standard form or spreadsheet. Include the information from your power bills: total costs, total amount of consumption, and demand charges. (The latter will be explained shortly in this chapter.)

2. *Energy audit.* This is explained in detail in the next section of this chapter.

TABLE 5–1

Yearly Value of Electricity Saved

YOUR ELECTRICITY COST PER KWH	VALUE OF THE ELECTRICITY SAVED IF THE NUMBER OF KILOWATT-HOURS USED PER YEAR IS REDUCED BY					
	500	1,000	2,500	5,000	10,000	25,000
5 cents	$25	$50	$125	$250	$500	$1,250
7 cents	$35	$70	$175	$350	$700	$1,750
9 cents	$45	$90	$225	$450	$900	$2,250
11 cents	$55	$110	$275	$550	$1,100	$2,750
13 cents	$65	$130	$325	$650	$1,300	$3,250
15 cents	$75	$150	$375	$750	$1,500	$3,750

Source: How to Reduce Your Energy Costs, 2nd ed. (Boston: Center for Information Sharing).

3. *Retrofitting.* In foodservice, about half of energy conservation comes from retrofitting to make existing appliances, systems, or buildings more energy-efficient. This effort includes everything from increasing the insulation in the building to insulating the hot-water tanks to installing timers on outside lighting and climate control systems. A retrofit project usually makes sense only when you know it will pay for itself in cost savings within a few years.

4. *Low-cost and no-cost ideas.* Before you do the retrofitting, which costs money, try changing the habits and routines of your staff. Turn off lights in unoccupied areas; don't leave appliances on when they are not being used; ask your staff members for energy-saving ideas, and then put them to work.

5. *Capital project.* Like home improvements that add value to a house, similar improvements add value to a business. New technology may have a high price tag, but consider its long-term usefulness. Computerizing the heating, ventilation, and air-conditioning (HVAC) system to turn on and off automatically and hold predetermined temperatures can result in huge savings over time. Cogeneration, an idea discussed near the end of this chapter, involves adding new equipment that captures otherwise wasted heat from appliances and uses it to heat water or generate steam. When appliances wear out and must be replaced, look for the newest energy-saving features.

6. *Continued surveillance.* In order to make any of the other five steps effective, they must be monitored, and their importance must be communicated to the staff. Soon you'll be documenting savings instead of dreading each month's utility bills.

5-2 ENERGY AUDITS

The good news is that because foodservice uses so much gas and electricity, you are automatically an important customer to your utility provider. You can leverage this clout to use the utility's expertise, which can help you manage your business. Utility companies often provide cash rebates, low-interest loans for equipment updates, and free design and technical advice. You might as well get to know them! In fact, you should have a contact person at every public utility (or, in deregulated markets, your retail electric provider, or *REP*) with which you do business.

An **energy audit** is the process used to determine how your facility uses energy and where savings might be obtained. Utility providers will perform simple walk-through audits at the request of customers. More detailed *analysis audits* may require the services of an engineer or consultant.

Walk-through audits are usually free of charge. The service provider's representative will already have information about your average bills when he or she comes out to inspect the property, look at your food preparation and storage equipment, ask about habits and procedures, and recommend improvements you could make. Some suggestions (such as replacing an appliance or adding insulation) may cost money. Of course, whether you act on the audit results is up to you.

A fee is typically charged for an analysis audit because it requires gathering more detailed information about the heating and cooling systems and appliances and even the illumination levels of the lights in your dining room. The recommendations may include any of the following:

■ Structural or design modifications to your building

■ Replacing or retrofitting some equipment

■ A target electric rate that is "best" for your particular business

Often, a financial analysis of each of the energy conservation measures (ECMs) is provided so you can compare the cost to the potential payoff over time. The recommendations are designed to give you lots of options. For example, let's say you have recessed lighting in the dining room ceiling and the audit suggests it is inefficient. You could:

■ Change the bulbs in the light fixtures to such energy-efficient choices as reflector (R) or ellipsoidal reflector (ER) bulbs.

■ Paint the room a lighter color, which would be more reflective and therefore require less artificial light.

■ Install more light switches or dimmers to allow better control of individual areas or rooms.

■ Lower the ceiling height.

■ Install skylights or light tubes.

Who performs these audits? A call to your utility company or REP should get you started. One advantage the utility has is that it can track the history of energy use at your site even before you got there. If it used to be another restaurant, this could be extremely helpful. Private firms are listed in telephone directories under "Energy Management and Conservation Consultants." When selecting a private contractor, it's important to ask:

■ What they charge

■ Exactly what the fee includes

■ Whether they represent a variety of equipment manufacturers (so they're not just trying to sell you their own brand of system or appliance)

■ If they are qualified as certified energy managers (CEMs) by the Association of Energy Engineers

■ To see sample reports and references

Often, equipment manufacturers' representatives offer a no-cost energy audit as part of their introductory service. Remember, although the advice is free, they are trying to sell you their products.

Some foodservice businesses challenge their own employees to self-audit, making their own suggestions for energy savings. Involving your staff in energy management is a wise move—not only for the potential savings but also because it is a responsible way to do business. Make energy efficiency the topic of a staff meeting. Many utility companies have developed their own walk-through self-audit forms for customers. Not every question on the form will apply to every restaurant or foodservice business, but it can be a good starting point.

Give employees an incentive, and let them try their own walk-through and report on their findings, as often as every 6 months. You might be surprised at what they come up with!

Keep in mind, especially with a brand-new business, that there is no substitute for the expertise of a utility company or contractor. If changes should be made, a thorough energy audit will help you prioritize them. It also can be the groundwork for future planning, including project financing, compliance with government regulations, and LEED certification (the Leadership in Energy and Environmental Design program), discussed in greater detail elsewhere in this chapter.

5-3 UNDERSTANDING AND MEASURING ELECTRICITY

Electricity is a source of power made up of billions of individual electrons flowing in a path, much like water in a stream. The stream is the *electrical current*. Current flows through *wires*, sometimes called **conductors**. Wires are coated with a layer of *insulation*, which prevents shocks and ensures the full amount of power reaches its destination. The type of insulation may vary depending on the type of appliance or whether its location is wet or dry. When several wires are bound together with additional insulation, you have a **cable**. Cables in foodservice, heavy-duty but flexible, are wrapped in steel and often referred to as *BX cables*. When the wires are encased in nonflexible steel pipe, the pipe is called a **conduit**.

The complete path of a current, from start to finish, is known as a **circuit**. Every circuit has three basic components: the *source*, or starting place; the *path* along which it flows; and the *electrical device* (the appliance) it flows into. A common type of circuit is found in the fluorescent light bulb (see Illustration 5-2).

A complete working circuit is called a *closed circuit*. An *open* (or *broken*) *circuit* is one that has been interrupted so no electricity flows. A **short circuit** occurs when current flows in an unplanned path due to a break in a wire or some other reason. Short circuits are unsafe because they can cause fire and/or other types of equipment damage.

Fortunately, safety devices can prevent the damage caused by short circuits. A **fuse** is a load-limiting device that can automatically interrupt an electrical circuit if an overload condition exists. Fuses protect appliances and humans as well as the wiring of the circuit itself. More about them in a moment.

A **circuit breaker** looks like an electrical switch. It automatically switches off when it begins to receive a higher load of electricity than its capacity. As the load decreases, the breaker may be manually switched on again without damage to electric lines or appliances.

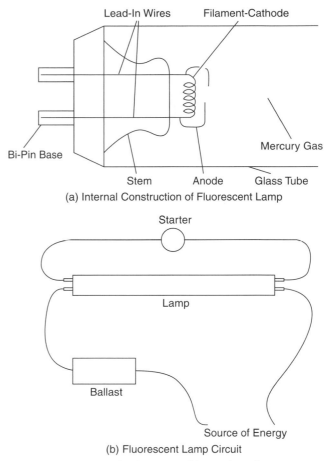

(a) Internal Construction of Fluorescent Lamp

(b) Fluorescent Lamp Circuit

ILLUSTRATION 5-2   The inner workings of (a) fluorescent lamp, and (b) simple fluorescent lamp circuit.

Source: Frank D. Borsenik, *Maintenance and Engineering for Lodging and Foodservice Facilities* (Lansing, Michigan: Educational Institute of AH&MA, 1977).

In addition to understanding kilowatt-hours (kWh) and kilovolt-amperes, you need to learn four basic terms to measure electrical energy. Why do you need to know these terms? You may need to figure amps, volts, or watts to help you decide what equipment to buy, to check the efficiency of existing appliances, or to see if you're about to overload a circuit by plugging in something new. Also, most kitchen equipment is rated (for output or energy efficiency) based on these terms:

1. *Ampere.* Commonly known as an **amp**, this is a term for how much electric current flows through a circuit. A 100-watt light bulb, for instance, requires 0.832 amps of electricity to light up. The larger the diameter of an electric wire, the more amperes it can safely carry, so it is vitally important that a licensed electrician determine the correct wire sizes for your needs. The National Electrical Code sets the trade standards for wires. The minimum size for copper wire, for instance, is #14; for aluminum wire, #12.

    Circuit breakers are also rated in amperes, most often 15 amps (a single circuit breaker) or 30 amps (a double).

    The typical home electrical outlet provides 15 amps to plug in appliances, lamps, and so on. As few as 0.02 amps can render an electrical shock. In electrical formulas, the letter I indicates amperes.

2. *Volt.* The **volt** is the driving force that pushes an ampere through an electrical wire. One volt is the force required to push 1 amp of electricity. For appliances, common voltages are 110–120 and 208–240. This means the amount of voltage necessary to operate the

appliance efficiently; most appliances have a minimum requirement of 120 volts. In electrical formulas, the letter E indicates volts.

3. *Watt.* The **watt** represents actual consumption of electrical energy—the amount of power in a circuit. One watt equals the flow of 1 amp of electricity at a pressure of 1 volt. Many electrical appliances are rated in terms of both watts and volts. In electrical formulas, the letter W indicates watts.

4. *Ohm.* This is a less familiar term; an **ohm** is a unit of electrical resistance. It is a measure of how well a substance conducts electricity or how effectively an insulator restricts electric flow. Examples of conductors are copper, silver, and drinking water. Examples of insulators are dry air, wood, rubber, and distilled water. In electrical formulas, the letter R (for *resistance*) stands for ohms.

Now that you know the terminology, you can use these simple formulas for converting the data you *have* into the energy measurement you *need* for a given situation.

There are two ways to find volts.

1. Divide watts by amps:

$$E = \frac{W}{I}$$

2. Or multiply amps by ohms:

$$E = 1 \times R$$

There are two ways to find amps.

1. Divide volts by ohms:

$$I = \frac{E}{R}$$

2. Or divide watts by volts:

$$I = \frac{W}{E}$$

To find watts, multiply volts by amps:

$$W = E \times I$$

HOW ELECTRICAL SYSTEMS WORK

The Edison Electric Institute estimates that 80% of all power-related equipment problems occur on the customer's side of the meter. When your repair technician shows up, you'll be thankful for your basic knowledge of electrical systems as you explain the problem and listen to his or her recommendations.

A building's electrical system uses either **alternating current (ac)** or **direct current (dc)**. Alternating current is by far the more common type of electrical current. It actually reverses its flow at regular intervals. Alternating current can be transmitted over long distances with minimal loss. The equipment costs to use ac power are lower than those for dc, and ac doesn't tend to overheat equipment like dc does. The most common ac energy in the United States is 60-cycle energy, which means the electrical current cycles 60 times every second. Why is this important? The number of ac cycles varies from country to country, and if you are purchasing equipment made in another country, it may run on 50-cycle power instead of 60-cycle power. If that's the case, you'll need to buy a converter to enable it to work in your facility. The nameplate on the appliance should state how many cycles it uses to operate.

The two main types of alternating current are single phase and three phase. **Single-phase current** is one current of alternating voltage. Most household appliances run on single-phase power. **Three-phase current** combines three streams of current that peak in a regular, steady pattern. Heavy-duty commercial appliances and industrial equipment run on three-phase power.

The common battery-powered flashlight is a good example of dc power. It has a constant flow of amps and volts, it doesn't travel a long distance, and it can overheat equipment. Dc power is not widely used in the United States, but it has limited applications such as lighted exit signs, elevators, and security systems.

Your utility company or REP can recommend what source of energy is preferable for your site. Often only one source is available anyway, and it must be brought into the building and channeled to points of use.

Because different voltages and types of electrical circuits are available, it is absolutely critical, when designing electrical service, to know exactly the types of electric service and circuits you have in an existing building and the types of utilities your equipment will need.

VOLTAGE COMBINATIONS AND LOAD FACTORS

Electrical equipment manufacturers design appliances with combinations of voltage and phase—that is, the force with which the amperes are pushed through the line and the number of alternating voltages available. Voltage and phase specifics should be part of any new appliance's service information; they are found on the nameplate of the appliance.

You should be familiar with these options because so many large appliances require **dual voltage**—that is, more than one type. Perhaps the controls or motors need 120 volts while a heavy-duty heating element requires 208 or 240 volts. Therefore, you must be sure your building's wiring can handle the challenges, whether the facility is brand-new or a remodel, or even if you are just replacing an old piece of equipment with a newer one. The voltage supplied to each electrical outlet directly affects the performance of the appliance plugged in there, so voltage must be correctly specified when ordering the equipment. The most common voltage combinations are:

- *Single phase: 110–120 volts.* This system includes two wires, one hot and one neutral. Together, the maximum amount of voltage they can generate is 125 volts.
- *Single phase: 220–240 volts.* In this two-wire system, both are hot wires. The maximum voltage between them is 250 volts.
- *Single phase: 110–120/220–240 volts.* This three-wire system includes two hot wires and one neutral wire. Between either of the hot wires and the neutral wire, the maximum voltage is 125 volts, but between the two hot wires, the maximum is 250 volts.
- *Three phase: 110–120/208 volts.* This four-wire system is made up of three hot wires and one neutral wire. The voltage between any single hot wire and the neutral wire is 125 volts, but the voltage between any two hot wires and the neutral one is 208 volts (single phase), and the voltage between all three hot wires without the neutral is 208 volts (three phase).
- *Three phase: 220–240 volts.* All three wires in this system are hot. The maximum voltage between any two of them is 250.
- *Three phase: 440–480 volts.* This is a three-wire system in which the maximum voltage between any two of the wires is 480.
- *Three phase: 110–120/220–240 volts.* This is a four-wire system, three hot and one neutral. The maximum voltage between any one of the hot wires and the neutral wire is 125; the maximum voltage between any two hot wires is 250.
- *Three phase: 440–480/277 volts.* This is a four-wire system, three hot and one neutral. The maximum voltage between each hot wire individually and the neutral is 277; between any two hot wires, 480; and between all three hot wires, 480.

Now let's put all this terminology into perspective. Generally, as you have learned, single-phase circuits are used for most small or light-duty appliances, such as a two-slice toaster. Heavy-duty appliances with large power and heating loads are usually wired at the factory for three-phase, three-wire electricity. They sometimes include smaller fuses and circuit breakers for different functions within the appliance. The manufacturer must supply a complete wiring diagram, which should always be kept with the appliance's service records. Countertop cooking equipment usually requires a two-wire, single-phase circuit. In short, the requirements will be very specific.

It is critical that the equipment voltage matches the service voltage of the building. To determine your building's electric service voltage, you can:

■ Call your electric utility provider and ask. (This is the surest and safest method.)

■ Look at the electric meter. Most of them are labeled to indicate if the meter itself is a Delta (120/240 volts) or a Wye (120–208 or 277–408).

■ Check the nameplates on your cooking equipment for the electrical ratings.

■ Ask an electrician to check the large electrical wall outlets with a voltage meter.

The other key piece of information you should learn is the restaurant's **load factor**. Expressed as a percentage, it is the amount of power consumed (in kilowatt-hours), divided by the worst-case (highest) consumption over a stated period of time (number of hours). The formula is:

$$\frac{\text{Kilowatt-Hours}}{\text{Highest Kilowatts} \times \text{Hours}} = (\text{kWh}) = \text{Load Factor (\%)}$$

The business should have a target load factor. Then, by tracking the load factor weekly or monthly, you can see how well the business is doing—or how well it *could* be doing. Luckily, with today's computer-based metering systems, this kind of analysis is possible. It's easy to calculate a target load factor:

1. Take into account all the hours the business is open (not just the business hours but also the before- and after-hours that employees are there). If a business site is being used 14 out of 24 hours a day, the target load factor is 14 divided by 24, which equals 58.3%. If you're calculating the target load factor for a full month, divide the total number of hours you're open that month by the total number of hours in the month. (A 30-day month has 720 hours.) A restaurant open 15 hours a day, 7 days a week, is open 450 hours a month. Divide 450 by 720 and you get a target load factor of 62.5%.

2. Now you are ready to compute the *actual* load factor. You will need these figures for the month:

■ The total amount of power consumption (in kilowatt-hours)

■ The demand meter reading (in kilowatts)

■ The number of hours you're open in a day (not the whole month, just one day)

Let's plug them into the formula and see what happens:

$$\frac{(\text{Total Consumption}) \ 2100}{\left[(\text{Demand Reading} \times \text{Hours in Operation}) \ 180 \times 15\right]} = \frac{2100}{2700} = .7777, \text{ or } 77.77\%$$

In the case of the restaurant open 15 hours daily, its actual power use is 15.27% higher than its target load factor. In the case of the business open 14 hours, actual power use is 19.47% higher than the target. In either case, energy-saving measures are overdue.

Tracking your load factors allows you to create a **load profile** for the business—that is, an accurate snapshot of how much power you need and what times of day you need it the most. Later in this chapter, when we discuss deregulation of the electric industry, you will see that having a load profile is necessary to get the best price for the electricity you buy.

ELECTRICAL INVENTORY

The more equipment you purchase and maintain, the harder it may be to keep track of if there is a problem. Problems may include:

■ Voltage transients (short, fast, high-voltage pulses of electricity that may damage equipment immediately or over time; also called *spikes*)

■ Momentary surges or sags in voltage

■ Momentary power losses

■ Electrical noise

■ Harmonic distortion (an imbalance in one wire of a single-phase ac line that can create an overload; seen primarily in fluorescent lighting systems and computers)

Some types of equipment are more likely to be the victims of power problems (phone systems, credit card scanners, personal computers), while others are more likely to *cause* problems for more sensitive equipment. Your heating or air-conditioning system, on a high-demand day, can cause such problems. Solving a problem may be as simple as making sure the victims and culprits are not on the same power circuit so they can't affect each other or upgrading wiring so it is large enough to handle a larger operating load.

The smart business owner will become acquainted with every piece of electrical equipment in the building; walk around and take inventory when the place is closed. From the largest motor to the smallest mixer, list everything. If a piece of equipment has an identification number or model number on it, write it down. Note the location of each piece of equipment and which electrical panel is feeding it. If the panels themselves aren't labeled, now is the time to do that, too, by shutting off each breaker and seeing what equipment turns off. One caution: Do not turn the large breakers (which control HVAC and computer equipment) on or off unless you know which part of your system they will affect.

Note any symptoms or problems. Be sure to ask employees if they notice anything unusual—flickering, surging, and so forth—and what time of day it seems to occur. When you purchase new equipment, put it on the list and write down the date it was put into service. (Sometimes the age of the equipment is a factor in its susceptibility to power problems—and older equipment almost always consumes more energy than newer appliances.) This way, you will be fully equipped to help the electrician or electric company help you. Most electric utilities have specialists who are familiar with the needs of foodservice businesses, so get to know the representative in your area. Remember, you are among their best customers.

HOW APPLIANCES HEAT

In electric equipment, heat is generated through a heating element. Wires inside the heating element accept the incoming electric current. It may be microchrome wire (round or flat ribbon-style wire), fiberglass-coated, or mica-coated. Fiberglass withstands temperatures up to 400°F, and more expensive mica-coated wire withstands up to 1000°F. Heating elements are also categorized by their **watt density**, or number of watts that they can produce per square inch—the higher the watt density, the greater the ability of the appliance to provide uniform heating.

If a piece of equipment doesn't have a simple on-off switch, it has a **thermostat** to control the heat. A thermostat also turns the equipment on and off, allowing it to heat to a desired temperature and then cycle on and off to hold that temperature for as long as needed. The cycle is often controlled by magnetic force. Magnets within the thermostat serve as contacts, keeping the power connected until the correct temperature is reached; then a bimetal control causes the magnets to snap open. This releases the power connection and the heat stops—until it cools down enough to cool the bimetal, which causes the magnets to snap shut and complete the power connection, and the cycle begins again. Solid-state controls—which use microprocessors to turn equipment on and off rather than moving parts or heated filaments—are now available in many types of kitchen equipment.

Kitchen appliances should also be equipped with their own set of cords that can be easily disconnected for cleaning and servicing. Neoprene cords are popular because neoprene is a tough but flexible substance that can withstand both water and grease. However, neoprene will soften under conditions of extreme heat. Rubber-coated cords are not recommended because they soften and deteriorate under grease and heat.

Today, many heavy-duty commercial cooking appliances can be ordered (at extra cost) for direct hookup to a 460-volt system. However, 460-volt equipment does not contain internal fuses or circuit breakers to isolate trouble. A malfunction in any part of the equipment will cause the whole thing to shut down completely until the problem is located and repaired.

When new equipment arrives in your kitchen, always read the nameplate before connecting it. If you do not follow the electrical requirements, you will notice problems. If the operating voltage is lower than it's supposed to be, the appliance will heat slowly or not heat fully, no matter how long it is left on. For instance, a 12-kilowatt fryer connected to a 230-volt power supply will be 25% less efficient and take longer to preheat.

If the operating voltage is higher than recommended, the appliance will become hotter, heat faster, and perhaps burn out the elements more quickly than it should. The same 12-kilowatt fryer, when wired to a 240-volt power outlet, will heat faster, but this will shorten the useful life of the appliance because its parts are working harder than they were designed to.

If you know some of the energy-related information for a particular appliance, but not all of it, remember that formulas can be used to calculate whatever you need to know. If an appliance is rated in kilowatt hours (kW), for instance, you can convert this rating to watts (W) and figure out how many amps (I) it will require. Let's say you have a single-phase 13.2 kW appliance. It's rated at 208 volts, but you need to know how many amps of electricity it will need. First, you need to convert the kilowatts to watts. This is a standard calculation because you know that 1 kilowatt equals 1000 watts:

$$13.2 \text{ kW} \times 100 = 13,200 \text{ watts}$$

Then use the formula you learned earlier to convert watts to amps:

$$I = \frac{W}{E}$$

For this appliance, that means:

$$\text{Amps} = \frac{\text{Watts}}{\text{Volts}} \quad \text{or} \quad \text{Amps} = \frac{13,200}{208} = 63.5$$

This appliance requires 63.5 amps of electricity.

The Electric Foodservice Council has calculated the energy requirements of several types of appliances, which you'll see in Table 5-2. This chart shows the approximate amounts of time and electricity required to preheat and maintain commonly used pieces of cooking equipment.

In this chapter, we focus mainly on electricity as a power source, but don't forget your gas-fired appliances when adding up energy costs. For example, a single gas pilot light (a necessary part of every cooking device approved by the American Gas Association) consumes about 750 Btu of gas per hour. Multiply that by 24 hours a day, and that's 18,000 Btu per day, or 18 cubic feet. (Gas is purchased in cubic feet.) If a cubic foot costs 5 cents, this equals 90 cents per day. The kitchen with a dozen gas appliances will spend almost $4,000 a year just to keep the pilot lights on. You'll learn more about gas usage in Chapter 6.

ELECTRICAL SERVICE ENTRANCE

Electricity is generated (converted into electricity from another source such as gas, coal, sunlight, or wind), then transmitted (moved in bulk to a wholesale purchaser), and finally distributed (delivered to end-users, such as homes and businesses).

The physical point at which a building is connected to the power source is called the ***electric service entrance***. Before you add a new piece of electric cooking equipment, you must find out

TABLE 5–2

Energy Use of Commercial Kitchen Equipment

TYPE OF APPLIANCE	KILOWATT INPUT	MINUTES TO PREHEAT TO 450°F	WATTS TO MAINTAIN AT 300°F	350°F	400°F
All-purpose oven	6.2	36	531	649	767
Baking oven	7.5	90	660	807	953
Pizza oven	7.2	45	410	507	599
Convection oven	11	10 or 11 (to 350°F)	1800		
Electric fryer (28-lb. fat capacity)	12	5 (to 350°F)	770		

whether it will require an increase in the size of the electric service entrance or the main circuit panel, a considerable expense.

The electrical service entrance may be overhead or underground, easily visible or not. This very important place has eight basic components, and, as shown in Illustration 5-3, you're already familiar with a couple of them. If there's a power outage, fire, or malfunction, you'll want to know enough to correctly describe it in these terms:

1. *Connection point.* The exact site at which the wires touch the building.
2. *Electric meter.* Records the kilowatt-hour consumption of the building.
3. *Demand meter.* Records the maximum demand for kilowatts.
4. *Master service switch.* Controls the flow of energy to the building; it automatically shuts off as a safety precaution if there's an overload or excessive demand. It's usually located on the outside of the building, and it is locked.
5. *Transformer.* As its name suggests, it changes incoming alternating current into whatever voltage is usable for a particular facility. Not all buildings have (or need) their own transformer; most very large ones do.

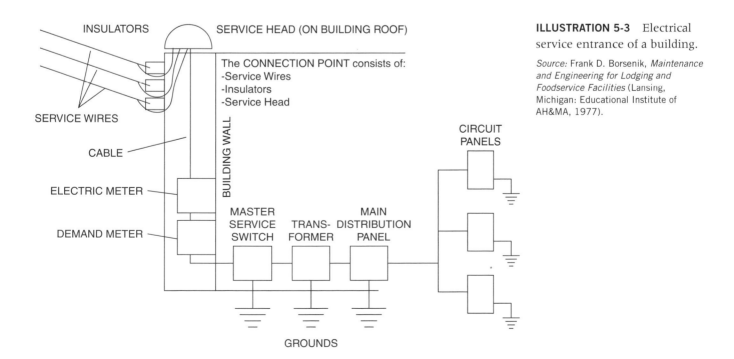

ILLUSTRATION 5-3 Electrical service entrance of a building.

Source: Frank D. Borsenik, *Maintenance and Engineering for Lodging and Foodservice Facilities* (Lansing, Michigan: Educational Institute of AH&MA, 1977).

6. *Main circuit panel.* Also called a *distribution panel*, this is the equivalent of the fuse box in your home. Separate switches, with circuit breakers, direct power to different areas of the building or pieces of equipment. Each panel also has a master fuse or circuit breaker that controls power to the entire panel.

7. *Secondary circuit panel.* Local codes may require separate circuits, in separate locations, for one particular appliance or function in the building. This provides a way to shut off the air conditioner, for instance, or all heavy equipment, without interfering with other power use. It's especially handy when repairs have to be made.

8. *Ground rod.* This low-impedance connection to the earth serves as a conducting body to which an electric circuit can be connected. A ground is necessary to operate electronic equipment.

Use of a **raceway** (sometimes called a *wireway*), as described in Chapter 3, allows for easier repositioning of receptacles, especially in the hot line area. Overall, think about more than wall space when you decide where electrical panels should be located. Consider accessibility for turning circuits on and off and for adding or modifying circuits as equipment needs change. Placing electrical panels in a back room, far from the equipment and lights they power, increases the length of wires and conduits—which, in turn, increases construction costs.

5-4 UNDERSTANDING ELECTRIC BILLS

Utility providers are not necessarily run by a higher power—no pun intended. They might make mistakes. They may misread a meter or estimate a bill without checking the meter at all. This is why foodservice managers should know how to read their own meters and to compare those readings with the bills when they arrive. Not only will you keep the utility companies honest, you'll be able to chart usage and see if the energy conservation measures you've implemented are working, week to week.

In a well-run operation, meter readings are taken regularly, usually weekly or every other week. Another purpose of these readings is to diagnose energy use problems. It is wise to designate one person as the "energy specialist" in your business so he or she can take responsibility for this task and for negotiating with utility providers for greater savings. In multiple locations, such as chain restaurants, it may be wise to hire an energy management coordinator who can handle these affairs on behalf of all locations. Independent companies also monitor and check utility bills for a percentage of the savings obtained. Bills should be checked for two types of error: calculation errors (incorrect addition or subtraction) and classification errors (e.g., being charged a residential rate instead of a commercial rate).

There are thousands of electric utility providers. Almost every one of them has a slightly different way of converting electrical usage into an electric bill, and almost every company bills restaurants differently than it bills residences. However, all restaurant bills have two major parts:

1. The **energy charge** (consumption charge) is based on the total amount of electricity used over the billing period.

2. The **demand charge** (capacity charge) is based on the highest amount of electricity used during peak use periods.

Demand is the amount of power required at a point in time to operate the equipment connected to the electrical system (also known as the **power grid**) of your business. It is measured in kilowatts. Demand should not be confused with electrical energy, which refers to power usage over a period and is measured in kilowatt-hours. To illustrate the difference, a piece of electrical equipment rated at 30 kilowatts (kW) requires 30 kW of power when it is operating. If it operates for 1 hour, it uses 30 kilowatt-hours (kWh) of electricity. If it operates for 20 hours, it will use 600 kWh of energy—but it's still rated at 30 kW.

Now let's look at the difference demand charges make in a utility bill. The power company must have enough generating capacity to produce whatever amount of electricity the customers need—not the number of kilowatt-hours but the amount of demand that may be put on the system at any given moment.

Let's pretend our fictitious Power Company, Inc., has only two customers. Customer A has a 50 kW load that is required 24 hours a day. Its total usage is 1200 kilowatt-hours (kWh), or 50 multiplied by 24. To serve Customer A, Power Company must have the capacity to generate, transmit, and distribute 50 kW of electricity.

Customer B also uses a total of 1200 kWh of electricity, but this usage is only 6 hours a day—200 kW for 6 hours equals 1200 kWh. To serve Customer B, Power Company must have the capacity to generate, transmit, and distribute 200 kW, or four times the load required for Customer A. Thus, the overall cost to serve Customer B is much greater than Customer A. It isn't fair to overcharge A for the greater peak demands of B, so Customer B pays more in demand charges than Customer A.

To bill customers accurately, both usage and demand must be measured, and this is done by meters. A meter is an electric motor with a large, circulating disk that rotates as energy passes through it. There are four types of meters. On three of them, you will see a row of dials that resemble clock faces. The dials record the number of rotations made by the disk.

You are billed for electricity in one of two ways: by the **kilowatt-hour (kWh)** or by the **Kilovolt-ampere (kVA)**. A kilowatt is 1000 watts, so a kilowatt-hour is the time it takes to use 1000 watts. A kilovolt is 1000 volts, so a kilovolt-ampere is the push it takes to propel 1000 volts through your wires. A meter keeps track of either kilowatt-hours or kilovolt-amperes.

A **kilowatt-hour meter** has only one row of dials. This type is used when the rate schedule is based only on kilowatt-hours used. Many utility companies are switching out these meters for digital meters. However, we include the meter-reading information because standard kilowatt-hour meters are still in use in some settings.

The second type of meter is called a **demand meter**. As you just learned with customers A and B, utility providers monitor (and can charge higher rates for) what they consider excessive use of electricity for businesses during peak periods. The demand meter may only measure for random 15- or 30-minute intervals. However, it has one hand that stays in the maximum-demand position to record the most energy that was used and another hand that moves as power needs increase or decrease. It is reset, using a key, by the meter reader on his or her normal rounds. Ask your meter reader exactly how to read your demand meter, as there are many models. Some are zeroed after every reading; others are cumulative.

Your meter may have two rows of dials. The top row is a kilowatt-hour meter; the bottom row is the demand meter. In commercial buildings, you may have two separate meters or this type of combination meter.

The **digital meter** is the most modern as well as the easiest to read. It looks like (and is read the same way as) the mileage indicator on a car speedometer.

No matter what kind of meter you have, subtract the first reading you take (e.g., at the beginning of the month) from the next reading (say, two weeks later) to get the total number of kilowatt-hours used during that time. Some meters have these words on them: "Multiply all readings by *x*." If yours does, you should multiply the readings you get by that number, or they won't be accurate.

When learning to read an electric meter, you may find it confusing that the hands on the dials rotate in opposite directions. To figure out which way each hand rotates, look at the dial itself. *Zero* is always at the top of the dial. If the number to the *right* of 0 is 1, we know this dial rotates clockwise, like a regular clock. If the number to the right of 0 is 9, we can assume the dial rotates counterclockwise.

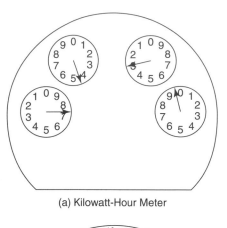

(a) Kilowatt-Hour Meter

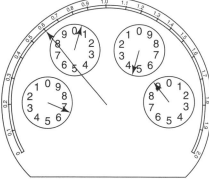

(b) Kilowatt-Hour Meter with
Kilowatt Demand Meter

ILLUSTRATION 5-4 (A) A kilowatt-hour meter (multiply all readings by 10 kilowatt-hours), and (B) A kilowatt demand meter (to get kilowatt-hours, multiply demand reading by 10).

Why are there several dials in a row? The one on the far right represents amounts from 0 to 10 kWh; the one to its left, from 10 to 100 kWh; the next one to the left, from 100 to 1000 kWh; and so on. The hand on the dial at the extreme right will complete one full circle, but the one to its immediate left advances only one number at a time—like the minute and hour hands of a clock. The same relationship exists between the other adjacent dials (see Illustration 5-4).

If the hand is pointing between two numbers, read it as the lower one.

Most utility providers require installation of a demand meter when the maximum energy use is 10 kilowatts or higher during their peak time period—usually from noon to 8:00 P.M. weekdays. This type of energy usage is called **peak loading**.

All utility companies vary their pricing structure somewhat, especially the demand charges. Some apply demand charges that vary according to the time of year. Others use a formula based on the relationship between the level of demand and the actual amount of energy used. Still others impose a minimum demand charge based on a high-energy- use period (midwinter in a very cold climate) and make customers pay it each and every month. This practice is known as a **ratchet clause**, and it is understandably unpopular, because high demand during one month can result in a higher bill during the rest of the year.

Because energy costs can vary drastically from season to season, the utility should prepare a **tariff analysis** for every restaurant, using its actual energy consumption as a guideline. This is a kind of overview of the billing, rates, and rules offered by the provider. The tariff analysis should be reviewed (and modified, if necessary) when major changes occur in the restaurant's operations or the utility provider's rate structure. Major changes include undertaking an energy conservation program or extensive remodeling. The document may be several pages long, but it is critical to have these details on paper.

Another item on your electric bill that should be reviewed periodically is the sales tax imposed by your local city or state. You may be able to apply for tax-exempt status for your utility bill if you use electricity (or gas) for processing foods. *Processing* is defined, for these purposes, as performing an operation in which a physical or chemical change is brought about. In foodservice, this includes most types of preparation and cooking: baking, broiling, chopping, slicing, frying, and mixing. It does not include keeping foods hot or cold (as no chemical change is happening to them at that point), heating or air conditioning, dishwashing, or lighting.

There are no separate meters for exempt and nonexempt utility uses, so it is up to you, as the business owner, to prove that more than half of your utility use is for processing (exempt) purposes. If you can, the entire bill may qualify for a sales tax exemption. In this chapter, you will find a list compiled by the Texas Restaurant Association to advise its members what that state's comptroller considers reasonable types of processing equipment that use energy as well as those that do not qualify in determining your sales tax exemption.

Your utility provider probably has written information on this topic as well as pamphlets explaining utility rates and how to read meters. Ask for similar information from gas and water suppliers.

To be truly energy proactive, it is important for the foodservice manager to know the relevant state and local codes governing the use of electricity, the calibration of meters, installation techniques for electrical equipment, and licensing requirements for electricians. The most widely accepted standard is the *National Electrical Code (NEC)*, which specifies how wiring must be done and what safety precautions must be taken. It is updated every 3 years, and major revisions made in 2008 involve the sizes of wires, conduit, and main electrical panels. These changes now allow the use of smaller sizes than would have

BUILDING AND GROUNDS

The Debate About Smart Meters

In the early 2000s, after decades of the same meters and meter-reading methods, utility companies began introducing the so-called smart meter. This uses electromagnetic frequency to transmit and receive data between the meter user and the electric utility, virtually eliminating the need for meter-readers who would otherwise take the readings in person.

At this writing, smart meters have caused almost as much controversy as they may have streamlined the meter-reading process for electric utilities. Groups like the National Campaign to Stop Smart Meters and the Wireless Radiation Protection Coalition say the meters emit "strong bursts" of a type of microwave radiation the World Health Organization has labeled a Class 2B carcinogen.

The groups also claim problems with internal wiring and shoddy installations of the new meters are to blame for hundreds of fires. And individual homeowners have expressed concerns about data privacy and security when they learn that information can be automatically transmitted in readings that may be accurate down to the individual appliance.

Utility companies say the meters are simply part of their ongoing effort to use technology to modernize energy transmission—and save energy by doing so. They point out the smart meters are installed outdoors, minimizing their contact with humans and animals, and the radiation they emit is a fraction of what a cell phone emits. And the European industry group smartmeters.com says no one can spy on customers or turn their power off as a result of smart meter use. This trade group calls smart meter detractors "fringe activists."

This debate is likely to continue to play out in the media, particularly because some utility companies making the switch to the new technology are charging customers a fee to keep their older analog meters.

---

been allowed under the old code, which will probably cost less. So, if you are constructing a new facility, be certain the architects and designers are using the most up-to-date NEC. Most local codes are based on the NEC, although they are not always updated in a timely fashion.

The code can also come in handy if you are considering adding new equipment to an existing kitchen. Equipment manufacturers also use the NEC when designing their products. (There is also a National Electrical Safety Code, or NESC.)

Another agency involved in the safe use and installation of electrical equipment is *Underwriters Laboratories* (*UL*). This organization establishes test procedures and tests equipment for compliance with its own rigorous set of standards. Its Standard 197, "Commercial Electric Cooking Appliances," pertains to the manufacture of most major appliances found in commercial kitchens (rated 600 volts or less) as well as coffee makers, food warmers, popcorn machines, and utensil warmers. In recent years, Standard 197 was updated to address safety concerns of the electric heating elements in appliances. The change requires that all "nonmetallic materials, including end seals, should not exceed their recognized temperature ratings." In the past, end seals (also called *terminations*) were not considered integral parts of a heating element; their role is to protect the element from moisture getting into it. Standard 197 is worth a closer look; it can be found online.

The UL certification label means the equipment meets the standards and is safe to use under normal conditions. However, be sure when you see the label that the entire piece of equipment has been certified—not simply one component, such as the fan of a convection oven.

CONSUMPTION CHARGES

Now that you know something about reading meters and converting quantities of energy, you're ready for the most daunting test: reading the electric bill.

BUDGETING AND PLANNING

Texas Sales Tax Exemptions for Foodservice Equipment

The following list is representative of the types of equipment the Texas Comptroller has determined qualify as processing and those considered nonprocessing.

For food processing, you may include all of the following listed equipment:

Rotisseries	Soft-drink dispensers
Ovens	Ice cream machines
Ranges (stoves)	Ice machines
Microwave ovens	Heat strips*
Fryolators	Steam table*
Toasters	Mixers/grinders
Food choppers	Broilers, charbroilers
Slicers	Proof boxes
Grills	Chicken boxes
Coffee makers	Chicken and fish breaders
Malt machines	

*Heat strips and steam tables cannot qualify if used to maintain a warm temperature after preparation, but may qualify if used to bring food from a cold or frozen stage to serving temperature—or, in the case of a heat strip, to keep food warm during preparation.

Your list may *not* include the following equipment:

Cash registers	Air conditioning
Coffee warmers	Lighting
Cream-o-matics	Exhaust fans
Milk dispensers	Music, signs
Heat lamps	Dishwashers
Refrigerators	Hot-water heaters
Freezers	Ice cream holding cabinets

While this list is not all-inclusive, it will by way of example help you understand the comptroller's definition of processing.

Courtesy of Texas Restaurant Association, Austin, Texas.

Most foodservice establishments are not large enough to qualify for lower industrial rates; instead, they are billed on a commercial or small general rate schedule. They are rarely charged a single rate for every hour of electricity they use. A single rate, or **uniform charge**, is more likely to be seen on a residential power bill.

Instead, for commercial customers, you'll probably find a **step-rate** or *declining-block schedule*—that is, a graduated schedule of rates that decreases as consumption increases. For instance, the first 50 kWh may cost you $0.100, but the next 50 will only cost $0.050, and the 100 after that will only cost $0.030.

A *time-of-day option* allows you to be billed based on the time of day your business is open. More power is used during weekdays, so that's when rates are highest. However, businesses that operate mostly nights, holiday, or weekends can qualify for lower, off-peak rates by choosing a time-of-day plan.

Another type of commercial rate is the *demand charge*. Remember the demand meter? The demand charge multiplies that reading—the average peak demand—by a charge per kWh. The demand charges may also be a declining step rate, depending on how much power is used; and if your facility exceeds its normal demand, there will also be an additional charge for consuming more than usual.

The utility provider may also impose additional charges on customers with low power factors. Your *power factor (PF)* is the ratio between what the demand meter says (*real power*) and what the utility company says is the total capacity available to you (*apparent power*). Most foodservice businesses don't run the kinds and amounts of equipment that result in a low-power-factor penalty. If it is a problem, however, capacitors can be installed to bring your voltage and current levels into closer sync, increasing your power factor and lowering your power bill.

Fuel adjustment charges (sometimes called *energy cost adjustments*) show up on your bill too. Because energy costs fluctuate constantly, utility companies in non-deregulated states are authorized to pass on these fluctuations to customers—sometimes as a charge, at other times a credit!

Finally, don't forget *late charges*, which are inevitably be tacked on if you don't pay the bill on time.

NET METERING

And what happens when your electric meter runs backward? This is not impossible if your business or home employs one of the alternative energy technologies described elsewhere in this chapter. *Net metering* is a special power measurement and billing agreement between a utility company and customers that generate their own power. For instance, a net metering agreement would allow you to use your own power first and rely on the utility for the remainder of your electricity, if necessary. If you end up generating more than you use, the meter dials actually spin backward. Some utility companies require you to have separate meters installed, one to track your homegrown production and one to track their power.

Every utility has different rules about net metering. For some, it is an annual agreement, and you're billed at the end of the contract for the power you have used; others allow monthly billing. Some pay you for the excess power you have generated, but others do not. If your utility provider won't buy your excess, however, chances are good that you can find another company that will.

CHARTING ANNUAL UTILITY COSTS

Beyond regular meter readings, foodservice operations can use two additional methods to track energy consumption. You may find them especially useful in preparing cost estimates to open a new facility. They are utility costs as a percentage of total sales and an *energy index* of Btus per meal served.

Let's look at them one at a time. First, to determine the percentage of utility costs, simply add all the utility bills (except the phone bill) for the month or year and divide the total by total sales for the same period. (The total sales figure is found in the income statement.) This method is simple, but it certainly doesn't give much detail about conservation measures.

To develop an energy index, you must convert all the types of energy used into a similar form so the figures can be combined. This is where the *Btu* comes in handy. The abbreviation stands for **British thermal unit**, which is the amount of heat needed to raise the temperature

BUILDING AND GROUNDS

Reading an Electric Bill

All managers should understand the information they receive on utility bills. This enables them to compare their own meter readings with those of the utility companies and to develop an awareness of how the billing figure is arrived at. Although utility bills come in a wide variety of shapes and forms, they usually display the same types of information shown on the sample electric bill (see Illustration 5-5).

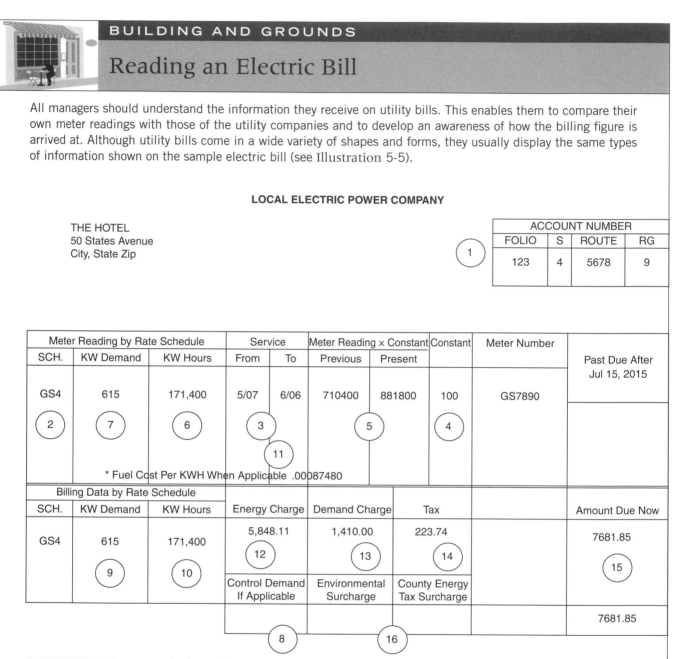

LOCAL ELECTRIC POWER COMPANY

THE HOTEL
50 States Avenue
City, State Zip

ACCOUNT NUMBER			
FOLIO	S	ROUTE	RG
123	4	5678	9

(1)

Meter Reading by Rate Schedule			Service		Meter Reading × Constant		Constant	Meter Number	
SCH.	KW Demand	KW Hours	From	To	Previous	Present			Past Due After Jul 15, 2015
GS4	615	171,400	5/07	6/06	710400	881800	100	GS7890	
(2)	(7)	(6)	(3)		(5)		(4)		
			(11)						

* Fuel Cost Per KWH When Applicable .00087480

Billing Data by Rate Schedule			Energy Charge	Demand Charge	Tax			Amount Due Now
SCH.	KW Demand	KW Hours						
GS4	615	171,400	5,848.11	1,410.00	223.74			7681.85
	(9)	(10)	(12)	(13)	(14)			(15)
			Control Demand If Applicable	Environmental Surcharge	County Energy Tax Surcharge			
			(8)	(16)				7681.85

ILLUSTRATION 5-5   A sample electric bill and key for interpreting it.

Source: U.S. Department for Energy, "Guide to Energy Conservation for Food Service," Washington, D.C.

NOTES FOR SAMPLE BILL

1. Your account number. Use it when communicating with the utility.
2. The rate schedule that applies specifically to your property. The utility company will furnish you with a copy.
3. Dates on which the meters were read. They represent the beginning and end of a billing period.
4. A constant by which the meter reading must be multiplied to get the actual usage.
5. The meter readings multiplied by the constant. If no constant is shown on your bill, this entry is the same as the meter reading.
6. Kilowatt-hours of electricity used.

7. The maximum kilowatt demand recorded during a demand interval in this billing period. The demand interval is measured in periods of 15, 30, or 60 minutes and shown in the rate schedule.

8. Control demand is shown if it is part of the rate schedule. In general, it is based on the maximum demand of the previous 12 months. The rate schedule shows how it is determined and its effect on the computation of the electric bill.

9. Kilowatt demand for billing purposes.

10. Kilowatt-hours for billing purposes; usually the same as item 6.

11. Fuel adjustment cost per kilowatt-hour. This varies as the utility company's fuel cost changes from a fixed normal cost.

12. Charge for kilowatt-hours used based on the rate schedule and the fuel adjustment cost.

13. Charge for maximum demand based on the rate schedule.

14. Local tax.

15. Total amount of the bill.

16. Surcharges that may be added to the bill to offset expenses of local governments.

The same types of information are shown on bills for other types of utilities, with obvious differences due to consumption in the applicable units: cubic feet (gas), gallons or cubic feet (water), and pounds (steam).

of 1 pound of water 1°F. This measurement is useful because all energy sources can be easily converted to Btus. Here's how:

Electricity. Multiply kilowatt-hours by 3413.

Natural Gas. Multiply cubic feet by 1000.

Oil. Multiply gallons by 140,000.

Steam. Multiply pounds by 1000.

When the resulting figures are added, the total is the amount of energy (in Btus) consumed in the restaurant overall. Don't be shocked—it will be in the millions!

Numerous software programs are available for you to use in tracking your business's energy use. If you know the amount of your electric bill and the number of customers you served during a particular period, you can easily determine how much you spent per customer on electricity.

A final note about electricity costs: Any experienced foodservice operator will tell you they're tough to track under the best conditions. Unusual weather and rate increases are just two of the factors that make forecasting difficult. In our experience, it is still preferable to monitor and pay your own utility bills than to agree to include utilities in a lease agreement and let the landlord keep track of them. Leases that include utility payments leave little or no incentive for conservation and the door wide open for rent increases based on higher utility costs you have no way of confirming.

5-5 POWER FAILURES

What happens when the power goes out? Power outage preparation should be part of emergency planning. Ask anyone in California, where utility customers encounter so-called rolling blackouts during soaring summer temperatures. The term *brownout* also has become familiar to many; it refers to a power interruption of less than 20 minutes. This typically occurs when the equipment being used requires more power than is available. Brownouts can occur on hot summer days or during winter cold snaps—whenever the equipment is running at full capacity and drawing a maximum load.

The U.S. Energy Information Administration (EIA) reports at least two dozen "major disturbances and unusual occurrences" per year that interrupt electrical service. Many of them are weather-related and last from two hours to several days. The EIA publishes an interesting monthly report, available online, about power prices, usage, and disruptions that covers both electricity and gas. Generac, a Wisconsin-based emergency generator manufacturer, estimates that 38% of power outages are the result of weather; 26% are from utility equipment problems or grid overload; 12% are from fallen trees on power lines; and 10% are from animal contact with the lines. The remaining 14% are prompted by human caused situations that vary from vandalism to construction errors.

A disruption in the power supply can have costly consequences to a restaurant if it lasts more than two hours. Most potentially hazardous foods (known as PHFs in food safety jargon) can be kept out of the temperature danger zone for that long, but no longer. Of course, during an outage, you have no way of knowing how long it will last. Managers on duty should take nine precautions:

1. Record the time the outage began.

2. Take periodic temperature readings (every 30 to 45 minutes) of the food that was prepared and being held at the time.

3. Keep all refrigerator and freezer doors closed as much as possible. (This is why, as you will learn in Chapter 9, it's handy to have a thermometer that measures a unit's internal temperature on the outside of the unit.)

4. Cover any open display cases in which food is iced or chilled.

5. Used canned fuel (such as Sterno) to keep hot foods hot. If the temperature of hot food on a steam table, for instance, drops below 140°F for more than two hours, it should be discarded. If the outage doesn't last that long but the food has dropped to less than 140°F, reheat it rapidly to 165°F, hold it at 140°F, and serve as quickly as possible.

6. Discontinue cooking and serving if it is not possible to wash, rinse, and sanitize utensils.

7. Discontinue cooking if light is insufficient for safe food preparation, including cleaning and sanitizing food contact surfaces.

8. Don't cook if there is no hot water or if water pressure is not sufficient.

9. Don't assume that cooking cold food stored at room temperature for too long will kill microorganisms and make it safe to eat. That is not the case. The rule is *always* "When in doubt, throw it out."

Table 5-3 includes guidelines for using refrigerated foods that have been subjected to a power outage.

In areas prone to weather-related power outages, the purchase of a power generator is an attractive option. If a foodservice business experiences more than two or three major power interruptions a year, having a backup generator makes sense. The generator can also be used for so-called **peak shaving**, or curtailing power use or shifting to another source to avoid peak loading and the higher costs it entails. The generator can be tied into the primary power grid and programmed to kick in not just when there's a power failure but whenever demand charges are highest. In other words, when the electric or gas company is charging its highest rates, you generate some of your own power and reduce dependency on the utility company, at least temporarily.

Noise used to be a common complaint about backup generators, but in recent years, manufacturers have focused on making them quieter and reducing exhaust emissions. They run on natural gas, come in a wide range of power capacities for restaurant use (from 7 kW to 150 kW), and can be matched to almost any voltage/amperage requirement. Some are designed for rooftop installation to save space. Generators can also be purchased that conduct their own self-diagnostic and start-up tests weekly to ensure they'll be ready to go to work at a moment's notice.

TABLE 5-3

Guidelines for Handling Potentially Hazardous Foods (PHFs) Stored in Refrigeration Units During a Power Outage

DURATION OF POWER OUTAGE	COLD FOOD TEMPERATURES		
	45°F OR BELOW	46°F TO 50°F	50°F OR ABOVE
0–2 hours	PHF can be sold.	Immediately cool PHF to 45°F or below within 2 hours	PHF cannot be sold.
2–3 hours	PHF can be sold, but must be cooled to 41°F or below within 2 hours.	Immediately cool PHF to 45°F or below within 1 hour.	PHF cannot be sold.
4+ hours	PHF Immediately cool to 41°F or below within 1 hour.	PHF cannot be sold.	PHF cannot be sold.

Source: Developed by Lucie Thrall, Director of Safety Management Services for Food Handler, Inc., Melville, New York.

In large facilities such as hotels, emergency power systems are incorporated into building design. They are required by code to have the capacity to handle a sufficient load to meet basic needs in case of a fire or other crisis event, often including emergency power for elevators.

RENEWABLE ENERGY

The rising cost of electricity, along with the unmistakable impact of climate change, has prompted more research into so-called *clean energy* or *green power*.

The U.S. Environmental Protection Agency (EPA) defines *renewable energy* as "electricity produced from resources that do not deplete when their energy is harnessed." The list includes water and ocean waves, wind and sunlight, geothermal energy derived from natural hot-water sources, and anaerobic digestion, or the use of natural decay to produce gas at landfills.

Proponents of renewable energy are quick to point out that if Americans yearn for less vulnerability to terrorism and less political instability in other parts of the world, a greater variety of methods should be used, and energy self-sufficiency should be a national goal.

In the United States, Congress has been slow to adopt a national energy policy that pays alternative energy sources more than political lip service. However, the federal government and many states offer tax credits, utility rebates, and low-interest loans for businesses or homeowners that make the switch to renewable energy. This significantly improves the cost-effectiveness of installing the equipment necessary to, for example, use solar power in additional to conventional power.

The U.S. Department of Energy sponsors an extensive list by state of the programs available. It is called the Database of State Incentives for Renewables and Efficiency (DSIRE), and can be found online.

Before you decide to use a renewable energy source, you must determine how much power is necessary and how much you could potentially generate with the alternative you are considering. At your location, do you get enough sun to use solar power effectively? enough wind for wind power? Just starting out, the technology can be expensive—so you had best determine if it is worth the outlay.

Depending on your target clientele, there may be a public relations benefit to renewable energy use as well. The effort your business makes to be environmentally friendly can be a selling point, and it might give you an edge over your competitors in customers' minds.

PURCHASING RENEWABLE ENERGY CREDITS. Most utility companies also have options for customers who want to support renewable energy development even if their homes or businesses are not suitable for solar panels or wind turbines. Signing up for a utility's "green power" program typically means paying a little more on the monthly power bill to support renewable energy sources in its portfolio.

If a "green power" program is not available, an alternative is to purchase Renewable Energy Certificates (RECs). The EPA suggests thinking about RECs as a type of currency. They are bought and sold, usually in increments of 1 megawatt-hour (1000 kilowatt-hours) that represent, in the EPA's words, "the right to claim the attributes and benefits of a renewable generation source."

Every REC is numbered and tracked to help ensure its authenticity. Buying a REC that is properly certified and verified indicates that you are doing your part to offset the greenhouse gas emissions associated with generating electricity. Sometimes, companies—like manufacturers that emit pollution in their processes—purchase RECs as a way to offset the damage they are doing to the environment. Utility companies sometimes purchase them to meet state-mandated goals that require a certain percentage of the utility's power come from renewable sources. An individual business can purchase RECs, or a group of businesses can.

RECs can be purchased based on a specific site where the "green" power was generated, on the date the generation occurred, or on the specific type of resource (solar, wind, geothermal) used to create the energy. The EPA has a Green Power Locator that allows anyone to locate green power suppliers that sell RECs.

SOLAR TECHNOLOGY IN FOODSERVICE. The tiniest solar photovoltaic (PV) cells are used in calculators and watches, but when larger cells are grouped together in modules of 40 or so, and multiple groups are mounted together (in configurations known as *arrays*), PV cells collect sunlight and convert it to sufficient amounts of electricity to power businesses and residences. It takes 10 to 20 arrays to power a household, according to the American Solar Energy Society.

It takes a lot of cells to make power because not all of the sunlight that strikes the cell is used to generate electricity. The industry is working hard to increase efficiency. In larger installations, the arrays collect and focus sunlight with mirrors to create a higher-intensity heat source; this process is known as *Concentrating Solar Power (CSP)*.

Solar power is as adaptable as electricity generated any other way: It can be used for heating water, heating or cooling air, and so on. It can be incorporated into building design—an advantage over wind turbines—and a solar system is relatively low-maintenance.

In 2011, the fast casual Chipotle restaurant chain committed to a plan to install solar panels atop 75 of its locations, which would make it the largest direct producer of energy in the restaurant business. By 2012, they had made 20 of the installations and confessed they'd run into plenty of challenges along the way.

"We have to consider several sites for every one that actually gets done," a Chipotle spokesperson told *QSR* magazine. "This is much more difficult and time-consuming than we originally thought."

One challenge is that solar panels must compete with vents and other necessary items already installed on the typical restaurant rooftop. Another is the cost—an estimated $15,000 to $20,000 for a system that may meet up to one-third of a quick-service restaurant's peak power needs. Batteries can be used to store some of the power, but they are expensive.

Initial cost also figures into solar installations at public institutions from schools to prisons. We've read about more than one promising proposal that can't get the necessary funding, even when proponents promise the plan will pay for itself within 10 to 15 years.

Almost any foodservice business can benefit from using a solar hot-water heater. Rather than generating electricity from the sun's heat, its thermal collectors heat the water. The chief difference between the types of solar water heating systems is how the water is circulated within them. Cold water tends to sink to the bottom of a water tank, while the hot water "lives" at the top of the tank. Simple explanations of the most common types are as follows:

■ *Batch heater*. Instead of collecting at the bottom of the storage tank, the water is directed to an insulated enclosure (the batch heater) where it is exposed to sunlight and heats naturally. It may not end up as hot as if it had been electrically heated to 140°F or more, but at least the sunlight does part of the job.

■ *Thermosiphon.* The solar collector is located at the bottom of the storage tank, where it heats water, which rises as it warms. Gravity naturally ensures the cold water is always at the bottom of the tank.

■ *Direct pump.* A small electric pump circulates the water from the collector to the storage tank, so the collector does not have to be at the bottom of the tank. The less distance the water must travel (in pipes), the more efficient the system.

WIND POWER IN FOODSERVICE. Wind power is another renewable energy source, which any farmer with a windmill has known for years. A single old-fashioned windmill produces from 1 to 5 kilowatts, just enough to pump water for livestock or home use, but new technology allows the progressive farmer to retrofit an existing pump system to generate 300 to 500 kilowatts—enough to power the property's irrigation system. They're not called *windmills* anymore, but *wind towers* and *wind farms.*

A few foodservice businesses have embraced wind technology. The Great Escape, a restaurant in Schiller Park, Illinois, has built a single wind turbine for its own use onsite. The company says it took 7 months to obtain Federal Aviation Administration approval (because it's 112 feet tall and could pose a navigation hazard for aircraft) and another 5 months for the local public hearing and zoning process. Today, it provides 108 kilowatts of power, serving an estimated 120% of the restaurant's needs. A 50-kilowatt backup generator is available in case of outages.

The wind conversion wasn't cheap; it cost the restaurant owners $375,000. They say they mortgaged their building in order to pay for it. But the excess power generated can be sold, and the turbine will eventually pay for itself in electricity cost savings.

Some people complain that the giant towers and turbines are noisy and unattractive, especially in large numbers, as well as hazardous to birds. Wind power advocates are making technical breakthroughs to bring down costs and make production more reliable, but natural gas-fired plants can produce electricity at 3 cents per kWh or less, while wind farms produce power at about 5 cents per kWh. Studies say this cost could be reduced enough to compete with gas-fired plants if wind farm developers could obtain the same favorable financing terms as utility companies or if the utilities would be willing to own the wind farms. Increasingly, the latter is the case. However, while some utility companies see the benefits of good public relations for adding renewable energy to their portfolios, others are fighting wind generation every step of the way. They claim it is unreliable because, of course, sometimes the wind doesn't blow steadily, or blow at all. Others are debating with small or remote wind farm owners over whose responsibility it is to pay for and build the additional power lines to connect their output to the grid.

You can learn more about the costs and feasibility of wind power from the American Wind Energy Association. The organization has European (EWEA) and Canadian (CANWEA) counterparts.

BIOMASS TECHNOLOGY. *Biomass* is the term for organic materials (from plants or animals) used to create a fuel source. It can generate electricity or fuels to power vehicles. Biodiesel and ethanol (ethyl alcohol) are examples of biomass fuels. Biomass is also being used to create new-generation recyclable plastics for use as utensils and food containers.

We mention biomass because foodservice grease traps and trash output are as good as gold to biomass energy producers. The Biogas Energy Project at University of California, Davis, made its first attempt at large-scale use of restaurant waste to create electricity starting in 2006. It has been a success and is still going strong. Waste collected from San Francisco–area restaurants is pumped into an anaerobic digester, a specialized vat typically used in wastewater treatment plants, which combines the sludge with bacteria to break it down. The process creates hydrogen and methane gas that can then be used as fuel or burned to produce electricity. Its creators say 1 ton of waste can power 10 houses for a day while keeping that ton of waste out of the landfill.

You will learn more about options for "green" waste options in Chapter 16.

GEOTHERMAL POWER. The capture and use of heat from naturally warm water (from 45°F to 55°F) found in some areas beneath earth's surface is described in greater detail in Chapter 7's discussion of HVAC systems. About one million American homes and businesses utilize geothermal heat pumps to both heat indoor spaces in winter and cool them in summer, requiring about 20% less electricity than they would without heat pumps. Federal legislation in 2005 increased the financial perks for this type of system.

The cafeteria, gymnasium, and school offices at Daugherty Elementary School in Garland, Texas, are being heated and cooled using geothermal energy starting in the 2013–2014 school year. This is part of a $17 million construction project expected to mean 20% to 40% lower utility bills for the first school of its kind in the Garland Independent School District.

Not everyone considers geothermal resources to be "green," however. In Hawaii, there has been public outcry against geothermal development since the 1980s. Accessing geothermal resources on the Big Island involves drilling in rainforests and tropical lowlands, an environmental risk some don't want to take.

Critics of the process say geothermal wells produce hydrogen sulfide, a gas with a distinctive odor that, when mixed with air, becomes sulfuric acid—a main component of acid rain. Fumes from geothermal development also have been blamed for nausea, headaches, and lung irritations of nearby residents.

If there is a lesson to be learned from our discussion of renewable energy, it is to approach any type of process or technology with an open mind and a willingness to see all sides of the debate.

5-6 ENERGY CONSERVATION

There are thousands of ways to conserve energy, but most can be grouped into four simple categories:

1. Improve the efficiency of equipment.
2. Reduce equipment operating time.
3. Recover otherwise wasted energy.
4. Use a cheaper energy source.

One-third of the daily energy use in foodservice operations is the result of food preparation activities. The term *cooking efficiency* means maximizing the quantity of heat transferred from the cooking equipment to the cooked product rather than the surrounding environment. This keeps your energy bills lower and also keeps the kitchen temperature more comfortable for employees.

The primary causes of inefficient cooking are:

■ Preheating equipment too long.
■ Keeping equipment turned on when it is not being used.
■ Using higher temperatures than necessary.
■ Opening appliance doors frequently, letting heat escape.

Sometimes the piece of equipment itself is at fault. It may be poorly insulated; its temperature may have been incorrectly calibrated; or perhaps it hasn't been kept clean enough to work at maximum efficiency. Energy-saving features should be considered in any equipment purchase, as it is almost impossible to retrofit an appliance to make it more energy efficient. How important is this? A single energy-efficient commercial appliance in a foodservice setting can save more energy in one year than an entire household.

For equipment, energy efficiency is measured in ratios (by percentage) and can be calculated any of three ways:

1. *Combustion efficiency* refers to fuel-fired equipment that uses gas, oil, or propane to produce heat. The key here is to control both the heat lost up the flue of the appliance and the radiant heat released into the atmosphere. There's a formula for this:

$$\text{Consumption Efficiency} = \frac{\text{Available Heat (the total amount of energy the appliance can produce)}}{\text{Purchased Heat (the amount of energy purchased for cooking with this appliance)}}$$

Typical combustion efficiency ranges from 40% to 70% for some gas-fired equipment.

2. *Heat transfer efficiency* refers to the amount of heat transferred to the cooked product compared to the total amount of available heat the appliance can produce. This percentage depends in part on the cooking process being used, the design of the equipment, and how efficiently it is being operated. Again, the formula:

$$\text{Heat Transfer Efficiency} = \frac{\text{Heat Transferred to Food (the amount of energy used cooking)}}{\text{Available Heat (the total amount of energy the appliance can produce)}}$$

Equipment design can increase this type of efficiency by adding insulation or recirculating exhausted air. Smart use of the equipment—keeping oven doors closed, closing the cover on a tilting braising pan—also minimizes heat loss.

3. *Overall energy efficiency* is the ratio of heat transferred to the cooked product in comparison to the amount of energy purchased for that appliance. The formula:

$$\text{Energy Efficiency} = \frac{\text{Transferred to Food}}{\text{Purchased Heat}}$$

This takes into account heat losses due both to combustion inefficiency and the surroundings.

Another way to shop for appliances, including computer equipment, is to look for the Energy Star rating. This program, developed by the EPA, began in the early 1990s as an effort to save energy and reduce pollution by rating some types of appliances and offering a third-party seal of approval of sorts. Since that time, businesses and consumers have saved more than $7 billion per year on energy costs by using Energy Star–rated appliances, and it is the standard by which commercial appliances are judged.

Appliances get Energy Star ratings when they score among the top 25% in their product category when it comes to energy efficiency, with potential savings of 25% to 60% on utility bills. As an example, an Energy Star–approved steamer can save $420 (electric model) to $820 (gas model), and it uses far less water than steamers that did not receive this rating. Ice machines typically are enormous users of water and power, but an Energy Star–rated ice machine uses 3.5 fewer gallons of water and 1.8 fewer kWh of electricity for every 100 pounds of ice it produces. And here's something to ponder: The EPA claims that if all the current reach-in refrigerators and freezers in the United States were replaced with Energy Star–approved models (which contain more modern refrigerants), we'd eliminate as much air pollution as we would by taking 475,000 cars off the road. In 2003, hot food cabinets, steam cookers, and fryers were added to the list of rated appliances. Learn more about the Energy Star program online at energystar.gov.

Reducing operating time is another way to save energy. In most kitchens, the hot-water heater is a perfect example. When used primarily for handwashing, the water is comfortable at a temperature of 105°F or less. However, the most common setting is 140°F. Do you know why? This is the original factory setting, and most people never bother to change it after it is installed. An expensive oversight! The water heater works harder to maintain the higher temperature, even when the business is closed.

For foodservice, it is more economical to install separate water heaters for high-temperature uses such as dishwashing and laundry and low-temperature ones for handwashing. You will learn in Chapter 6 about boosters that can be installed near the hot-water appliances. As its name suggests, a *booster* can boost the temperature of the water as it enters the appliance, lowering the heat loss that would occur if it had to be piped in from a more remote source. And how about checking for leaks and using spring-loaded taps in sinks to minimize wasteful dripping?

Another typical restaurant oversight is the fryer. Today's fryers take 10 minutes or less to heat to full temperature, but in many foodservice settings, they are on and fully heated all day. This not only wastes energy but also minimizes the life of the fryer oil. As you'll learn in Chapter 11, it is best to turn an unused fryer down to a lower temperature (if not off altogether) when it is not needed. Griddles and hot tops (see Chapter 10) can be purchased with heat controls in sections—heat only what you need. There are plenty of ways to save energy by using just as much appliance as you need, and no more.

One way to minimize the demand charges discussed earlier in this chapter is to stagger the preheating of equipment. By sticking to a start-up schedule and not turning everything on to preheat at once, the peak demand reading will be lower on your electric meter. How much difference can this make? Take a look at the charts (Tables 5-4A and B) from the Electric Foodservice Council. Remember, it's not the overall amount of energy used that varies; it's the peak, or maximum demand, for which you are also being charged.

In foodservice, energy conservation goals must also be seen from the point of view of the guest. If you're only looking at the bills, you're not looking far enough. Nowhere is this more evident than in your heating, ventilation, and air-conditioning system. If your guests are complaining that they are uncomfortable, your lower utility bill may actually be having an adverse impact on business. Set the thermostat at 68°F in winter and 76°F in summer, and you combine pleasant temperatures with savings of up to 18% in winter and up to 30% in summer.

Lighting can be another power drain. Fluorescent lights are certainly the most economical, but they may also be the least flattering unless you investigate some of the more modern types, which we discuss in greater detail in Chapter 7. Also remember that lighting creates heat; therefore, when you reduce lighting, you lose heat, and if you lose heat, your central

---

TABLE 5-4A

Simultaneous Start-Up

15 MIN. INTERVALS	15 MIN.	30 MIN.	45 MIN.	60 MIN.	75 MIN.	90 MIN.
Oven 6.2 KW	36 min.					
Fryer (1) 12 KW	5 min.					
Fryer (2) 12 KW	5 min.					
Griddle 6.5 KW	7 min.					
EST. DEMAND	18.7 KW	8.7 KW	5.4 KW			

Start-Up Time—36 Minutes

Maximum Demand—18.7 KW

Total Energy Used—7.7 KWH

ON ▬▬

IDLE ▬ ▬ ▬

heating system must work harder (and use more energy) to make up for it. You will learn more about both HVAC and lighting basics in Chapter 7.

In the typical 24-hour coffee shop, nearly 60% of the energy used flows through the kitchen. We realize it has become popular to sear food quickly or to prepare dishes individually. From a power consumption standpoint, however, you realize the most savings by:

- Cooking in the largest possible volume at a time
- Cooking at the lowest temperature that can still give satisfactory results
- Carefully monitoring preheating and cooking temperatures
- Reducing peak loading, or the amount of energy used during peak demand times (usually between 10 A.M. and 8 P.M.)
- Using appliances (ovens, dishwashers) at their full capacity
- Venting dining room air into the kitchen to meet kitchen ventilation requirements
- Using a heat pump water heater, which heats water as it cools (or dehumidifies) the kitchen
- Using an evaporative cooling system
- Recovering heat from equipment—refrigerators, the HVAC system, even kitchen vents—for reuse
- Increasing the hot-water storage tank size
- Keeping equipment clean and serviced

Later in this text, in the chapters on individual types of equipment, we discuss how to maintain each to maximize energy efficiency.

TABLE 5-4B

Staggered Start-Up

15 MIN. INTERVALS	15 MIN.	30 MIN.	45 MIN.	60 MIN.	75 MIN.	90 MIN.
Oven 6.2 KW	36 min.					
Fryer (1) 12 KW				5 min.		
Fryer (2) 12 KW					5 min.	
Griddle 6.5 KW						7 min.
EST. DEMAND KW	6.2 KW	6.2 KW	2.9 KW	5.1 KW	5.9 KW	5.7

Start-Up Time—82 Minutes

ON ▬▬▬

Maximum Demand—6.9 KW
Total Energy Used—7.6 KWH

IDLE ▬ ▬ ▬

COGENERATION

Another way to save energy dollars is to recapture and reuse heat that would otherwise be wasted. This is called **combined heat-power (CHP)**, **heat recovery**, *heat reclamation*, or **cogeneration** (the latter because it uses the same power for two jobs). Waste heat can also be recaptured from HVAC systems, refrigeration compressors, and computers, and the idea is adaptable enough that:

■ Hot exhaust air can heat incoming cold air.

■ Hot exhaust air can heat water.

■ Hot refrigerant can heat air or another liquid.

■ Heat from a boiler can be used to make steam or to heat air or water.

The trend in CHP is toward clean, quiet, reliable microturbine generators that neither require special permits to install nor generate environment-damaging emissions.

Both gas and electric utility companies have done extensive research on cogeneration. The Gas Research Institute conducted a study of utilities for 1000 restaurants in 28 U.S. cities, suggesting a cogeneration system in each case. This particular system, which impacted both gas and electric appliances in the restaurant, cost $128,000 to install. However, the average utility cost savings was an impressive $22,000 per year, meaning the system would pay for itself within 5 or 6 years.

Another study was conducted by the Electric Power Research Institute in conjunction with Pennsylvania State University. Reuse possibilities in four types of restaurants (cafeteria, full menu, fast food, and pizzeria) were assessed in cities with a variety of climate conditions; the study revealed that each of the following technologies reduced energy consumption by at least 10% (and as much as 25%).

REUSE OF DINING ROOM AIR FOR KITCHEN VENTILATION. Because air in a dining area is already conditioned, it can be used to meet hood ventilation requirements above the hot line. This reduces the amount of (naturally hotter) kitchen air that must be conditioned. Air can be transferred from the dining area to the kitchen using a single-duct system or transfer grilles. After use in the exhaust hood, the heated air can even be used to preheat incoming makeup air. However, the cost of this method makes it feasible only in restaurants that use the largest amounts of hood ventilation air.

HEAT PUMP WATER HEATERS. A *heat pump* is a mechanical device that moves heat from one area to another.

In all restaurant types, a heat pump was cited as an effective way to use less energy to heat water. An additional benefit of **heat pump water heaters (HPWH)** is the cool air from the evaporator, which can be routed to spot-cool the hottest areas of the kitchen. The heat pump contains an evaporator, a compressor, a condenser, and an expansion valve (see Illustration 5-6), much like the refrigeration systems you'll learn about in Chapter 9. Instead of cooling, however, the heat pump recovers heat from an area and uses it to heat water.

The location of a heat pump water heater is important. To work at its best, it should be in a space where the ambient temperature is between 45°F and 95°F. If operating properly, and if its air intake filters are kept clean, this handy appliance is capable of producing 30% to 70% of the hot water needs of an average restaurant.

RECOVERY OF REFRIGERATOR HEAT. Hot air released by refrigerators and walk-ins can be used to heat water. However, because refrigerators and walk-ins run 24 hours a day, storage is necessary to retain the heat during the hours when there's no demand for hot water. This is possible with a two-tank system that consists of a preheating tank that keeps water warm using air from the refrigerator, then feeds it into a conventional water heater tank, which doesn't use as much

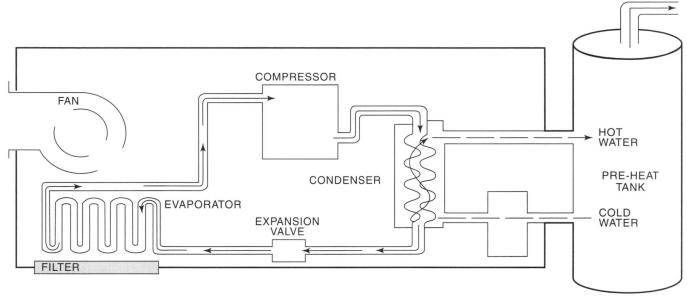

ILLUSTRATION 5-6 The inner workings of a heat pump water heating system.

energy because it doesn't have to work as hard. A side benefit of this system is that in warm months, the building's HVAC system doesn't have to work as hard to cool the space. The heat is being captured and used for other purposes.

HVAC HEAT RECOVERY. As the restaurant is being cooled (air conditioned), the heat from the cooling process can be captured to heat water elsewhere in the building. Typically, more heat is generated than needed, so correctly sizing the storage tank is critical to the success of this method. This method also uses a two-tank system similar to the one just discussed.

Online and in foodservice trade magazines, you can find plenty of examples of restaurants that are using "co-gen" successfully.

5-7 CONSTRUCTING AN ENERGY-EFFICIENT BUILDING

If you really intend to "go green" and create an environmentally friendly building inside and out, there are two major advantages to building a new structure instead of moving into an existing one.

1. You are free to select the most energy-efficient systems and designs on the market (as long as you can pay for them.)
2. You can design the building itself to minimize its energy requirements, using daylight and other lighting techniques to minimize the need for heating and cooling. Realize up front, however, that the idea of *sustainable construction* has become quite chic—or as some put it, eco-chic. The specialized experts, consultants, materials—everything will cost more than traditional building design and construction.

Here is the trade-off: By designing and selecting wisely, your future energy bills can be 30% to 50% lower than they would otherwise have been. In fact, you might even go beyond your city or state's minimum standards for construction, adding a bit more insulation, self-closing valves on water faucets, or any number of extras to enhance your savings. So no matter how "green" your architect or builder claims to be, before construction begins, look at the blueprint and designs and check this list of items. Are you saving as much energy as you could be?

- Plenty of daylight; minimizing the use of electric lighting
- Lots of light switches and dimmers, allowing for flexibility in turning off lights that are not needed
- The most efficient types of lamps and fixtures; energy-efficient ballasts for fluorescent lights
- High-pressure sodium lights in parking areas
- Efficient exit signage
- Timers or computerized or photoelectric controls for indoor/outdoor lighting
- Occupancy sensors for storerooms
- Glazing for windows to reduce incoming heat and increase daylight penetration
- Use of sufficient insulation for roof and walls
- Use of light colors, both inside and outside
- Positioning of building so that, if possible, trees or sloping land provide an insulating shield from wind and weather
- Awnings or overhangs to shield windows from direct sunlight
- Use of spectrally selective window film that cuts incoming heat in hot-weather areas
- Adjustable shades or blinds and, if appropriate in your climate, windows that open
- Caulking and weatherstripping around doors and windows
- Double doors or revolving doors at entrances
- Energy-efficient hot-water system, with tank located near main point of use and insulated pipes
- Low-flow and dripless faucets
- Efficient HVAC system, organized in zones so only areas in use are heated or cooled, with programmable wall thermostats and adjustable vents
- Locking covers on wall thermostats
- Heat pumps, where appropriate
- Restroom exhaust fans wired to go on/off with lights
- Installation of a computerized Energy Management System for optimum energy control

Many of the items on this list can also be used in improving the efficiency of existing buildings.

Think of the building itself as a shell that is the primary barrier separating a controlled, temperate indoor environment from the often harsh and unpredictable outdoor conditions. Better yet, think of the building as a filter, and use it to allow selected bits of the outdoors inside to make the indoor environment more comfortable: light, fresh air, and humidity. If the building works well as both a barrier and a filter, you're on the right track to using energy wisely.

GREEN CERTIFICATION

Today, both the government and various facets of the construction industry are on your side in attempts to improve a building's energy efficiency. The U.S. Green Building Council's trademarked LEED certification process (LEED stands for Leadership in Energy and Environmental Design) is the national benchmark for green construction.

The Green Restaurant Association (GRA), based in Boston, Massachusetts, has its own certification program. This group was established in 1990 as a nonprofit clearinghouse for foodservice businesses interested in sustainable practices and technology, from food purchasing practices to waste disposal to safer commercial cleaning and pest control products.

In order to be GRA-certified, a restaurant must have a full-scale recycling program, not use Styrofoam, and commit to fulfilling an annual green education requirement. The GRA also has a checklist of concerns the foodservice business must address in order to meet one of three categories of GRA certification, based on the number of points earned by tending to these concerns. The list is as follows:

- Water efficiency
- Waste reduction and recycling
- Sustainable food
- Energy
- Disposables
- Chemical and pollution reduction

In addition, as technology advances and more "green" alternatives become available, the business must commit to showing progress and earning more points to maintain or improve its certification.

Certification fees start at $700 and depend on business size. The GRA also offers consulting service contracts of 1 to 5 years in length, does environmental audits, endorses products (some of which are sold on its website), has an extensive database of sustainable restaurant operation information, and publishes a Certified Green Restaurant Guide for consumers, which is available online.

INSULATION AND AIR QUALITY

Air that leaks out of your building and air that leaks in have one thing in common: They both place an additional burden on your heating and cooling system. It's your job to minimize these losses and gains by sealing and insulating. Doors and windows are prime culprits, but dampers, skylights, and utility and plumbing entrances also must be secured.

Weatherstripping your doors is a good example of an often-ignored insulation priority. Let's say you have double doors at the exterior entrance to your cafeteria. Where the doors meet is an opening of about ¼ inch. It isn't an eyesore, and you're busy, so you never quite get around to doing the weatherstripping. However, on a pair of 80-inch doors, this quarter-inch gap adds up to a 20-square-inch opening! Most homes and offices are full of these kinds of seemingly insignificant leaks that truly can become energy drains.

Insulation minimizes the transfer of heat by some of the same principles as cooking, conduction, and convection. We know that heat flows naturally from a warmer environment to a colder one. All materials conduct heat, but some conduct it so poorly they actually resist it. Insulation is measured by its heat resistance, or R-value. The higher the **R-value**, the more resistant (the better insulator) it is. Your city building codes probably specify minimum levels of insulation, most commonly R-19 for walls and R-30 for roofs.

Even through layers of insulation, buildings "breathe"—that is, they take in a certain amount of fresh air and vent stale air back outside through windows, fans, and ducts. The amount of fresh air delivered is often dictated by health ordinances or building codes. The current recommendation by the American Society of Heating, Refrigeration and Air Conditioning Engineers (ASHRAE) is 15 to 20 cubic feet of air per minute per occupant of the building.

Especially in older structures, air quality may suffer at first when the building is tightened to make it more energy efficient. Humidity may condense on walls or windows, and existing ventilation systems may no longer be adequate. As you work at energy savings, ask about ways to do it without compromising air quality.

MEASURING YOUR PROGRESS

You've read elsewhere in this text about the need for keeping good records, and nowhere is it more important than in tracking utility bills and savings. You can set goals, complete surveys, determine priorities, and embark on numerous impressive conservation efforts, but how do you know how well they have really worked unless you can compare the before and after figures?

Ideally, you should begin by charting costs and consumption figures from the utility bills for the previous 12 months. If you have not been in business that long but the building was occupied by others, contact your utility companies for energy use records for the address. Setting an accurate base figure or starting point is critical to your success. How well you do from here is completely up to you!

■ SUMMARY

Energy use per square foot in foodservice is greater than in any other type of commercial space, although research indicates utility costs can be trimmed by 20% to 30% with an eye on energy conservation.

Having an energy audit of the building is a good way to begin. Audits can be simple walk-throughs—often done free of charge by your utility providers—or more detailed analysis audits in which every energy-use item in your building is scrutinized and suggestions and cost estimates for improvements are offered.

You should have a contact person at each local utility to discuss questions or problems. It is possible to chart your utility costs weekly, monthly, and annually, or even break them down per meal.

This chapter explains how electricity is measured (in amps, volts, watts, and ohms) and what materials are good or bad conductors of electricity. The two types of electrical current are alternating current (ac) and direct current (dc). Alternating current is the more common. Smaller appliances use single-phase electrical wiring; larger or more heavy-duty equipment uses three-phase electrical wiring. Combinations of voltage and phase are listed in specifications for every piece of electrical equipment you purchase. You should be able to determine if an individual circuit has enough electrical capacity to accommodate the appliances you will plug into it.

Every business owner, no matter how large or small the business, should consider opportunities to employ alternative energy technologies. Solar and wind power are available in many markets, and if it is impractical to install them directly for your facility, you have the option of purchasing "green power" or Renewable Energy Certificates (RECs). Before you dismiss renewable energy as too expensive, remember the tax advantages and low-interest loans available to encourage this type of project. There also is the eventual payoff of lower utility costs, plus a public relations boost for "going green." A few examples were included in the chapter.

Energy conservation methods can be grouped into four simple categories: improve equipment efficiency, reduce equipment operating time, recover energy that would otherwise be wasted, or use a cheaper energy source. The more you know about your utilities, the more you can track down and correct water, gas, or electricity wasters.

Energy professionals may preach conservation, but it is the foodservice operators and the general public who will have the final say in how they use these resources—and whether they choose to conserve them.

■ STUDY QUESTIONS

1. List three questions you would ask if you were giving an energy audit to a restaurant.
2. What is the difference between a *kilowatt-hour* and a *kilovolt-ampere*? Why do you need to know that?

3. What would be your primary considerations when making a decision to install solar panels on a foodservice business?

4. List three reasons it is necessary to be familiar with the terms *amps*, *volts*, and *ohms* as they relate to foodservice.

5. Explain briefly why a business might be billed differently for different hours of electricity use.

6. What is an energy index, and what does it do for you as the owner of a foodservice business?

7. What is the fuel adjustment charge on your electric bill?

8. List four ways you can save energy in a commercial kitchen.

9. In case of an emergency, is it okay to pull or flip the master service switch of your electrical system? Why or why not?

10. Under what conditions would a restaurant owner want to purchase a power generator?

Savoy Bar & Grill

Albuquerque, New Mexico

SITUATION

Two brothers, already well known in the area for their other successful restaurants, want to open a third location that is "notably more upscale" than its sister establishments, a "blend between a steakhouse and a seafood restaurant," with an average check of $30–$40 per person. They chose to remodel an existing restaurant. Their priorities: intelligent product flow, impeccable sanitation, and a sophisticated but unpretentious dining experience with a focus on wine.

CHALLENGES

- An extremely tight space for the display hot line.
- The owners want to bake all their own bread and eventually supply bread to other restaurants.
- The hood was not where the original plans said it was located! The existing configuration took up precious aisle space, requiring that an area where customers pick up food be redesigned.
- The decision to use heavy-duty equipment, which is larger (deeper) than comparable café-style equipment, to meet the restaurant's potential volume needs.
- Placement of an ice cream freezer not affected by proximity to high-heat equipment. (The restaurant makes its own ice creams and sorbets.)
- Placement of remote cooling equipment (the refrigeration compressor) on the roof of the building.
- Owners wanted to use the existing walk-in cooler and keep restrooms in current locations of the building.
- A separate kitchen for the bar and lounge area.

COMMENTS FROM THE DESIGN TEAM

- "It is compact and we can produce items quickly. And it's easy to clean, because the equipment is sitting on curbs and the fryers are on casters. This is the first kitchen I worked in where everything is seamless. Equipment is placed so there are no gaps between equipment, and no food particles can accumulate on the floor or between pieces of equipment."—Bob Peterson, executive chef
- "This area [the display hot line] is about 18 feet long by 12 feet wide. Because we had to use the existing hood space, this area is extremely tight. There might be as many as four people working in this area, so the cooking system must be precise in every way. No one can cross over another person."—Jim Lawrentz, kitchen designer, National Restaurant Supply, El Paso, Texas
- "We had some difficulty getting the refrigeration lines in for remote refrigeration. They had to be brought in through a back corner and run through the wall, and they come out in areas we designed into the counter where electrical and plumbing lines come in too. Once the counter is set, you can't pull it up to make adjustments. This is the kind of major challenge I enjoy."—Jim Lawrentz

HIGHLIGHTS OF THE DESIGN

- A 45-seat lounge, a 60-seat covered patio, two meeting rooms (36 and 50 seats respectively, one with audio-visual equipment), a 25-seat wine tasting room, and a 140-seat main dining room all were incorporated into the 10,000-square-foot space. The kitchen is 1800 square feet.
- A wood-fired oven for bread baking and vegetable-roasting has wood storage beneath the door. It is located in the prep kitchen and is visible from the dining room.
- Glass-encased, refrigerated wine storage cabinets double as the walls of the wine tasting room.

RESULTS

The equipment investment was $325,000. Savoy Bar & Grill hosts about 2000 customers per week. Three-month sales: $900,000.

TEAM MEMBERS

- *Owners:* Keith and Kevin Roessler, Executive Chef Bob Peterson, Lynn Devejian, and Catarhina Forsting
- *General manager:* John Mark Collins
- *Architects:* Claude Vigil Architects, Albuquerque, New Mexico
- *Interior designers:* Internal, plus consultant from Claude Vigil
- *Kitchen designer:* Jim Lawrentz, National Restaurant Supply, El Paso, Texas

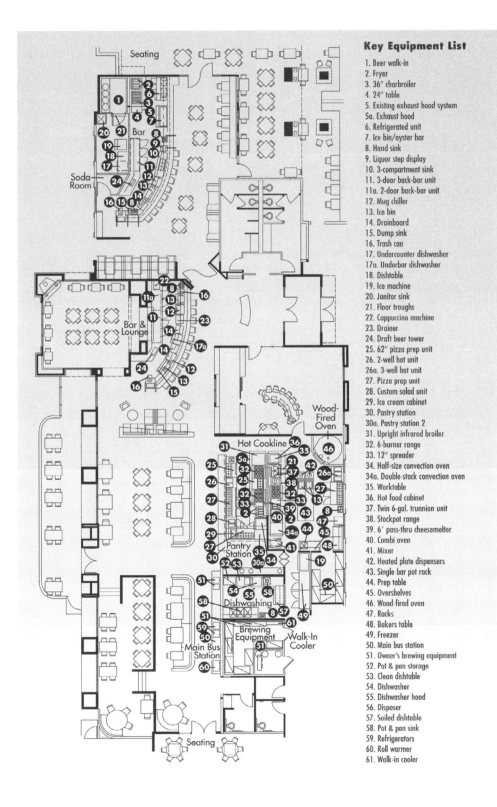

Key Equipment List

1. Beer walk-in
2. Fryer
3. 36" charbroiler
4. 24" table
5. Existing exhaust hood system
5a. Exhaust hood
6. Refrigerated unit
7. Ice bin/oyster bar
8. Hand sink
9. Liquor step display
10. 3-compartment sink
11. 3-door back-bar unit
11a. 2-door back-bar unit
12. Mug chiller
13. Ice bin
14. Drainboard
15. Dump sink
16. Trash can
17. Undercounter dishwasher
17a. Underbar dishwasher
18. Dishtable
19. Ice machine
20. Janitor sink
21. Floor troughs
22. Cappuccino machine
23. Drainer
24. Draft beer tower
25. 62" pizza prep unit
26. 2-well hot unit
26a. 3-well hot unit
27. Pizza prep unit
28. Custom salad unit
29. Ice cream cabinet
30. Pastry station
30a. Pastry station 2
31. Upright infrared broiler
32. 6-burner range
33. 12" spreader
34. Half-size convection oven
34a. Double-stack convection oven
35. Worktable
36. Hot food cabinet
37. Twin 6-gal. trunnion unit
38. Stockpot range
39. 6' pass-thru cheesemelter
40. Combi oven
41. Mixer
42. Heated plate dispensers
43. Single bar pot rack
44. Prep table
45. Overshelves
46. Wood-fired oven
47. Racks
48. Bakers table
49. Freezer
50. Main bus station
51. Owner's brewing equipment
52. Pot & pan storage
53. Clean dishtable
54. Dishwasher
55. Dishwasher hood
56. Disposer
57. Soiled dishtable
58. Pot & pan sink
59. Refrigerators
60. Roll warmer
61. Walk-in cooler

- *Project manager:* Eduardo Jones, National Restaurant Supply, Albuquerque, New Mexico
- *Equipment dealer:* Rick Levis, National Restaurant Supply, Albuquerque, New Mexico

Source: The full article about this project first appeared in the June 2007 issue of *Foodservice Equipment & Supplies* magazine, a Reed Business Information publication. The original article was written by Donna Boss; our encapsulation and the layout appear with permission.

6

GAS, STEAM, AND WATER

Gas and steam provide power for many of the major kitchen appliances you will use and may also run your heating and/or cooling system. A basic knowledge of water quality and plumbing is necessary for foodservice managers.

After reading this chapter, you will be able to:

- Describe the uses of gas and steam in foodservice.
- Identify the basics of how gas and steam equipment work.
- Identify energy-saving use and maintenance tips.
- Identify water quality problems and how to deal with them.
- Identify the basic foodservice plumbing requirements.
- Identify the basics of installing and maintaining a drainage system.
- Explain hot-water needs and how water heaters work.

6-1 GAS ENERGY

Gas has many uses in foodservice. You may use it to heat or cool your building, to heat water, to cook food, to chill food, to incinerate waste, and to dry dishes or linens.

It's called *natural gas* because, indeed, it is not man-made. It was formed underground several million years ago by the decay of prehistoric plants and animals and is now pumped to the earth's surface for use as a fuel. Illustration 6-1 shows how natural gas is extracted from wells, then processed and transported through a series of pipelines to its final destination: your home or business.

The American Gas Association credits the Abell House, a stagecoach stop in Fredonia, New York, as the first commercial establishment to use natural gas for cooking. Back in 1825, the pipes were hollowed-out logs. Gas was propelled through the logs into the building to a single-flame stove with a reflector plate. We've come a long way since then.

There are different types of gas for different uses. The one most commonly known as natural gas is mostly **methane**. When it is highly compressed for storage, under incredibly cold conditions (below 260°F), it becomes **liquefied natural gas (LNG)**.

When it is manufactured—in a process that mixes methane with hydrogen and carbon monoxide—it is known as *synthetic gas*. Other gas combinations—propane, butane, isobutane—may be called **liquefied petroleum gas**, *LP gas*, or *bottled gas*.

In the United States, restaurants use natural gas to operate as much as two-thirds of their major cooking equipment.

Its chief benefit is its instant ability to provide intense heat. Chefs like cooking with gas because it allows them to work quickly and to use different types of burners to direct heat where it is most needed. Modern-day foodservice ranges may feature **power burners**, high-input gas burners that burn twice the amount of gas (and supply twice the heat) of a conventional gas burner.

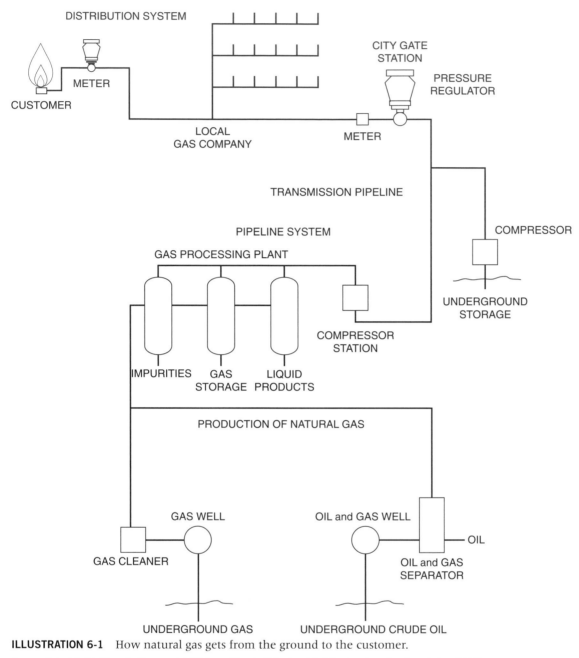

ILLUSTRATION 6-1 How natural gas gets from the ground to the customer.

Source: Robert A. Modlin, ed., *Commercial Kitchens*, 7th ed. (Arlington, Virginia: American Gas Association, 1989).

Other appliances that may be gas fired include broilers, fryers, griddles, steamers, coffee urns, and ovens. These appliances may have automatic *pilot lights*, which stay lit and indicate the gas is ready if you turn on the appliance, or may require the user to light the pilot manually with a match.

GAS TERMINOLOGY

Let's look at a single gas range burner as it is being lighted. Every burner assembly has an **orifice**, or hole, through which the gas flows. The orifice is contained in a unit that may be known as a *hood, cap,* or *spud*. The diameter of the orifice determines the gas flow rate. Some orifices are fixed; others are adjustable (see Illustration 6-2). The gas flows from the orifice in

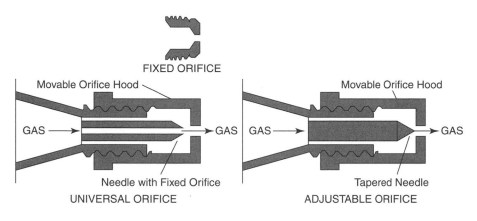

FIXED ORIFICE

Movable Orifice Hood

GAS → → GAS

Needle with Fixed Orifice

UNIVERSAL ORIFICE

Movable Orifice Hood

GAS → → GAS

Tapered Needle

ADJUSTABLE ORIFICE

ILLUSTRATION 6-2 The orifice, or point at which gas flows into the range burner, may be fixed or adjustable.

the form of a *jet*, which causes a rush of air (called *primary air*) to flow into the *burner tube*. It is in this tube (sometimes called the *mixer tube*) that the gas mixes with air and ignites.

Most burners are also equipped to adjust the amount of primary air that comes into them by adjusting the size of the air opening. This piece is called the *air shutter*. The air–gas mixture then flows through *burner ports*, a series of round holes from which the flames burn. As they burn, the flames also use *secondary air*—the air around them. (More about burners later in this chapter.)

When air and gas are mixed, the speed at which the flame shoots through them is called the *burning speed*. Burning speeds vary with the amount of air and the type of gas being used. The ideal ratio is 10 parts air to 1 part gas, but this is hard to achieve in the real world. So you may encounter *incomplete combustion*, where the fuel doesn't burn fully because something's not quite right: not enough air in the air–gas mixture, poor ventilation, or improper flame adjustment. You can spot incomplete combustion in the flames themselves, which are tipped with yellow.

Yellow-tipped flames are caused by a lack of primary air to the burner, possibly because lint or grime has collected to block the primary air openings. This means you must clean your burners, not adjust them.

Yellow flames create black *carbon soot*, which makes cleanup harder and can eventually clog vents and orifices, making your appliances less efficient. A buildup of carbon soot will affect your exhaust canopies too, getting them dirtier and making them work harder. If cleaning the burners does not give you the desired results, an air shutter adjustment may be needed, which requires a qualified service person.

Incomplete combustion also gives off varying amounts of carbon monoxide, which is odorless, colorless, and tasteless but harmful nonetheless.

What you are looking for is *flame stability*—a clear blue ring of flames with a firm center cone—indicating the air–gas ratio is correct and you are using the fuel under optimum conditions. When gas fuel burns completely, you get heat energy, harmless carbon dioxide, and water vapor. Nothing is wasted, and no harmful pollutants are released into the atmosphere. Once again, you can alter the flame stability by changing the burning speed (adjusting the orifice so less gas flows in) or by changing the primary airflow into the mixture (adjusting the air shutter).

Don't confuse yellow-tipped flames with the red or orange streaks you sometimes see in a gas flame. These color streaks are the result of dust in the air that turns color as it is zapped by the flame; they should not be a problem. Also remember you are wasting gas when you use high flames that lick the sides of your pots and skillets. In fact, when a completely unheated pan is placed on the gas range, it is best to begin heating it on medium heat so the tops of the flames do not touch the surface of the pan. Carbon monoxide and soot are produced when intense heat hits the cool metal surface. Increase the heat only after the pot or pan has had a chance to warm up.

THE GAS FLAME

The flame of the gas burner represents the ultimate challenge: to mix gas and oxygen in just the right amounts to produce combustion, yielding controlled heat with minimum light. The simplest, most effective example of this is the old-fashioned *Bunsen burner*. This type of burner premixes air and gas prior to reaching the flame, resulting in a highly efficient flame that burns intensely but cleanly, without smoke.

The shape and size of the burner are the two factors that place the flame exactly where direct heat is needed most. In a toaster or broiler, for instance, the gas flame is directed at a molded ceramic or metal screen, which is heated to a deep red color and emits infrared heat rays that penetrate the food being cooked. The latest innovation in the industry, as mentioned earlier in this chapter, is the high-input gas burner, which burns twice the amount of gas (for greater intensity of heat) as a conventional burner of the same size.

Amazingly, the natural gas–air combination can produce a stove temperature of up to 3000°F. One of the functions of the gas-fired appliance is to limit and distribute the available heat so as to reach the correct temperature to cook foods properly.

You'll notice a gas burner sometimes lets out a *whoosh* or roaring sound as it lights. This is called *flashback*, and it happens because the burning speed is faster than the gas flow. This type of flashback occurs more often with fast-burning gases such as propane.

Another type of flashback happens when the burner is turned off, creating a popping sound that is known as the ***extinction pop***. Sometimes it's so pronounced that it blows out the pilot light flame. You usually can correct both types of flashback by reducing primary air input to the burner. If you're unsure about how to do it yourself, remember that burner adjustment is a free service of many gas companies.

Flashback is not hazardous, but it is annoying. It creates soot and carbon monoxide and often means you have to relight your pilot. Inside the burner, repeated flashback occurrences may cause it to warp or crack. It makes more sense to get the burner adjusted than to live with flashback.

Several other conditions may require professional attention and adjustment (see Illustration 6-3). You may notice that the flames seem to lift and then drop on some parts of the burner head at irregular intervals, as if some unseen hand were playing with the control knob. This burner may seem a bit noisier than the others, making a roaring sound whenever the flames increase. ***Flame lift***, as this is sometimes known, is not a stable, normal burner condition and should be corrected immediately.

Incomplete combustion may also cause ***floating flames*** that are lazy looking and are not shaped as well-defined cones. This is a dangerous condition, and you'll usually notice it in the first minute or two after a burner has been turned on, before it achieves the proper air-flow. If the flames don't assume their normal, conical shapes quickly, have the burner checked.

Finally, the most serious condition is ***flame rollout***. When the burner is turned on, flames shoot out of the combustion chamber opening instead of the top of the burner. Flame rollout is a serious fire hazard and must be repaired immediately. The burner may not be correctly positioned, or something may have obstructed its inner workings. Either way, call the service person—fast.

GAS BURNERS

The burner assembly is the basic unit in a gas-fired piece of equipment that mixes gas with oxygen and thus produces the heat required for cooking. Today we see many types of burners, each designed to meet the demands of a particular appliance. All burners have ports, a series of round holes from which the flames burn. There are wide and narrow ports, and they can be arranged in various patterns (see Illustration 6-4). A few of the most common burner types are discussed next.

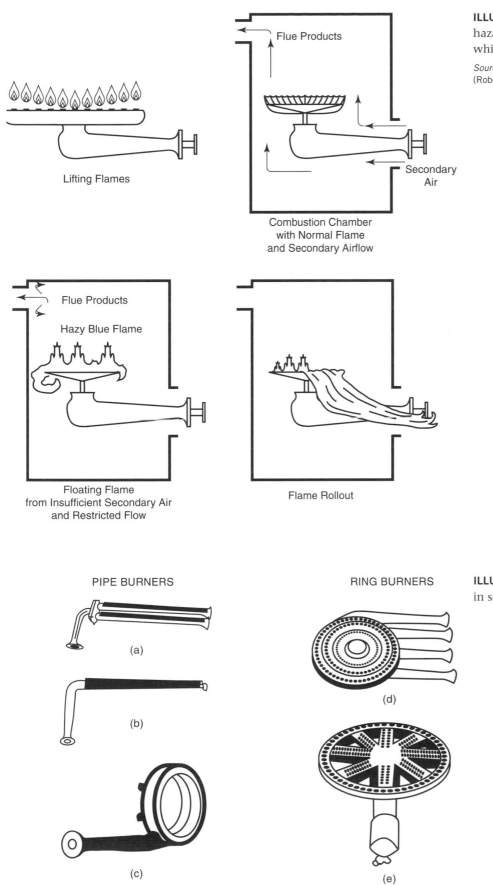

Lifting Flames

Flue Products

Secondary Air

Combustion Chamber
with Normal Flame
and Secondary Airflow

Flue Products

Hazy Blue Flame

Floating Flame
from Insufficient Secondary Air
and Restricted Flow

Flame Rollout

ILLUSTRATION 6-3 The four most hazardous gas flame conditions, all of which require adjustment.

Source: Anne Marie Johnson, *Cooking for Profit* (Robert Hale, Ltd.: London, 1991).

PIPE BURNERS

(a)

(b)

(c)

RING BURNERS

(d)

(e)

ILLUSTRATION 6-4 Gas burners come in several shapes and sizes.

PIPE BURNER. This is a pipe (usually made of cast iron) with two or more rows of ports drilled along its length. You'll find pipe burners in ovens, griddles, and broilers. Although most pipe burners are straight, some are loop burners, in which the pipe is bent into a circular or oval shape.

RING BURNER. These are widely used on rangetops, steam tables, boilers, and coffee urns. The standard ring burner has one or two rows of ports arranged in a circle. It's made of cast iron and comes in a variety of sizes. To increase the capacity of a burner, several ring burners of different diameters can be nestled one inside the other so that one, some, or all of them can be turned on as needed. Some ring burner ports face sideways instead of straight up so food cannot be spilled into them accidentally.

Ring burners are sometimes known as *atmospheric burners* because the secondary air comes from the atmosphere around them. However, the flame tips often suck in too much oxygen, making them less efficient and requiring frequent adjustment of the primary air–gas mixture.

SLOTTED BURNER. This is really a type of pipe burner because it is made of the same cast-iron pipe and can be straight or circular. The ports in a slotted burner are all aimed in the same direction to form a single large flame. Some have only one wide slot, with a corrugated ribbon of heat-resistant metal alloy located inside the slot to allow a wider (but still safe) path for the flame. Slotted burners are typically found in hot-top ranges and deep-fat fryers.

FLAME RETENTION BURNER. This is a slotted burner with additional ports drilled into the pipe. It allows more heat to flow into the burner, which improves combustion and reduces flashback. Flame retention burners are considered very efficient, with a wide range of heat settings and the ability to fine-tune the primary air adjustment.

RADIANT (INFRARED) BURNER. The usual infrared burner is a set of porous ceramic plates with about 200 holes per square inch on its surface. Air and gas flow through these holes and burn very hot (about 1650°F), which makes this type of burner ideal for broilers. These burners can be located at the sides of a fry kettle for maximum heat transfer or suspended inside a protective cylinder located at the bottom of the fry tank.

Fryers with infrared burners boast 80% energy efficiency, compared to about 47% for conventional fryers. Their heat recovery time (the time it takes to return to optimum cooking temperature after a new batch of cold food has been loaded into the kettle) is less than 2 minutes. The same benefit—heat intensity—makes the infrared burner popular for griddles. A sturdy, 1-inch-thick griddle plate can be used instead of a thinner one that is not able to retain heat as well.

Infrared burners work so well because they use *radiant heat*, and the best example of radiant heat is the sun. Have you noticed how the sun can warm your face on a winter day even though the air around you is cold? In the same way, the infrared burner sends its high-frequency waves directly from the heat source to the food. The rays turn into thermal (heat) energy when they hit the food; they do not heat the air. When you're cooking, you get the most energy efficiency from an appliance that heats the food, not the air that surrounds it.

RANGETOP POWER BURNER. This burner premixes gas and combustion air in correct proportion to produce high heat and efficiency. Unlike the standard ring burner, the power burner does not rely on the atmosphere to supply its secondary air. The burner head is enclosed in a sealed metal ring through which no excess air can enter. Flames are spread evenly through the ring over the bottom of the cooking pot so less heat is wasted, more heat is delivered to the food, and the kitchen stays cooler. The burner head acts as a shutter mechanism, readjusting the premixed air and gas whenever the controls are turned up or down.

A recent appliance development is the power burner range, with two (front) power burners and two (back) conventional burners, the front ones for speedy cooking and the back ones for keeping food warm.

INFRARED JET IMPINGEMENT BURNER. The IR jet, as it is known, is a type of high-efficiency burner that also uses less gas than conventional burners. It is a power burner that premixes gas and air in a separate chamber before burning. This mixture is fed into the burner by a blower and ignited at the burner surface. The perforated ceramic burner plate holds the flames in place and allows them to impinge (hit hard) on the bottom surface of the pan.

PILOT LIGHTS AND THERMOSTATS. The pilot light is an absolute necessity in the gas-fired commercial kitchen. There are several kinds of pilot lights, some automatic and some manual, and most gas appliances make it possible for you to easily adjust the pilot light, if necessary, with the turn of a screw. Illustration 6-5 shows the inner workings of the two most common types of gas pilot lights. The newest technology, however, replaces the gas pilot with electronic spark ignition of the gas flame, reducing gas consumption because the pilot is not continuously lit.

Electronic spark ignition offers the benefit of knowing that standing pilots are not burning when the facility is not operational. Pilots that unintentionally go out during down times allow gas to escape into the environment. This safety feature is especially important in schools that do not operate year-round.

The standard gas pilot light is about ¾ inch high. This small flame is located next to the main burners of the appliance. In some cases, each burner has a separate pilot; in others, a single pilot light is used to light more than one burner.

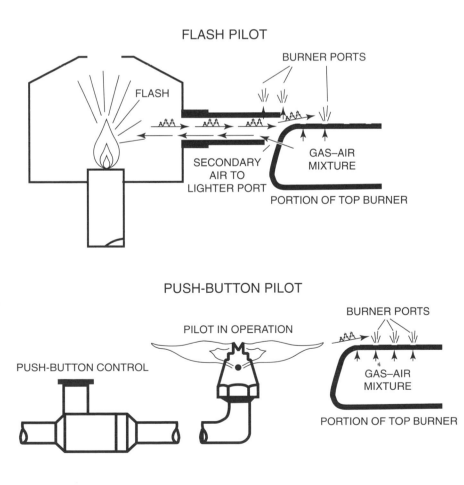

FLASH PILOT

PUSH-BUTTON PILOT

ILLUSTRATION 6-5 The main parts of a typical pilot light system. The flash pilot lights on its own; the push-button pilot requires someone to push the button.

Source: Robert A. Modlin, ed., *Commercial Kitchens*, 7th ed. (Arlington, Virginia: American Gas Association, 1989).

If the pilot light goes out, it is a signal that the gas has been shut off. Like other gas flames, pilots can also burn yellow, which means dirt or lint may be blocking the opening. You can remove the dirt by brushing the orifice clean again.

Pilot lights have safety features, the most common of which is the **thermoelectric control**. When a junction of two metal wires (called a **thermocouple**) is heated by the pilot flame, a very low electric voltage is generated—just enough to fire an electromagnetic gas valve and hold it in an open position. If the pilot light fails, the thermocouple cools, the electric current stops, and the gas valve is closed by spring action. To resume the flow of gas, the pilot must be manually relit.

The **thermostat** is the control used on most gas-fired equipment to maintain the desired burner temperature. By far, the most common thermostat is a knob or dial called a *throttling* or *modulating control*. It allows the flames to rise or fall quickly, then regulates the gas flow to keep the burner temperature constant.

Some types of cooking appliances, such as deep-fat fryers, require quick heat recovery in less than 2 minutes. In these cases, a **snap-action thermostat** opens fully to permit maximum heating until the desired temperature is reached. Then it shuts off just as quickly.

Remember that the function of a properly working thermostat is to turn down, or shut off, the supply of gas as soon as the burner reaches the desired temperature. Should you need to reduce the burner's temperature, turning down the thermostat will not be sufficient. Heat is already stored in the appliance, and it will take time for it to dissipate and cool down. On a gas oven, for instance, you should set the thermostat to the new, lower level and then open the oven doors to allow it to cool more quickly.

Many people have the mistaken impression that ovens and other appliances heat more quickly from cold when the temperature is set first to HIGH and then turned down. In fact, the oven won't heat any faster, and you risk forgetting to turn it down and damaging the food by cooking it at an excessive temperature.

MAINTAINING GAS-POWERED EQUIPMENT

Preventive maintenance is as important with gas appliances as any others to prevent equipment malfunction. We have adapted these maintenance suggestions—all of which still apply—from an article in *Equipment Solutions* magazine's September 2002 issue:

- Perform routine checks and/or maintenance both weekly and monthly, and document what you've done.
- Weekly tasks include cleaning burner parts and orifices; checking the primary blower speeds; vacuum-cleaning the entire blower system.
- Monthly tasks include adjusting and cleaning air inputs and pilot lights; checking and calibrating thermostats; checking the burner valves. If the latter are difficult to turn or move, lubricate them (sparingly) with high-heat valve grease. Also, check monthly for a good balance of exhaust and make-up air. Remember, gas-operated equipment depends on a uniform exchange of new air to replace the air used in the combustion process.
- Whenever another piece of equipment is added to the cooking line, have a flow test performed on the main gas line (by a professional repair person or gas company representative) to ensure your hot line has sufficient pressure when all the cooking equipment is at peak gas consumption levels. This, in turn, will ensure the efficient operation of each appliance.

READING GAS METERS AND BILLS

If you have gas appliances, you also have gas meters, which look like Illustration 6-6. The single upper dial on a gas meter is used only to test the meter and is not part of your reading. The dials are sometimes called **registers**.

Although gas is measured in cubic feet instead of kilowatt-hours, the meter dials work the same way (and are read the same way) as electric meter dials, even when the gas company reads them by radio signal as smart meters.

In large commercial buildings, the gas meter may be more sophisticated, with two sets of dials much like the combination electric meter. Called a *compensating meter*, this meter also adjusts (compensates) when gas pressure or temperature at the location varies from normal conditions. Of the two sets of registers, one will be marked "Uncorrected" or "Uncompensated" and the other will be labeled "Corrected" or "Compensated." Read the "Corrected" meter to determine your gas usage. The "Uncorrected" dials are used by the utility company for checking the meter.

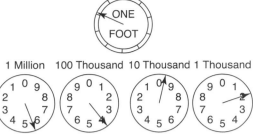

ILLUSTRATION 6-6 Gas meter dials are read the same way as electric meters, but they measure cubic feet instead of kilowatt-hours.

Illustration 6-7 is a sample gas bill. In all instances, gas consumed is measured in cubic feet, or cubic feet per hour (cfh). However, the rates may be based on therms. (One therm equals 100,000 Btu, or approximately 100 cubic feet.) In our sample bill, the cubic feet are converted to therms to determine the cost of the gas fuel.

The gas utility (in this case, Nicor, Inc. of Chicago, Illinois) estimates the natural gas a customer uses makes up 75% to 80% of the gas bill over one year. As a regulated utility, it does not profit from the actual cost of the gas and must pass it on without markup. Some deregulation affecting heavy users of natural gas allows the larger consumer options to bid their natural gas service. A heavy user would, for example, use 200,000 therms per year or more.

These notes describe items that appear on the Nicor bill:

METER READING SECTION

The "Current Reading" is determined by one of three methods:

■ An *Actual Meter Reading* means your reading was taken by a company meter reader or recorded by an automated device.

■ *Customer Reading* means you reported your meter reading to the utility company.

■ An *Estimated Meter Reading* means the gas company estimated the reading based on previous usage and the weather. Nicor says it estimates most meter readings every other month; then, on the next bill, any difference from actual use is automatically corrected when an actual meter reading is taken.

The "Previous Reading" is the reading of the gas meter in the last reading period. The "Difference" is the difference between the previous reading and the current reading.

To determine therms used, the gas company multiplies your use by the Btu factor. The Btu factor may vary from month to month, and is listed on the bill.

"Delivery Charges" are Nicor's costs to operate and distribute gas to you. These charges make up 15% to 20% of your gas bill over one year.

MONTHLY CUSTOMER CHARGE

This is a minimum charge for most customers, and it is the same each month, even if you do not use natural gas that month. For the gas company's Rate 4 business customers, this charge is based on meter size and potential maximum hourly use of natural gas in cubic feet per hour (cfh). Most business customers are in the "less than 1,000 cfh" category. Customer charges are:

■ Less than 1,000 cfh: $20.37

■ 1,000 cfh to 10,000 cfh: $59.82

■ More than 10,000 cfh: $105.56

ILLUSTRATION 6-7 A sample gas bill.

Courtesy of Nicor, Inc., Chicago, Illinois.

1-888 NICOR 4U

Services for
Good Eats Restaurant
123 Main St.
Anytown

Account: 1-23-456-78
Commercial

Payment Information
Previous acct. balance   = $ 917.29
Payment rec'd. Thank you = $ 917.29

Meter Reading
Current: 2522 Estimated Meter Reading—February 4, 2013
Previous: 1435 Actual Meter Reading—January 5, 2013
Difference: 1087 (30 days)
Conversion to Therms: 1087 X 1.013 BTU Factor = 1,101.13 Therms

Delivery Charges

Monthly Customer charge:		$ 20.37
First 150 Therms	150 @ .1329 =	$ 19.94
151–952 Therms	951.13 @ .0682 =	$ 84.87
Environmental Recovery	1,101.13 @ .0024 =	$ 2.64
Franchise Cost	=	$ .24
Governmental Agency Adj.	=	$ .08 $ **106.14**

Natural Gas Cost
1,101.13 Therms X .6700 $ **737.36**

Taxes

State Tax	1,101.13 @ 0.024	
Utility Fund Tax	845.90 @ .10%	
Municipal Tax	845.90 @ 5.15%	$ **70.84**

Total current bill—est. $ 916.74

*Have you considered our Budget Plan? You can equalize your monthly
payments to cushion the impact of winter bills.*

*Thank you for your prompt payment record. We are pleased to have you
as our customer.*

Total Amount Due $ 916.74

THERM USAGE/DISTRIBUTION CHARGES

These charges cover the gas company's fixed and variable operating and distribution costs. A portion of these charges carries from month to month based on the amount of natural gas used. There are price variations at certain levels of therm usage. Volumetric delivery charges are:

- First 150 therms @ $0.1329
- Next 4850 therms @ $0.0682
- More than 5000 therms @ $0.0482

ENVIRONMENTAL RECOVERY COST

This charge covers costs for the environmental monitoring and possible cleanup of former manufactured gas plants in the service territory. This charge changes periodically; the gas company does not profit from it.

FOODSERVICE EQUIPMENT

Maintenance Tips from the Pros

Frank Murphy—now with Five-Star Energy, Inc. in Glenville, Georgia—supplies these maintenance tips for specific types of equipment.

BROILERS AND CHARBROILERS

Most problems with broilers and charbroilers are the result of grease and food particles clogging the burners, pilots, or shutters. These components must be cleaned weekly and properly adjusted to restore optimum operation.

A good way to prevent this buildup is to make sure broiler grates are positioned to direct the excess grease flow for burn-off, or collect it in grease drawers. Clean the broiler grates at the end of each day, which is easy to do—just place the grates flat on the broiler and set the gas valve on HIGH for 45 minutes. Then turn off the broiler and allow the grates to cool. Remove them from the broiler when they've cooled down, and clean them (both top and bottom surfaces) with a wire brush, damp cloth, and mild detergent.

The grate channels and burner radiants should be thoroughly cleaned as well. Brush the burner's heat reflector to remove dust or debris, and clear all the burner portholes at least once a week.

FRYERS

A major maintenance problem for gas-fired fryers may be its location in the kitchen. It is essential that there are no restrictions for new air entering the burners or blower motor. If the airflow is restricted, the fryer sidewalls and internal control components will be abnormally high. This will cause the electrical controls to overheat, and soon the equipment's performance will diminish.

To prevent this problem, make sure the gas connections to the fryer are tight. Be sure that enough fresh make-up air is available. Clean the hood filters each day, checking again for airflow restrictions there.

Another maintenance concern for fryers is slow temperature recovery, which is related to having a reliable and controllable heat source. A fryer that takes too long to recover its temperature when cold food items are dropped into the kettle is losing its capacity to conduct or radiate heat efficiently. In high-efficiency fryers, the burner seals may be leaking, the blower motor speed may be too low, or a broken temperature probe could be the problem. In tube-heated fryers, internal heat-baffle wear causes recovery problems. Your owner's manual should offer guidance in these situations.

CONVEYOR OVENS

The cooling fan exhaust grilles should be wiped clean daily. Check that the cooling fan is turning when the oven is operating. Remove the entire conveyor every month so you can clean the jet-air "fingers," being careful to replace all components in their original positions. Every 2 or 3 months, clean the combustion motor's blower air intake, which is usually located behind a closed panel.

RANGES

Most maintenance problems with gas-operated ranges occur because bits of food and greasy particles settle into the gas lines or ports, preventing a smooth flow of gas and air to the pilot lights and burners. To identify this condition, check the burner pilot lights and flames for clear, even combustion. Clean the burner grate surfaces every day using a wire brush, damp cloth, and mild detergent. Clean and adjust the range pilots monthly; they should sit level in their mounts and be situated so the pilot light flame can easily ignite the burner. Check the range burners at least once a month for heat flow. If foods are taking too long to cook, ask a factory-authorized service person to check the gas pressure and flue.

GRIDDLES

Every month, check the underside of the griddle to make sure grease is not running where it's not supposed to go—such as into the air vent for the gas pressure regulator. If the vent is blocked, the griddle's heating performance will become erratic. Each griddle section and its burner/orifice must be kept clean to maintain a consistent flame and heat correctly. Also be sure the thermostat mechanism is securely in place.

Source: Reprinted with permission of *Equipment Solutions* magazine (Talcott Communications, Chicago, Illinois, September 2002).

FRANCHISE COST

This is the cost for municipal franchise agreements. The gas company does not profit from this charge.

GOVERNMENTAL AGENCY ADJUSTMENT

This adjustment covers governmental fees and added costs, excluding franchise costs. This cost changes periodically; the gas company does not profit from it.

TAXES

Gas bills include a number of taxes; in the case of Nicor, these taxes make up 5% to 10% of a customer's total bill over one year. In Illinois, where Nicor is located, there are three types of state taxes and two municipal taxes on utilities.

SAVING ENERGY WITH GAS

There are many simple, practical ways to take full advantage of the instant heating power of gas. Choose equipment that is enclosed and insulated, keeping the energy within the appliance (or absorbed by the food). Cook at the lowest temperature, or in the largest volume, possible. Especially for solid-top ranges, use flat-bottomed cookware that makes full contact with the cooking surface. Curves and dents in pots and pans waste money. The bottoms of the cookware should be about 1 inch wider than the diameter of the burner.

Although big flames and kitchen showmanship have their place, for most cooking duties it is sufficient that the gas flame tips barely touch the bottoms of the cookware and do not lap the sides. Burners should be adjusted accordingly. Don't keep pots at a boil when simmering would be sufficient, and cover them to hold in heat.

A common tendency is to turn equipment on early to let it heat up. Again, this is a waste of fuel and time. For open-top ranges, preheating is simply not necessary; for griddles, low or medium flames are sufficient for just about any kind of frying. Broilers don't require much, if any, preheating; gas ovens, solid-top ranges, and steamers can be preheated, but no more than 10 minutes.

Energy saving is another good reason many ranges and griddles are built as adjoining, temperature-controlled, multiple-burner sections. During slow times, learn to group food items on the least possible number of sections, which eliminates the need to keep the entire cooking surface hot.

Regular cleaning and maintenance of the appliances are two important keys to wise use of natural gas, but other energy-saving innovations are in the works. One is a concept called **heat transfer fluids (HTF)**. The idea is to power several pieces of equipment with a single burner using a series of pipes and a heated fluid that runs through the pipes. (The heated fluid cannot be water because its pressure would become too high and create steam.) The fluid may also be run through a heat exchanger, if necessary, to boost its temperature along the way.

GAS PIPES

Natural gas for commercial kitchens flows through large pipelines at pressures of 600 to 1000 pounds per square inch (psi). This high pressure is reduced by a series of valves to arrive at your gas meter at about 25 psi. Both the size and the quality of pipes used are critical in setting up a gas system for your business.

By totaling the Btus required when all gas equipment is on, you can estimate the total amount of gas required and calculate the size of the pipes needed. Divide the total number of Btus needed per hour by 1000. This figure is the total number of cubic feet of gas needed. Let's say your place would use 400,000 Btus per hour. That's 400 cubic feet per hour. Then use the pipe sizing table (see Table 6-1) to estimate the diameter of pipe to install. You'll notice this depends, in part, on how far the gas has to travel from the meter to the kitchen. For our example, let's estimate that the distance from meter to kitchen is 40 feet. The table says a 1¼-inch

TABLE 6-1

Pipe Sizing Table

This shows the maximum capacity of pipe in cubic feet of gas per hour (based on a pressure drop of 0.3-inch water column and 0.6 specific gravity gas)

NOMINAL IRON PIPE SIZE (IN INCHES)	INTERNAL DIAMETER (IN INCHES)	LENGTH OF PIPE (IN FEET)									
		10	20	30	40	50	60	70	80	90	100
½	0.622	132	92	73	63	56	50	46	43	40	38
¾	0.824	278	190	152	130	115	105	96	90	84	79
1	1.049	520	350	285	245	215	195	180	170	160	150
1¼	1.380	1050	730	590	500	440	400	370	350	320	305
1½	1.610	1600	1100	890	760	670	610	560	530	490	480
2	2.067	3050	2100	1650	1450	1270	1150	1050	990	930	870
2½	2.469	4800	3300	2700	2300	2000	1850	1700	1600	1500	1400
3	3.068	8500	5900	4700	4100	3600	3250	3000	2800	2600	2500
4	4.026	17500	12000	9700	8300	7400	6800	6200	5800	5400	5100

Source: Steven R. Battistone, *Spec Rite: Kitchen Equipment* (Cincinnati, Ohio: Food Service Information Library, 1991).

pipe should be more than adequate. We might install a 1½-inch pipe to compensate for additional equipment installed in the future.

Most gas utility companies have commercial representatives who will help determine proper pipe sizing. Their consulting service is usually free.

Because the gas pipes will be a permanent part of the building, they might be constructed of a durable material, such as wrought iron. Copper or flexible stainless steel is also now used for natural gas piping, although care must be taken to install it properly. And some utilities use heavy-duty plastic piping in their distribution systems—but it can only be installed underground and is never used inside buildings.

It is always preferable to install gas pipes in hollow partitions rather than solid walls to minimize their potential contact with corrosive materials. Gas pipes should never be installed in chimneys or flues, elevator shafts, or ventilation ducts.

Gas appliances are attached to their gas source with a **gas connector**, which is a flexible, heavy-duty brass or stainless steel tube, usually coated with thick plastic. One end is fastened permanently to the building's gas supply, the other to the back of the appliance. The appliance end should be a **quick-disconnect coupling**, an easy shutoff device that instantly stops the flow of gas with an internal assembly of ball bearings and a spring-loaded plug (see Illustration 6-8). Most quick disconnects have built-in polymers or wax seals that melt at temperatures above 350°F, immediately shutting off the gas supply even if the hose is still connected. An additional accessory well worth the cost is the restraining device, which is essentially a stiff steel cable attached to both the appliance and the wall. The cable ensures the connector is not damaged if the hose is accidentally stretched too far. In most municipalities, this is mandatory safety equipment. Shutoff occurs automatically when kitchen air reaches a certain high temperature (as in a fire) or when an employee disconnects the coupling to move or clean behind the appliance. As a further safety step, a mechanical shunt-trip valve must be installed in the main gas feed line to equipment under the exhaust hood. This device shuts off the gas supply to the equipment when the fire suppression system is engaged.

Quick disconnects are not only safety precautions; they are also handy when rearranging and cleaning the kitchen or servicing the appliances. Like pipes, quick disconnects come in different lengths and diameters. You order them based on the connection size, the gas pressure, and the length needed. Quick disconnects also are available for steam and water appliances.

ILLUSTRATION 6-8 The components of a connector kit used to attach an appliance to its gas source.

Source: Dormant Manufacturing Company, Pittsburgh, Pennsylvania.

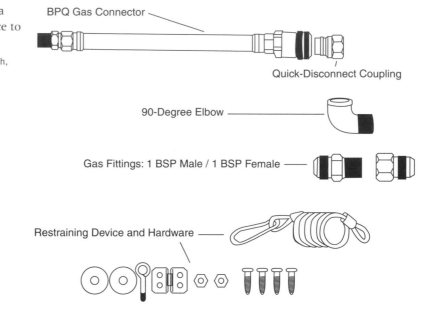

BPQ Gas Connector

Quick-Disconnect Coupling

90-Degree Elbow

Gas Fittings: 1 BSP Male / 1 BSP Female

Restraining Device and Hardware

6-2 STEAM ENERGY

Steam is water vapor, which occurs when water molecules are suspended in air by the heat added to them. Steam molecules carry large quantities of heat, and they return to their original form (condense) when they come into contact with a cooler surface.

When we discuss steam and its uses in foodservice, the terms *heat* and *temperature* (sometimes used interchangeably) take on completely different meanings.

■ *Heat* is the total amount of energy contained in steam or water at a given temperature.

■ *Temperature* is used to describe how hot a particular object is.

It takes 180 Btus to heat 1 pound of water from freezing (32°F) to boiling (212°F). However, to change this same pound of water to steam requires an additional 790 Btus. This means steam contains about six times the energy of boiling water.

The temperature of steam is generally related to its pressure. In short, the hotter the steam is, the higher its pressure, as shown in Table 6-2. The higher its pressure, the more steam molecules it contains. As these molecules condense, most of the Btus they contain are transferred quickly to the food being cooked, and the condensation creates room in the airspace for even more steam molecules to take the place of the ones that just condensed, in a cycle that continues until the heat source is turned down or off.

Steam is simple, clean, and quick, and it has been around longer than either electricity or gas as a heat source. In foodservice, steam is used extensively in the dish room to heat water and sanitize and dry dishes. In cooking, steaming is a healthful alternative to rangetop cooking that holds in nutrients and can be done quickly. Most foods can be cooked in a steam appliance with three significant advantages: greater control over the food quality; less energy use than other types of cooking equipment; and minimal handling, as the food often can be prepared, cooked, and served in the same pan. Steam is also a more efficient way to thaw frozen foods than immersing them in boiling water.

Steam is a major component of these popular kitchen appliances:

■ The **steam-jacketed kettle** is a large bowl within a bowl used for making sauces, soups, and stocks. The kettle has a sturdy outer layer. Between the two bowls is an area about 2 inches wide into which steam is pumped, which provides high but uniform cooking temperatures. The water used to create the steam can be heated with either gas or electricity (see Illustration 6-9).

TABLE 6-2

Steam Pressure and Temperature

STEAM PRESSURE (POUNDS)	TEMPERATURE (°F)
0	212
1	215
2	218
4	224
8	235
15	250
20	259
25	267
40	287
45	292
50	298

■ A *steamer* is a rectangular ovenlike appliance with an insulated door. It can be used for steaming vegetables, braising meats, cooking rice, thawing frozen food—any process that would benefit from the addition of moisture (see Illustration 6-10).

■ Convection steamers contain a fan or blower that circulates the warm, moist air for quicker, more even cooking.

■ Steam tables, often seen in cafeterias or on serving lines, hold food above a reservoir of hot water to keep it warm; similarly, a *bain-marie* is a hot-water bath in which an urn of gravy or a delicate sauce is immersed, also to keep it warm.

Steam systems and appliances work in one of three ways:

1. *Steam generators* use electricity to heat water and make their own steam. Small generators, called *boilers*, can be located right in the kitchen, under or near the steam equipment.

2. *Heat exchangers* take steam already made from one source, circulate it through a series of coils to clean it, and use it to heat another source. The steam from a building's heating system could, for example, be captured and recycled by a heat exchanger and then be used to heat the same building's hot-water tanks.

3. *Steam injectors* shoot pressurized steam directly into an appliance to produce heat. This is the least efficient way to use steam, because it's a one-time use. Condensation is drained away, not reheated and reused.

The push for water and energy savings has led to the development of *boilerless steamers*, also known as *no-hookup steamers*. These are not plumbed with a water source; water is added to

ILLUSTRATION 6-9 A steam-jacketed kettle is among the most versatile pieces of cooking equipment.

Courtesy of Vulcan-Hart, Baltimore, Maryland.

ILLUSTRATION 6-10 Any type of cooking or thawing that requires moist air can be done in a steamer.

Courtesy of Vulcan-Hart, Baltimore, Maryland.

them as needed, and they use it very efficiently—only 10% of the water used by a conventional, fully plumbed steamer. This means the difference between 8 or 10 gallons a day versus 30 gallons an hour. They don't cook as quickly as conventional steamers, but unless you must cook large quantities of product in a short time, a no-hookup steamer will more than fulfill your steam cooking needs, perhaps saving so much in water and energy costs that the unit can pay for itself within a year. In addition, if you use boiler-dependent equipment, by law the hot steam drained out of the unit must be followed with a cold-water chaser; thus, more water and cash go down the drain. Today's boilerless models can be set on standby, too, saving energy when not in use.

Steam equipment can further be classified into pressurized and unpressurized. The amount of pressure in a steam appliance is related to the temperature of the steam: The higher the temperature, the higher the pressure. Pressurized steamers cook food quickly because the steam can be superheated and comes into direct contact with the food. Unpressurized steamers are not as efficient. Unpressurized steam may not become as hot, and, as it touches the colder foods or cookware, its temperature is lowered even more. Eventually, it condenses back into water and is vented away into a drain.

Most states have specific protocols for boiler-based steam systems, whether they are designed for whole buildings or for individual pieces of equipment. Small boilers in equipment are held to the same requirements as large boilers. They all require yearly inspections, which some operators wish to avoid.

STEAM REQUIREMENTS FOR EQUIPMENT

Steam is measured in boiler horsepower (BHP). As a general rule, 1 BHP creates 34.5 pounds of steam per hour and is equivalent to about 10 kilowatts of electricity. The boiler is the piece of equipment that boils the water to make it into steam. If a boiler is rated by the manufacturer as producing 5 BHP, this means it produces 5 × 34.5 = 172.5 pounds of steam per hour.

To calculate the size of boiler you'll need for a whole kitchen, you must find out how much steam flow is needed by each piece of equipment; add these numbers and divide by 34.5. For instance, a kitchen with eight pieces of steam equipment may require a total steam output of 187.5 BHP. Divide 187.5 by 34.5, and you discover you need a boiler with an output of at least 5.43 BHP.

However, you should know more than how much steam will be produced. You must also know how much force, or pressure, the steam will have. In most English-speaking countries, steam pressure is measured in *pounds per square inch*, or **psi**. Once again, remember the temperature of the steam goes *up* when the pressure goes *up*.

The other factor that affects steam pressure is altitude—not drastically, but the equivalent of a 2°F or 3°F drop in temperature for every 1000 feet above sea level.

Finally, both steam temperature and pressure are affected by the distance the steam must travel to get from the boiler to the appliance. Heat loss is determined by the number of feet of pipe traveled plus every valve and fitting through which the steam must flow.

Foodservice industry research indicates the most expensive way to set up a kitchen is to install individual boilers for each piece of equipment, so it is ironic that that's the most common way it is done. Self-contained boilers have higher maintenance costs than a single large unit, and they add more heat to the already sweltering kitchen environment.

Like other appliances, boilers have efficiency ratings to consider. A boiler that requires 140,000 Btus and has a 50% efficiency rating will deliver 70,000 Btus of heat to its water supply to make steam. (This is the equivalent of a little more than 2 BHP.)

Steam can be an economical energy source, especially if your building already has a *clean steam* system built in. (When steam is referred to as *clean*, it means it is pure and has not been contaminated by chemicals.) If the building is not already fitted with steam pipes, you must decide if you will be using enough steam appliances to justify the expense of installing them.

STEAM TERMINOLOGY

Here's the way a steam system works. Steam is made by boiling water in the boiler, which may also be called a *converter*. It is then piped to the appliance where it will be used. At the appliance, the steam hits a *coil* (coiled copper or stainless-steel tubing), which condenses the steam and transfers its heat to be used in the appliance. As this transfer occurs, the steam cools and becomes water again. This *condensation* is removed from the appliance through a *steam trap*. The condensation usually returns to the boiler through another set of pipes, called the *return piping*, to be reheated and made into steam again (see Illustration 6-11).

The steam trap is one of the most vital parts of a steam system because it helps regulate the overall pressure. Oddly enough, steam traps are not placed directly in the main lines of a steam system. Steam does not flow through the steam trap. Instead, the trap is placed near the ends of the steam lines. It operates almost like an overflow valve, opening now and then to discharge water without affecting the rest of the steam or the steam pressure. Your steam appliance (or system) has one of two kinds of steam traps: an *inverted bucket* or a *thermodynamic disc*. Either way, the results are the same. Condensation plus air and carbon dioxide are collected in the trap, then discharged as the trap becomes full.

Valves control the steam pressure based on the amount of flow, size of pipe, and intensity of pressure needed for the appliance or system. There are several types of valves: *Electric solenoid valves* control the steam flow; *pressure-reducing valves* regulate steam pressure within the main supply lines; *test valves* allow you to test steam pressure at a point close to the appliance;

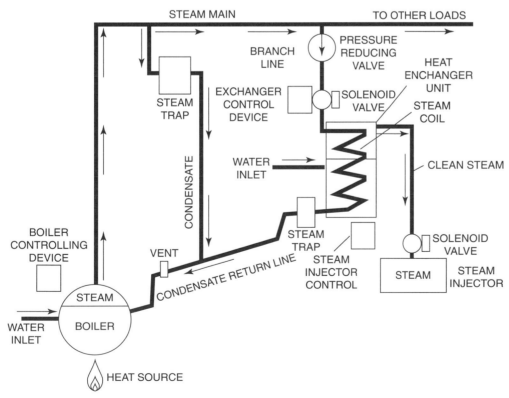

ILLUSTRATION 6-11   The basics of steam system operation.

Source: Hobart Corporation, "Steam in Perspective" (Troy, Ohio).

and *manual shutoff valves* ensure the steam can be safely turned off by hand any time the system must be serviced.

When purchasing or installing steam equipment, you'll need to decide if the unit is adequate for the job. If the unit does not generate its own steam, can the steam system in your building make enough steam to operate it?

Your installation contractor should be able to calculate the ***allowable pressure drop*** for the appliance. This is the total amount of available pressure minus the required amount of pressure to run that appliance. As you have already learned, the length and diameter of the pipes can affect this calculation—and, ultimately, the performance of the equipment. Friction and condensation in the pipeline naturally cause a drop in pressure, as do elbow joints and valves, and you have to allow for those.

Most equipment sales representatives are familiar with the particulars of installation. However, you still need to know the basics when replacing an old machine, purchasing a new one, or troubleshooting a steam-related problem.

COMMON PROBLEMS AND DIAGNOSES

Safety is the main consideration in dealing with steam equipment malfunctions. The mandatory first step is to shut off the steam supply and depressurize the steam line before attempting to disassemble the equipment.

The problem may be as simple as inadequate steam pressure. Take pressure readings as close to the appliance as possible, while steam is flowing through the line. Trace the steam flow through the entire appliance, by checking all valves, strainers, and coils for visible leaks or damage.

On most steam equipment, you will find evidence of hard-water mineral deposits, as the traces of minerals dissolved in the water settle and form *scales* inside tanks and pipes or lime buildup inside boiler tanks. Even if chemical additives are used periodically to control scale buildup, you should disassemble equipment now and then and remove scales manually. Excessive or frequent buildup is a sign the problem is not being properly treated, and perhaps a professional water treatment expert should be consulted. We'll talk more about water quality problems later in this chapter, but it is safe to say that water supplies in most cities are hard enough to cause significant problems in commercial steam use for foodservice. Steam equipment manufacturers cover themselves in these situations by specifying a minimum water hardness acceptable for their appliances. Equipment failure caused by unacceptable water quality is not covered under their warranties. Manufacturers may also recommend the installation of a water-softening system or, at least, a filter to remove silica and chlorine from water used to make steam. A water-softening method known as Zeolite is often recommended for hard-water areas; Zeolite specifically attracts and filters minerals and salts out of the water. Some manufacturers offer their own descaling kits.

6-3 YOUR WATER SUPPLY

Safe, plentiful water is often taken for granted by guests, employees, and managers in foodservice—which is ironic when you consider what a truly scarce resource it is. In fact, salt water makes up 97% of all the water on earth. Another 2% is inaccessible, frozen in remote ice caps and glaciers. More than half of the single percentage that remains worldwide is now diverted for human use, and yet the combination of increased population, industrial technology, and irrigation have pushed people to use an amazing 35 times more water than their ancestors did just three centuries ago. The United Nations estimates that by 2025, at least 800 million people will live in areas with what it terms an "absolute scarcity" of water.

According to the U.S. Environmental Protection Agency (EPA), Americans use 100 gallons of water per person per day. The EPA began its Water Alliances for Voluntary Efficiency program (WAVE) for hotels, restaurants, and other businesses in the 1990s, promising conservation of up to 30% for undertaking a series of water-saving measures. Overall, however, we still appear to be playing catch-up as a species when it comes to water conservation.

The other major water-related problem, at least in the United States, is the delivery system itself. Many of our water mains and pipes are more than a century old and, long neglected, are reaching the end of their useful life. The nation sees more than 237,000 water main breaks annually. When pipes break, water pressure drops and dirt and debris are sucked into the system and jeopardize water quality. At this writing, there is widespread agreement among experts that the nation's water system is in need of an enormous and expensive overhaul, and fixing it may change the way Americans use, and pay for, water.

In recent years, desalination (removal of salt from salt water to make it drinkable) has received much attention. Most current plants employ a technique called ***multistage flash distillation***, which removes contaminants from seawater by boiling it, then condensing (distilling) the steam.

Another technique is called ***reverse osmosis (RO)***. Highly pressurized seawater is pumped through a semipermeable membrane that allows only the freshwater molecules to flow through, leaving the mineral ions behind. In foodservice, reverse osmosis equipment is becoming popular as a way to purify water for steam, drinking, cooking, and humidification. RO technology can address the problems of both hard-water scaling (caused by calcium, magnesium, and manganese salts) and soft-water scaling (caused by sodium and potassium chloride) in water pipes. Because it removes solids better than normal filtration, RO offers the advantages of reduced water-related maintenance and better equipment life in addition to improvements in water quality. Illustration 6-12 shows the major components of a reverse osmosis water filter.

Water is a major expense, and water in foodservice establishments is given much more scrutiny than you probably give your home tap water. Samples are checked for bacteria, pesticides, trace metals, alkalinity, and chemicals. Before you lease or purchase a site, a water test is in order. And before the water is inspected in a laboratory setting, your own senses can offer clues to a few important basics.

> *Taste or odor.* Sometimes you happen to live in an area where—there's no other way to put it—the water tastes or smells funny. The locals may be accustomed to it, but visitors to the area notice it right away. It can affect the flavor of coffee, hot or iced tea, and any beverage in which you place ice cubes. Taste and odor problems are typically

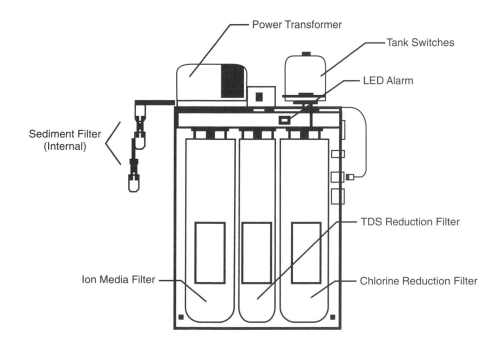

ILLUSTRATION 6-12 A reverse osmosis water filtration system includes a prefilter for sediments (not shown); a filter to reduce chlorine; a filter for total dissolved solids (TDS) that cause scale formation in pipes; an ion media filter that replaces the minerals with non-scale-forming particles; and a storage tank for the filtered water.

caused by the presence of organic materials in the water. You may need to find an outside source of ice to purchase, make ice and beverages with (more expensive) bottled water, or install crushed carbon filters to minimize customer contact with off-tasting water.

Color. Expert water quality advice is needed for this one. Iron or manganese in the water supply can result in odd colors that, through clear water glasses, look unpalatable. Filtering may help, but getting rid of this condition is a surprisingly technical problem.

Turbidity. When solids are suspended in water, it looks cloudy or murky—a definite turnoff in foodservice. Filtration, with a water-softening system, is a reasonably priced, low-maintenance alternative.

Two other water quality concerns are common:

Corrosion. This condition is the result of oxygen or carbon dioxide becoming trapped within the water supply; it is often caused by the level of acidity in the water system. It affects the useful life of pipes and equipment and can be corrected by installing a filtration system. What kind, and how extensive the problem is, can be determined by a water quality specialist.

Water hardness. Hard water contains a high proportion of minerals or salts. As you've learned, this condition causes an eventual buildup of scales on equipment, which requires constant preventive maintenance to prevent clogging of tanks and water lines and malfunction of the equipment. One-half inch of scale in the interior linings of a restaurant's hot-water heater can increase the appliance's energy consumption by as much as 70%. Scale also attacks ice machines, coffee makers, dish machines, and more.

THE HOW AND WHY OF HARD WATER

To a certain extent, hard-water scaling affects about 85% of water users in North America. Hard water is actually an unavoidable, natural process. As rain falls, it dissolves carbon dioxide gas from the air and becomes a weak solution of carbonic acid. As this acidic rainwater passes through the ground, it erodes and slowly dissolves limestone rocks. Limestone, found virtually everywhere on earth, typically consists of calcium carbonate, which is responsible for the hardness that causes scale. Scale is a result of the abnormal behavior of calcium carbonate, which becomes less soluble as water temperature increases. This means that as hard water is heated, the calcium carbonate can no longer stay dissolved and precipitates (separates)—or falls out of the water—as scale. We can soften hard water by adding lime to achieve a pH factor of around 10, followed by a treatment process of adding CO_2 gas to bring the pH level down to 9.5.

Hard water is also troublesome because it encourages the formation of soap scum and makes it more difficult for the *surfactants* (foaming agents) in soap to produce lather. In these cases, you have to compensate by using more dish detergent. You may also notice that dishes or cooking utensils washed in hard water become slightly discolored over time. Table 6-3, used with permission of the Hobart Corporation, lists common maladies and possible causes of malfunctioning dish machines. Notice how many of them are related to water quality.

In manufacturers' equipment specifications, water hardness is usually measured in **grains**, the amount of solids (calcium carbonate and other minerals commonly found in water) expressed in parts per million (ppm) and reported by the amount found in 1 gallon. One grain of hardness is equivalent to 17.1 ppm. If water contains more than 65 ppm, or 3.8 grains of hardness per gallon, it is a good candidate for water-softening treatment.

1–3.5 grains per gallon = slightly hard water

3.5–7 grains per gallon = moderately hard

7–10.5 grains per gallon = very hard

TABLE 6-3

Common Problems and Symptoms When Dish Machines Malfunction

STEAM DISH MACHINES SOME COMMON SYMPTOMS	LOW STEAM PRESSURE	STEAM TRAP OPEN	STEAM TRAP CLOSED	BACK PRESSURE IN CONDENSATE RETURN	STEAM PIPE TOO SMALL	MINERAL BUILDUP ON COIL	CONTROLS MALFUNCTION	LINE STRAINER FOULED	STEAM INJECTOR BROKEN OR FOULED	SOLENOID VALVE FOULED	PRESSURE RELIEF BAD	INCONSISTENT STEAM PRESSURE	BLOWER DRYER VENTS MISADJUSTED	IMPROPER CHEMICAL ADDITIVES TO TREAT STEAM
LOW TANK TEMPERATURE	×	×	×	×	×	×	×	×	×	×	×	×	×	
HIGH TANK TEMPERATURE							×						×	
VIBRATION				×			×			×	×			
LEAKING STEAM PIPES										×				×
FREQUENT VALVE FAILURE										×	×			×
SCALE/CORROSION in TANK														×
SCALE/CORROSION in COILS														×
HIGH FINAL RINSE TEMP.							×			×				
LOW FINAL RINSE TEMP.	×	×	×	×	×	×	×	×		×	×	×		
LOW BLOWER DRYER TEMP.	×	×	×	×	×	×	×	×		×		×	×	

Courtesy of Hobart Corporation, Troy, Ohio.

It is expensive to treat hard water and, because it doesn't pose a health hazard, many municipalities don't bother to do so, leaving it up to individual users to complete the job. Even if the local water utility won't treat your water for you, it might offer specialized assistance in helping you diagnose problems and determine solutions.

WATER QUALITY FACTORS

Water quality is not always a major concern. Indeed, in the United States, modern purification techniques have virtually eliminated such water-borne illnesses as cholera, typhoid, and dysentery. However, our water supplies are not without problems. Day-to-day human activities—farming, construction, mining, manufacturing, and landfill operation—affect water quality, affecting wildlife and marine life as well as humans.

Strange but true: Most water available for drinking is unfit for human consumption before it is treated. In 1974, Congress enacted the Safe Drinking Water Act, which authorized the federal government to establish the standards and regulations for drinking water safety. The EPA now sets and implements those standards and conducts research. State governments are primarily responsible for implementing and enforcing the federal mandates. The EPA reports that about 90% of community water systems comply with its standards. This figure is always controversial because some experts assert the EPA's standards are not tough enough. They claim there are as many as 1000 potential pollutants but the EPA has established enforceable limits for only 100 of them.

For water to be declared potable, it must meet federal drinking standards, whether the water system is public or privately owned. Regulations are strict, and water is tested at least four times each year. However, groundwater is not tested with the same intensity as surface water because it generally is not exposed to as many contaminants and pollutants. When the source of water is both surface and ground, then the water is tested as if all was surface water. In general, city residents and businesses get their water from a public water system, which is piped to users through a common water supply system. Public water systems are defined as those serving at least 25 persons or more and having at least 15 customer connections. In rural areas, residences and businesses may be on a private water system (i.e., individual well).

Water quality debates are often in the news, especially when a system is found to have higher-than-normal lead levels. But plenty of other contaminants and pollutants make headlines:

- Arsenic occurs naturally in some groundwater and is also a residue of mining and industry. At low doses, it is linked to cancer and diabetes; at high doses, it is poisonous.

- Pathogens are bacteria that can cause gastrointestinal illnesses. You've heard some of their names in news reports: Escherichia coli (*E. coli O157:H7*), cryptosporidium, and Yersinia enterocolitica are just a few. Farm waste runoff and sewage discharge can result in their accidental introduction to a water system.

- MTBE is a fuel additive designed to reduce air pollution. When spilled or leaked from storage tanks, it can contaminate water and cause liver, digestive, and nervous system disorders.

- Perchlorate is used in making fireworks, weapons, and rocket fuel. It interferes with thyroid function in humans.

- Trihalomethanes (THMs) are among the most common groundwater contaminants. They form when chlorine reacts with organic material such as fallen leaves. THMs may contribute to miscarriage risks and bladder cancer.

- Ammonium perchlorate is an additive the National Aeronautics and Space Administration (NASA) and the Defense Department used for rocket fuel and munitions. For disposal, perchlorate often was dissolved in water and poured on the ground because defense officials did not consider low levels hazardous to humans.

Perchlorate remains in use today and has been unregulated for years. However, in 2012 the EPA announced its intent to develop a National Primary Drinking Water Regulation for perchlorate. This is significant because officials and scientists have long disputed whether the trace amounts in groundwater, usually 4 to 100 parts per billion, are enough to suppress hormone levels in humans, which fluctuate slightly anyway. The EPA contends that although healthy adults are probably not at risk from ingesting minute amounts of perchlorate, it may affect the development of unborn babies and young children.

In addition to its inherent controversies, the water treatment process itself doesn't sound all that appetizing. Chemicals and gases—including lime, ferric sulfate, chlorine, ammonia, carbon, polymers, ozone, carbon dioxide, and fluoride—are added to drinking water to remove impurities, kill harmful viruses and bacteria, eliminate odd tastes and odors, and even help prevent tooth decay. These substances are mixed into the water, which is then sent through huge basins called *flocculators* where large paddles mix the water while the additives do their jobs and prompt the "bad" particles to group together and sink to the bottom of the tank. After this, the water passes into a settling basin, flowing slowly for 4 to 8 hours as the enlarged particles continue to settle to the bottom. A secondary treatment phase—more chemicals, more mixing, more settling—removes most of the chemicals that were originally put in, not just the bad stuff. The water is then filtered through anthracite coal, sand, and gravel, a process that catches any remaining particles. Finally, it is disinfected to kill bacteria.

One of the most troubling developments for the foodservice industry in recent years is the addition (at the secondary stage) of a disinfectant known as **chloramine**, a combination of ammonia and chlorine. Municipalities are using it because it remains in the water longer than chlorine alone. In fact, in 2012 chloramine use was mandated as part of the water treatment process by the EPA in what is called the Stage 2 Disinfection Byproducts Rule. The EPA says chloramine does a good disinfecting job while creating far fewer unwanted byproducts, known as **chlorocarbons**, which may cause cancer.

Why, then, has the same water treatment method been banned in many European nations? The problem with chloramine is that it wreaks havoc with foodservice appliances, particularly steam equipment. When the water is heated to such high temperatures, it becomes highly corrosive, pitting stainless-steel surfaces almost immediately and causing rubber and polyurethane parts and fittings to wear out quickly.

Because it is difficult to know when or if chloramine is being added to your water supply, it's important to check with your local water facility. At this writing, the most effective way to protect equipment is by installing what is called a *hollow carbon filter assembly*. Makers of other types of carbon filters may promise chloramine reduction, but in our experience the hollow carbon filter assembly lasts the longest and works the most effectively.

Another substance that had routinely been added to water systems is now undergoing an increasing consumer backlash. Water fluoridation began in the 1950s as a public health initiative to strengthen Americans' teeth and prevent dental decay. Since then, however, further research has shown that fluoride also stains some people's teeth (a condition called **fluoridosis**), causes digestive problems, and may exacerbate some allergies. Detractors point out that fluoridating the water system means everyone who drinks the water gets the same dose of fluoride—whether or not it is the appropriate dose. Some believe the process affects such water-intensive pursuits as brewing beer.

As a health precaution, or in areas where the local water has persistent mineral content that results in taste or odor problems, many foodservice facilities opt to filter their own water. You can purchase different types of filters to counteract different problems: a carbon filter for odor and taste problems, an integrated UV plus activated carbon filter to kill viruses and remove particles. There are filters for chemical absorption, turbidity reduction, and heavy metal reduction. Filters have load capacities and are sized by flow rate. Capacities of water filtration systems can range from a small, single-cartridge unit that can be attached to a single machine like a coffee maker to a multi-cartridge system capable of filtering all the water that enters a building, more than 100 gallons per minute. No matter what their size, the principle is simple: Cold water enters the filter at an inlet valve, where it is directed through the internal filter chamber. It exits the body of the filter through an outlet connection, generally at the bottom of the chamber. Changing the internal filter element is usually as easy as opening the chamber, removing the old element, and putting in a new one. The manufacturer recommends the frequency with which filters should be changed.

The effectiveness of the filtration systems probably vary as much as their manufacturers' claims, but all of them should adhere to two important NSF International Standards: 42 ("Aesthetic Effects," which governs taste, odor, chlorine content, and particulate reduction) and 53 ("Health Effects," which governs turbidity, Giardia cyst content, and asbestos reduction). Properly filtered water can extend the life of your most expensive equipment, such as steamers, combi-ovens, dish machines, ice machines, and beverage dispensers, by eliminating scale and slime buildup. Better energy efficiency and fewer maintenance calls can translate into cost savings. And, of course, using filtered water for customers—in beverages, ice-making, and cooking—is also a plus.

TRENDS IN WATER TECHNOLOGY AND CONSUMPTION

As businesses grapple with the safety hazards of strong chemicals used for cleaning and sanitizing, some new technologies and processes have emerged. **Electrolyzed water** (*E-water* or **ozone water**) has been found effective in eliminating food-borne pathogens. E-water is produced by applying an electrical current to a weak solution of water and salt, which produces a superacidic, sanitized water that contains powerful oxidants—although it doesn't look, smell, or taste any different from tap water. It can be used for handwashing and on food contact surfaces; to sanitize cutting boards, utensils, and more—even to sanitize raw foods, because of its power as a bacteria-killer. E-water is especially easy to use because it doesn't leave soapy residue on food contact surfaces, doesn't have to be rinsed off, and works well on floors and stainless-steel surfaces. Grease exhaust hood filters can be soaked in it overnight. In tests, less dish detergent was needed when E-water was used in the dish machines. Equipment to produce electrolyzed water is now in use in the foodservice industry, but the technology has not yet become cost-effective enough for home use.

Kin to E-water is *activated water*, also created by applying a small amount of electricity to water, but without using salt. The charged water acts as a conductor, attracting dirt particles and suspending them in the water, which allows them to be wiped away more easily. The advantage of activated water over E-water is supposed to be that it sanitizes more quickly.

Yet another technique has been developed for identifying hazardous particles in water. Laser beams are shot through a stream of water to check for microorganisms. They can detect anthrax, *E. coli*, or any other particle not previously identified in a water supply. Each type of microorganism looks different, and the lasers are precise enough to differentiate them. So far, this type of system scans for live organisms (like bacteria) but still sees chemicals only as unidentified particles. It can also detect and report any type of increased particle activity. The newly patented technology can be used as part of security measures, to monitor water safety at large public events.

Whether it is safety, flavor, or marketing, bottled water appears to have staying power and sales appeal in any foodservice setting. Bottled water accounts for 15% of the total nonalcoholic beverage market, making it the second largest commercial beverage category in the United States, outdone only by soft drinks (at 23%). Free of sugar, calories, and alcohol, it outsells beer, wine, juice, and coffee and is a beverage for all day-parts of a foodservice operation. Most customers feel it enhances the dining experience when they are offered still or sparkling water by the bottle, and this tactic certainly increases check averages. Bottled water sales continue to grow in restaurant settings in fine-dining, casual, and quick-service operations. Water provides a sales opportunity every time a table is seated or before the next course arrives.

Lately, the selling point is not necessarily the water but what is in it; line extensions now include a vast array of "enhancers." We now have vitamin water, energy water, fitness water, and fruit-flavored water, all of them in various flavors and colors. If water is artificially flavored with passion fruit, sweetened with Splenda, and dyed green, is it still water? Some contend these products are more like diet sodas without the carbonation; an increasing number of consumer groups are challenging the nutritional claims and asserting that the public is simply being duped into paying a high price for convenience. Flavors aside, it is true that those who purchase bottled water expecting it to be purer than tap water may be wasting their money. Legally, bottled water does not have to be any cleaner than tap water. The same Safe Drinking Water Act provisions cover it—and yet it costs, on average, 625 times more than tap water.

Some environmental advocates believe the safety and quality of tap water is better regulated than most bottled water and that the bottles themselves have created additional environmental problems. According to Food and Water Watch, the production of plastic water bottles consumes an estimated 47 million gallons of oil a year, and about 1.5 million tons of plastic waste per year ends up in landfills—assuming that 80% of the bottles are discarded rather than recycled.

The International Bottled Water Association, an industry trade group, counters that water bottles "are 100% recyclable," make up only one-third of 1% of the U.S. waste stream, and that water bottle manufacturers have reduced the weight (and therefore the plastic content) of the bottles by 32% in the last decade.

Mineral water is exempt from the Safe Drinking Water Act because it contains a higher mineral content than allowed by the U.S. Food and Drug Administration (FDA) regulations. "Manufactured" waters, such as club soda and seltzer, are also exempt from the act because they are considered soft drinks. For other types of bottled water, however, the FDA now requires additional labeling on individual bottles to identify better the source of the water. Common terms include:

- *Spring water.* Collected as it flows naturally to the surface from an identified underground source or pumped through a bore hole from the spring source.

- *Artesian water.* Tapped from a confirmed source before it flows to the surface.

- *Well water.* Tapped from a drilled or bored hole in the ground.

- *Mineral water.* Collected from a protected underground source; contains appreciable levels of minerals (at least 250 ppm of total undissolved solids).

- *Sparkling water.* Contains the same level of carbon dioxide in the bottle as it does when it emerges from its source.

- *Purified (distilled) water.* Produced by distillation, deionization, or reverse osmosis.

BUYING AND USING WATER

Water is purchased in much the same way as electricity. A meter measures the number of gallons that enter the water system, either in cubic feet or in hundreds of gallons. The meter, which isn't equipped to record the huge numbers used by most foodservice locations, will show a basic number. The meter reader multiplies it by a constant figure, known as the **constant multiplier**. For instance, if the meter shows 1200 and the multiplier is 100, 120,000 gallons have been consumed.

When you turn on a sink in your kitchen, the water rushes out at 50 to 100 psi. This pressure is more than enough to get it from the city's pipes into the building, which only takes up to 20 psi. The excess pressure is used to move water into numerous pipes throughout the facility. This is called the **upfeed system** of getting water. In fact, 50 to 100 psi is strong enough to supply water to the upstairs area of a building four to six stories high. If your facility is in a taller building than that, you will probably need water pumps to boost the pressure and flow.

Pumps can be used to increase water pressure; valves (called **regulation valves**) can be used to decrease it. Your goal is to control the water coming into your facility to avoid fluctuating pressure or an uneven flow rate. And whenever there's a possibility that contaminated water could backflow into the potable water system, a *backflow preventor* should be installed. **Backflow** might occur whenever a piece of equipment, such as a commercial dishwasher, is capable of creating pressure greater than the incoming pressure of its supply line.

The segments of the typical upfeed system are:

Water meter. The device that records water consumption. It is the last point of the public water utility's service. Anything on your side of the water meter, including all pipes and maintenance, is your responsibility, not the water company's.

Service pipe. The main supply line between the meter and the building.

Fixture branch. A pipe that carries water to a single fixture. It can be vertical or horizontal and carry hot or cold water.

Riser. A vertical pipe that extends 20 or more feet. It can carry hot or cold water.

Fixtures. The devices (faucets, toilets, sinks) that allow the water to be used by guests and employees.

Hot-water heater. The tank used to heat and store hot water (discussed later in this chapter).

Pipes. The tubes fitted together to provide a system for water to travel through. They can be copper, brass, galvanized steel, or even plastic. Building codes determine what materials are acceptable for different uses. Copper is the most expensive type of pipe, but it's considered easy and economical to work on. Plastic pipes are allowed only for limited uses. The most common type of plastic pipe is made of **polyvinyl chloride (PVC)**. It is inexpensive, corrosion resistant, and has a long life, if you're allowed to use it.

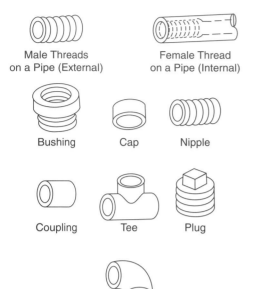

Male Threads
on a Pipe (External)

Female Thread
on a Pipe (Internal)

Bushing   Cap   Nipple

Coupling   Tee   Plug

90-Degree Elbow

ILLUSTRATION 6-13 Examples of the most common pipe fittings. Fittings allow pipes to connect and bend.

Fittings. The joints of the pipe system. They fit onto the ends of pipes, allowing them to make turns and to connect to each other and to other appliances or fixtures. Some of their names describe their shapes; the most popular fittings are the bushing, cap, coupling, elbow, plug, and tee. Some fittings have threads (either internal or external) to be screwed into place; others are compression type (see Illustration 6-13).

Valves. Valves control water flow and are made of brass, copper, or cast iron. Use of the correct valve minimizes plumbing problems. Gate valves are used to vary water flow and allow water to go in either direction. Check valves allow water to flow only one way. They are marked with an arrow indicating the direction of flow. Safety valves are spring-loaded valves operated by temperature or water pressure to relieve excess pressure if they sense a buildup.

READING WATER METERS AND BILLS

Like gas, water consumption is usually measured in cubic feet, but occasionally you will see a meter that measures in gallons. If so, that will be printed on the meter. These meters have two common faces. One has a simple readout in the center that indicates the number of cubic feet that have been used [see Illustration 6-14(a)]. The hand that makes its way around the dial is used only to indicate that water is flowing through the meter.

The other type of meter has a series of small dials that are read like gas and/or electric meter dials. The 1-foot dial is not part of the reading; it only indicates whether water is flowing through the meter. Start your reading with the 10-foot dial [see Illustration 6-14(b)].

A water bill lists several charges. General use is billed at a flat rate for every 1000 gallons used. Sewer services are also billed in 1000-gallon increments. You can often get a small break on your bill by paying early. Some cities apply a variety of nonmetered charges, including a fee for maintaining the town's firefighting equipment, a water line repair service fee ("leak insurance"), or a water treatment or "water quality" fee. Illustration 6-15 is a sample water bill.

You might ask your utility company if your business can install a submeter, a separate meter to track water that does not go into the sewer. Use of a submeter is not common, but it allows you to subtract the submeter count from your total gallons used and, therefore, pay less for your sewer bill. A resort hotel, for example, uses water to fill its swimming pools or irrigate a golf course—uses that do not flush water into the sewer system.

ILLUSTRATION 6-14 Two types of water meters: (a) simple readout, and (b) dial meter.

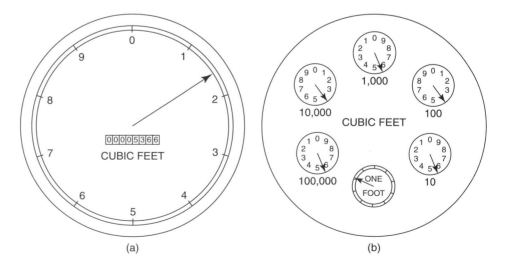

CUBIC FEET

(a)

1,000

10,000   CUBIC FEET   100

100,000   ONE FOOT   10

(b)

Your Restaurant, LLC
XYZ Restaurant Group
1234 McKinney Avenue
Your City 85123-4567

****TO QUALIFY FOR PROMPT PAY AMOUNT, PAYMENT MUST BE
RECEIVED IN OUR OFFICE BY THE DUE DATE OF JULY 14, 2014.**

Customer: YOUR RESTAURANT, LLC
Service Address: 1234 MCKINNEY AVE.

COMMERCIAL
Account Number:
001-12345678-010

Billing Date: 06-30-14
Water Used This Month: 268,700 Gallons   Days Served: 31

SERVICE PROVIDED	METER NUMBER	READ PREVIOUS	READ 6/30/2008	USAGE IN 100 GALS	USAGE CHARGE	CUSTOMER CHARGE	TOTAL
Water	123456	89181	91868	2687	586.14	6.64	592.78
Sewer				2687	658.31	5.93	664.24
Surcharge				2687	103.93	0.00	103.93
					CURRENT CHARGES:		**$1,360.95**

NOTES:

Water charged at $1.70 per 1000 gallons up to 10,000; $2.20 per 1000 gallons above 10,000 gallons.
Sewer charged at $2.45 per 1000 gallons.
Untreated Wastewater charged at $0.3868 per 1000 gallons.

ILLUSTRATION 6-15   A typical commercial water bill.

Also be aware that you aren't just paying for water; you're also paying to rid your restaurant of water and waste. It is common for the water utility to assume that all water used by your restaurant is discharged as sewage and to charge you accordingly. However, not all the water you use ends up in the sewer system. Ask your utility company to help you determine what percentage of water you buy actually reaches the sewer and to adjust your bill accordingly.

WATER CONSERVATION

All our best efforts aside, Americans still use 35 billion gallons of water a day. Restrooms, kitchens, and landscaping are the three most water-intensive areas, and most restaurants have all three! You may be surprised at how many water utility companies offer water conservation tips, often on Internet sites. A mountain of information is out there for the restaurateur who wants to train employees to practice conservation. Most suggestions are simply common sense; a few are truly inventive. Some of the ones that follow were adapted from the Massachusetts Water Resources Authority in Boston.

KITCHEN AND SERVICE AREAS

- Turn off the continuous flow feature of drain trays on coffee/milk/soda beverage islands. Clean them thoroughly as needed.
- Many newer appliances have water use requirements set under the federal Energy Policy Act of 2005. For instance, new commercial dish machines have perfected designs

that take water usage well below 1 gallon per rack. New water-saving devices, such as nozzle design and power-wash features, can reduce water use by as much as 50%. On older machines, check the manufacturer's instructions to see if the dish machine spray heads can be reduced to lower-flow ones.

■ Do your food thawing and utensil presoaking in tubs or basins of water, not running water. Better yet, allow for longer thawing times in refrigeration so you won't have to use water to speed the process.

■ Boilerless steamers are the equipment of choice for water savings.

■ Adjust ice-making equipment to make and dispense less ice when less is needed. The big mistake is purchasing an ice machine that is too small or just the right size for the operation, because ice machines run on an economy of scale: The larger the machine, the more efficiently it runs. If an operation has a machine that overproduces, it can simply be put on a timer and set to be on during off-peak hours (when water or electricity rates are cheaper) and off during peak hours.

RESTROOMS

■ Repair leaky toilets and faucets. One leaking toilet can waste 50 gallons of water a day; a dripping faucet can waste at least 75 gallons a week.

■ Install aerators, spring-loaded valves, electronic sensors, or timers on all faucets.

■ Replace worn-out fixtures with water-saving ones.

■ Apply water conservation stickers on mirrors to remind both employees and customers to turn off taps.

LANDSCAPING

■ One inch of water per week is sufficient to sustain an established lawn or landscape. Gauge rainfall, and augment with only as much water as is needed to equal 1 inch per week. This means monitoring the watering schedule—what you'll need in the summer is not the same as in the fall, for instance.

■ After a heavy rain, wait at least 10 days to water again.

■ Don't water on windy, rainy, or hot days, when more water evaporates than gets to your landscaping.

■ Investigate a drip irrigation system for flowers, shrubs, and new plantings. Drip irrigation saves 30% to 70% of the water used by an overhead sprinkler system.

■ Sweep sidewalks, loading docks, and parking lots instead of hosing them down.

The water-guzzling capital of the United States is Las Vegas, with its massive (and highly landscaped) casinos, backyard pools, and green boulevards transforming what was once desert land. Las Vegas uses an estimated 325 gallons of water per person per day. At that rate, the region is expected to run out by the mid-2030s, according to some experts. However, the city's hospitality industry is doing its part with conservation measures, from water-saving plumbing fixtures to lawn-watering restrictions to new types of water purification technology.

6-4 CHOOSING PLUMBING FIXTURES

Your plumbing fixtures are among the hardest-working items in your business. Fortunately, many guidelines can assist you in selecting them. The Uniform Plumbing Code sets fixture requirements for the public area of your restaurant, primarily the restrooms. For the kitchen, and food preparation in general, NSF International has extensive guidelines. Here are some considerations when selecting your fixtures and designing your restrooms.

WATER CLOSET. Yes, that's the fancy name for a toilet. It should be made of solid glazed porcelain, with a flush tank that discharges water when a lever or button is pushed. Another way to flush the tank is with pressure valves; however, they use more water. The toilet should have a self-closing lid; some have no lids at all.

There are plenty of ways to save water in the restrooms. Dual-flush-option toilets, used in Europe and Asia for more than a decade, let customers use as little as 0.8 gallons per flush, or 1.6 gallons, depending on need. Pressure-assist toilets use a pressure vessel inside the tank to create a combination of water line pressure and compressed air to flush. Any commode that uses less than 1.28 gallons per flush is considered a *high-efficiency toilet (HET)*; today's legal standard for new construction is 1.6 gallons per flush, but many older facilities still have the old-fashioned toilets that use 3.5 gallons or more per flush.

Your city's plumbing code specifies the number of toilets and urinals in public buildings. In foodservice settings, some cities require more if you serve alcoholic beverages. The general rule is two toilets for every 100 female guests and one or two urinals and one toilet for every 100 male guests, as discussed in Chapter 4.

URINAL. This companion fixture for men's restrooms should also be solid glazed porcelain. There are stall, wall, and pedestal-style installations; the wall-mounted urinal is the best because it makes cleaning easier beneath the urinal. The flush valve is the most common mode for flushing urinals.

LAVATORY. The *lavatory* is also called a *hand sink*. The preferred material for this important part of every restroom is, again, glazed porcelain. The hand sink is required in most cities to supply both hot and cold water, with a common mixing faucet for temperature control. Aerators are a must for your restroom sinks; these simple attachments to the faucet head will reduce water flow from 1.5 gallons per minute (gpm) to 0.5 gpm. On average, they save about $268 per sink per year without compromising water pressure.

The sink should have an overflow drain. Other health code requirements include soap dispensers (not bar soap) and disposable towels for hand drying. Although heater-blowers can dry hands with warm air, they are not particularly energy efficient.

Regulations specify the minimum number of hand sinks in restrooms, and you must be certain at least one sink is installed such that a person in a wheelchair can use it, to meet the guidelines of the Americans with Disabilities Act. We'll discuss hand sinks in the kitchens in just a moment.

OTHER CONSIDERATIONS. Generally, it is advisable to have one floor drain in each restroom stall and at least one in the urinal area of the men's room. If the restrooms are large, consider installing additional floor drains to make mopping easier as well as to catch any potential plumbing overflows.

A working exhaust fan may be required by local health code. Even if it's not, it is a good idea to circulate the restroom air. Install spring-loaded doors on restrooms to prevent people from leaving them open. Finally, another crucial consideration: Restrooms must meet both local and federal requirements of accessibility for guests with physical disabilities.

At the back of the house, the plumbing fixtures must withstand heat, grease, heavy-duty cleaning products, and all the rigors of cooking. They include sinks and drains, discharge systems, venting systems, and hot-water tanks. As a rule, the architectural drawings of your building will include plumbing, electrical, and mechanical connections: Ask that the drawings be rendered in ¼-inch scale and include a schedule of equipment to be plumbed.

SINKS AND HANDWASHING SYSTEMS

Before we discuss the multiple types of sinks used in foodservice, let's talk for a moment about the importance of the hand sink. The FDA reports that 40% of all food-borne illness is the result of poor handwashing practices by employees and cross-contamination from touching

ILLUSTRATION 6-16   The components of a hand sink station.

Courtesy of The Eagle Group, Clayton, Delaware.

the faucets or other surfaces that may not be clean. Proper handwashing is a combination of appropriate water temperature, the duration and type of scrubbing, and the use of soap. It's a matter of teaching employees the right way to do it and then, human nature being what it is, monitoring to make sure they do it right. *You must not wash hands in the food sinks or wash food in the hand sinks—it is absolutely against health codes!*

Hand sinks are at the core of today's increasingly important hygiene stations; an example is shown in Illustration 6-16. They are now being designed and located for convenience and frequent use rather than for minimal compliance to local health codes. Locating hand sinks where they will be used demands more than the traditional minimum quality. The closer to the action, so to speak, the greater the need for improved functionality and appearance. Sink stations must be kept clean, which includes regular sanitizing of sinks, faucets, and dispensers on the daily list of routine maintenance. A basic hand sink selection checklist should include:

- Select a manufacturer with an established food-service reputation.
- Select a manufacturer with easy access to technical service in your area.
- Use Type 304 stainless steel, shaped for enhanced strength and proper drainage.
- Choose seamless construction for better hygiene and more effective sanitizing.
- Select a deeper sink bowl for better drainage, with splashguards to prevent cross-contamination.
- Consider antimicrobial surfaces or a highly polished finish.
- Tailor your selection to the needs of the location.

Food contamination is a problem serious enough that manufacturers now produce automated hand-washing systems or stations that reduce hand contact with equipment. Some can be installed using existing plumbing; others are self-contained units. For use in either restroom or kitchen, the unit is activated by motion, not touch, and leads the user through the wash (dispensing soap and warm water) and dry (with warm air) process in less than a minute. The unit can also "read" the handwashing frequency of each individual. And you, as a child, thought your mother was strict about this? Think again! Here are the seven steps in a typical computerized handwashing operation:

1. A beep in the food production area reminds you it's time to wash your hands. It can be programmed to sound either at timed intervals or after a task is completed.

2. You approach the sink and, without touching anything, water streams out at just the right temperature. Get your hands wet...

3. ...and then, about 7 seconds later, a built-in device dispenses soap. (Depending on the system, you may have to punch in your employee code number to get the soap.) The lathering and scrubbing is up to you, but it should take about 20 seconds.

4. At that moment, more water comes on for rinsing—again, about 20 seconds' worth.

5. Dry your hands using the hot-air dryer.

6. Some units provide an optional antibacterial spray for sanitizing.

7. The computer software records your accomplishment of a proper wash. In some systems, your employee code number even qualifies you for a small gift for frequent washes.

The most sophisticated systems, most often used in hospitals, allow the handwasher to insert the hands past the wrists into two separate cylinders. The machine provides a low-volume (but high-pressure) spray of water and sanitizing solution that lasts from 12 to 20 seconds. It requires electric power as well as standard plumbing connections. More about handwashing practices in Chapter 8.

Even if you have just plain hand sinks, most health departments have rules about how many and where they must be located. (These requirements usually apply to bar areas as well.) At this writing, here are the norms:

- A hand sink should be within 15 feet (in a straight line) of any food prep area.
- One hand sink is required for every five employees or every 300 square feet of facility space.
- One hand sink is required for every prep and cooking area.

In addition to hand sinks, you need several other types of sinks in your kitchen. In the dish room, there is the pot sink (for washing pots and pans), the warewashing or *scullery sink*, and the three-compartment dish sink. The three compartments are for washing, rinsing, and sanitizing. Elsewhere, there's the prep sink (for scrubbing and peeling vegetables), the utility sink (for mops and cleaning), and the bar sink (for the bar area).

Sinks should always be made of stainless steel, which is durable and easy to clean. Manufacturers use two types of stainless steel, known as Type 430 and Type 304. Both are approved for foodservice use, but Type 304 is considered more durable because of its content: 8% nickel in addition to the standard 16% chromium. Rounded corners (called *coved* corners) make sinks easier to clean. You can also clean more thoroughly under the sinks if they are installed so the water faucets come straight out from the walls (and the water pipes are located behind the walls instead of beneath the sinks).

Other requirements are a swiveling, gooseneck-style faucet that can reach each compartment of the sink; an overflow drain for each compartment; and ample supplies of both hot and cold water. You can choose from many faucet types, but aerators and stream regulators save the most water.

For pot sinks, add a drainboard to the list of requirements. If the drain board is more than 36 inches long, it will need its own support legs. NSF International now requires that drainboards be welded to the sink bowls. Here are basic sink installation guidelines:

Pot sink. The height of the sink edge should not exceed 38 inches. The sink itself should not be more than 15 inches deep (most are 12 to 14 inches deep) on legs or a pedestal no more than 24 inches tall. The depth of the sink, from front to back, should not exceed 28 inches (see Illustration 6-17).

Warewashing (scullery) sink. Local health codes dictate the number of compartments or bowls these sink units must have, plus their backsplash height, water depth, drainboard size, and so on.

Dish sinks. These are used mostly in small, limited-menu operations. They are three-compartment sinks with a minimum bowl size (for each compartment) of 16 × 20 inches and a water-level depth of 14 inches. A dish sink also usually requires a double drainboard—on each side of the far left and far right sink bowls. Illustration 6-18 is a handy three-compartment design made to fit into a corner.

Combination pot-dish sink. Also in small restaurants, a three-compartment unit can be installed with slightly larger sinks to do double duty for washing both pots and dishes. The minimum bowl size here is 20 × 20 inches, with a water depth of at least 14 inches.

MINIMUM OPENING DIMENSION FOR EACH SINK 12" X 12" WITH MINIMUM WATER
CAPACITY OF 6 GALLONS BELOW THE OVERFLOW LEVEL.

MULTIPLE SINKS TO BE USED IN THE MANUAL WASHING OF EATING AND DRINKING
UTENSILS. AT LEAST 3 UNITS OR BASINS SHOULD BE PROVIDED.

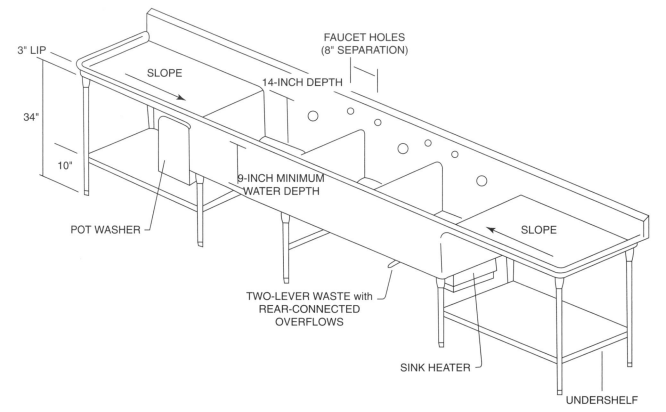

ILLUSTRATION 6-17 The pot sink is organized to permit the addition of faucets, shelves, water heaters, and a pot washer.

Source: Adapted from Carl R. Scriven and James W. Stevens, *Manual of Equipment and Design for the Foodservice Industry* (Albany, New York: Thomson Learning, 1989).

ILLUSTRATION 6-18 For small spaces, consider three-compartment sink designs that fit into corners.

Courtesy of The Eagle Group, Clayton, Delaware.

You can also order these with 24 × 24-inch bowls and, if you'll be washing a lot of full-sized baking sheets, you'll want 24 × 28-inch bowls.

Hand sink. The hand sink most often required by city ordinance is either wall mounted or pedestal style. Again, the wall-mounted sink is easier to clean beneath. The typical hand sink is 20 inches long by 16 inches wide, with a depth of 8 inches, but smaller sizes are also acceptable. In choosing one, remember it will get a lot of use.

Bar sink. This is a four-compartment sink, either 8 or 10 feet in length. A minimum of 24 inches of drainboard space is recommended on both sides of the bar sink. The bar sink is generally 18 to 24 inches wide and 1 foot deep. Most have a special overflow drain; make sure this drain is at least 1 inch in diameter.

Prep sink. This is usually a two-compartment sink unit, although some health codes mandate a third bowl. A heavy-duty garbage disposal may be installed in one of the compartments. Size will vary depending on the amount of prep work done in the kitchen; the most common size is a 20-inch-square sink with a 10-inch depth.

Utility sink. This is typically the big, deep, rectangular sink in the back of the kitchen that always looks so beat up and untidy. At least it is useful, if not attractive. A wall-mounted sink will allow storage (of buckets, etc.) beneath it.

At least one manufacturer has introduced mobile sinks—movable, on casters, with quick-disconnect water lines so the sink unit can be completely relocated (temporarily) so you can clean more thoroughly behind it. This is a difficult area to reach and, over time, can harbor lots of bacteria and other hazardous gunk. No tools or special skills are required to disconnect the lines.

DRAINS AND THE DISCHARGE SYSTEM

Your myriad sinks are drained into the drainage or **discharge system**, which receives the liquid discharge created by the food and beverage preparation area. The first component of the discharge system is on the sink itself: the **trap**, which is a curved section of pipe whose lowest part traps (or retains) some water. The trap is called a *P trap* when the drain pipes go into the wall; it is called an *S trap* when the drain pipes go into the floor.

In addition to these traps, it is a good idea to have floor drains located directly beneath your larger sinks. The drains in a commercial kitchen must have a **dome strainer** (or *sediment bucket*), much like a perforated sink stopper that traps bits of dirt and food as liquids go down the drain. For the heaviest-duty jobs, a floor drain with a much larger strainer compartment (called a **sump**) is recommended. The sump is at least 8-inches square. Type 304 stainless steel is the preferred material for drain fabrication, and coved corners make them easier to clean. This type of drain is commonly referred to as a **floor sink**. Floor sinks can be set flush or below finished floor height; however, many jurisdictions require that they be set ½ to 1 inch above the finished floor. Check your local code requirements before giving direction to the plumber. The reason the drain is recessed slightly is to prompt water to flow toward it. The drainpipe should be 3 to 4 inches in diameter, and its interior walls must be coated with acrylic or porcelain enamel that is both nonporous and acid resistant. A nonslip floor mat, with slats for drainage, should be a standard accessory beneath every sink.

How many floor drains should you have in your kitchen? Let's count the areas in which drains are a must to catch spills, overflow, and dirty water from floor cleaning:

1. Hot line area
2. Prep and pantry area
3. By the pot sinks
4. Dishwashing area
5. Dry storage area

6. Outside the walk-in refrigerator

7. Waitstations/service areas

8. Near steam equipment

9. By the bar sinks

10. Under the ice maker

The ice maker has another unique drainage requirement: a recessed floor. One smart idea is to install several drains in a trough 1 to 2 feet wide and several feet long, covered with a rustproof metal grate. This is effective along the length of the hot line area or in the constantly wet dish room. Floor troughs are generally constructed of stainless steel with a subway-type grating. They come in various widths and lengths to fit your needs. You can also decide if you want stainless-steel bar grates or flat fiberglass grates in yellow or gray. These troughs are set just a bit below floor height to allow water to be directed to them when a spill occurs.

When we talk about draining away waste, we're not just discussing water. The water often contains grease, and grease disposal is an enormous (and messy) problem in foodservice. Most towns and cities require a **grease interceptor**. It is commonly known as a *grease trap*, although the professional plumbing industry discourages the use of this terminology. Your area's building code lists which kitchen fixtures must be plumbed to the interceptor; typically, the water/waste output of the garbage disposal, dishwasher, and all sinks and floor drains must pass through the interceptor before it enters the sewer. Employee restrooms and on-premise laundry appliances generally do not have to be connected to the interceptor.

The role of the grease interceptor is to prevent grease from leaving the restaurant's drainage system and clogging the city sewer system. Foodservice wastewater is a big problem for sewers designed primarily for residential waste. Thus, fines and surcharges may be imposed on restaurants if their effluent (outflow) exceeds the local standards for its percentage of fats, oil, and grease (FOG, in industry jargon).

As waste enters the interceptor, it separates into three layers: The heaviest particles of food and dirt sink to the bottom; the middle layer is mostly water, with a little bit of suspended solids and grease in it; and the top layer is grease and oil. The interceptor traps the top and bottom layers while allowing the middle layer to flow away into the sewage system. Interceptors come in different sizes, and you should choose one based on the gallons of water that can run through it per minute, the number of appliances connected to it, and its capacity to retain grease. See Illustration 6-19 and Table 6-4, a sample size chart from the Uniform Plumbing Code.

Cleaning the interceptor regularly is necessary because the bottom layer can clog pipes if allowed to build up, and the top layer can mix with, and pollute, the middle layer. Most foodservice facilities hire a trap-cleaning service company to handle this unpleasant task. It is a costly

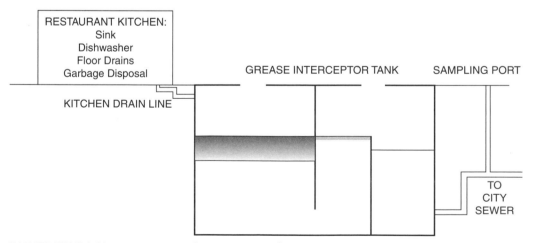

ILLUSTRATION 6-19 How a grease interceptor works.

TABLE 6-4

Grease Interceptor Requirements

TOTAL NUMBER OF DRAINAGE FIXTURE UNITS (DFUS) CONNECTED	REQUIRED RATE OF FLOW PER MINUTE (GALLONS)	GREASE RETENTION CAPACITY (POUNDS)	REQUIRED RATE OF FLOW PER MINUTE (LITERS)	GREASE RETENTION CAPACITY (KILOGRAMS)
1	20	40	76	18.2
2	25	50	95	22.7
3	35	70	132	31.8
4	50	100	189	45.4

Source: International Association of Plumbing and Mechanical Officials (IAPMO), 2012 Uniform Plumbing Code.

activity, and not without legal ramifications. The service company must be licensed to haul the grease waste to specially approved treatment areas. It's no longer enough for a restaurateur to trust that the grease is being taken care of. The smart ones take a proactive approach. Once in a while you'll see news reports about service companies that skirt the law by dumping waste into creeks or unapproved areas. You would be wise to thoroughly research your area's grease removal requirements and to interview several service companies. Ask for, and contact, their references.

There are two types of interceptor cleaning: skimming (removing the top layer) and a full pump-out of the tank. For most foodservice facilities, skimming is not sufficient. The heavy lower layer of particles must also be filtered away. You might decide on a combination of services—frequent skimming, with a full pump-out at regular intervals. The types of foods you serve and your volume of business should be your guidelines, along with a scientific measurement of the effluent to see how much FOG or chemicals it contains. In some cities, the penalties are so strict that restaurateurs include a pretreatment step, adding fat-dissolving chemicals or filtering the waste before it even gets to the grease trap. Electric undercounter units recover grease for discarding as trash, not sewage.

Outside installation of the grease interceptor is recommended, at a level several feet below the kitchen to use gravity in your favor in grease elimination. Building inspectors seldom allow an interceptor to be located anywhere inside the building, but if it happens to be inside, it should be flush with the kitchen floor. Early in the building process, a call to your local plumbing inspector will provide the particulars for your city and probably save you a lot of trouble.

We must also discuss the dry part of the discharge system, which is known as the **venting system**. Its main purpose is to prevent siphoning of water from the traps. Vents (called *black vents*) on both sides of the grease trap equalize the air pressure throughout the drainage system, circulating enough air to reduce pipe corrosion and help remove odors. Vent pipes for kitchens and restrooms extend up and through the roof.

DRAINAGE TERMINOLOGY AND MAINTENANCE

Drainage and vent pipes have specific names. Knowing them will make it easier for you to discuss your discharge system.

Black vent. A vent pipe that connects the venting system to the discharge system. These vents are found near the grease trap, allowing air to enter the trap and preventing contaminated water from flowing out of it.

Building drain. This is the main drain, which receives drainage from all pipes and carries it to the sewer for that particular building.

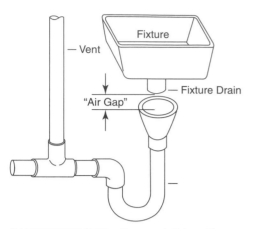

ILLUSTRATION 6-20 Proper sink installation prevents backflow.

Source: Frank D. Borsenik, Maintenance and Engineering for Lodging and Foodservice Facilities (Lansing, Michigan: Educational Institute of AH&MA, 1998).

Building sewer. This pipe carries the drainage beyond the building and into the public sewer.

Drainpipe. Any pipe that carries away discharge from plumbing fixtures. Usually drainpipes are horizontal; if they are vertical, they are called **stacks**. Sometimes drainpipes are referred to as *soil pipes*.

Vent. A pipe that allows airflow to and from the drainage pipes.

The Uniform Plumbing Code contains specifics about choosing sizes of pipes and vents. The general rule is that smaller pipes flow into larger pipes—never the reverse.

The flooded kitchen floor is every restaurant manager's nightmare. Typically, the floor floods because the drains are clogged, and the drains are clogged because everybody thought they were somebody else's job to take care of.

Drainage systems require periodic maintenance to keep them open and working efficiently. Simply because they depend on gravity to work, they occasionally become clogged when debris blocks the natural flow of the system. You can use flexible metal rods, called **augurs** or *snakes*, inserted into pipes to break up the debris, or you can pour in chemicals formulated to dissolve grease and soap buildup. Either way, the clogged material should be removed and not flushed back into the system. There are all types of augurs, including heavy-duty ones with gas-operated motors and 300 feet of line.

DRAINAGE PROBLEMS

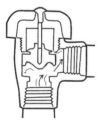

ILLUSTRATION 6-21 A vacuum breaker works to drain excess water from a pipe or line and prevent back siphonage.

The two terms most frequently heard when there's a water backup in the kitchen are **backflow** and *back siphonage*. Backflow results when dirty water (or other unsafe materials) flows into the drinking water supply. Back siphonage occurs when negative pressure builds up and sucks contaminated water into the freshwater supply. Either way, you're in trouble.

Avoiding backflow is not complicated if you've hired an experienced plumber who will plumb your system to include a space (never less than 1 inch) between each pipe and its drain, which prevents contaminated water from flowing back into the water supply (see Illustration 6-20). This space is often referred to as an **indirect waste**. Floor drains that receive condensate from refrigerators are also required to have an indirect waste.

An **atmospheric vacuum breaker** is another smart addition to your water line, especially on any hoses you use in the kitchen to clean the floor or flush out drains. The vacuum breaker (see Illustration 6-21) is a small shutoff valve that allows the water to drain completely after the faucet is turned off and minimizes the chances that fresh water can be contaminated by whatever the hose has touched.

Back siphonage is the term for reversing the normal flow of your water system because of a vacuum (or partial vacuum) in the pipes. This sometimes happens after firefighters use a hydrant in your area or the water system is shut off temporarily for repairs. Like sipping a soft drink through a straw, when water comes back on after being shut off it may create pressure to move it in the opposite direction than is desired. The vacuum breaker is, again, the most popular preventive measure.

Restaurants in areas with winter temperature extremes must also deal with frozen pipes, which create their own set of challenges for your staff. Heed the advice from the Restaurant and Hospitality Association of Indiana in the Building and Grounds section.

BUILDING AND GROUNDS

Frozen Pipes? Yipes!

In the heart of winter, the time when facilities already need extra maintenance—wiping up wet floors and shoveling snow at entrances, for example—don't let frozen pipes become another headache. Follow these prevention tips from maintenance experts.

1. In autumn, shut off valves that supply outdoor water pipes. Because water might be trapped in the pipes, leave outdoor faucets open to prevent pipes from bursting.

2. Set the thermostat at a minimum of 65°F. Higher is fine, but if your establishment is scheduled to close for an extended period, don't turn the heat down too far.

3. Periodically check the water flow in faucets. When the temperature drops to 0°F or below, open wide both the hot- and cold-water taps on all faucets for 30 seconds. Then close them to a point where they drip slightly. This flow should reduce the threat of freezing in the line.

4. Heat hidden pipes. During a cold snap, open cabinet doors that house hidden bathroom and kitchen pipes to let heat in.

5. Raise the temperature to 75°F if pipes freeze. Then open both the hot and cold taps so they can drip slightly and give water a place to go when the pipe begins to thaw.

6. Take care when heating a frozen pipe. You can be electrocuted if you try to use a hair dryer on a pipe near standing water. Never use a butane torch or open flame on a pipe, either.

Source: Newsletter of the Restaurant and Hospitality Association of Indiana, Indianapolis.

6-5 HOT-WATER HEATING

Most hot-water heaters aren't given much attention—until there's no hot water. Because you'll spend more money heating water than you spend on the water itself, you should know the basics of managing this valuable resource.

The average restaurant guest prompts the use of 5 gallons of hot water. This figure decreases in the fast-food arena, of course, where disposable utensils and plasticware are the norm. In a table-service restaurant, however, the 5-gallon figure includes water to wash, dry, and sanitize dishes, glassware, utensils, plus pots and pans and serving pieces.

Water must, by health ordinance, reach certain temperatures for certain foodservice needs. These temperatures are listed in Table 6-5. They correspond to the thermometer chart in Chapter 8's food safety discussion about minimum temperatures for safe handling and storage of food.

A small restaurant may have a single hot-water tank with a temperature of 140°F. Near the dishwasher, a booster heater will be installed to raise the final rinse water to the required 180°F. In larger operations, you may install two or more separate water heating systems for different needs.

TABLE 6-5

Required Water Temperatures for Foodservice Operations

Restrooms	110°F – 120°F
Pot sinks	120°F – 140°F
Dishwashers	140°F – 160°F
Final dish rinse	180°F

To determine what size water heater you will need, multiply the 5-gallons-per-guest average by the maximum number of guests you would serve at a peak mealtime. For example, if 200 guests are likely to be served, multiply by 5 gallons; you will need 1000 gallons of hot water per hour. The other figure you must determine is the maximum amount the water temperature will have to rise to become fully heated. For instance, in the winter, the water may be as cold as 35°F when it enters the building. Your water heater must work hard to get it to 140°F. The temperature rise in this case is 140°F minus 35°F, or 105°F. Manufacturers' charts tell you how much gas or electricity your water heater will consume for different temperature rises.

Even if your water heater is large enough and its output is hot enough, one more variable affects the availability of sufficient hot water: the way it is piped. If the pipes are too small, the water heater doesn't empty and refill fast enough. Also, heat is lost along the way when the water must travel long distances to reach appliances or faucets. Insulate the pipes against this heat loss. If the distance cannot be shortened between the source and appliances, you may need to install recirculation lines and a pump.

TYPES OF WATER HEATERS

The most common type of water heater is the *self-contained storage heater*. It heats and holds water up to 180°F, delivers on demand, and requires no external storage tank. This type of heater comes in sizes ranging from 5 to 100 gallons. At a 100°F temperature rise, the self-contained storage heater can heat 500 gallons per hour. A closely related type of water heater is the *automatic instantaneous heater*, designed to heat water immediately as it is drawn through the tank. Again, no external storage tank is needed. Finally, the *circulating tank water heater* heats the water and then passes it immediately to a separate storage tank using gravity or a pump.

Exciting technological developments in the field of water heating may help restaurants save money. Some companies are experimenting with recovering **waste heat**. This means reusing the heat given off by air conditioners or kitchen appliances (such as big walk-in refrigerators) that is normally wasted by capturing it with a heat pump and using it to heat water.

The **heat pump water heater (HPWH)** should be located wherever such waste heat is available. An air-conditioning system gives off as much as 16,000 Btus per hour of waste heat, so it's easy to recover the 3000 to 5000 Btus necessary to heat water. In hot climates, when the air conditioners run all day, it is not uncommon to be able to heat all the water you need at no cost. Because heat pumps can also capture heat from outside air, they work best in locations where the temperature is more than 50°F year-round. Illustration 6-22 shows how the HPWH works. A survey of eight U.S. quick-service restaurant chains found their HPWHs saved from $851 to almost $4,000 per year in water-heating costs, and the systems paid for themselves in 4 to 20 months. A side benefit of the system that siphons off waste heat is that it makes places like kitchens and laundry rooms more comfortable and easier to cool.

The use of solar energy to heat water has also been introduced in foodservice, and has become much more cost-effective as technology has improved. In a *batch water heater*, solar collectors are built right into the storage tank. They are described as simple and efficient, but in cold climates they must be drained during the winter to prevent freezing. They are probably smarter options for more temperate areas.

A **thermosiphon** is another type of solar water heater. It heats water by circulating it through the solar collectors (called a *direct system*) or by circulating antifreeze in sealed tubes through the water (known as an *indirect system*). Either way, an advantage to this system is that the circulation does not require an additional heat pump.

Tankless water heaters are systems that not only provide an endless supply of hot water but also are more energy efficient than the cylindrical tank-style water heaters. Tankless water heating systems were pioneered in Europe and Asia, and they've caught on in America (despite higher up-front costs) because of increased energy prices. They heat water only on demand. The traditional electric or gas hot-water heater cycles on and off all day, keeping the capacity of the tank at around 120°F, but tankless units heat only the water flowing through them, and only when someone opens the faucet. A sensor detects the demand for hot water, signaling a

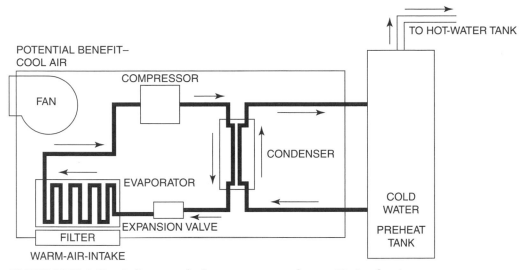

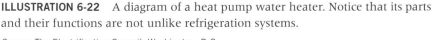

ILLUSTRATION 6-22 A diagram of a heat pump water heater. Notice that its parts and their functions are not unlike refrigeration systems.

Source: The Electrification Council, Washington, D.C.

heating element or heat exchanger, which turns on to a preset temperature based on the flow rate of the water and other parameters. The water flows across the internal heating elements/heat exchanger and exits the unit at the desired temperature, a process that takes only a couple of seconds.

The unit remains on until all the hot-water faucets are closed. As soon as the sensor detects that water has stopped flowing, the power to the unit is turned off completely. Because tankless water heaters have no refresh rate (they are instantaneous), there is no need to overheat water to 130°F or 140°F, as is the case with conventional hot-water heaters. You select an output temperature that matches your actual needs (usually 105°F to 110°F), which also saves considerable energy.

Tankless water heaters are available in both electric- and gas-powered models; your choice depends on a few important factors: installation costs, availability and cost of power sources, water usage history of your operation, and so on.

Gas tankless models are 80% to 85% more efficient than traditional water heaters; electric tankless models are 98% more efficient than traditional water heaters. The electric models tend to cost less up front than the gas models, but gas is a cheaper long-term fuel source than electricity. Best of all, tankless water heaters are estimated to last 20 to 25 years versus half that time for the traditional models, according to the U.S. Department of Energy.

Both gas and electric models can be installed in groups, the gas models with a single manifold, to serve the needs of a large facility. Before you decide to purchase and install a tankless water heater, ensure your facility has sufficient electrical capacity for the additional demand. It may require upgrading your electrical service connections.

■ SUMMARY

This chapter includes a great deal of information about two of the most necessary substances in any commercial foodservice facility: gas and water.

In foodservice, gas energy heats buildings, cooks food, and dries dishes. This chapter explains how a gas burner works and describes the potential problems if it is not kept clean and properly adjusted. There are several types of burners, chosen depending on cooking needs.

You also learned about the working parts and significance of the pilot light on gas appliances. Pilot lights may be manual or automatic, and most can be adjusted easily by hand. Pilot lights also must be kept clean and properly adjusted.

Although many people assume it is necessary to preheat gas equipment, this is a waste of time and fuel. It is also wasteful to cook with large flames. In fact, the gas flame tips on a range burner should barely touch the bottom of your cookware and should not lap the sides.

Steam energy is water vapor that carries a large quantity of heat and can also be used to cook food, heat water, and more. The hotter the steam is, the higher its pressure. The size and capability of the boiler, and the sizes and lengths of pipes through which the steam must pass, all affect the output of the steam system. Water quality is another component in clean, efficient steam output.

Water quality is also critical for drinking, cooking, and dishwashing. This chapter includes a discussion about possible causes of water taste or appearance problems, and why (and how) some foodservice businesses filter their water. It explains the reasons behind water treatment in public systems and mentions recurring water quality hazards and controversies.

Bottled water continues to be a popular option in foodservice and retail despite critics' persistent charges that it is simply treated tap water with a high price tag and despite environmental concerns about plastic waste. The kinds of water—mineral, sparkling, artesian, and so on—are defined in this chapter, with additional information about the Safe Drinking Water Act.

The development of special forms of activated water for cleaning and sanitation is also explained. The chapter includes simple, commonsense ways to conserve water.

The National Uniform Plumbing Code contains guidelines for equipping restrooms and specifies the proper sizes for fixtures, pipes, and vents. Hand-to-food contamination is responsible for nearly 40% of all food-borne illnesses, so providing enough of the correct types of handwashing facilities is critical. Automated handwashing systems are available to track employees' progress in this task.

In addition to several types of sinks for different purposes, no kitchen is complete without a grease trap to prevent sewer blockage by intercepting grease and solids before they enter the sewer system. The grease trap must be cleaned regularly by a reliable company that disposes of the grease correctly, and it is not enough to rely on the company to do so. A business can be fined for putting out too much solid waste. Your locality may have strict enough laws that, to avoid fines, you must pretreat restaurant waste before it even enters the grease trap.

■ STUDY QUESTIONS

1. Briefly describe how a burner on a gas range works. What two components mix to make the fire, and how do you adjust each of these components?

2. What is flashback, and how do you prevent it?

3. List three ways to save energy when cooking with natural gas.

4. How do you decide whether or not to use steam cooking equipment in a commercial kitchen?

5. How do you compensate for hard water in a foodservice setting?

6. Why must you determine the length and diameter of pipe to use when installing steam equipment?

7. What's the difference between a P trap and an S trap? As a foodservice operator, why do you need to know that?

8. List three features to consider when choosing sinks for your kitchen, and explain why they are important.

9. What is a grease trap, and how do you empty one?

10. How do you unclog a kitchen floor drain? Why might it become clogged in the first place?

11. When discussing water pressure, what is an upfeed system?

12. How do you determine the size and type of water heater you need for a foodservice facility?

A CONVERSATION WITH...

Dr. Arvid Osterberg

ACCESSIBLE DESIGN EXPERT AND ARCHITECTURE
PROFESSOR, IOWA STATE UNIVERSITY

Dr. Arvid Osterberg developed and teaches one of the few classes in the nation devoted to inclusive design. He helps students—and architects, building contractors, and inspectors—interpret and implement the Americans with Disabilities Act Standards for Accessible Design (ADASAD).

In addition to being a professor of architecture at Iowa State University, Dr. Osterberg is the author of the manual Access for Everyone: A Guide to the Accessibility of Buildings and Sites with References to 2010 ADASAD. *The latest edition was published in 2010, the same year the latest ADA standards were released.*

Q: How did you wind up specializing in accessibility of buildings?

A: I grew up in the Chicago area and went to the University of Illinois at Champaign/Urbana for my undergraduate and masters degrees. In the first couple years there, I had a colleague who used a wheelchair and was also in the architecture program. Sometimes we had to go from one end of campus to the other between classes, and I was impressed that he got there faster than I did! But when we got to the architecture building, he had to go in through the back door and up a wooden ramp, while I went up the steps and through the front door. Inside, I climbed three flights of stairs, and he used an elevator he had a key for. I saw examples of discrimination against people with disabilities, but also some that could be considered discrimination against the average, normal person. About that time, each region of the country had only one university campus mandated to be accessible, and ours was the one selected.

Even further back in life, though, I grew up at a time when lots of people got polio. I had it when I was a kid, and so did many people I knew. For me, it was like a bad flu and I came out of it without a problem, but I saw others—like my brother, who was paralyzed for a time and came out of it with one leg shorter than the other, and had to wear a leg brace—and I'm sure that made me more aware of living with a disability.

Q: What were some of your first jobs?

A: When I was an undergraduate, I'd get summer jobs in Chicago. My first was for the Greyhound Bus Company. At the time, they were a very big company and had restaurants all over the country at their bus stops, known as Greyhound Post Houses. These were a very big part of the organization, and they had their own division to design them. My first job was to draw up some of these restaurant and kitchen arrangements, and I got familiar with all the stainless-steel equipment. There's a lot of it!

I also studied dining halls at Air Force bases for the Army Corps of Engineers Research Laboratory. One thing we did was set up time-lapse cameras to observe people's behavior in the dining halls, and I sat there and observed, too. I'd eat my lunch and take notes about things like seating patterns. One thing I noticed was that some tables were extremely popular—as soon as someone got up to leave, someone else would sit there.

Q: What were you looking for in the research?

A: Ways to make the designs more efficient and reduce some of the congestion and problems. At the time, there was a lot of interest overall in evaluating facilities post-construction to see how well they actually worked,

and then learn from that to make better designs. Unfortunately, we don't do that enough in the design professions today. It really should be part of the typical design process—for architecture, interior design, and so on. We ought to be evaluating how things work and trying to make improvements as we go along.

Q: You've also done some work in the area of housing accessibility.
A: Yes. After graduation I worked for an architectural firm, but I got interested in research and in teaching, so I went back for a masters degree. There was a lot of funding for housing-related research at the time, and I got particularly interested in how elderly people and folks with disabilities were just not thought of very carefully in terms of design issues. I became interested in finding out how architects could do a better job of designing environments that were more supportive. For example, if there was a "handicapped" apartment—I don't like using that word, but that was what they called it at the time—it was clumsy and unappealing to other renters because it looked institutional. That was when I started developing the attitude of what we now call *inclusive design*—to try to make those kinds of facilities supportive, but not awkward and cumbersome.

I went to a conference on gerontology at the University of Michigan in Ann Arbor because I was interested in designing housing for older people. They had a doctoral program in architecture and also an Institute of Gerontology, so that's where I ended up going for my doctoral work. There were so many interesting projects going on there—I even studied the effects of vision loss on older drivers. But my major focus was on retirement communities. I wanted to take a more progressive approach, to design facilities that would encourage older people to stay active. There's a balancing act between being supportive and making environments too easy, too comfortable, so there's no incentive to go out and walk or do anything.

Q: How did the "Design for All People" course and *Access for Everyone* book come about?
A: I took a position at Iowa State University and figured I'd be here 2 or 3 years—and it's been 35 years! The university started asking my advice about access issues on campus, and that has developed into a good long-term relationship with what is called Facilities Planning and Management. They still send me the plans for buildings that are being built on campus as well as plans for additions and remodelings so I can review them and provide comments.

When the Americans with Disabilities Act came out, the university decided it should do a very thorough assessment of the campus to find out where the problems were and how to correct them. It's hard enough to get it right on new construction and even more challenging to get it right on existing buildings. So, my goal ever since has been to try to improve accessibility on campus, and it's fortunate that it is also part of my teaching and my research.

The course I developed, called "Design for All People," is taught yearly, and it fills up every time. We role-play, and students get in wheelchairs and evaluate different environments around campus. We look at many issues, like vision loss and hearing loss.

When we started dealing with the ADA's accessibility guidelines, I found them difficult to work with—lots of problems in terms of how they were organized and how to access the information. And the *Access for Everyone* guide grew out of that.

Q: When you walk into a foodservice facility, do you automatically think about accessibility issues because of your background?
A: I guess it has become kind of an obsession! It is interesting how you become a better observer all the time, and as time has passed, I've become sensitive to different issues. In the beginning, it was more about having maneuvering space, not having stairs, and being able to reach countertops and equipment—those sorts of things. Now, I think about lighting, the sound, the acoustical environment, the congestion, how hard a space is to navigate and figure out overall. For instance, a lot of current dining areas have these islands or stations that you move around to, to order food or serve yourself. It's great for people in terms of having variety and choice, but it can also be confusing and lead to congestion problems.

And it amazes me how many things are built and designed that don't seem

to have any common sense behind them. The standards the government put out are really important, and they're a good help, and they're fundamentals. But what I'm finding is we need to keep adding to that. We need to have additional recommendations to try to help people think through some of these other issues that aren't very well covered.

Q: In which areas?

A: There's really nothing in the standards to address or accommodate vision loss, hearing loss, sensory loss. The primary concern is still spatial arrangements, having enough space and having things at the right height so people are able to reach them. That's very important, but there's a lot more to it. So, it's an ongoing issue. I think we've made tremendous progress. The environments today are far superior to those built 30 or 40 years ago, but I think we also have a long way to go.

Q: Anything else you notice when you're dining out?

A: Wherever I travel, I love to find the small, independent dining places in any city. They are so much fun to try—although I can't help but notice that a lot of them have serious accessibility problems. The owners have not seemed to think even about the most basic safety aspects of their

buildings—the doorways, the steps to get in and out. I think we need to reach out to small business owners and make them understand that these are important issues—not only for business but also to protect themselves, so they don't get sued when people fall.

Q: Falls are the major problem in the back-of-house areas, too, aren't they?

A: They certainly are. In both public spaces and back areas, I think owners need to be proactive about this and take it seriously instead of trying to deal with it after the fact. They need to assess their facilities from the viewpoint of an older person, or a person with balance problems or vision problems. They need to consider things like if you walk from a bright exterior to an extremely dark interior, your eyes have to adjust, and older people's eyes take longer to adjust. Just one step at the entrance, or a high threshold at the door—these seem like little things, but they can be big problems. And they don't have to be. Oftentimes there are easy solutions.

So, I really applaud the richness and variety of small foodservice businesses—I would hate to see the country end up with nothing but big chain restaurants—but they need to be safe and accessible for everyone, and the owners need to be made aware of these issues.

Q: If you could give advice to students in the foodservice management or design field, what would it be?

A: I think a lot of students get very focused on what their own area of interest is when they could learn a great deal from being exposed to other areas. I do that naturally, being curious about everything, but a lot of undergraduate programs are pretty restrictive. There are so many required courses that you really have to work to reach out and develop your interests and your passions and not just stay strictly on one path.

To me, it is lifelong learning that never really stops. Even as a teacher, I've sat in on other professors' courses, and I learn from my students all the time from the projects that we do. So, I would encourage students to reach out for hands-on experience and to get summer jobs and part-time jobs. If they're interested in foodservice, they should try working in all facets of foodservice. In school, take an interior design course; or try a sociology course for insights about human behavior. There are opportunities like this on every campus if you have an open-minded attitude. Don't limit yourself to the prescribed program. Take advantage of the resources around you.

7 DESIGN AND ENVIRONMENT

■ INTRODUCTION AND LEARNING OBJECTIVES

The *environment* of a foodservice establishment is, for the person eating there, the characteristics that make the person feel welcome, comfortable, and secure. For the employee, environment means a safe, comfortable, and productive work setting. And today's foodservice business owner has yet another critical consideration: the environment of the planet. More attention than ever is being paid to energy conservation, waste reduction, and pollution control in our industry. In some cases, businesses that do not heed these important trends face legal penalties or even closure.

Luckily, with forethought and expert advice, every component of the dining environment can be designed with conservation as well as safety and atmosphere in mind.

After reading this chapter, you will be able to:

- ■ Identify the fundamentals of proper lighting.
- ■ Select appropriate colors for foodservice spaces.
- ■ Identify the fundamentals of noise and sound control.
- ■ Identify the fundamentals of temperature and humidity control.
- ■ Identify the fundamentals of managing heating and air-conditioning systems.
- ■ Identify the fundamentals of maintaining ventilation and indoor air quality.

7-1 LIGHTING

In a commercial building, lighting represents up to 40% of its energy usage. In foodservice, lighting is also the single most important environmental consideration, although lighting rules are probably the most difficult to define. When lighting works properly and does what it was designed to do, most of us don't think much about it; when it doesn't, we definitely notice. Correct lighting enhances the mood of a dining area, the appeal of the food, and the efficiency of a kitchen. And yet, in each of these situations, "correct" means an entirely different thing.

First, consider that no one actually sees light. Instead, what we call *light* is a reflection off an object. Its intensity can be measured (in foot-candles, which you'll learn more about in a moment)—and yet, the human eye cannot see a foot-candle. What the eye can sense are differences in brightness, color, and other characteristics. Try striking two matches, one in darkness and one in daylight. Their appearance is very different, and it is not because the two matches were struck any differently. It's because our perception of them differs with the surrounding conditions.

Seen in this light (pun intended), it is a complex sensory challenge to illuminate a dining area. Lighting is an integral part of creating ambience, with artsy fixtures and new technology that allow smaller fixtures to do bigger lighting jobs. A lighting design consultant should be on your planning team if the budget allows. If not, be sure to choose an interior designer with extensive experience in lighting. This person will be determining not only the placement and appearance of *light fixtures* but also their intensity, their direction, and the contrast of light levels in different parts of the dining area.

The lighting environment should match the type of facility. Using restaurants as our examples, a quick-service place is brightly lit to help move guests through the ordering and pickup process and to discourage lingering (sometimes given the unflattering term *camping* by restaurant folk). Big windows take advantage of daylight to lower utility bills and give customers the impression of a quick, casual meal, almost like eating on a patio. This type of restaurant typically spends from 12% to 15% of its annual budget on lighting.

An intimate eatery, however, requires subdued light levels to encourage a leisurely or romantic dining experience. An establishment that features fine artwork on the walls will want to play up its collection by spotlighting it. In most dining situations, experts try to think of each table as a single space and design the lighting accordingly. This results in more sources of light, but at lower levels. Light also can be used to cast shadows, adding texture to an otherwise dull or underutilized space.

Restaurants that are multipurpose—serving business lunches during the day and intimate dinners in the evening—should install dimmers on the lighting system to enable a change of light levels to reflect different moods. Dimmers also come in handy after closing time because they allow you to crank up the lights for thorough cleaning. Dining areas that also serve as meeting rooms have special needs: spotlight capability for speakers at podiums; perhaps track lighting with multiple circuits so the lights can be moved as needed for panel discussions, note-taking, use of chalkboards, and so on. All fixtures that could interfere with the use of a projection screen (for slides or videos) should be dimmer controlled.

The Daniel Boulud Brasserie in Las Vegas's Wynn Resort Hotel (see Illustration 7-1) shows how lighting design can be both beautiful and useful. The ceiling serves not only a decorative function but also as the source of downlights directed at individual tables. Wall sconces also add decorative touches.

Even within a dining space, light levels vary. An entryway, waiting room, cocktail lounge, and dining room each require separate treatment. This may require *lighting transition zones*, the technical term for shifting people comfortably between two types of lighting, giving their eyesight a moment to adjust to the change. A lighting transition zone is necessary in a lobby, for example, when guests come from bright sunlight into a darker indoor space.

Another common lighting design term is *sparkle*, which refers to the pleasant glittering effect, not unlike a lit candle, of a lighting fixture. Sparkle is a must in creating a leisurely or elegant dining atmosphere.

Light sources and levels of intensity should cause both the guest and the food to look good. Although every design expert has a theory about how to accomplish this ideal, all agree that the use of incandescent light is the best way to achieve it. First, however, you must control any incoming natural light, which can affect utility bills as well as diner comfort. If you have ever sat at a luncheon with sunlight streaming through the window right into your eyes or strained to read something because of glare, you know the dilemma. In new structures, architects can locate windows to avoid the harshest daylight or even take advantage of daylight if it benefits a particular concept. Other types of external lighting also might be intrusive, from parking lots or exterior streetlights. Many problems can be remedied with window blinds, shades, curtains, or awnings. These should be carefully chosen to blend with the mood of the facility and for ease of cleaning.

The most common ways to use light are indirect and direct.

Indirect lighting washes a space with light instead of aiming the light at a specific spot. Indirect lighting minimizes shadows and is considered flattering in most cases. Wall sconces are one example of indirect lighting. Often the light fixture itself is concealed.

Direct lighting aims a certain light at a certain place to accent an area such as a tabletop. When that area is directly beneath the light fixture, it is referred to as *downlighting*. A chandelier is an example of direct lighting.

ILLUSTRATION 7-1   The lighting for Daniel Boulud Brasserie is an excellent example of the use of both direct and indirect lighting.

Courtesy of Daniel Boulud Brasserie, Wynn Las Vegas Resort Hotel, photo by Eric Laignel.

One lighting caution that we add from experience: If yours is a truly dark dining room, be prepared for the occasional complaint that it is "too dark to read the menu," usually from older diners. They are not just being ornery—the average 60-year-old receives only one-third as much light into his or her eye as the typical 16-year-old. Equip the waitstaff with a few pocket-size flashlights, and they'll be able to offer immediate enlightenment, so to speak, in these cases.

MEASURING LIGHT

Now we'll introduce you to some of the terms you will use when making lighting decisions.

Lumen is a measurement of light output; it is short for *luminous flux*. One lumen is the amount of light generated when 1 foot-candle of light shines from a single uniform source.

Illumination (also sometimes called *illuminance*) is the effect achieved as light strikes a surface. We measure illumination in *foot-candles*; 1 foot-candle is the light level of 1 lumen on 1 square foot of space. In the metric system, foot-candles are measured in terms of *lux*, or 1 lumen per square meter. A lux is about ¹⁄₁₀ foot-candle.

All light sources except natural light are classified in terms of their *efficiency*. Efficiency is the percentage of light that leaves the fixture instead of being absorbed by it, and efficiency is measured in *lumens per watt*. You may also hear the term *efficacy*, which refers to how well a lighting system converts electricity into light. Efficacy is a ratio of light output to power output and is also measured in lumens per watt.

The *color rendering index (CRI)* is yet another measurement of light. The CRI, on a numerical scale of 0 to 100, indicates the effect of a light source on the color appearance of objects. It isn't as confusing as it sounds. The index measures the "naturalness" of artificial light compared to actual sunlight. A higher CRI (from 75 to 100) means a better color rendering. Simply put, things look more true to their natural color when the light on them is strong enough to get a good look at them. So a CRI of 75 to 100 is considered excellent, while 60 to 75 is good and below 50 is poor. Incandescent and halogen lamps (bulbs) have the highest CRI ratings; clear mercury lights usually have the lowest CRI ratings.

You can't discuss lighting without deciding on the color of the light. Most of us are familiar with warm tones and cool tones, although we don't think much about why different types of lighting appear to be different temperatures. Interestingly, you may assume warm, reddish tones are the hottest colors on the spectrum, but scientists rank **color temperatures** from low (red) to orange to yellow and, finally, to bluish white, which is the highest or hottest color temperature. Whether a light source appears warm or cool is its *correlated color temperature (CCT)*, or **chromaticity**. Color temperatures, CCTs, and chromaticity are all measured in **kelvins (K)**. Kelvins are named for Baron William Thomson Kelvin, a British physicist and mathematician in the 1800s who developed this color temperature degree scale based on the Celsius temperature scale—the lower the degrees on the Kelvin scale, the warmer and cozier the effect of the light. In hot climates, designers try to light with high Kelvin output to make the environment seem cooler.

Light loss factors take away from a light source's potential. (The word *lamp*, in professional lighting terms, refers to what consumers usually call a *bulb* or *light bulb*.)

Lamp lumen depreciation is the gradual reduction in a bulb's light output as the bulb ages, losing some of its filament or phosphorus.

Lumen dirt depreciation is the reduction of light output due to the accumulation of dirt, dust, or grease on the bulb.

The **ballast factor** explains the lamp's output using any particular ballast. As you'll learn in a moment, some types of lamps require a separate piece of equipment called a **ballast** to be able to work, and some ballasts are more effective than others at allowing the lamp to emit its maximum light capability. The ballast factor is the ratio between the light output using a specific ballast and the maximum potential output of the light under perfect conditions. Most ballast factors are less than 1.

Finally, the **power factor (PF)** is a measurement of how efficiently a device uses power. A lamp that converts all the power supplied to it into watts without wasting any in the process has a power factor of 1. Often lamps that require ballasts have a PF less than 1 (e.g., 0.60 or 0.90) because some of the electric current is used to create a magnetic field within the ballast, not to produce light.

Lighting devices are referred to as *high power factor (HPF)* or *low power factor (LPF)*. This is important because sometimes utility companies penalize customers when their electric loads have a low PF—that is, if you're using too much electricity to generate the amount of light or heat that you need. Utilities may impose standards on the types of lamps you can buy to qualify for energy-saving discounts or rebates, requiring that they be rated HPF (see Illustration 7-2).

These factors are so critical that lighting experts generally recommend you use only one lamp manufacturer for reorders instead of buying from several sources and getting mixed results.

Kitchen lighting is another animal altogether. For employees' safety and comfort, and to allow them proper attention to detail as they work, kitchen lighting must be bright and long-lasting, and it should be selected to give off the least possible amount of heat. The most popular all-purpose kitchen lighting is still fluorescent lamps—inexpensive, durable, and bright. A new generation of smaller, more efficient fluorescents, known as T-5s, can be paired with reflectors for better use of fewer lumens per watt than the popular T-8 fluorescents.

The trick in lighting a kitchen is to make things bright without causing glare (and the resulting eyestrain). Recommendations from the Illuminating Engineers Society of North America (IESNA) include matte or brushed finishes on countertops and careful lighting around highly reflective surfaces, such as mirrors and glazed walls. In addition, kitchen light fixtures and lamps must be able to withstand the rigors of professional cooking and cleaning. Lamps that can withstand humidity for areas like dish rooms and walk-in refrigerators are called *damp-labeled* luminaries. Both the U.S. Food and Drug Administration and Occupational Safety and Health Administration (OSHA) now require shatterproof lamps. This so-called **protective lighting** usually includes a coating to prevent bits of glass or chemical from flying out in case of breakage.

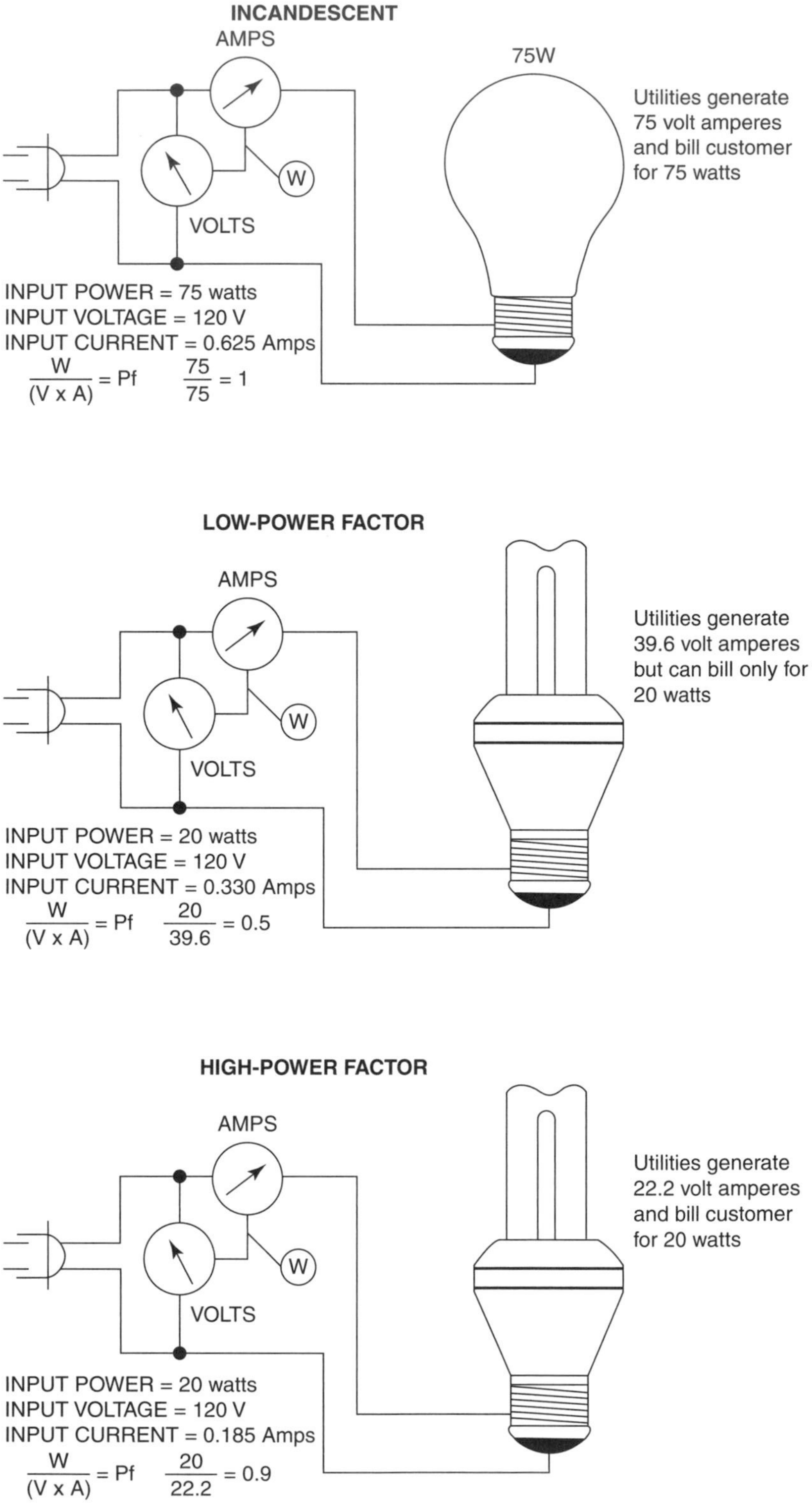

ILLUSTRATION 7-2 Utility companies prefer HPF products to LPF products. These diagrams show how power is used in each type of lamp.

INCANDESCENT
AMPS

75W

VOLTS

Utilities generate 75 volt amperes and bill customer for 75 watts

INPUT POWER = 75 watts
INPUT VOLTAGE = 120 V
INPUT CURRENT = 0.625 Amps
$$\frac{W}{(V \times A)} = Pf \quad \frac{75}{75} = 1$$

LOW-POWER FACTOR
AMPS

VOLTS

Utilities generate 39.6 volt amperes but can bill only for 20 watts

INPUT POWER = 20 watts
INPUT VOLTAGE = 120 V
INPUT CURRENT = 0.330 Amps
$$\frac{W}{(V \times A)} = Pf \quad \frac{20}{39.6} = 0.5$$

HIGH-POWER FACTOR
AMPS

VOLTS

Utilities generate 22.2 volt amperes and bill customer for 20 watts

INPUT POWER = 20 watts
INPUT VOLTAGE = 120 V
INPUT CURRENT = 0.185 Amps
$$\frac{W}{(V \times A)} = Pf \quad \frac{20}{22.2} = 0.9$$

LIGHTING TECHNOLOGY

The foodservice industry is fortunate that in recent years, lighting technology has focused on energy conservation and cost savings, making lighting more affordable and adaptable. The three basic types of artificial lighting are **light-emitting diodes** (commonly known as LEDs), **incandescent lamps**, and **electric discharge lamps.** The fluorescent and compact fluorescent lamps (CFL) common in foodservice use are types of electric discharge lamps; more about them in a moment.

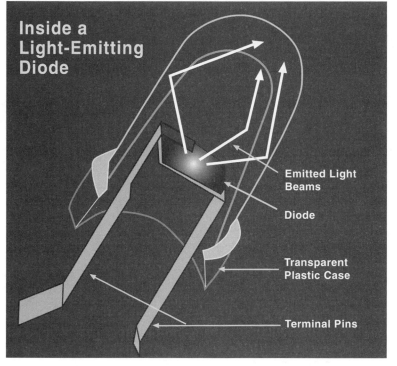

ILLUSTRATION 7-3 Diagram of a light-emitting diode, or LED.

LEDs are revolutionizing the industry. First used in the 1990s to illuminate appliances, remote control devices, and digital alarm clocks, they were soon adapted for traffic lights and signage and then for home and office lighting. They also are now used in the brake-light systems of most automobiles.

An individual LED looks like a tiny light bulb that fits into an electrical circuit, but, in fact, it operates much differently. Its inner workings, shown in Illustration 7-3, include a simple semiconductor, a microchip called a *diode*. In most applications, diodes don't give off much light at all. But a light-emitting diode is specially constructed to release photons in large numbers, and the bulb that houses the diode concentrates or aims the light in a certain direction. Illustration 7-4 shows what a modern LED bulb looks like. We think of LED lights as being bright green or amber, the most common colors used in clocks and appliances, but by grouping them or using different chemical compositions in the diodes, LEDs can produce other colors as well.

The chief advantages of LEDs are substantial. They put out brighter light than incandescent bulbs, but they consume up to 80% less energy and generate very little heat. They can be safely used indoors or outdoors, as they don't require high-voltage electricity. They last up to 10 times as long as a regular lamp, and, when they're ending their useful life, they gradually get dimmer instead of burning out with an unexpected pop. And, unlike fluorescent lights, LEDs do not contain mercury.

The downside of LED use is that it is still more expensive than traditional lighting methods, yet it lacks their overall finesse—they're bright but not warm. Like so many other types of technology, the price of LEDs has come down substantially since they were introduced. In commercial settings, LEDs are likely to be found in displays, some architectural and landscaping uses, where lighting quality doesn't matter as much as brightness.

ILLUSTRATION 7-4 A modern LED bulb.

Courtesy of iStockphoto LP.

What most of us think of as a traditional light bulb is actually an *incandescent lamp*. It is a filament encased in a sealed glass bulb. Electricity flows into the filament through the base of the bulb; the glass is usually coated to diffuse the light. While they are inexpensive, most light bulbs have a relatively short life (1000 hours), which means they have to be changed fairly often in a commercial environment. Long-life bulbs, even up to 20,000 hours, are available, but we can't attest to the accuracy of that claim. If incandescent bulbs are used at a higher voltage than originally intended, their lives are shorter. (The suggested voltage is usually stamped on top of the bulb.) They also have rather poor efficiency (only 15 to 20 lumens per watt), which means they give off heat.

The typical foodservice operation uses incandescents where the particular glow of this type of light is necessary to create a mood or if dimmers are used to vary the light levels. One option for commercial use is to try 130-volt bulbs rather than 120-volt. They produce a little less light but last longer in the normal (120-volt) light socket.

There are several variations of the incandescent lamp—including spotlights, floodlights, and elliptical reflector (ER) lamps—all of which may be endangered species by the time you read this. See the text box, "A Cloudy Future for Incandescence," for the details.

The *electric discharge lamp* is one that generates light by passing an electric arc through a space filled with a special mixture of gases. That's why you will sometimes hear them called *gaseous discharge lamps*. In this category, you find these types of lamps:

FOODSERVICE EQUIPMENT

A Cloudy Future for Incandescence

Conservation groups worldwide have seized on the energy inefficiency of incandescent lighting in recent years and, in 2007, Australia and Canada became the first two nations to announce phase-out plans for incandescent lighting by 2012.

In the United States, Congress has been trying to regulate the energy efficiency of light bulbs for years. In 1992, it passed the National Energy Policy Act (EPACT), setting minimum performance standards for some types of light bulbs and fluorescent tubes—and sending the lighting industry into a panic by giving manufacturers just 12 months to make the changes.

In 2007, Congress passed the Energy Independence and Security Act under the premise that the nation needs to save energy in order to be less dependent on foreign oil. However, some conservative lawmakers tripped up enforcement of the 2007 legislation, claiming the government is infringing on individuals' rights to make household consumer decisions. Late in 2011, they attached another bill onto federal budget legislation, prohibiting the U.S. Department of Energy from enforcing the efficiency standards through September 2012.

What did go into effect are the law's new packaging requirements. Light bulb packaging is now labeled, front and back, with information about the bulb's brightness, expected lifespan, and annual cost to operate.

Manufacturers are seeing the handwriting on the wall, even if Congress continues stalling. The law says new incandescent bulbs must be at least 25% more energy-efficient than their traditional counterparts. Meeting the new efficiency standard means, for instance, that a 100-watt bulb could use no more than 72 watts to accomplish its level of brightness (in this case, 1,600 lumens).

The U.S. standards don't apply to some specialty lighting types: colored bulbs; the smaller bulbs used inside home refrigerators and ovens; impact-resistant lights for medical, stage, and studio purposes; and even three-way light bulbs.

The original law set the phase-out of traditional incandescent bulbs in this order: 100-watt bulbs (January 1, 2012); 75-watt bulbs (January 1, 2013); 40- and 60-watt bulbs (January 1, 2014). It allows retailers to sell off their excess inventory as long as the bulbs were not imported or manufactured after those deadlines.

The controversy may continue for years about the incandescent bulb created by Thomas Edison more than 130 years ago. Just as the giant Swedish home products retailer IKEA announced it will sell only energy-efficient LED lighting by 2016, lighting manufacturer Osram Sylvania released a consumer survey saying one in eight Americans intend to stockpile 100-watt incandescent bulbs before they go off the market entirely because they simply prefer them to the newer alternatives.

- Fluorescent
- Mercury vapor
- Halide or halogen
- High- and low-pressure sodium

Unlike the incandescent lamp, the electric discharge lamp cannot operate directly by threading or screwing it into a fixture. An additional piece of equipment, called a *ballast*, is required. The ballast is a current-limiting device that acts as a starting mechanism. It is generally mounted on top of the fluorescent light fixture and is replaceable. (Replacement costs are lower for separate ballasts than those permanently attached to the fixture.)

The two basic types of ballasts are the magnetic ballast and the electronic ballast. Magnetic ballasts are the older technology; they cannot be used with dimmers. An energy-efficient magnetic ballast is available; manufacturers say it saves 20% over the regular ballast.

Electronic ballasts cost more than magnetic ones, but the additional expense buys a ballast that is more energy efficient and has less of the humming and flickering sometimes associated with fluorescent lighting. Electronic ballasts can also be used with dimmers.

Multiple types of electronic ballasts complicate the choices. The *rapid-start ballast* can operate up to four light fixtures at a time, with parallel wiring so if one burns out, the other three continue to shine. Rapid-starts may take a few seconds to fully start. The *instant-start ballast* uses less energy than the rapid-start and starts up immediately, but it may reduce the life of the fluorescent bulbs. Manufacturers of the newest entry on the market, the *program-start ballast*, say it saves energy and extends bulb life.

When electronic ballasts were first introduced, their life spans were very short, but recent improvements have increased them. However, buyers should still specify a warranty that covers replacement materials and labor for up to 3 years.

When comparing ballasts, look for those rated with the letters *P* (which indicates it is thermally protected and self-resetting) and *A* (which makes the least humming noise).

Both magnetic and electronic ballasts may generate **harmonics**, a distortion of power frequency that can create electromagnetic interference on power circuits. This wreaks havoc with sensitive electronic equipment, such as computerized cash registers or order-taking systems. Ballasts even have **total harmonic distortion (THD) ratings**—the lower the percentage of THD, the less likely it is to distort the power line. Low THD is less than 32%, which is the recommendation of the American National Standards Institute. Therefore, high power factors and low THDs are the rule for whatever type of ballast you buy.

An **integral ballast lamp** is a lamp and ballast in a single unit. It may be magnetic or electronic and has a screw-in base that allows it to fit directly into almost any standard light fixture. Combining lamp and ballast makes for easy retrofitting. Replacement costs are higher, however, because the whole unit must be replaced. You can also purchase adapters that contain a ballast. These last through the lives of three or four lamps and are generally more economical than the integral ballast lamp.

Fluorescent lamps are the most common form of electric discharge lamp used in the hospitality industry. The fluorescent bulb converts more of the energy it uses to light than an incandescent bulb, making it a more energy-efficient choice. The fluorescent tube lasts from 7,000 to 20,000 hours, with an efficiency of 40 to 80 lumens per watt. Most fluorescent tube-shaped bulbs come in three standard sizes: 24-inch, 48-inch, and 64-inch.

All of the several types of fluorescent bulbs come in a variety of colors and tones. Most people are familiar with straight, tube-shaped bulbs, although they also come in U shapes. The numbers indicate the diameter of the tube, measured in eighths of an inch. The slimmer the tube, the less energy it requires.

- The T-12 (1½-inch diameter) is the least efficient (but also least expensive) fluorescent bulb, although it does have an energy-efficient T-12 cousin that uses 15% less energy— and emits 15% less light. T-12s require a magnetic ballast.
- The T-8 bulb (1-inch diameter) is an upgrade from the T-12. It uses half as much energy as the T-12 and eliminates that annoying flicker sometimes associated with fluorescents.

It is also easy to retrofit a T-12 system to T-8s. You may find government or utility company incentives for doing so.

▪ A high-performance T-8 system may be referred to as a "premium" or "super" T-8. It includes bulbs and ballasts that promise additional energy savings of 10% to 20% over the regular T-8 system.

▪ The T-5 (5/8-inch diameter) is the next-generation fluorescent. It produces the same amount of light as a T-8 bulb but is smaller and uses less energy. T-5 and T-8 systems require electronic ballasts.

Compact fluorescent lamps (CFLs) feature smaller-diameter tubes bent into twin tubes, quad tubes, or even circular shapes in many sizes. The smaller diameter of the CFL tube makes it economical to manufacture them with higher-quality phosphorus, which improves light output and makes the color seem more natural.

Lighting designers swear that technology has greatly improved the light quality of CFLs—that often, you wouldn't know you were seeing fluorescent light unless you actually looked at the bulb. However, they have been a tough sell in the American market for their initial cost, harsher light quality, and unusual appearance. They also simply don't fit into some standard fixtures that were designed for incandescent bulbs.

CFLs typically are three to four times more efficient than regular light bulbs. They are available in several color temperatures (from a warm 2700 kelvins to a cool 4100 kelvins) to achieve different effects. In dining areas, lighting experts recommend the warm white bulbs.

If you're not ready to make the change, you can still realize substantial savings by using CFLs in hallways, offices, walk-in coolers, and exhaust hoods. They don't emit heat, which saves on air-conditioning costs, and they don't have to be changed as often as other types of bulbs.

On the downside, CFLs generally cannot be used outdoors or in any situation where they might get wet. They work best in mild temperatures, not in cold weather, and their lives are shorter if you use them in recessed or enclosed fixtures. Their lives are also shorter if you turn them on and off a lot. They don't dim except with special dimmable ballasts (they are not unsafe when used in regular light fixtures; they just don't dim).

Finally, fluorescent lamps in general pose their own environmental problems: They contain small amounts of mercury and must be disposed of as hazardous waste. T-8 tubes contain less mercury than T-12s.

High-wattage compact fluorescents (HW-CFLs) probably don't have many foodservice uses, but we'll mention them. They are made to replace the traditional bulbs in super-bright industrial, high-ceiling spaces like warehouses.

Another breakthrough in lighting technology is the ***E-lamp***, which uses a high-frequency radio signal instead of a filament to produce light. Inside the sealed globe, the rapidly oscillating radio waves excite a gas mixture that, in turn, gives off light. The light hits a phosphorous coating on the inside of the globe and glows.

The E-lamp lasts even longer than the CFL—at least 20,000 hours—because it does not contain an electrode that can burn out. However, the phosphorous coating wears out, and the E-lamp gradually dims over time. E-lamps fit most fixtures, will operate whether they're installed upside down or sideways, and are not susceptible to cold climates like some CFLs.

For consumers, the push for manufacturers to improve age-old lighting technology has been beneficial. Now that lighting undergoes performance testing, some of today's bulbs, coated inside with a substance called rare earth (RE) triphosphor, achieve a CRI of 70 to 89, compared to a CRI of only 50 to 65 for the older-style halophosphorous lamps. Rare earth lamps produce 5% to 8% more light with no increase in energy consumption, and they are a popular choice in retrofitting older fluorescent lighting systems. However, they cost at least twice as much as conventional lamps.

Some interesting combinations have also been created, such as the *compact halogen lamp*. This is actually an incandescent bulb with halogen gas inside, which prolongs the life of the filament. Compact halogen lamps give off brighter, whiter light than regular incandescents and last longer too. They are used in downlighting, accent lighting, and retail display. Many restaurants are beautifully lit with MR16 halogen spotlights. An alternative that saves even more energy (and money) is the MR16-IR; the *IR* stands for "infrared restricted." A 50-watt MR16-IR can replace a 75-watt MR16. Again, initial costs are higher, and you might assume 10 or 15 watts here or there isn't a big enough difference to bother with—but think about how many bulbs you use overall. The savings due to long life and lower energy use pay off over time.

Mercury vapor lamps are also known as *high-intensity discharge (HID) lamps*. They are used for lighting streets and parking lots, and their intensity varies from 15 to 25 lumens per square foot. The lifetime of these lamps is between 12,000 and 20,000 hours. Mercury vapor lamps are made clear or white, of which the white ones have better color rendition.

HID lamps are considered hazardous waste because of their mercury content, however, and must be disposed of accordingly. The DOE estimates HID lamps are responsible for about 17% of total lighting energy consumption in the United States.

In fact, the **metal halide lamp** is a variation of the mercury vapor lamp, with metallic halide gases added to improve color rendition and efficiency. These lamps have an efficiency level of 80 to 100 lumens per watt and lifetimes of 7,500 to 15,000 hours.

Sodium pressure lamps are highly efficient light sources used primarily in parking garages, driveways, hallways, and anywhere lighting is used as a security measure. Their color rendition is poor, but these bulbs are very efficient, with 85 to 140 lumens per watt and a life of 16,000 to 24,000 hours. There are high-pressure sodium (HPS) and low-pressure sodium (LPS) types.

No matter what type of bulb you use, you must also make a choice of light fixtures, or *luminaries,* to hold the bulbs. A fixture is a base, with wires that connect to an electric power source and a reflective surface that directs the light to whatever surface you want to illuminate. A fixture can be highly visible and decorative, such as a chandelier or stained-glass tabletop lamp, or recessed into a wall or ceiling where it is virtually invisible. For recessed fixtures, you can purchase a special bulb called an **elliptical reflector (ER) lamp**. This is an incandescent bulb with a unique shape that focuses its light outside the lamp envelope. None of the light is wasted by heating up the inside of the fixture. Using ER lamps saves 50% of the energy you'd use with regular incandescent bulbs, but only if they are used in recessed fixtures.

For electrical discharge lamps, the fixture also holds the ballast in place. Before you buy fixtures, think about how easy (or hard) they will be to clean and replace. Also, consider the costs of replacement items such as decorative globes or safety covers.

If you are retrofitting an existing lighting system, professional advice from a lighting consultant is not a waste of money. Although the price tags for state-of-the-art fixtures, ballasts, and lamps may be so high you immediately assume you can't fit them into your budget, after the initial expense they could save you more than 40% on lighting costs month after month and replace old fixtures that were just about worn out anyway. A consultant can help you evaluate what you've got and decide if the improvements are worth the money.

CONTROLLING LIGHT LEVELS

How about the use of natural light? One interesting way to light your dining area is to install **solar tubes** on your roof (see Illustration 7-5). Each aluminum tube is a mere 13 inches in diameter (so small it doesn't disturb the rafters or ducts between roof and ceiling) but can light a 10-square-foot indoor area, providing the equivalent of 1500 watts. One end of the tube is covered with a clear acrylic dome and peeks out several inches above the roofline, facing the best direction (usually southwest) for maximum sunlight exposure. The other end opens as a hole in the dining room ceiling. The tube is sealed with special flashing to prevent water leaks. Because aluminum is highly reflective, the tube can generate light even on cloudy

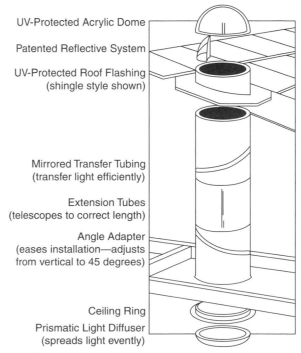

UV-Protected Acrylic Dome
Patented Reflective System
UV-Protected Roof Flashing (shingle style shown)
Mirrored Transfer Tubing (transfer light efficiently)
Extension Tubes (telescopes to correct length)
Angle Adapter (eases installation—adjusts from vertical to 45 degrees)
Ceiling Ring
Prismatic Light Diffuser (spreads light evenly)

ILLUSTRATION 7-5 Only a few inches of the solar tube's total length can be seen above the roofline. *Source:* Solatube, Carlsbad, California.

days. Inside the building, the opening is sealed by a frosted globe that diffuses the light and looks like just another light fixture attached to the ceiling.

Reflection is one good way to make light go farther. A ***reflector*** is a shiny, custom-fit surface that mirrors the light that bounces off it. Reflectors fit between the bulb and fixture. Because most are shaped like an inverted bowl, they also help aim the light, reducing glare and focusing the light downward. On the downside, walls can sometimes appear darker, and it's harder to make illumination seem uniform throughout a room.

Reflectors cannot be used in most recessed light fixtures but are often recommended as a retrofitting technique for other types of lamps because of their potential energy savings. By installing a reflector, you can remove up to half of the bulbs in a fixture, saving half on your lighting costs and a little on your heating and cooling costs too. For example, a four-bulb lamp operated with two bulbs and a reflector still has 60% to 75% of the light output it would have with all four bulbs burning. Also, for spaces that are too dark, use of a reflector with a full contingent of bulbs can actually boost existing light levels by 15% or more.

Most fluorescent fixtures, to minimize glare, use either a ***lens*** or a ***louver*** to cover the bulb itself and prevent direct viewing of it at most angles. In fact, there's a standard ***glare zone***—at just about any angle of more than 45° from the fixture's vertical axis, most naked bulb lighting is uncomfortable to look at and causes reflections on work surfaces and computer screens. A sturdy, clear plastic lens helps diffuse glare and makes the light more uniform.

Louvers do the same thing in a slightly different way. Instead of covering the bulb, they are like small, partially open doors that aim the light to eliminate glare at certain angles. Louvers are more effective than lenses at reducing glare, but because they shade some of the light, they allow less light output overall. If louvers are not properly adjusted, people may complain that ceilings look dark.

An ***occupancy sensor*** can be used in a conference room, storage area, restroom, or walk-in refrigerator. It detects motion and activates a controller device that turns on light fixtures. Then, if no motion is detected within a specific period, the lights turn off until motion is sensed again. Compared to manual on–off switches, occupancy sensors can reduce the use of lights by 15% to 30%. Most sensors allow for adjustments of motion sensitivity, periods the lights stay on or off, and a bypass switch that overrides the system when necessary. They are also handy at dark back-of-the-building loading docks and entrances.

Occupancy sensor costs often include installation of mounting devices in walls or ceilings and wiring to hook them up to the building's power supply. In simple systems, regular light switches are merely replaced by sensors.

These types of lighting control systems can help save energy, but only if you think them through carefully before installation. Many expensive systems have been disconnected because they caused too much inconvenience, or because a system wired to alert a security company of possible intruders resulted in false alarms. You can also control lighting with *timing switches*, light-sensitive ***photocells***, and programmable ***sweep systems***. An inexpensive timer can be used to turn lights on and off at the same times every day. They work well with lights that have predictable operation periods—for example, outdoor signage, corridors, and parking lot flood-lights. Similarly, outdoor lighting can be hooked to a photocell that turns the light on when darkness falls and off again when the sun rises. The only problem with photocells is that on dark, rainy days, the outdoor lights may come on unnecessarily.

If your lighting needs change from day to day, a more complex programmable system is better. Sweep systems can be set up to turn lights on and off sequentially in one area, on a whole floor of a building, or in the entire building.

Dimmers allow light output to be adjusted for the needs of a particular space, either manually or by sensors that detect light levels, daylight, or occupancy. Precise adjustment of lights is some-times called ***tuning***. Most dimmers are installed during new construction, but they can also be successfully retrofitted. Daylight sensors use photocells to detect light levels and dim fixtures auto-matically near windows or skylights. In a kitchen, a dimmer can be used to adjust light in work surface areas; in a dining room, lights can be adjusted to full output to assist the cleaning staff.

Perhaps the biggest problem with dimmers is they still require incandescent bulbs. As mentioned earlier in the chapter, CFL lamps simply don't dim—they're either on or off.

7-2 THE USE OF COLOR

In the dining room, the lighting system and color scheme must work together to enhance the environment. Colors depend on their light source because, as you know, the same color can look completely different when seen under different types or intensities of light. Color and light together can also be used to make space seem large or smaller and more intimate.

Color can be used to convey a theme, a style, a geographic region, a way of life, or even a climate. The next time you're out studying your competitors, look at the colors of their walls and try to decide what that color is saying. Color psychology is the study of color's impact on people's moods and buying habits, based on the principle that certain colors evoke certain feelings. Warm colors—orange and yellow tones, such as peach and terra cotta—are considered inviting; white denotes cleanliness and purity; black is sophisticated and dramatic; and so on. Some basic color rules follow.

- Light tones and cool colors make small spaces look larger.
- Darker tones and warm colors shrink rooms, creating a greater sense of intimacy in a large space.
- Dark colors can be used to mask structural features, such as ductwork, that may otherwise interfere with the design of a room.
- Bright, primary colors can be combined in contrasting combinations for a super-modern feel. Think bright blue or lime green and white, or black and red.
- Muted colors have a soothing effect on a room.
- If more than one color is used for walls, one must be dominant and the other subordinate (an accent).

Don't just consider walls when you think about color. Tabletops, chairs, and any other surfaces can be the sources of important accents to the color palette. Authors and scholars Baraban and Durocher assert there are *color cycles* in the restaurant industry—popularity trends that seem to last about 8 years at a time. It would be smart to ask your decorating and design team what cycle you're in. As you make lighting decisions, remember they will serve you (and your guests) in several ways: visually, psychologically, thermally, and socially as well as aesthetically. The food should look appetizing and wholesome; the dining area must look pleasant and appropriate to the image, price range, and type of cuisine.

USE OF ARTWORK

Making a statement with your décor by using murals and other forms of art is an idea worth exploring. The walls aren't going to be bare, are they? What you choose to put there will add personality and flair to your public areas.

A *mural* is a wall-size painting, often created directly on the wall itself. It's a big, gutsy artistic statement that usually becomes the focal point of a room, and it can affect your overall design in several ways:

- It can be a signature that communicates the restaurant's concept.
- It can be used to personalize a large dining space, adding color and motion to an area that might otherwise feel more like a warehouse than a restaurant.
- Conversely, a mural may be used to open up small dining spaces by adding extra dimension to the wall space.
- A mural can add sophistication. It can be wild, vibrant, and ultramodern, or soft and subdued. Either way, it cannot help but make a statement.

Adding a mural is as big a step as the size of the mural itself, and the price tag for a large original work can easily top $10,000, so it requires careful consideration (see Illustration 7-6).

ILLUSTRATION 7-6 For many years, Gino's Italian Ristorante in Boise, Idaho, has featured huge, lush murals on almost every wall. Boise artist Fred Choate designed each of them with surprises for guests who study them to make discoveries.

Courtesy of Gino Vuolo, Gino's Italian Ristorante, Boise, Idaho.

When hiring an artist, ask to see many photos of their work, and visit some of the other mural sites in person. A mural can be an impressive conversation piece, but if not properly planned and professionally painted, it can be distracting and downright annoying instead of mood enhancing. Either way, it's an investment that you, and your guests, must live with for a long time. Be certain you have a contract with the artist or designer that specifies the size and content of the finished work (nothing that might be considered offensive), including preliminary sketches; a timeline for completion of the mural; the materials that will be used; the procedure and price if touch-ups or alterations are needed; and how payment is expected—usually some up front and the rest upon completion and approval.

A less permanent and less expensive way to add color to your dining area is to display the work of local artists, galleries, or art dealers. Depending on your arrangements with the art supplier, the works are priced and the restaurant may receive a small commission when one is sold. This allows you to change the décor with the seasons, by the month, or whenever you think the place needs a change. Whenever a new artist or selection is featured, you have a nice reason to invite guests and members of the local news media to a reception for the new display.

As when purchasing a mural, when selling the works of others it's a good idea to sign a short, basic contract with the artist or gallery representative. It should specify the number of days or months the work will appear in the restaurant, with starting and ending dates; specifics about commissions; rules about who has responsibility for safely hanging and lighting the artwork; accepting money for the pieces that sell; and picking up the pieces that don't sell within a certain time frame (so you won't be stuck storing them). It should also contain a statement that waives your liability in case any of the work is damaged or stolen while in your possession. The gallery owner or artist should give you a neatly typed sheet that lists every piece being given to you for sale, along with the prices and a stack of business cards. You'll need this on hand in case prospective buyers ask for it.

Showcasing local talent is not without risks, but it has a satisfying, community-minded spirit as a side benefit. It decorates your walls, boosts your business, and nurtures goodwill—all inexpensively.

Finally, when thinking about artwork, don't forget the restrooms. Many restaurants decorate all other areas to the max but leave their restroom walls completely bare. Why?

KITCHEN LIGHTING AND COLOR

When lighting a foodservice kitchen, trends and ambience take a backseat to efficiency. The idea here is to reduce employees' eyestrain by minimizing glare while making sure light levels are bright enough to allow a safe working environment. Planners must consider:

- Square footage of the area
- Ceiling height
- Contrast and colors of the products being processed at the workstation
- Colors of the surrounding walls

A back-of-the-house lighting system is usually fluorescent, with shields on the fixtures to protect food and workers from falling glass in case a bulb should break. Fluorescent fixtures may be placed parallel to workers' line of sight, as this results in less glare. In storage areas, light fixtures should be located over the centers of the aisles for maximum safety. Be sure to counteract the reflective properties of stainless-steel surfaces, such as worktables and refrigerator doors. A **satin finish** instead of a shiny one reduces the glare considerably.

Some kitchen tasks—cake decorating and garde manger duties (cold-food work, preparation of hors d'oeuvres, etc.)—are detailed enough to necessitate brighter lighting than other tasks. The Illuminating Engineering Society recommends a light intensity of at least 30 foot-candles throughout the restaurant and 70 foot-candles at so-called points of inspection, such as pass windows and garnishing areas, where the food gets a close look before being served. You may find that, in your area, the city or county government has set minimum lighting requirements as part of your licensing or permit process. We know of some cities that require a minimum of 50 foot-candles throughout the kitchen; others specify these requirements wherever food is prepared or handled, or where utensils are washed.

Kitchen lights require considerable maintenance because grease can build up on the fixtures and shields, so choose them for easy cleaning. In addition, fixtures that are labeled *vapor proof* (moisture resistant) must be used in exhaust hoods and dishwashing areas. *Protective lighting* is used in many prep areas and refrigerated spaces; these shatterproof lamps are designed so employees do not have to touch bare glass and food cannot be contaminated in case of breakage. Protective lighting options include:

- *Standard fluorescents.* Energy-saving or full-wattage, rapid start, preheat, and more.
- *High-output (HO) fluorescents.* For areas with temperature extremes, hot or cold. HOs have the added benefit of maintaining 95% of their original output during their entire life span; standard metal halide lamps tend to fade by half toward the end of their useful life.
- *Very-high-output (VHO) fluorescents.* Most often used for walk-in refrigerators and freezers.
- *Black lights.* Ultraviolet lights help identify contamination in food processing areas.
- *Teflon-coated incandescents.* Used wherever an exposed light fixture requires extra shatterproof protection. These also are safe at very high temperatures and resist chemical contamination.

Suggested maintenance of protective lighting is a method known as **group relamping**, in which all lamps in a fixture are replaced at the same time. This eliminates annoying flickering and protects the ballast.

The colors of walls and furnishings don't seem to be a major concern in most kitchens. However, a well-designed combination of light and color can reduce eyestrain and increase worker efficiency, possibly even reducing accidents and boosting morale. Light, cool colors are probably the best choices. Using one lighter color and a darker version of the same hue creates a mild contrast that is easy on the eye.

Because white is highly reflective, it should be avoided for kitchen walls; however, an off-white for kitchen ceilings can be a smart choice. Its natural brightness helps boost light levels.

The use of colors to signal special equipment and areas is also recommended. Danger is usually red, marked with paint or reflective tape on moving parts of equipment, swinging doors, and so on. Yellow is used to mark the edges of steps and landings. Green is used to identify first-aid kits or areas.

A final note about lighting: Local building codes, as well as your insurance company, have specific requirements for emergency lighting. Most of them come from either the National Electrical Code or the Life Safety Code. An emergency lighting system must be able to provide at least 1 foot-candle of brightness for 90 minutes, and it must have its own power source, independent of the main power system, in case of an electrical failure.

7-3 NOISE AND SOUND CONTROL

Both the dining area and the kitchen pose numerous challenges when it comes to noise reduction. Unfortunately, the challenges usually are not obvious until the space is occupied and someone, either employees or guests, starts complaining.

Typical restaurant sounds are many and varied: people conversing; waiters reciting the day's specials or picking up orders; cleaning of tables and busing of dishes; kitchen equipment grinding, whirring, and sizzling; the hum of the lights or the HVAC system. If you are located on a busy street, add traffic noise; at an airport—well, you get the point. Music is also an integral part of many restaurant concepts. Where would the Hard Rock Cafés or Joe's Crab Shacks be without it?

It is interesting that, while too much noise can cause discomfort, an absence of noise is just as awkward. It feels strange to sit at a table and be able to hear every word of your neighbors' conversation and to know that your own conversation is probably being eavesdropped on as well. In short, a restaurant's noise level should never be accidental. It is an important component of the environment and mood.

In restaurant settings, some businesses pump up the sound system volume in hopes of creating a sense of "happening" in the dining area. The challenge here is not to have a loud restaurant but to achieve a sound level of high enough quality and volume that guests will notice and enjoy it yet also comfortably converse through it. Juggling these priorities is harder than it sounds. Even if yours is a fabulous sound system, a stimulating noise level is bound to turn some customers away.

THE NATURE OF NOISE

The first step in controlling noise is understanding its two basic characteristics: *intensity* and *frequency*. Intensity, or loudness, is measured in *decibels*, abbreviated dB. The lowest noise an average person can hear close to his or her ear is assigned the level of 1 decibel, while a 150-decibel level causes pain to the average ear. A noisy restaurant averages 70 to 80 decibels, as shown in Table 7-1.

TABLE 7-1

Comparative Noise Levels for Foodservice

DECIBELS	NOISE EQUIVALENT
10–20	Broadcast studio
20–30	Average whisper at 5 feet, very quiet residence
30–40	Average school, average residence, library reading room, museum
40–50	Quiet restaurant dining room, quiet office
50–60	Average office, noisy residence, quiet street
60–70	Average restaurant dining room, noisy office, automobile at 30 mph
70–80	Noisy restaurant dining room, automobile at 50 mph, restaurant kitchen
80–90	Noisy street, police whistle at 15 feet
90–100	Fire siren at 75 feet, subway train, riveter at 30 feet
100–120	Train passing at high speed, airplane propeller at 10 feet
130	Threshold of pain

Source: Reproduced with the permission of the National Restaurant Association.

In Europe, where noise pollution is becoming a concern in larger cities, the World Health Organization estimates about 40% of the population lives with an ambient daily noise level above 55 decibels, comparable to a commercial dish machine running day and night. This is significant because, even at tolerable volumes, a relentless drone of noise can raise blood pressure, interfere with sleep patterns, cause stress, and hinder children's development and classroom learning.

Frequency is the number of times per second a sound vibration occurs. One vibration per second is a hertz (abbreviated Hz). Humans hear vibrations that range from 20 per second (low frequency) to 20,000 per second (high frequency). In a restaurant setting, a high-frequency sound is more objectionable than a low-frequency sound. Even sounds that are low intensity may be objectionable to guests if they are also high frequency. Sound travels from the source to the listener and back again in very speedy fashion—about 1100 feet per second. Inside a building, sound can be either absorbed or reflected by all the other things in that space: walls, ceilings, floors, furniture, and equipment. If it reflects or bounces off surfaces (called **reverberant sound**), the area can build up a sound level much higher than if the space were not enclosed.

It is not just dining areas where noise is problematic. At the back of the house, everything seems to conspire to *create* noise. Dishes clatter, chefs bark orders, dishwashers *whoosh,* toilets flush, refrigerators hum, and the kitchen exhaust system drones on and on. It can be a real cacophony. Your goal is not only to keep it from reaching the guests but to keep it from driving your staff crazy.

A study on restaurant noise levels by the University of California at San Francisco raises the issue of potential health hazards to employees of consistently loud eateries. Although the findings are inconclusive, the study authors measured everything from a low of 50 decibels in a Chinese bistro to almost 91 decibels in a busy microbrewery. A noise volume of 75 decibels or more requires that most people raise their voices to be heard in conversation—which, ironically, just creates more noise—and OSHA guidelines require employees in other fields of work to wear earplugs with noise levels of 90 decibels or more.

The primary concerns of the designer are those points at which noise is most likely to emerge from the kitchen—particularly the pass window, where expediters and waitstaff are positioned to call out orders and pick up food. Computerized ordering systems have already improved and quieted this process. The other problem site is the dishwashing area, where doors that are too thin or opened too frequently permit the clatter to emanate into the dining room.

One simple noise control strategy, often overlooked, is to make equipment operate more quietly by maintaining it properly, as we discuss in greater detail in Chapter 9. This includes careful examination of basic mechanical systems. HVAC ducts can act as chambers for sound transmission, amplifying the motor noise from the system's fan and condenser and sending it throughout the building. Metal plumbing pipes also transmit noises, as rushing water makes them vibrate and trapped air creates knocking sounds.

Now let's look at ways you can choose "sound smart" alternatives, in and out of the kitchen. General sound control can be accomplished in two ways: Either suppress the sounds from their source or reduce the amount of reverberant sound by cutting the travel direction of the sound waves. Sound engineers use *hard concave surfaces* to concentrate sound. The hard surface traps the sound and then sends it in the desired direction. If a *soft convex surface* is used, the sound is absorbed and deadened. These principles can be adapted to the ceilings, walls, and furnishings of any room.

CEILINGS. The ceiling of a room is a natural choice for sound control treatment because there is not much on the ceiling that would get in the way. Spray-on acoustic surfaces, acoustical tile, fiberglass panels padded with fabric, wooden slats, or perforated metal facings are less noisy alternatives to plain plaster or concrete ceilings. Panels and slats should be at least ¾ inch to 1 inch thick and suspended instead of being directly attached to the ceiling. Be sure not to paint acoustical tile, or it will lose its sound-absorbent effect.

WALLS. Covering walls with padding, fabric, or carpet helps a great deal to muffle sound. Organizing the dining room on several levels also helps somewhat. Movable partitions, which can be purchased up to 5 feet in height, can contain sound as well as create intimacy by breaking up a large dining area into smaller spaces. Walls and partitions can even be decorative and see-through to maintain the open feeling of a room.

DRAPERIES AND FURNISHINGS. In some types of foodservice businesses, fabrics are appropriate in the dining area and should be chosen for their sound absorption qualities. Window coverings can muffle sound if they are made of heavy material. Tables can be padded and covered with cloth, minimizing clanking dish noise. Chairs can also be padded and covered; although a porous fabric allows sound waves to penetrate, it also absorbs dirt more quickly. High-backed booths absorb noise.

Even something as simple as choosing a slightly smaller table can have an impact on sound, as it prompts diners to talk closer and more quietly to each other.

CARPETS. Floor coverings have a major impact on noise. If your dining room is carpeted, choose a carpet with high pile. It will be more expensive to purchase and maintain, but it won't wear as readily as cheaper carpets. Learn more about carpet and flooring choices in Chapter 8.

USE OF MUSIC

Finally, because music is so often viewed as a marketing tool, we cannot overlook its role in the restaurant environment. Many casual and theme restaurants make a considerable investment in sound systems, making music a part of their atmosphere. It can be used to muffle the ambient noise of guests and kitchen clatter and to add to the desired spirit of the place.

Make music selections based on the demographic mix of your guests and your overall concept. A direct satellite music feed is probably the best way to ensure employees don't hear the same song several times during a work shift—a surprisingly common employee gripe. Multiple music subscription services have sprung up in today's competitive market, all available for a monthly fee. A subscription takes the hassle out of keeping your music selections fresh and current and offers a wide range of choices. You can change the style of music by time of day with the touch of a button. Of course, some casual eateries still opt to play a local radio station on their sound systems, but it is unwise. Why would you want customers to hear advertisements, for your competitors or anything else, while they eat?

Perhaps the best thing about a music-subscription service is that the annual licensing fees are included in the cost. This is important because businesses must pay these up-front fees for the legal right to play almost any song in a public setting. The American Society of Composers, Authors, and Publishers (ASCAP) and Broadcast Musicians Incorporated (BMI) are the two major music-licensing organizations in the United States; for European artists, it's the Society of European Stage Authors and Composers (SESAC). They represent the people who create the music, ensuring they receive royalties for its use. Music licensing is a multimillion-dollar industry. The good news is that purchasing the annual licenses entitles you to freely play millions of songs. The bad news is that licensing is expensive. The cost is determined by a number of factors: the size of your establishment, whether you include dancing or a cover charge or use a jukebox, and others, from a minimum of about $300 to a maximum of more than $8,000 a year for each license. Discounts may be given if your restaurant is a member of a recognized state restaurant association.

Music licensing is nothing to neglect. If you are not using a subscription-based satellite service, you can be fined and even sued for playing tunes without paying the license fees. The licensing companies, which refer to themselves as *performing rights organizations*, have enforcement people whose job entails going into stores, restaurants, bus stations, office buildings—anywhere music might be playing in the background—to check for current licenses and cite business owners who don't have them. Learn more about music licensing specifics at the websites of these companies: ascap.com and bmi.com.

Make sure you get expert advice in selecting the right-size amplifier—that is, one designed to work with your space and size (by watts and number of channels)—to complement your number of speakers. The newest amplifiers are sophisticated enough to allow programming sound levels based on time of day or day of the week. Some have built-in microphones that sense the ambient sound level and automatically adjust the music level to fit. Proper placement of speakers is another job for experts. In foodservice, ceiling-installed speakers are more common than wall-mounted ones.

KITCHEN NOISE CONTROL

Controlling loudness in the kitchen begins with doors that separate it from the dining room. Some kitchens don't have doors but rather passageways; acoustical treatments on both walls and ceilings are especially important in these transition zones.

Of course, some restaurateurs believe that having customers witness kitchen sights, sounds, and smells is part of the mystique and excitement of eating out. In the open or *display kitchen*, noise is part of the atmosphere. The Romano's Macaroni Grill restaurant chain uses the open kitchen and its associated noise level as part of the design and concept.

Inside a kitchen, consider these noise abatement options:

■ Installing acoustical tile ceilings

■ Undercoating all work surfaces

■ Putting the dish room in an enclosed area

■ Installing all refrigerator compressors in a separate mechanical room or area outside the kitchen

■ Properly installing and maintaining exhaust fans to reduce humming and vibration

■ Using plastic or fiberglass dish carts and bus containers instead of metal ones

7-4 HEATING AND AIR CONDITIONING

Heating, ventilation, and air conditioning, collectively and commonly known as **HVAC**, are used to maintain a level of comfort for both guests and employees. Today, environmental considerations are top of mind in selecting a system—the type of refrigerants used, the potential energy savings, and so on. Your building's HVAC system must be carefully selected, properly operated, and continuously maintained to do its job effectively. The key environmental comfort factors at work here are:

■ Indoor temperature

■ Humidity

■ Air movement

■ Room surface temperature

■ Air quality

To modify and control these factors, these types of equipment are part of most HVAC systems:

■ Furnaces (to produce hot air)

■ Boilers (to produce hot air)

■ Air conditioners (to produce cold air)

■ Chillers (to produce cold air)

■ Fans (to circulate and remove air)

■ Ductwork (to move air)

■ Filters (to clean air)

Individual comfort is a simple matter of balancing a person's body temperature with that of the surrounding environment. The body gives off heat in three ways: *convection, evaporation*, and *radiation*. An example of heat loss by convection is when air moves over a person's skin. The movement creates a temperature difference between skin and air. An example of heat loss by evaporation is perspiration, when heat causes liquid to turn to vapor. Heat loss by radiation happens when two surfaces of differing temperatures are placed right next to each other, like being seated by a window on a cold day. The right combination of temperature and relative humidity to make people comfortable is referred to by experts as the **comfort zone**. The parameters of this zone are highly subjective; let's just say that, in finding your facility's comfort zone, you're trying to balance the environmental conditions with the natural heat loss of the bodies of your guests or workers. Comfort is important in the kitchen too.

The challenge to foodservice managers and planners is to find the comfort zone while paying attention to odors, noise, and air quality. The HVAC system must also be flexible enough to change if environmental demands change, whether that means a change in season, a change in crowd size, or a change in building size. Select a system that has a quick response time and can be controlled automatically and operated at low cost. Another option, popular in Europe and catching on in the United States, is the use of separate, smaller systems in different areas of the building. These are decisions that can best be made with the help of engineers.

When discussing an HVAC system, the air that runs through it is known by different names depending on location and use. In alphabetical order, common terms include:

Conditioned air. Air that has been cooled or heated mechanically (by HVAC) and released into the building's interior.

Desiccant air. **Desiccants** are drying agents, which may be included in the HVAC system to reduce humidity.

Exhaust air. Air that must be removed from cooking sources (e.g., ranges and fryers) or enclosed spaces (e.g., restrooms). Once exhaust air has been removed from the building, it should not be reused. The volume of air removal (how fast it is removed) is measured in cubic feet per minute (cfm).

Makeup air. Air that must be supplied to an area to replace the exhaust air that has been removed.

Outdoor air. Air that is taken from outdoors.

Return air. Air that is removed from an interior space, then returned to the HVAC system for recirculation or exhaust. Sometimes called *recirculated air*.

Supply air. Air that is delivered to an area by the HVAC system. It may be used for ventilation, heating, cooling, humidifying, or dehumidifying.

Transfer air. Air that flows from one part of a building to another. In a restaurant, transfer air comes from the dining area into the kitchen, which helps keep cooking odors from drifting back into the dining area.

Once the HVAC system is installed, several of your staff should always be aware of how the system operates. Now, let's talk briefly about each part of an HVAC system and how it contributes to the overall heating and cooling process.

HOW HVAC SYSTEMS WORK

The two major HVAC components that heat cold air are the heating plant and the heating system. The **heating plant** is where fuel is consumed and heat is produced. The **heating system** is the means by which the heat is distributed and controlled. The heating plant uses one of several heat sources: electricity, natural gas, liquefied petroleum gas (LPG), fuel oil, or steam. Your options depend on the availability of the fuel, its cost, the cost of equipment and upkeep, environmental regulations, and safety concerns, some of which were discussed in Chapters 5 and 6.

Electricity is available everywhere. While it is safe and clean, it tends to cost the most. If the need for heat is low and other fuel sources are not reliable, electricity may be your first choice. Natural gas, fuel oil, and LPG all produce heat by combustion, a combination that requires special equipment, such as boilers and furnaces. So, while fuel costs are lower, equipment and maintenance costs are higher. Also, the availability of these fuels is not as widespread in some areas as electric power.

Parts of the heating plant (shown in Illustration 7-7) are:

▪ The *preheater*, which takes completely cold air and begins the heating process.

▪ The *heater* itself, which is a type of motor that runs on fuel or electricity to create heat.

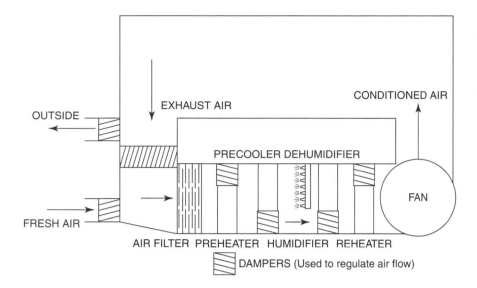

ILLUSTRATION 7-7 Layout of a year-round HVAC system.

Source: Frank D. Borsenik and Alan T. Stutts, *The Management of Maintenance and Engineering Systems in the Hospitality Industry,* 4th ed. (Hoboken, NJ: John Wiley & Sons, Inc., 1997).

■ The *humidifier*, which adds water to the air (often necessary because heat dries the air so much).

■ The *dehumidifier*, needed when outside air brought into the building for heating (or cooling) contains too much moisture. *Desiccants* are drying agents used to dehumidify air.

■ The *fan*, which blows the air from the motor or brings in air from outside. Fans have their own electric motors.

■ The *reheater*, used in some systems with humidifiers or dehumidifiers. A reheater boosts the heat of air that is otherwise too wet to blow into a room.

Parts of the heating system are:

■ *Ducts* and *vents*, which are the pipes and openings through which the air travels into parts of the building. Vents are also called *diffusers*; they are covered with *grilles* that can be manually opened and closed to restrict or encourage airflow.

■ *Dampers*, which are used inside the ductwork to regulate airflow. They can either heat or cool the air that passes through them, depending on the season.

■ The *filter*, which purifies the air by catching dust particles that would otherwise circulate.

■ The *thermostat*, which is the control unit that allows you to regulate the air temperature of a building or area.

Most climates require both heating and air conditioning. When you think of air conditioning, you probably think of cold air. However, the typical air-conditioning system does both cooling and heating, plus all the same filtering and humidifying a heating system does. Mechanical cooling equipment works by extracting heat from air or water, then using the cooled air or water to absorb the heat in a space or building, thus cooling that space.

Refrigerated air conditioning is similar to commercial refrigeration in that air is cooled in the same way—in a room or inside a refrigerated cabinet. The components of the system may be assembled in several ways, but each accomplishes the same goal: to produce refrigerated air that cools a given space. The main components are a metering device, an evaporator, a compressor, and a condenser. Illustration 7-8 shows how *vapor compression* works to refrigerate air.

Here's how the system works. The *refrigerant* (which is, at this point, about 75% liquid and 25% vapor) leaves the metering device and enters the *evaporator*. This is called the *low side* (meaning low pressure) of the refrigeration system.

The refrigerant mixture moves through the coiled tubing of the evaporator. As its name suggests, the liquid evaporates as it moves along. By the time it leaves the evaporator, it should be 100% vapor. (This is known as the *saturation point* of the refrigerant.) Typically, a fan is used to cool the evaporator area.

ILLUSTRATION 7-8 How vapor compression works to refrigerate air.

Source: Frank D. Borsenik and Alan T. Stutts, *The Management of Maintenance and Engineering Systems in the Hospitality Industry*, 4th ed. (Hoboken, NJ: John Wiley & Sons, Inc., 1997).

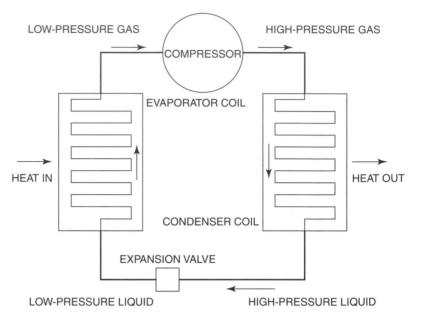

The vapor is drawn into the ***compressor*** by the pumping action of pistons and valves. It is superheated by the time it leaves the compressor, or the *high side* (meaning high pressure) of the system. The hot vapor travels through more tubes to the ***condenser***, where it begins to turn back into liquid. Another fan is used here to help the condensing process, cooling the hot gas to help it return to its liquid state. By the time the refrigerant reaches the end of the condenser coil, it should be 100% liquid again. From here, an *expansion valve* controls the flow of the liquid refrigerant. Its pressure drops as it is forced through the valve, which causes its temperature to drop. The cold, low-pressure liquid refrigerant then cycles back into the evaporator, and the process begins again.

An air-conditioning system is chosen primarily based on the size of the space you intend to cool and on the overall climate of your area. You will purchase a unit that provides a certain number of ***refrigerated tons***. A ton sounds like a lot, doesn't it? For measurement purposes, 1 ton of refrigeration is equivalent to the energy required to melt 1 ton of ice (at 32°F) in a 24-hour period.

Remember, in Chapter 5, we defined a Btu as the amount of heat needed to raise the temperature of 1 pound of water 1°F. Well, the melting of 1 pound of ice absorbs 144 Btus, so the melting of 1 *ton* of ice (2000 pounds) takes:

144 (Btus) × 2000 (pounds per ton) ÷ 24 (hours) = 12,000 Btus per hour

How many tons of refrigeration your system requires depends on many factors, including the size of the space, the number of people you'll serve at one time, and the type of cooking you'll do. A refrigeration specialist will do a heating and cooling ***load calculation*** to determine the size of unit you will need. In the United States, federal law also requires manufacturers to label air conditioners with an ***energy-efficient rating (EER)*** so you'll know how much power it will use. The EER is figured by dividing the Btus used per hour by the watts used per hour. The higher the EER, the lower the energy consumption.

The most popular HVAC system is known as a ***packaged air-conditioning system***, so named because it is self-contained. A fuel, such as gas or electricity, provides heat, and a refrigeration system provides cooling. Heating and cooling share the ductwork, dampers, and thermostat.

Speaking of air ducts, they must be properly sized to allow free passage of air. The fan and motors in a system must work too hard if the ducts are too small.

HVAC TECHNOLOGY ADVANCES

Depending on the climate in your area, you can help your packaged rooftop HVAC system save energy, as well as prolong its life, by installing an ***economizer***. This helpful unit is a set of dampers, sensors, and controls that introduces cool outdoor air into a building when it senses the outdoor air temperature is right. In mild climates, it takes the burden off the air conditioner to cool the building mechanically. The process is sometimes called *free cooling* in the HVAC industry. Some economizers sense temperature only; others sense humidity too. It's worth looking into and can be purchased as part of a new rooftop HVAC unit or retrofitted to an existing system. In some cases, it can reduce mechanical cooling costs by half or more.

In some regions of the country, you need **heat pipe exchangers** when summer weather brings both high temperatures and high humidity, especially if the doors to your eatery are opened and closed often. The heat pipe, pioneered by the National Aeronautics and Space Administration to cool the electronic components of spacecraft, uses no electricity. Installed as a part of your HVAC system, it evaporates and condenses a fluid in a continuous loop that transfers heat from the supply air to the return air stream—that is, it pre-cools warm air before it reaches the evaporator coil, takes some of the moisture out of that warm air, and discharges it into the return air. Air flows through the heat pipe simply because of where it is placed; no power is required to force it to work. When several Burger King restaurants in Clearwater, Florida, experimented with heat pipes, they decreased the relative humidity of their buildings from 85% to 60% and saved more than 20% on their electric bills.

Yet another option is to use the earth's constant temperature to heat and cool air. The *ground source heat pump* may be the most energy-efficient, clean, and cost-effective HVAC system available, but *geoexchange technology*, as it is called, is still fairly new. In the United States, fewer than a million systems are installed nationwide. Geoexchange technology combines traditional plumbing equipment and air-conditioning ducts with large *ground loops*, flexible piping filled with water and nontoxic antifreeze. The loops of pipe are buried 4 to 6 feet underground, either vertically or horizontally, where they adopt the natural underground temperature of 55°F. A simple electric compressor circulates the liquid and cools or heats the air going into a home or other building. The advantage is that, because it's already at a temperate 55°F, it requires less energy to heat or cool the indoor air to the desired temperature. Another plus: no noisy condensers and no pollution. Geoexchange or geothermal technology can be expensive to install, but some utility companies provide grants or rebates to encourage use of this nonpolluting energy source.

In climates known for their high humidity, everyone knows comfort involves more than temperature. It requires reducing the moisture in the air, which also helps eliminate humidity-related problems like mold, mildew, and fungus growth and their associated odors. For this purpose, *desiccant HVAC systems* were created. A blower, usually located on the roof of the building, pulls fresh air through a rotor coated with either silica gel or titanium silicate. The combination of these substances and the rotor motion dries the inbound air. From that point, the drier air passes through a heat exchanger (a pipe or another rotor) to either heat it or cool it. The air is then forced through ducts into the building's interior as is or into a conventional air-conditioning system for further cooling or heating. At the end of the chain, another blower sucks in more air, into which the excess moisture from the system is dumped and vented back outdoors. The system can maintain a relative humidity level of 60% or less and has the added benefit of keeping coils and filters cleaner.

Desiccant HVAC is an expensive proposition, and field studies have shown that it's probably not worth the expense for quick-service eateries, where guests do not linger long in the building. But for full-service dining establishments, it's worth a look. If only 15 guests a day increase their purchases by $4, staying longer for that extra drink or dessert in the air-conditioned comfort, you'll pay for your desiccant unit in less than a year while making life a little more comfortable for both customers and employees.

Getting the right mix of temperature and relative humidity is especially difficult in foodservice, when the seated, relaxing customers have different needs—and perceive temperature and humidity differently—than the frantically busy staff members serving them. The American Society of Heating, Refrigeration and Air Conditioning Engineers (ASHRAE), long regarded as the expert in this field, suggests a dining area comfort zone of 76°F with 45% relative humidity. How you accomplish this is best left up to a qualified HVAC contractor.

In new construction, the orientation of a building can enhance natural ventilation, thus reducing the overall need for mechanical cooling. In temperate climates, attention can be paid to the direction of prevailing breezes. The use of ceiling fans and control of heat and glare from sunlight can keep buildings comfortable in such climates with minimal HVAC use.

VENTILATION AND AIR QUALITY

The most important reason for ongoing maintenance of your HVAC system is its impact on air quality. Imagine, for a moment, a poorly designed ventilation system. In this restaurant, the guests are treated to each and every smell emanating from the kitchen—even the food that was cooked *yesterday*! Every time the front door opens (which is tough, because incoming customers must fight the pull of a badly balanced exhaust system), a rush of outside air pushes in, along with dust, insects, and uncomfortable drafts. The cleanup crew spends much of its time wiping off a greasy film that seems to settle on everything. Utility bills are higher than they need to be.

In short, ventilation is part of your overall ambience. At best, it is hard to do perfectly. At worst, it may jeopardize the health of your workers and your customers. Indoor air quality (IAQ) has become a huge concern, even prompting lawsuits and legislation on the state level. The EPA cites poor indoor air quality as a top public health risk. Americans spend 80% of their time indoors. Many of today's buildings are more tightly sealed than ever. Windows don't open, meaning that people inside must depend on adequate ventilation from the HVAC system. All too often, they are not getting it.

Over the years, we have all heard the news stories about so-called sick buildings, with medically identifiable diseases and symptoms that affect people. Toxic molds and Legionnaires disease are examples of **building-related illness (BRI)**. Doctors also now recognize *sick building syndrome (SBS)* as a medical condition, when 20% or more of a building's occupants exhibit symptoms of discomfort that disappear when they leave the building. What could possibly cause BRI or SBS? *Legionella*, for example, grows in stagnant water that has a source of iron, as in a building's air-conditioning system. Mold spores and dust mites can also seriously compromise air quality. Chemical contamination may be a culprit. A class of chemicals known as *volatile organic compounds* is often to blame, including cleaning solutions, paints, pesticide products, and some construction materials. Because most HVAC systems take in air from the outdoors, consider the locations of all air intake vents. Keep them clear of:

- Automotive exhaust from streets or parking areas
- Air being exhausted from other nearby buildings
- Your loading dock, where delivery truck fumes could be a factor
- Garbage dumpsters and unsanitary debris
- Standing water

The primary IAQ issue for most commercial kitchens is kitchen-related odor and smoke. Clean indoor air (CIA) laws have banned smoking in most public buildings, so that is no longer a major concern and is not discussed in this book.

The ASHRAE standards for proper ventilation are shown in Table 7-2. The ventilation rate is the amount of *cubic feet of air per minute* that must be replaced per person, and it varies somewhat with the function of the room itself. The four overall ways to improve indoor air quality are:

1. Bring in a sufficient amount of outdoor air.
2. Manage the flow of air within your interior spaces.

TABLE 7–2		
Air Quality Standards		
AREA	VENTILATION RATE	OCCUPANCY LEVEL
Dining rooms	20 cfm/person	70 persons/1000 square feet
Cafeteria, fast food	20 cfm/person	100 persons/1000 square feet
Bars, cocktail lounges	30 cfm/person	100 persons/1000 square feet
Public restrooms	50 cfm/water closet or urinal	
Kitchens (cooking)	15 cfm/person	20 persons/1000 square feet

Source: American Society of Heating, Refrigeration, and Air Conditioning Engineers, Atlanta, Georgia.

3. Use high-quality filtration systems.

4. Keep the HVAC equipment clean and in good working order.

In working with HVAC professionals, you may hear a few catchy terms. *Energy recovery ventilation (ERV)* is a system in which energy is used to change both temperature and humidity of incoming air. The desiccant HVAC system is a type of ERV. *Demand control ventilation (DCV)* is a technology option that allows you to adjust the amount of outdoor air that comes into your system, much like a thermostat allows you to control temperature. DCV systems are popular in foodservice because the number of guests varies greatly at different times of the day. Computerized sensors vary the amounts of air supplied and exhausted depending on the occupancy of the space. The sensors are stationed in each zone of the building, and they sense occupancy by detecting carbon dioxide levels in the air. Other sensors monitor the air pressure of the system. Installing this type of ventilation system is expensive, and the sensors and controls must be calibrated regularly, but it will save energy and money in the long run by freshening and exhausting air automatically and only as needed.

Your HVAC system is designed to provide **ventilation**—that is, to blow fresh or properly treated, temperature-controlled air into a room and to send a corresponding amount of used air out of the room. All airflow has a direction, and all HVAC systems have areas of positive pressure (to force air out) and negative pressure (to invite air in). Any system's positive and negative pressure and airflow can be drawn as a simple diagram called a **pressure footprint**, as shown in Illustration 7-9. Plus signs on the chart indicate positive pressure. A **positive air pressure zone** means more air is supplied to an area than is removed from it. Minus signs indicate negative pressure. A **negative air pressure zone** is created when more air is removed from an area than is supplied to it. The arrows on the drawing indicate airflow direction.

Remember, air naturally flows from areas of positive pressure to the areas of negative pressure. Slight pressure differences are fine, but large pressure differences are almost like miniature indoor wind gusts. They can cause problems.

Let's look more closely at this footprint for a small restaurant with a single air supply system located in the dining area. Exhaust fans are used in the kitchen to eliminate cooking smoke, fumes, and grease and in the bathrooms to eliminate odors and stale air. The smoking and nonsmoking sections could be in either corner of the dining room because both corners are close to the incoming air supply and to exhaust areas.

For restaurants of this size, the outdoor air supply is sometimes eliminated based on the assumption that negative pressure on the kitchen exhaust system will end up drawing in outdoor air passively as doors and windows are opened. A small air conditioner or heater may be installed to recirculate interior air in the dining room, but this would be a mistake. The kitchen needs outside air to keep venting its grease and smoke away from the building, and the dining room needs outside air as a fresh air source.

Controlling airflow looks easy on paper. However, while it should be relatively simple, adjustments must be made for how crowded the restaurant gets at different times of day. It's a balancing act that never ends.

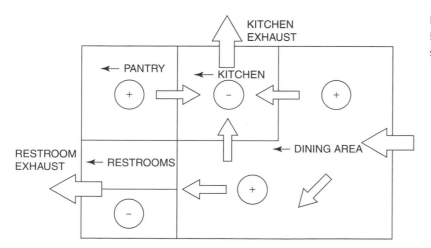

ILLUSTRATION 7-9 A pressure footprint that indicates the airflow of a room and its HVAC system.

A design trend to improve the energy efficiency of commercial HVAC systems is **underfoot air distribution (UAD)**, where ducts are placed so air enters the room at floor level rather than ceiling level. In the case of air conditioning, this means the air can enter the room at a higher temperature than if it came in at the ceiling level by coming in at points closer to where people are congregating.

HVAC SYSTEM MAINTENANCE

No matter what type of HVAC system you choose to install, you must consider one more important design factor—that is, think about how difficult it will be to get to the components when they must be inspected, cleaned, or serviced. Preventive maintenance is an essential part of keeping your system working at its best, saving you the most money in the long run and prolonging the life of the system by as much as 10 years. After a good cleaning and tune-up, your system will also automatically be up to 7% more energy efficient.

It is best for a professional HVAC service company to put your business on a regular inspection and maintenance schedule. Experts say the most common problems with HVAC systems are:

- Loose belts
- Dirty air filters
- Poorly lubricated bearings

Be certain the inspections address these items. Make sure air filters are cleaned or replaced, motors lubricated, heat transfer surfaces and drain lines cleaned, and switches and thermostats checked for accuracy. These service calls should be scheduled in the month immediately before your area's cooling season begins and again just before the peak heating season.

You might also ask the service technician about computerizing your HVAC controls if they aren't already. This allows you to control the system remotely and provides valuable diagnostic tools that track power usage, temperature limits, refrigerant levels, air filter conditions, maintenance records, and more. After spending more than $1 million a year on HVAC-related service agreements, the casual dining chain Applebee's installed a companywide computerized system. With input from the individual location managers, Applebee's headquarters can now remotely control the climate functions in each restaurant over the Internet without having to put lockboxes on the thermostats.

Basic control options for simpler HVAC systems are on–off switches, which look and work like light switches, and time locks, which are programmable timing devices to turn fans on and off at certain times of day. Unless you truly understand how to operate a time lock and how to change it when conditions change, you may become so frustrated you'll just disconnect it. Be sure the installer fully familiarizes you with whatever controls you decide on.

The foodservice HVAC system must not only circulate air but also remove dirt and chemicals. This creates the need for air filtration. Air filters remove airborne dust particles. The most widely used standard for filter effectiveness is the ASHRAE *dust spot method*, a rating that labels the filter with the average percentage of dust it removes. There are many bargain filters rated as low as 25%, but the suggested minimum rating is from 65% to 85%, and you can even get 95% ones. These higher-rated filters are known as *High-Efficiency Particulate Air (HEPA)* filters. For serious odor removal situations, an activated carbon filter adds extra effectiveness. Superior filters cost more but can pay for themselves within a few months with the cost savings from a cleaner system that uses less energy. No matter what filter you select, if it is not changed regularly, it won't do its job.

The smoking versus nonsmoking issue has also created a need for products such as the *electronic air cleaner*, which rids the air of gases and odors along with dust particles. This ceiling-mounted unit sucks room air through an electrically charged field (called an **ionizer**) and into a series of metal plates. The plates attract the newly charged airborne particles and collect them, ridding the air of 95% of its impurities (see Illustration 7-10). An optional charcoal filter at the end of the cleaning process helps remove odors and some gases. Some units require regular filter changes; others do not. A quiet fan returns the cleaned air to the room.

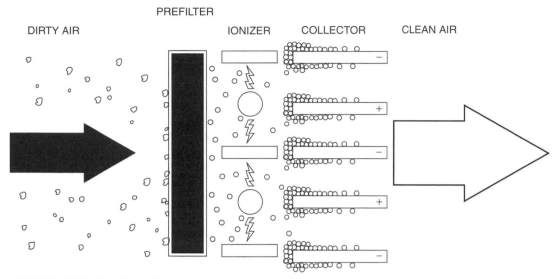

DIRTY AIR   PREFILTER   IONIZER   COLLECTOR   CLEAN AIR

ILLUSTRATION 7-10 How electronic air cleaners work.

As a foodservice business, you have three basic responsibilities in setting your air quality standards:

1. To provide sufficient ventilation for basic indoor air quality, including control of odors, grease, and smoke
2. To provide a comfortable environment for diners
3. To minimize energy use and operating costs

To accomplish this, we offer a number of suggestions, from design to installation. Some have been adapted from information in the technical bulletin *Ventilation for a Quality Dining Experience*. Today, it is ironic that this 2003 bulletin was distributed by what was then Philip Morris (now Altria), as part of the tobacco company's push to keep restaurants from banning cigarette smoking by helping them improve their indoor air quality.

BEFORE CONSTRUCTION. On a floor plan, define how each area of the building will be used as well as how it may be converted to another use over time. (If you'll be leasing existing space, ask if the lease includes regular maintenance by a professional HVAC technician.)

DURING DESIGN. Review the ventilation zones and airflow patterns of the pressure footprint to see they make sense and comply with standards for that type of use. Evaluate the ventilation specifications of ASHRAE Standard 62-2010 as well as local codes and ordinances. Check the locations of outdoor air intakes to be sure they are not near pollution sources. Check the control sequence of intake and outgo for its ability to stay within your standards under different operating conditions. Finally, check the design of your controls; make sure it is simple enough to be easily operated.

DURING CONSTRUCTION. Select construction materials and finishes for low toxicity and easy maintenance. Choose filters that will help you achieve your air quality standards. Evaluate plans for operation, maintenance, and cleaning of your system. Consider painting or rustproofing outdoor units if your climate warrants it.

DURING START-UP. Inspect the new construction to ensure it complies with design specifications. Bring in a certified contractor to test, balance, and review your airflow patterns. Review the operation and maintenance manuals for instructions and service information. (Put this on your maintenance cards or charts.) Follow the formal commissioning (start-up) procedure for the system to be sure it operates correctly under a variety of conditions.

DAILY MAINTENANCE. Ensure adequate outdoor air ventilation during and after cleaning and pesticide applications. Avoid using strong deodorant, disinfectant, and pesticide products. Use high-efficiency, filtered vacuums to clean all carpets and upholstery. Regularly check and clean exhaust filters and grease traps.

MONTHLY MAINTENANCE. Check controls and ventilation equipment once a month to be sure they are working well and in compliance with your area's regulations. This means checking and changing air supply filters and cleaning condensate pans and humidifiers as necessary. Always use the correct size of filter and keep a written record of filter replacement. Keep the area around the unit itself free of debris, stacked boxes, and anything else that would block airflow.

QUARTERLY MAINTENANCE. Check and clean the fans, air intakes and return grilles, vents and exhaust ducts, and the blades of ceiling fans. Check the dampers, tightening or lubricating them as needed. Check inside the air conditioner for leakage, which should be repaired promptly.

ANNUAL MAINTENANCE. Look closely at all the routine cleaning and pest control products you use to be sure they are the least toxic to do the job. How many do you really need? Are they safe? Keep a notebook of information about them. Check and clean cooling coils and condenser coils in the spring; check and clean heat exchangers and ducts and determine overall furnace efficiency in the fall. Make sure any service person you hire to check your system is looking at belts, fans, filters, thermostats, and system calibration.

COMMERCIAL KITCHEN VENTILATION

By law, all commercial kitchen equipment that produces any type of smoke, fire, or grease must be ventilated. Professional advice is needed to create a ventilation system; in fact, many cities require that licensed mechanical engineers design these systems. ***Commercial kitchen ventilation (CKV)*** requires changing the ambient air to remove heat, odors, grease, and moisture from the workspaces, primarily the hot line and dishwashing areas. Without proper ventilation, heat and humidity become unbearable, grease slowly but surely builds up on walls and other surfaces, and odors waft into the dining area nearby. Because the doors between kitchen and dining room swing constantly, an improperly cooled kitchen also means more work for the dining room HVAC system, which must compensate for the blasts of hot air. Pilot lights and flame settings are affected by negative pressure. Perhaps most important, grease buildup in hoods and ducts is the most common cause of kitchen fires. As you can see, there are lots of reasons to get it right the first time.

Early kitchens were equipped with propeller-type exhaust fans that were mounted on walls. As kitchen equipment became more complex, designers developed the ***hood*** or *canopy* mounted directly over the cooking equipment to draw smoke, moisture, heat, and fumes up and away from the kitchen. This led to the installation of ducts, completely separate from other HVAC system ducts, connected only to the kitchen exhaust fans. For many years, the national building codes mandated a fixed exhaust volume determined by the size and type of hood and the cooking equipment beneath it, whether cooking was taking place or not.

Today's CKV system can respond automatically to the amount of fumes and heat being generated below it. Exhaust quantities can be adjusted from 33% to 50% with variable speed fans. The ducts, hoods, and fans do double duty by bringing in fresh air (*replacement air*) and removing contaminated air (*exhaust air*). Known as a ***balanced system***, it requires expert design and upkeep to prevent wasting costly treated air and to prevent drafts. The system regulates the amount of outside air introduced into a kitchen area based on a predetermined minimum level of carbon dioxide. If that level is exceeded, the system increases incoming air velocity. Illustration 7-11 is a diagram of a variable-speed hood.

Hoods and fire protection systems have the most stringent legal requirements of any food-service equipment. The health department is concerned about sanitation; the fire inspector wants to know fire hazards have been abated; environmental agencies want assurances that you're not spewing pollution into the air. In most states, insurance regulations also govern the design and operation of ventilation systems. A commonly accepted procedure nowadays is that a third party (not you, in other words) must inspect your system every 6 months to make sure it's clean and working properly—or you may be denied fire insurance.

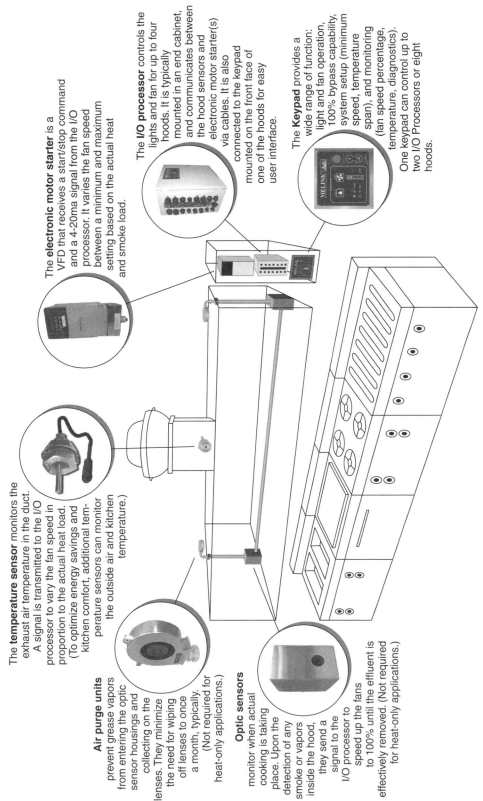

The **temperature sensor** monitors the exhaust air temperature in the duct. A signal is transmitted to the I/O processor to vary the fan speed in proportion to the actual heat load. (To optimize energy savings and kitchen comfort, additional temperature sensors can monitor the outside air and kitchen temperature.)

Air purge units prevent grease vapors from entering the optic sensor housings and collecting on the lenses. They minimize the need for wiping off lenses to once a month, typically. (Not required for heat-only applications.)

Optic sensors monitor when actual cooking is taking place. Upon the detection of any smoke or vapors inside the hood, they send a signal to the I/O processor to speed up the fans to 100% until the effluent is effectively removed. (Not required for heat-only applications.)

The **electronic motor starter** is a VFD that receives a start/stop command and a 4-20ma signal from the I/O processor. It varies the fan speed between a minimum and maximum setting based on the actual heat and smoke load.

The **I/O processor** controls the lights and fan for up to four hoods. It is typically mounted in an end cabinet, and communicates between the hood sensors and electronic motor starter(s) via cables. It is also connected to the keypad mounted on the front face of one of the hoods for easy user interface.

The **Keypad** provides a wide range of function: light and fan operation, 100% bypass capability, system setup (minimum speed, temperature span), and monitoring (fan speed percentage, temperature, diagnostics). One keypad can control up to two I/O Processors or eight hoods.

ILLUSTRATION 7-11 In modern exhaust hoods, temperature sensors and photoelectric eyes monitor smoke, fumes, and heat and adjust air levels automatically. The kitchen staff can manually override the system if necessary.

Source: Melink Corporation, Cincinnati, Ohio.

The National Fire Protection Association (NFPA) sets the national norms; Standard 96 applies to the construction and installation of hoods, canopies, exhaust fans, and their fire protection systems. The NFPA does not make exhaust volume recommendations, but Underwriters Laboratories (UL) and the Building Officials and Code Administrators (BOCA) do. In 2003, in an attempt to reconcile all these codes and recommendations, ASHRAE created a new Standard 154 that has become the industry benchmark for CKV design. Revised in 2011, the latest version is ANSI/ASHRAE Standard 154-2011. (ANSI is the American National Standards Institute.)

There are two basic types of exhaust hoods:

Type I is used for collecting and removing flue gases, smoke, and some of the grease generated by the cooking process. The two subtypes in this category are *listed* and *unlisted*. Listed hoods meet the requirements of the UL 710 Standard.

Type II is used where steam, vapor, odors, and heat are generated, such as above dishwashers and steam tables. It is not intended to handle removal of smoke or grease-laden air. The two subclassifications here are condensate hoods and heat/fume hoods.

UL and NFPA standards require that all exhaust hoods be constructed of minimum-gauge stainless or galvanized steel, with a liquid-tight, continuous external weld, and that they include an approved filter or removal device to get rid of grease. Hoods, their filters, and ductwork must all be separated from any combustible material by at least 18 inches. (Smaller clearances are allowed, but only if an approved fire barrier is installed.) The exhaust system's point of discharge at the top of the building must be at least 40 inches above the roof. The standards also require a fire suppression system that includes both manual and automatic shutdown of cooking equipment and extinguisher nozzles above each piece of surface cooking equipment (broilers, fryers, ranges, etc.).

Hood design has improved greatly in recent years as manufacturers have refined the shapes of hoods to introduce makeup air as they draw away smoke and pollutants. The three basic types of hood installation are wall, corner, and island. A hood mounted on a wall is often more efficient because it is not affected by cross-drafts, as a hood mounted above an island in the center of the kitchen would be. The curtain of makeup air on an exhaust hood is called a *plenum*; there are internal (under the hood) plenums and external (mounted on the outside of the hood) plenums.

The amount of exhaust air that must be removed, and the rate of its removal, depends on many variables: hood size and design; the amount of hood overhang (how well the hood covers the appliances installed below it); the height of the hood in relation to the appliances; and the amount of steam, smoke, or vapor generated by the appliances as they are used. Some suggestions for maximizing hood efficiency are found in the "Foodservice Equipment" text box.

An expert will use one of two basic ways to determine exhaust requirements. The first common method is to design a CKV system to provide a specific number of changes of air in

FOODSERVICE EQUIPMENT

Ventilation Toolbox: Going for Greater Efficiency

- Use deeper and taller hoods for fuller capture and containment.
- Push back equipment to maximize overhang and minimize rear gap.
- Lower hoods.
- Add side and end panels.
- Place heavy-duty equipment such as broilers in the middle of the lineup.
- Place light-duty equipment such as ovens on the end.
- Don't waste hood space over noncooking areas.
- Introduce makeup air at low velocity.

Source: Fisher-Nickel, Inc., PG&E Food Services Technology Center, San Ramon, California.

a kitchen space each hour the system is in operation. Industry experts recommend that a ventilation system provide 20 to 30 total air changes per hour or that it exhausts a total of 4 cubic feet per minute (cfm) per square foot of floor space. To calculate the volume of air in the kitchen, use the total dimensions of the room without deducting square footage for the space occupied by equipment.

The second method is to maintain a fixed velocity of air across all of the hood openings. In other words, the air has to move a certain number of feet per minute across the face of the hood. Most hood manufacturers indicate in their specifications the air volume necessary for proper operation, which may vary a bit if the hood is installed in a corner or against a wall. The BOCA recommendation is an airflow of at least 100 cfm per square foot of cooking space beneath your hood for a wall-mounted unit, and 150 cfm for an island unit. Underwriters Laboratories suggests 150 cfm for wall units and 300 cfm for island units but makes its calculations using *linear* feet, not square feet.

For example, let's say we have an island canopy that is 20 feet long and 4 feet wide for a total of 80 square feet. Using the BOCA standards, we'd need 12,000 cfm: 80 square feet × 150 cfm. Using the UL standards, however, we'd only need half that capability: 20 feet × 300 cfm = 6000 cfm. Again, state or local codes may govern your minimum requirements.

Earlier in this chapter, we talked about positive and negative air pressure zones and the need to create air balance. In the kitchen, because exhaust needs are so great, a system to add makeup air is required. The critical design consideration for kitchen exhaust and makeup air is to build up enough air velocity to capture vapors, grease, and heat at the cooking surfaces without creating wind gusts that blow out pilot lights and make doors difficult to open or close. To achieve the slight difference that keeps air moving without gusting, makeup air generally replaces about 85% of the exhaust air. It is normally introduced into the kitchen from supply grilles (also called *diffusers*) located as far across the room as possible from the main exhaust and as close as possible to the ceiling. A combination of diffusers that introduce and vent air away from a space can be used to provide specific, directional airflow within that space regardless of its air pressure relative to other, surrounding spaces. When choosing a location for the diffusers, first decide where to put your pass window or pickup counters. Cross-ventilation is usually a good thing, but not if the air blows directly across the food and cools it off before serving.

According to the *Whole Building Design Guide*, there are three types of exhaust fans:

1. *Up-blast fans* are centrifugal fans, made of aluminum and mounted on the roof, directly on top of the exhaust stack.

2. *Utility fans* are also roof-mounted, used where high-static pressure losses exist. The inlet and outlet are 90° from each other.

3. *Inline fans* are located in interior ducts and used where it is impossible or impractical to mount an outdoor fan.

The fans themselves may be *propeller-type fans* or *blower-type fans* (also called *centrifugal fans*). You have certainly seen *ceiling fans*, which operate on the same principle as the propeller-type fan, with a group of blades connected to a motor. They don't need much horsepower to move large volumes of air, but they can be noisy and accumulate grease and dirt on their blades, which reduces their effectiveness. The blower-type fan consists of a motor inside a specially formed housing and blades that incline forward or backward. Blowers cost more than propeller fans to install, but they work better for kitchen exhaust systems because they stand up better to the rigors of heat and static pressure. The newest fan technology is the *direct-drive fan*, which eliminates the problem of broken or worn-out fan belts. Sensors automatically change the fan speed based on smoke and heat output. These can be retrofitted into existing hood systems and can pay for themselves quickly in energy savings alone.

The fan's job is to remove smoke from the cooking surface. The amount of exhaust air must be equal to or greater than the amount of air generated naturally by the cooking equipment. Without sufficient **capture velocity** to pull the stale air up and away, it will roll off the cooking line into the rest of the kitchen. This **rollout** condition is what you are trying to avoid by having exhaust fans in the first place.

Kitchen ducts are also subject to a host of special requirements so the hot grease and particles they carry cannot start fires. Ducts must be made of a minimum 16- or 18-gauge steel, welded into a single sheet (not screwed together), and enclosed in a fire-resistant shaft throughout the building. More about fire safety in Chapter 8.

When exhaust fans are located on roofs, grease buildup over time can damage the roof itself. Several manufacturers offer *grease guards* to prevent this; these are installed at the base of the fan to catch grease before it hits the roof.

Inside the hood canopy itself, grease buildup can also be a problem. A grease filter is an integral part of any hood installation, trapping the grease before it enters the exhaust duct. The traditional grease filter is a set of interlocking metal baffles that remove grease by centrifugal force, collecting it and using gravity to drain it downward into a trough and then into pans attached to the hood. These pans, and the filter system itself, must be easily accessible and removable for cleaning. Some codes specify the minimum distance allowed between the lowest edge of the filter and the cooking equipment below. Frequent cleaning is absolutely necessary to maintain a filter's effectiveness. If grease is left too long, it can harden, bake on in the heat of cooking, and hamper the airflow in the exhaust system.

Another method of grease removal, known as *extraction*, takes place as the exhaust air moves through a series of specially designed baffles. The air moves at high speed and must change direction in order to go through the baffles, throwing out the grease particles by centrifugal force along the way. The particles collect into a grease cup, which holds more than 6 pounds of grease and can be cleaned and replaced daily, or the particles are collected in grease gutters, which are also cleaned daily.

New technology has introduced packed bead bed filters that grab grease particles and ultraviolet (UV) lights that attract and pulverize particles. Large extraction systems for high-volume cooking operations often include a wash cycle that can be activated by pushing a button on the system control panel. Hot water and detergent are sprayed inside the system to wash the day's grease accumulation from it. Afterward, the wash cycle shuts off automatically. This type of cleaning is handy if the chore of cleaning large numbers of filters, fans, and ducts is too expensive or time-consuming.

When selecting a CKV system, you may hear the terms *tempered* and *untempered*. A *tempered* system is one that has the capability to heat or cool the replacement air, depending on the season; an *untempered* system does not. Untempered systems are touted for their energy-saving features. They work well in moderate climates but may have drawbacks in areas with seasonal temperature extremes. In recent years, system designers have recognized the potential for beneficial reuse of the heat generated in a commercial kitchen. Illustration 7-12 is one example of heat recovery technology to capture kitchen heat as it is vented away and use it to temper incoming (replacement) air during cold weather.

Overall, the kitchen exhaust system can be the most expensive single purchase in a kitchen, so be sure one person or company can be held accountable for its performance. In too many cases, a budget-minded entrepreneur pieces together a hood from one company, ductwork from another, a fan from another, and then has somebody's brother-in-law install it all. When the system does not work properly, each supplier blames the other, and the facility is stuck with a system that is a continual problem.

VENTLESS HOODS

In some situations, space limitations, building configuration, or prohibition of structural changes prevent the use of a standard exhaust system. In these cases, a ventless exhaust system is an interesting option. The Skydome in Toronto, Canada, has a ventless system, and its owners estimate it cost $500,000 less to install it than designing and installing ductwork for this moving, dome-shaped structure would have. Ventless systems are also an option if you are leasing space because you can take the whole package with you when you leave, which certainly is not possible with standard ductwork.

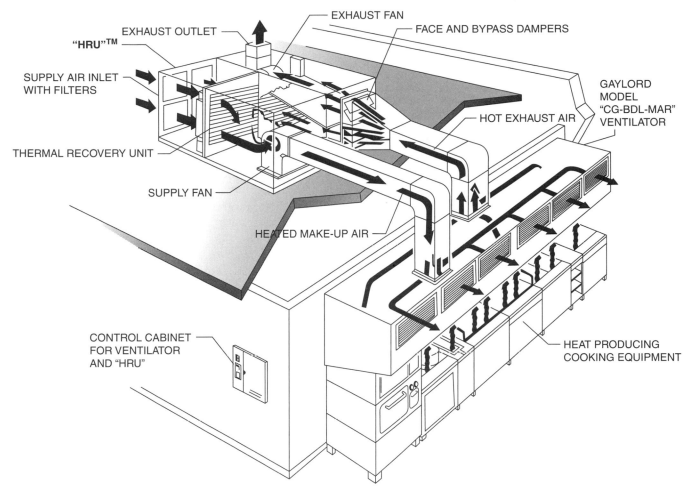

EXHAUST FAN
FACE AND BYPASS DAMPERS
EXHAUST OUTLET
"HRU"™
SUPPLY AIR INLET WITH FILTERS
GAYLORD MODEL "CG-BDL-MAR" VENTILATOR
HOT EXHAUST AIR
THERMAL RECOVERY UNIT
SUPPLY FAN
HEATED MAKE-UP AIR
CONTROL CABINET FOR VENTILATOR AND "HRU"
HEAT PRODUCING COOKING EQUIPMENT

ILLUSTRATION 7-12 Gaylord Industries says its Heat Reclaim Unit (HRU) has an average heat recovery efficiency of 65%, using the heat generated in cooking to heat replacement air during cold weather months.

Source: Gaylord Industries, Inc., Tualatin, Oregon.

The ventless exhaust system is self-enclosed and uses every possible type of filtration system instead of the usual air intake and outgo: the traditional baffle filter for grease particles, stainless-steel mesh screens that trap smaller particles; disposable paper or fiberglass filters (including high-efficiency HEPA filters) for the smallest particles; and charcoal filters for odor removal. Powerful (and sometimes noisy) fans suck the air through the filters and discharge newly cleaned air back into the cooking area. All these filters require cleaning and regular replacement; many ventless systems include sensors that light up to alert the need for a filter change. Makeup (replacement) air is not part of these systems, so your HVAC system will be the only source of fresh air to counteract the heat of cooking. Most manufacturers recommend the same amount of external air circulation as the ventless hood's filtration capacity—for example, if the hood operates at 400 cfm, you'll need 400 cfm of air from your HVAC system—to keep cooking conditions comfortable.

The ANSI/ASHRAE Standard 154-2011 covers ventless hoods, which must meet additional safety standards in order to receive UL and NFPA listings, including built-in fire suppression systems. UL Standard 197 says a ventless hood cannot emit more than 5 milligrams of particulates per cubic foot of air in an 8-hour period. It also must be equipped with a safety interlock system that shuts off the hood and all cooking equipment if the filters are either incorrectly installed or too dirty to operate properly.

You can purchase individual pieces of equipment with their own, self-contained ventless systems, or ventless systems already assembled from the factory, to install above existing equipment. One important note: Most cities have regulations that restrict the types of equipment that can be used with a ventless hood. Often this means they cannot be used with gas-power equipment, only electrical.

7-5 AIR POLLUTION CONTROL

Many of the technological improvements to foodservice equipment in recent years are designed to reduce pollution, and thank heavens. A Southern California study found that restaurants, particularly those that do a lot of frying, produce approximately nine times more air pollutants than a city bus. Griddles and deep-fat fryers in the four-county Los Angeles area were found to emit a total of 13.7 tons of particulate matter and 19 tons of volatile compounds per day—as much as an oil refinery! Most of these harmful compounds come from the fats that drip onto open flames during cooking.

This research was done in conjunction with efforts by Los Angeles to reduce air pollution by passing strict new ordinances that would include mandatory pollution control for any eatery that cooks more than 50 pounds of meat and/or 25 pounds of other foods per day. Other large U.S. cities are also interested in this issue, so you might as well be too.

What exactly is air pollution? Smoke and gases, grease, airborne dirt, and germs begin as individual particles, too small to see until thousands of them are grouped together. One particle is measured in microns—one micron is 1/25,400 inch long—and that's what air pollution control equipment must remove from the air system.

Let's use smoke as an example. A single particle of restaurant kitchen smoke is 0.3 to 0.8 microns in size, so it is much easier to measure it in terms of its density or *opacity*—the degree to which the smoke blocks the flow of light in a room. An opacity level of 100% would be solid black; 0% would be completely smoke-free. The cities that have adopted smoke pollution ordinances usually require an opacity level of no more than 20%.

That sounds reasonable until you consider that the average charbroiler emits enough smoke for a 60% to 70% opacity level. Opacity levels can actually be measured by instruments (called *opacity meters* and *cascade impactors*); more often, however, the inspector's own eyeballs and experience are used to determine what's in compliance and what's not.

Grease particles, also measured in microns, are generally larger than smoke particles—approximately 10 microns and up. That's why your grease extractor has to be top of the line, capable of removing 95% of the grease it processes.

Cooking odors are molecules generated by the combustion of animal or vegetable products during the cooking process to make an extremely complex mixture of ***reactive organic gases*** (referred to by experts as *ROGs*). Grease absorbs a small percentage of odor ROGs, but most of them float around in the air, much too small to be removed by any type of filter. One popular method to remove odor ROGs is to push them through a layer of activated charcoal, which absorbs and retains them. Unfortunately, charcoal is highly flammable, and, when it becomes fully saturated, it begins to release odors—sometimes worse than the original odors! A second common method of odor removal is to use potassium permanganate, which oxidizes and actually solidifies the molecules, then retains them. Either charcoal or oxidation can remove 85% to 90% of odor molecules from the air.

Much like electronic air cleaners, some air pollution control units work using electrostatic precipitation (ESP). Electrostatic cells inside the unit are made of aluminum plates, some grounded and others charged with 5000 volts of electric power. As air enters the unit, it moves through a series of thin electric wires, which charge it with 10,000 volts of power. Smoke and grease particles in the air receive a positive charge from the wires. They are repelled by the positive plates but attracted by the negative plates, so they collect on the negative plates. The system is washed regularly with hot water and detergent to remove the accumulation of grease and smoke particles. Often the fire protection system is wired to the air pollution control unit so both can go to work quickly in case of a smoky fire. The least sophisticated models only abate smoke; others also control odor.

■ SUMMARY

Many people believe that lighting is the single most important environmental consideration in foodservice. If your budget allows, hire a lighting design consultant to help you plan the location and light levels you need, both natural and artificial. Coordinate the lighting design with the color scheme you choose, both for safety and for aesthetics. Take advantage of new lighting technology like LEDs and CFLs, which may be more expensive at first but save money in the long run because of their longevity. The colors you use for walls and furnishings will depend in part on the lighting sources you choose.

Noise levels in a restaurant are part of its ambience, but noise control is a key consideration. Pay attention to both the frequency and the intensity of the noise level—not just what the guests hear but also the kitchen clamor. Architectural modifications and the placement of equipment, furnishings, and draperies can all help modify noise levels. The judicious use of music in your dining area is usually an asset, but only with a good sound system. Remember to pay the licensing fees to use the music, or you will face fines from the performing rights organizations.

Plan your heating, ventilation, and air-conditioning (HVAC) system so it provides a consistent, comfortable atmosphere during inevitable temperature changes and varying occupancy levels.

Your kitchen exhaust system is part of the HVAC system, and it should be kept clean and maintained rigorously. Building codes, health codes, and insurance regulations all have major effect on your HVAC and kitchen ventilation decisions, which are best left to experts for design and installation.

Perhaps the best thing about today's technology is that practically any environment-related system can be programmed to save energy by sensing occupancy or activity and adjusting accordingly. Lights can be placed on automatic timers or dimmers; music levels can adjust themselves automatically to fit the crowd noise. The addition of an economizer in some climates can automatically let outdoor air into a building to assist with cooling on pleasant days. Even your kitchen exhaust system can sense when, and how much, cooking is being done, and increase (or decrease) its performance in response. These automated functions are not inexpensive, but they pay for themselves quickly with time and energy savings.

■ STUDY QUESTIONS

1. Explain three ways to save on the cost of lighting.
2. In lighting terminology, what is the difference between *efficiency* and *efficacy*? Which would you say is more important in choosing a lighting system?
3. Why do fluorescent lamps work better now than they used to?
4. What are some advantages of using LEDs? Is there a situation in which you would not use them?
5. List two things you can do to reduce reverberant sound in a dining area, and explain why they work.
6. List three ways you can reduce noise in a busy kitchen.
7. If your restaurant's air-conditioning system didn't seem to be working well, what is the first thing you would check?
8. Briefly describe four methods mentioned in this chapter of achieving a heating and cooling system that is environmentally friendly.
9. What is a desiccant HVAC system, and why would you need one?
10. What are the negative effects of a poorly ventilated kitchen? Name three.
11. What is a balanced ventilation system, and why is it important?
12. What's the difference between regular ducts and kitchen exhaust ducts?

8

SAFETY AND SANITATION

■ INTRODUCTION AND LEARNING OBJECTIVES

Safety is an integral part of the design of a building, and this includes food safety as well as employee and guest safety. In kitchen design, everything must be considered with an eye for easy and effective cleaning. Refrigeration must be directly accessible to the prep areas where it is required to keep foods correctly chilled. Counters and other food display equipment must balance the challenges of effective merchandising versus safe holding temperatures. Making it all work together can be complex and difficult—and expensive, especially if it is not done correctly.

Improving safety in and around your foodservice business also can positively affect your staff's morale and productivity, decrease your insurance costs and legal liability, and even attract more customers.

After reading this chapter, you will be able to:

- Identify and explain the fundamentals of fire safety.
- Describe ways that ergonomics affect design.
- Identify training methods to ensure employee comfort and safety.
- Select the most appropriate flooring and carpeting for food service spaces.
- Identify and explain the fundamentals of sanitation.
- Identify and explain the fundamentals of food safety.
- Identify and explain the fundamentals of waste management.

Safety is a state of mind. It requires making a conscious decision to conduct yourself, and your business, so as to do everything possible to prevent accidents and injuries. It means identifying all potential hazards in your work processes or in the facility itself, developing safety measures to minimize these hazards, and training employees with an effective, ongoing safety program.

8-1 FIRE PROTECTION

Because we ended the preceding chapter with a discussion on kitchen hoods and exhaust systems, let's continue in that vein. About one-third of all restaurant fires originate in the kitchen, and they are generally flash fires on cooking equipment. Prevention of these incidents requires two important steps: control of ignitable sources and control of combustible materials. The most common source of the kitchen fire is grease, a natural by-product of many cooking processes. When fats are heated, they change from solid to liquid. They are then drained off in the form of oil, or they become atomized particles in the air, propelled upward by the thermal currents of the cooking process. Low-temperature cooking creates more liquid grease; high-temperature cooking creates more grease-laden vapor. The vapor is sucked into the exhaust hood where, as it cools, it settles on surfaces and becomes a fire hazard inside the exhaust system.

If the kitchen staff has had the proper training and the right safety equipment is available, the rangetop fire can be extinguished within moments. If not, it can quickly expand into the duct-work, reaching 2000°F as it comes into contact with highly flammable grease and lint particles. Therefore, an *automatic fire protection system* is a necessity. In fact, most state insurance departments require a fire safety inspection from an exhaust hood expert before insurance companies can issue a commercial fire insurance policy. As we've mentioned, the site must usually be reinspected every 6 months to keep the insurance in force. Even if the six-month rule doesn't apply in your area, it is a good idea to have your system professionally cleaned and checked twice a year anyway.

The National Fire Protection Association (NFPA) is the authority on this topic and sets the stringent regulations for commercial kitchen installations. Most canopy manufacturers offer fire protection systems as part of their package, including installation, but you can also hire an independent installer.

An automatic fire protection system, such as the one shown in Illustration 8-1, consists of spray nozzles located above every piece of external (not ovens) cooking equipment on the hot line. Rules about the numbers of nozzles and their locations are specific:

■ Rangetops require one nozzle for every 48 linear inches.

■ Griddles require one nozzle for every 6 feet of linear space.

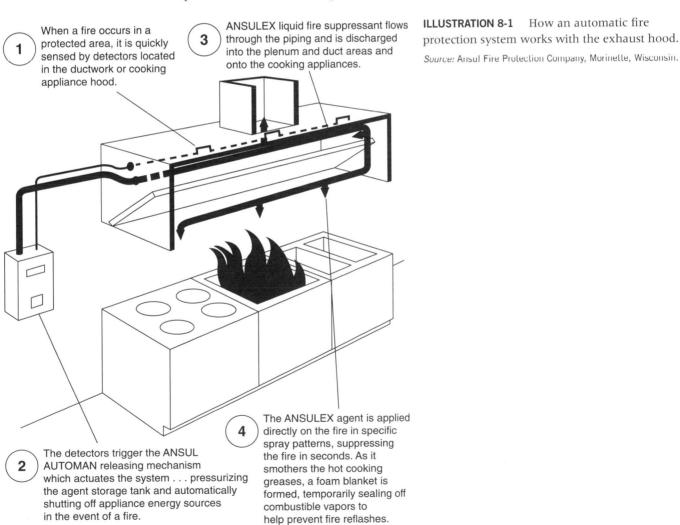

1 When a fire occurs in a protected area, it is quickly sensed by detectors located in the ductwork or cooking appliance hood.

3 ANSULEX liquid fire suppressant flows through the piping and is discharged into the plenum and duct areas and onto the cooking appliances.

ILLUSTRATION 8-1 How an automatic fire protection system works with the exhaust hood.

Source: Ansul Fire Protection Company, Marinette, Wisconsin.

2 The detectors trigger the ANSUL AUTOMAN releasing mechanism which actuates the system . . . pressurizing the agent storage tank and automatically shutting off appliance energy sources in the event of a fire.

4 The ANSULEX agent is applied directly on the fire in specific spray patterns, suppressing the fire in seconds. As it smothers the hot cooking greases, a foam blanket is formed, temporarily sealing off combustible vapors to help prevent fire reflashes.

- Open broilers (gas, electric, or charcoal) require one nozzle for every 48 inches of broiler surface.
- Tilting frying pans require one nozzle for a surface 48 inches in width.
- Fryers require one nozzle apiece or one nozzle for every 20 inches of fryer surface.

Nozzles are placed between 24 and 42 inches above the top of the equipment. (This varies depending on the type of appliance.) The nozzles activate automatically to shoot water or a liquid fire retardant at the cooking surface when the temperature reaches 280°F to 325°F. The heat detector may be located in the ductwork or in the hood.

If you feel you will need to move equipment around under the hood, you can request that it be designed for full-coverage suppression. This means the nozzles are installed on the hood every 12 inches, from end to end, allowing appropriate fire protection no matter where the equipment is placed. For some types of equipment, the nozzles must be specially aimed in particular ways—so if you do make changes to the hot line, you'll want to make sure a specialist checks your nozzle placements.

Inside the ductwork is an internal fire protection system whereby a fuse link or a separate thermostat is wired to automatically close a fire damper at the ends of each section of ductwork. The exhaust fan shuts off, and a spray of water or liquid fire retardant is released into the interior. Similar systems can be operated by hand instead of automatically. Some keep the exhaust fan running to help remove smoke during a fire.

In addition to the exhaust system's fire protection, several handheld fire extinguishers should be mounted on the kitchen walls, and employees should know how to use them. The automated system, when it is triggered, is so thorough that you must close the kitchen and begin a major cleanup, so often a handheld extinguisher is sufficient for minor flare-ups, and a lot less messy. Today, most insurance coverage requires **Class K** fire extinguishers in commercial kitchens. The NFPA classifies fires by the type of material that is burning; K (for *kitchen*) was added to the list in 1998. These fire extinguishers work on the principal of **saponification**, the term for applying an alkaline mixture (such as potassium acetate, potassium carbonate, or potassium citrate) to burning cooking oil or fat. The combination creates a soapy foam that quenches the fire.

Finally, as with any other public building, ceiling-mounted sprinkler systems are also worth investigating; their installation may significantly reduce your insurance costs. There is a

---

BUILDING AND GROUNDS

10 Commandments of Fire Safety

1. Make certain portable fire extinguishers are properly located and mounted; of the correct type and size, and properly identified; and fully charged and maintained.
2. Train all employees in the proper use of portable fire extinguishers.
3. Train employees in conducting a patron evacuation of your restaurants.
4. Post emergency telephone numbers in appropriate locations.
5. Keep paths of exit unobstructed, well lit, and equipped with emergency lighting and exit signs.
6. Do not permit the accumulation of refuse or rubbish. Follow recognized housekeeping procedures.
7. Clean your kitchen hood, filters, and duct of grease accumulation.
8. Have your kitchen hood's automatic fire protection system inspected and serviced by qualified persons every 6 months.
9. Have your automatic sprinkler system inspected annually.
10. Inspect your electrical system for frayed wiring, prohibited use of extension cords, overloading of circuits, clearance around motors to prevent overheating, and ready accessibility to all electrical panels.

Source: Reprinted with permission from *Restaurant Business*, a publication of Ideal Media, LLC, New York.

common misperception that, if it detects even one wayward flame, the entire sprinkler system will douse the whole building, but this is generally not the case. In fact, most restaurant sprinkler systems have heads that activate only when a fire is detected directly beneath them.

Ask your local fire department for suggestions and fire safety training tips for employees. And by all means, keep up with those fire inspections. In recent years, insurance companies have challenged kitchen fire claims, and courts find the restaurateur is at fault—and cannot collect the insurance money for fire damage—when routine maintenance and cleaning has not been performed.

8-2 ERGONOMICS

The idea of *environmental design* is derived from the field of human engineering, also known as **ergonomics**. Ergonomics is an applied science that involves studying the characteristics of people and designing or arranging their activities (sports, on-the-job duties, etc.) so they are done in the safest and most efficient manner. This includes designing machines and work methods with safety, comfort, and productivity in mind.

Here are just a few ways that ergonomics impact design in foodservice:

■ Planning easy entrance and exit from the facility

■ Placement of public areas to make service easier

■ Design of specific work areas to facilitate safer or more sanitary work

■ Finding alternatives to repetitive tasks or heavy lifting

To illustrate the many options, let's walk through a well-designed, "human-engineered" facility that includes public dining space. First, notice that no unnecessary barriers impede the flow of guests into the lobby area. Comfortable, well-arranged seating eliminates cross-traffic, creates intimate dining areas, and minimizes the clutter of too much movement by wandering guests or staff. The customers' feelings of comfort and security have been taken into account in the arrangement of tables and the control of noise, temperature, and humidity. From the use of lighting and color, it is clear the space planners paid attention to psychological and physiological comfort factors, and it is also evident this can be accomplished without a commitment to expensive remodeling projects.

As we proceed to the back of the house, we notice that kitchen equipment has been chosen for easy cleaning and maintenance as well as durability. There are convenient access panels for plumbing and electrical connections. Whenever possible, equipment is mobile and can roll between locations. In some cases, even ovens and refrigerators can be rolled on wheels. Aisles are wide enough to accommodate people and equipment. Carts for rolling heavy supplies around are available so people don't have to lift and carry them.

When food is delivered, handling time is minimal because the storage areas are close and organized efficiently. The big walk-ins are nearest the receiving areas, with smaller, ancillary refrigeration units nearer the hot and cold production areas.

The dish room is arranged to vent heat away and to eliminate unnecessary employee movements. Noise from the dish room doesn't spill into the dining area because it is well insulated.

The entire back of the house has been designed for ease of cleaning. Employees can work comfortably because their work areas are well lit and free of grease and fumes. Work surfaces are poised at the correct height for minimal back strain and to avoid reflecting glare from lights.

In the late 1990s, ergonomics became more than an interesting idea. The U.S. Occupational and Health Administration (OSHA) began requiring some employers to adopt full-scale ergonomics programs to change—or at least limit—working conditions that contribute to **musculoskeletal disorders**, or MSDs, such as carpal tunnel syndrome, back pain, and tendonitis. OSHA lists the risk factors for MSDs as:

■ Repetitive work (typing, chopping, etc.)

■ Using power tools

■ Being seated for long periods in the same position

- Maintaining awkward posture while working
- Reaching (either below the knees or above the shoulders) to grasp objects
- Bending or twisting the torso while lifting heavy objects

You can see how quite of few of these activities are part of life in a busy foodservice operation. OSHA-approved plans to minimize MSD risks include several specific steps:

- Have a set procedure for employees to report MSD symptoms or injuries.
- Inform employees of the workplace hazards that may cause MSDs.
- Provide training for all new employees to avoid MSD risks, and repeat it for existing employees at least once every 3 years.
- Analyze each job and task to identify the MSD risk factors, and reduce them to the extent feasible.
- Ask employees for their input in the job/task analysis.
- Provide workers with medical care (at no cost to the worker) for their job-related injuries.

Most of these steps seem like common sense for a conscientious employer, and, according to OSHA, the average annual cost of modifying an otherwise risky job so it is safer is only $150. However, at this writing, about half of U.S. workers are still not covered by any type of company ergonomics program. Our suggestion is that you undertake such a program voluntarily. In the long run, you will save money on workers' compensation insurance, employee medical claims, and time lost due to work-related injuries.

8-3 EMPLOYEE COMFORT AND SAFETY

The manager of a foodservice facility should constantly be thinking about the productivity of his or her workforce. Since the early 1900s, when Frederick Taylor began studying workers on American factory assembly lines, productivity and human engineering have fascinated scientists. The concept is not complex. Anyone who's spent time in a kitchen appreciates the importance of having the right tools and equipment placed in convenient locations in a comfortable environment.

Exactly what are we avoiding by identifying potential hazards and developing safety measures? Here are just a few of the concerns your staff members are exposed to daily:

- Slips and falls on a wet floor, icy sidewalk, or ladders or steps. (Just this category of mishaps results in more than 3 million injuries to employees and another 1 million injuries to guests every year.)
- Cuts from knives, electric slicers, and other sharp kitchen tools.
- Burns from hot oil, hot water, the entire cooking process, and dishwashing.
- Back injuries from moving and handling heavy equipment and bulk supplies.
- Bumps and bruises from dropping things, bumping into things, mid-aisle collisions on busy evenings, and so on.

With all that to think about, when employees are first hired, safety should be taught as a way of life and a part of your business culture. This means training them about how to use equipment correctly, explaining why these safety procedures are in place, and posting the procedures in prominent locations—perhaps as "Safety Checklists" for tasks or appliances. The training, any follow-up discussions, and especially any violations should be noted in writing as part of each employee's master file.

What makes for a safe kitchen work environment? We've already discussed the necessity for noise control, adequate lighting, temperature control, and removal of stale air. Now let's talk about basic equipment needs. Foodservice workers spend about 90% of their time on

BUILDING AND GROUNDS

Slips, Falls, and the Foodservice Industry

Slips and falls are a costly fact of life in the foodservice industry. Consider these statistics:

■ OSHA says slips, trips, and falls account for between 12% and 15% of all Workers' Compensation costs, and that the average cost for a single serious injury is near $30,000.

■ Slip-and-fall accidents accounted for 57% of foodservice general liability insurance claims, according to CNA Insurance.

■ A study by the Food Marketing Institute found that 53% of all Workers' Compensation claims and public liability suits against supermarkets resulted from injuries sustained as a result of slip-and-fall accidents.

■ The Bureau of Labor Statistics (BLS) annual census of occupational injuries and fatalities reported that 3.4 of every 100 foodservice workers hurt themselves from a slip or fall bad enough to require at least a day off from work.

■ The BLS also found that 22% of slip-and-fall injuries resulted in more than a month off work.

their feet. It's not the standing that's the real problem but rather the back and muscle strain that results from having to bend at uncomfortable angles to do repetitive work in a standing position. In an ideal situation, the work surface could be raised or lowered to a height where the person could work without having to bend. More realistically, you can adjust the worktables to the correct height for the persons who use them most often. However, general height recommendations exist; these are listed in Table 8-1.

Another industry rule of thumb is actually a rule of elbow—the height of the worktable should be 4 inches below the worker's elbow. At this height, the person should not have to raise or lower the arms to uncomfortable angles to accomplish tasks. The exposed edges of the table should be smooth or rounded to prevent cuts, bruises, and snagged clothing.

A person works best when tools or materials are within 24 to 36 inches of the center of his or her waistline. A good combination of elbow room without much reaching can be achieved if each worker at a table has no more than 5 linear feet of work space—3 feet if space is tight and supplies can be situated directly in front of the worker. The surface should be brushed or *matte* (not glossy or shiny) stainless steel, and lighting should be designed to minimize glare off both worktables and walls.

Storage of hand tools, small equipment, and supplies is best done on two or three shelves, located in front of the people who will use them, or no farther than 28 to 38 inches from the standing worker. Frequently used items should be stored at eye level; under-tabletop shelving could be added for flat pans, cookie sheets, and so on.

In areas such as walk-in pantries and freezers, store the most frequently used items between eye level and waist level to minimize the risk of back injury. Overall, try to avoid

TABLE 8–1

Height Recommendations for Kitchen Tables

EQUIPMENT USE	SUGGESTED HEIGHT
Worktable, for light work done standing	37–39 inches (for women)
	39–41 inches (for men)
Worktable, for heavy work (kneading, chopping)	35–36 inches
Equipment table (for mixers, etc.)	26–28 inches

using ramps, stairs, and ladders in your kitchen, as these are chronic causes of accidents. If you must use them, equip stairways and ramps with handrails, and mark them with yellow caution tape or paint at top and bottom or on each stair step. Individual steps should be between 5 and 8 inches tall (the height of a step is called its *riser*), and the step should be from 9 to 11 inches deep (from back to front). Some experts suggest the safest steps have depths and risers that are both 8 inches. They also suggest a larger landing area be provided every 10 to 12 steps.

If lifting is required, back braces or lifting belts should be available near storerooms and in the receiving area. Wheeled carts or hand trucks should be used to move heavy or bulky supplies.

When using kitchen equipment, the rules are mostly a combination of common sense and attention to detail. Keep it clean, and keep it well maintained. Other good rules:

- Don't leave equipment where someone might trip on it.
- Don't try to use a piece of equipment for something it was not designed to do.
- Read the instruction manual or owner's manual, and pay attention to warning stickers placed by the manufacturer on the equipment itself.
- Keep electrical cords out of the way and in good condition.
- Check the batteries and chargers of battery-powered equipment.
- Pay attention when the equipment is in use. Keep both hands on it when it is operating. Use safety guards if they are part of the equipment.
- Always unplug it before you adjust or repair it.
- Have a red tag system or another clear way to mark equipment that needs to be repaired so others will know not to use it.

Some restaurants have a policy of "clean it after each use." This is partly to get equipment ready for the next use but also because cleaning it offers a good chance to notice and document any repair that may be needed.

Hot pads and mitts should be stored at all cooking stations to discourage employees from using a good old kitchen towel to lift hot plates, pans, or utensils. The best ones are designed with a liquid-and-vapor barrier to prevent burns. These items should be washed regularly and discarded when they show signs of excessive wear.

Depending on your building's layout, in corners or hallways where people may be emerging from doors into a common corridor, the installation of convex mirrors might help avoid collisions between scurrying workers.

Do not ignore the receiving area when it comes to safety concerns. Your dumpsters should be placed inside a locked enclosure, especially if you have mechanical equipment (such as trash compactors) as part of the setup. Sloped-front dumpsters should be equipped with safety legs to keep them from tipping forward accidentally. Floodlights at the receiving door should be equipped with motion sensors that activate whenever employees are working there and that minimize the number of insects attracted to a light that's always on. Another way to discourage flying bugs is with air screen or plastic strip curtains on open loading docks. Again, lifting belts, back braces, hand trucks, and dollies should be stored in this area.

FLOORING AND FLOOR MATS

OSHA standards address important aspects of floor safety. Standard 1910.22(a)(2) states that all floors must be kept clean and dry, and that where water spills are likely, precautions must be taken. These may include mats, platforms, floor drains—anything that can provide a dry place for workers to stand.

In our view, the least you can do for your kitchen staff is provide a bit of cushioning for the long hours they spend on their feet. Kitchen mats, when properly selected and used, provide traction for employees, minimize accidental breakage, and keep floors cleaner as well.

Rubber mats are arguably the most comfortable, but the longest-wearing mats contain nylon cords melded to the rubber. You can also buy mats made of neoprene rubber (the same material wetsuits are made of), solid vinyl, and sponge vinyl. Beveled or tapered edges provide an extra measure of safety against tripping and are useful in heavy-traffic areas or if you're rolling carts. The mats should not be hard but textured, which forces people to change posture when standing on them for long periods.

Before you buy mats, decide where you will put them. Kitchen mats are usually 3-foot squares that can interlock and between ½ and ⅝ inches thick. Thicker mats provide more cushioning but thinner ones are easier to clean. There are special greaseproof mats for areas near fryers, grills, and griddles; slightly raised bar mats, which drain spilled liquid beneath them to prevent slipping; and vinyl loop pile mats for kitchen entrances and exits, which can be specially treated to make them germ-, mold-, and mildew-resistant. For public entrances, mats are only ¼ inch thick (so people won't trip on them) and should be able to trap incoming snow, ice, moisture, or mud. Your logo or a welcoming message can be emblazoned on the mat. If a mat is to be placed over a carpeted surface, make sure molded nubs on the bottom of the mat grip and adhere to the carpet.

Slippery floors are a safety issue in every kitchen, but mats improve employee comfort, and, in dish rooms, they minimize breakage of dishes and glassware. Where wet floors are a persistent problem, as in dish rooms, you also can keep a fan blowing directly onto the floor and require that workers in these areas wear shoes with nonskid soles. Insisting on slip-resistant shoes is increasingly common in foodservice as a way to minimize restaurant owners' liability. Some purchase the shoes for their employees as part of their uniforms; others spell out acceptable footwear styles and have discipline policies for those who show up repeatedly without the correct footwear. Slip-resistant shoes can reduce the number of slip-and-fall accidents by 70% to 80%.

Even with protective mats, it is the manager's responsibility to see that floors are cleaned often. Today's mats are light, weighing 3 to 5 pounds apiece, for easy lifting and cleaning. They can be taken outdoors to a loading dock or back parking lot and hosed down with high-pressure water and cleaning fluid that melts the grease buildup. Or you can use a motorized floor scrubber/vacuum (sometimes called a *wet-vac*), which has rotary brushes and uses hot water and detergent to loosen grime from mats and floors and then whisks away most of the moisture with an absorbent squeegee located at the back of the machine.

The floor covering you select will go a long way toward ensuring a safe working environment. You will hear the term ***aggregate*** to describe nonskid floor surfaces. This means something material has been added, usually to a top or second layer of the floor material, to give it more friction. It might be fine sand, clay, tiny bits of gravel or metal, carborundum chips, or silicon carbide.

In addition to selecting flooring material for its slip resistance, consider whether it is easy to clean. Some aggregate floors are almost impossible to mop; a floor-scrubbing machine is needed to effectively clean them.

Flooring choices are numerous, as shown in Table 8-2, but few of them are truly capable of withstanding the rigors of the foodservice kitchen. Among veterans, unglazed quarry tile is the undisputed favorite. Quarry tile contains bits of ceramic (usually carborundum chips), clay, and/or silicon carbide for slip resistance. Even when it wears down in a high-traffic area, it maintains its skid-proof quality. Quarry tile should be at least ⅜ inch thick; in areas with heavy foot traffic or where appliances are frequently rolled, a ½ inch or ¾ inch thickness is preferred.

The tile is only part of the puzzle, however. It must be grouted into place, and that grout must be able to withstand significant differences in pressure, temperature, and moisture as well as exposure to grease and chemicals. The recommendation is a so-called thickset mud bed, using either an epoxy or a special cement with latex added for elasticity. Epoxy is generally considered more durable, but either type of grout is susceptible to deterioration if it is constantly wet. The tile is laid onto a setting bed of cement, which should be sloped toward the floor drains. This facilitates steam or pressure cleaning and keeps water from standing beneath the tiles and damaging the floor over time.

Another option is a one-piece floor, poured all at once and made of epoxy, polyester, urethane, or magnesite cement. This is also called a ***monolithic floor***. An epoxy composite floor consists of several layers: the epoxy or resin, then an intermediate sealer layer, an aggregate layer to make the floor slip-resistant, and a clear sealing finish (also epoxy). The resulting floor is ³⁄₁₆ to ¼ inch thick and can withstand extreme temperatures around fryers and other cooking equipment. An epoxy composite floor is roughly half the strength of a quarry tile floor, but its chief

TABLE 8-2

Types of Flooring

	ADVANTAGES	DISADVANTAGES
Asphalt tile	Low cost	Least resilient
	Large color selection	Affected by oils, kerosene
	Moderate maintenance	Possible cracking or deterioration from temperature extremes
Vinyl tile	Wide color and pattern range	Prone to scratching and scuffing
	Resistant to oils and solvents	
	High resiliency	
Carpeting	Sound absorbent	Greater-than-average maintenance needs
	Decorative	Moderate to high initial cost
	Comfortable	Affected by spills
Linoleum	Available in sheets or tiles	Lacks resistance to moisture and alkaline solutions
	Wide color range	
	Low cost	
	Good resilience	
Rubber tile	High resilience	Lacks resistance to oils and solvents
	Wide color selection	
Quarry tile	Resistant to moisture	High initial cost
	Low maintenance	Grout subject to deterioration
Terrazzo	Decorative	High initial cost
	Durable	Unsuitable for areas subject to spillage
	Low maintenance	
Wood	Decorative	Unsuitable for areas subject to excessive moisture and oil spillage
	Comfortable	Moderate care and maintenance required
Concrete	Durable	Dusting and blooming
	Low maintenance	
	Low to moderate initial cost	

Source: Reproduced with the permission of the National Restaurant Association.

advantage is that it is seamless—there is no grout to clean or maintain. It is skid- and scuff-resistant, resilient, and unharmed by moisture, food spills, acidic substances, and the like. Epoxy is also easy to repair—just grind down the worn or damaged part and pour a new layer there.

Composite sheet vinyl is another type of seamless flooring, laid in sheets on a bare floor with a layer of adhesive. For foodservice applications, an extra top layer of aggregate is mixed with the vinyl for slip resistance. Heat welding is used to seal the perimeter of the sheets onto the floor. Sheet vinyl is attractive, durable, and has excellent resistance to moisture and food acids. It is superior for use in wet areas, although this type of floor is susceptible to scratches, dents, and some stains. Today's sheet vinyl comes in all kinds of funky colors and patterns that mimic everything from wood to slate to jungle prints, so it is suitable for front-of-house areas as well as kitchens.

Hubbelite is a compound mixture of cement, copper, limestone, magnesium, and a few other ingredients. It is about ½ inch thick, strong, slip-resistant, and comfortable to walk on, and it is gaining popularity in remodels because it can be installed over any existing floor type, including quarry tile. Hubbelite has excellent stain and moisture resistance, and the addition of copper and magnesium makes it somewhat insect-repellent and prevents bacteria and mold growth. Cleaning it is also simple; just mop it. The downsides are that Hubbelite comes in only one color, brick red, and you must choose an installer who is familiar with the product to do a good job with it.

Ceramic tile can be used for both floors and walls, in kitchens or dining areas. It is durable and decorative, and it can immediately bestow charm and character or add artsy, ethnic touches to any space. Commercial-grade tile is usually baked clay covered with a glaze to make it sturdier. New technology has given new life to this ancient material, with many more varieties, colors, styles, and patterns of tile readily available. It is simple to clean, too, with soap and water. The more stylish and decorative the tile, the more expensive it is, so budget-minded restaurateurs may opt for tile only on half-walls, some countertops, or as accents in display kitchens.

The lowest-cost types of flooring are vinyl tile and sealed concrete. Vinyl tile is considered fairly high maintenance and does not offer maximum traction or heat resistance, although it is easy to install and clean. We'd recommend it only for use in dry storage areas, never in food prep or dishwashing areas. In some jurisdictions, health regulations do not allow its use for commercial foodservice.

A concrete floor sealed with a water-based epoxy coating is a good choice. Be sure the dry thickness of the coating is 1.5 to 2.5 millimeters. Whether you choose vinyl, tile, or a one-piece poured floor, it is critical that it be properly bonded to the concrete slab beneath it. This is called *curing the floor*, and if it isn't done correctly, the floor will eventually crack. No artificial curing agents or sealers should be used; they will destroy the adhesives used to set the floor and keep it in place.

For public spaces, terrazzo flooring is an attention-getting option. Terrazzo begins with a highly durable cement or epoxy base into which chips of granite, marble, tile or even seashells may be mixed. It can be pricey and it takes a long time to install, but it is durable and very attractive.

Hardwood floors are a popular option in dining areas, and technology has produced more durable products and finishes than ever before. Known as *engineered wood products (EWP)*, these materials also are considered environmentally friendly because they are made from scrap wood. Softer woods, such as white pine, do show their age—and that's exactly why some designers choose them, to take on a comfortable, worn look with use. Mats or area rugs can be used for especially high-traffic spots. Cleaning is a simple routine of sweeping, dust mopping, and vacuuming; the precaution here is to use cleaning products formulated for wood floors. Avoid products made for dusting wooden surfaces—they'll only make a floor slippery.

Carpeting is never an option in the kitchen, but in dining areas, beautiful carpet can make a lasting impression. Worn and soiled carpets, however, can make an equally strong impression: that the place is dirty and the staff is inattentive at best, oblivious at worst. In fact, dust mites, fungi, and bacteria can live and thrive in carpets that are not well cared for. In busy commercial settings, it is probably best to install carpet by the roll rather than squares—fewer edges to lift up and cause people to trip. Carpet has a definite life cycle and must be replaced periodically, but it is often preferred in dining areas for its ability to suppress noise.

FLOOR CLEANING AND MAINTENANCE

For floors that are carpeted, you'll need an action plan to keep them clean, safe, and looking terrific. Here are five practical tips for carpet maintenance:

1. Make sure the carpet is properly installed. If it's not done right, it will buckle and wear out more quickly. Choose patterns and blended colors over a single, solid color, which shows more dirt.

2. Vacuum frequently so dirt will not accumulate. We'll talk more about vacuums in a moment. When vacuuming is not an option, keep dirt at a minimum by using a sweeper.

3. Place mats at all entrances; about 70% of the soil on carpets is brought in by foot traffic from outdoors and from the kitchen. Entrance mats should be 12 to 15 feet long; 30% of soil is trapped on the first 3 feet of mat. Of course, keep the mats clean, or they'll only contribute to the problem.

4. Implement a spot removal plan. Ask the installer or manufacturer for the correct method (scrape, spray, blot, etc.) and products to clean up those inevitable spills.

5. Clean by extraction. Contract with a professional carpet cleaner to deep-clean and remove any harmful bacteria buildup. Most dining rooms could use this service every other month.

To select the vacuum cleaners to care for your carpet, ask yourself these questions: Is the area to be cleaned a large, open area or a small, enclosed space? Will the vacuum be carried, wheeled, or rolled on a cart to points of use? Should the vacuum be able to dust and do edge cleaning as well as carpet cleaning? Are there special requirements for indoor air quality, noise levels, or power requirements?

The most popular commercial-use vacuum cleaners are the *traditional upright* with one motor and *direct airflow* and the *two-motor upright* with *indirect airflow*. The first is a vacuum that draws air and dirt directly from the carpet through the motor fan and into a soft cloth dust collection bag that is emptied or replaced when full. The single-motor vacuum is called traditional because it has been around for years and has one basic use: to vacuum carpets. Its nozzle comes in standard widths of 12 to 16 inches. The two-motor label identifies upright vacuums that have one motor to drive the brush rollers and a second motor to provide suction. These units are called indirect because dirt and air must be pulled from the carpet, through the dust bag, and into the suction motor. These vacuums are sometimes touted as **clean-air vacuums** because the incoming dust is filtered to the dust bag before entering the motor. Two-motor models are usually larger than direct-airflow vacuums, have a hard-shell case in which the dust bag fits, can vacuum bare floors as well as carpeted areas, and have standard nozzle widths of 14 to 18 inches.

A special version of the two-motor model, called the **wide-area vacuum**, has a nozzle width of up to 36 inches, a longer dust bag, and no hose for attachments. You may see them in use in airports or ballrooms.

Don't bother using ampere or horsepower ratings to determine how much suction or power your vacuum is capable of producing. Instead, ask about its *cfm rating*. This is the number of cubic feet per minute pulled through the 1¼-inch orifice that is the typical size of a vacuum's nozzle opening. The higher the cfm number, the greater the airflow and, therefore, the better its cleaning ability. Units with a high cfm rating measured at the nozzle opening and a revolving brush roller have the best overall ability to clean a carpet.

Like any appliance, vacuums require regular maintenance to keep them working at their peak. They have a distinctive motor noise you'll become familiar with, and your first sign of trouble is if the vacuum sounds different than usual. This may mean a belt is wearing thin, the brush roller is slipping, the dust bag is full, or the suction motor's fan should be replaced.

A good rule to teach your cleaning crew is: If it is not commonly found on the floor, don't vacuum it up, pick it up. This certainly prevents many vacuum cleaner breakdowns. Of course, some manufacturers offer magnetic bars in their vacuums to snag paper clips, thumbtacks, and the like, but these lull the user into a complacency that usually results in the malfunction of the machine.

When vacuuming is not possible (usually because guests are present), it's good to have a couple of nonmotorized carpet sweepers. Theses have rotary blades that spin with the simple action of being pulled or pushed across a floor. Sweepers work on carpet or bare floors, picking up everything from dust to glass particles to wet or dry food bits and depositing them all in an internal dustpan. They are easy to use and easy to clean.

A wet-vacuum capable of sucking up water is a necessity for noncarpeted areas. The appliance of choice is the *dry steam vapor cleaner*, a portable unit that includes a water tank and assortment of hoses, brushes, and wandlike attachments for reaching tough-to-access spots. The hot steam (230°F to 330°F) cuts through just about any type of kitchen grime and grease. This hardworking machine can clean floors, restrooms, ceramic tile walls, tabletops, vinyl or plastic chairs, and the grout between tiles. It handily tackles the greasy challenge of the exhaust hood and stainless-steel back-panel walls of the hot line, steaming the grease and melting it enough to be wiped away with a cloth or squeegee, then mopped and allowed to dry.

Modern commercial floors are cleaned with soap and warm water. The cleaning products are formulated to attack oily, fatty residues and soften them for easier removal. Choose products that work on both concrete and quarry tile, and use them with the correct types of cleaning tools. There are floor-scrubbing brushes designed to clean grout lines between tiles. Brushes with epoxy-set nylon bristles and aluminum brackets and handles are more sanitary than cotton mop heads and wooden handles that absorb dirt and foster bacteria growth.

An important fact overlooked in many commercial kitchens is that simply mopping a floor is not sufficient to clean it. Mopping actually spreads the soil around, depositing some into the grouting and most into the mop bucket—to be reapplied to the floor. Therefore, several sets of brushes, mops, and buckets are necessary. This also prevents cross-contamination between different parts of the restaurant, where different cleaning chemicals may be used. It is best to use a two-compartment bucket, or separate wash and rinse buckets. The correct cleaning procedure is:

1. Sweep the floor before mopping.
2. Scrub the corners and hard-to-reach places by hand, with a scrub brush.
3. Use a squeegee to move the soiled water from scrubbing into a central area.
4. Use a wet-vacuum to remove the water from the floor.
5. Use the mop to touch up the newly wet-vacuumed area.

Mop buckets are offered in sizes from 26 quarts to 44 quarts, with wringers to accommodate mop heads from 8 ounces to 36 ounces. Mop sizes are based on the weight of the cotton used to make the mop strands. What we call a broom for home use is known as a *floor brush* or *deck brush* in the commercial kitchen. These are 10 to 12 inches wide and come cut in several angles for cleaning flat surfaces, reaching beneath appliances, and cleaning along baseboards; wider (36-inch) brushes are called *lobby brushes*. Whenever floors are damp-mopped or scrubbed, you must also have an adequate supply of "Danger, Wet Floor" cones or caution signage to alert people that the floor is wet. In fact, OSHA requires this in Standard 1910.145(c)(2).

You might also want to reread the discussion in Chapter 6 about electrolyzed water (E-water) for its use in commercial cleaning and sanitizing.

For dry cleanup, good deck brushes are essential. Some manufacturers offer them with boar's-hair bristles, nicknamed *Chinese pig brushes*, which are useful for cleaning up fine particles such as flour and sugar. Synthetic bristles are good for all-around use. Whatever your choice, an angled broom is most useful to reach beneath tables and into tight corners.

CLEANING STAINLESS STEEL

Regular maintenance has a green component: It reduces the need for harsh chemicals and aggressive cleaning procedures. You should select cleaning products only after you understand your restaurant's needs. What type of surface are you cleaning? Preserving it is as important as cleaning it. What types of substances will you be cleaning up? Is the water hard or soft? What is the cleaning process, and how often will you use this particular product? Two organizations, Green Seal and the Green Restaurant Association, certify cleaning products as eco-friendly. Recent additions to the mix include a federally approved disinfectant made from thymol, a derivative of the herb thyme, and products that contain promicrobials, beneficial enzymes that work to dissolve bacteria. Green Seal established a standard (GS-37) for restroom chemicals, glass cleaners, and general-purpose cleaners.

Because many of the surfaces and appliances in a commercial kitchen are made of stainless steel, your employees need to know how to keep it looking its best. Most people assume stainless steel is rustproof, but this is not so. Some background information: Metals that corrode in a natural environment are known in the scientific world as *active*; those that do not are called *passive metals*. Stainless steel is an alloy containing chromium and nickel, two passive metals that shield it against corrosion by acting as a sort of protective coating. Over time, however, three types of wear can break down this protective film:

1. *Mechanical abrasion.* Steel wool pads, wire brushes, and metal scrapers can abrade stainless steel.
2. *Water.* Hot, hard water can leave behind mineral deposits, often called *scales*, **scaling**, or *lime buildup*. On steel, these look like a white film.

3. *Chloride.* This is a type of chlorine found naturally in water, salt, and many foods. If you boil water in a stainless-steel pot, the discoloration you may see is most likely caused by chlorides. Many cleaners and soap products also contain them.

Rust buildup makes it impossible to sanitize a surface, so it's important to prevent it by using nonabrasive cloth and/or plastic scouring pads, with alkaline or nonchloride cleaning products at their recommended strengths. After washing, the stainless steel should be rinsed thoroughly, wiped to dampness, and allowed to air-dry completely. (Oxygen helps maintain the passivity of the coating.) Some equipment may need to be descaled regularly with a deliming agent. If scaling seems to be a persistent problem, soften the hard water with a filtration system.

Before you wipe down a stainless-steel surface, look for the *grain*— the tiny lines that show the direction in which it was first polished by the manufacturer. Wipe parallel to the grain, in the direction of the lines.

CLEANING SUPPLIES AND CHEMICALS

Ammonia, chlorine, ethanol, and formaldehyde are just a few of the chemicals that may be found in products in a commercial kitchen. They are also on OSHA's list of hazardous substances, a lengthy compilation of substances the agency has classified as irritants, corrosives, carcinogens, neurotoxins, and so on.

If any product on your premises is or contains a hazardous chemical, federal law requires you to have a Material Safety Data Sheet (MSDS) on the premises. OSHA says an MSDS must contain the following information:

- The names of the chemicals in the product and any common names that may be used for them.
- The chemical manufacturer's name and contact information.
- The physical and chemical characteristics of the product's hazardous ingredients (such as appearance, odor, pressure).
- The health hazards potentially posed by the chemical (any symptoms it may cause or medical conditions that may be worsened by exposure to it.)
- The physical hazards it may pose (flammable, combustible).
- How it would come into contact with a person to cause these problems (if it can be ingested, inhaled, etc.)
- The permissible exposure limit set by OSHA.
- Emergency first aid procedures pertinent to this type of chemical exposure.
- Precautions for safely storing, using, and disposing of it.
- The date the MSDS was prepared.

Most often, the manufacturers of these products have already created these sheets and can provide them.

Of course, all this information is of no use unless your staff knows where to find these sheets and how to use them. This should be part of any new employee's orientation process as well as the subject of periodic refresher courses.

GREEN CLEANING. The fast-growing green cleaning trend is an attempt to achieve satisfactory cleaning and sanitizing results using fewer harsh chemicals as well as cleaning techniques that conserve water, energy, and even such items as paper towels.

The EPA has jumped on this bandwagon with its Design for the Environment (DfE) Safer Product Labeling Program. As part of the agency's Office of Pollution Prevention and Toxics, its goal is to reward manufacturers for formulating products (including cleaning products) that are safer for humans and the environment, through a process that it refers to as "informed substitution." To be considered "safer" choices, the EPA says the new product formulations should exhibit as many of the following characteristics as possible:

■ They should be technically feasible.

■ They should deliver a product that is the same or better value as the original in terms of performance and cost.

■ They should provide an "improved profile for health and the environment."

■ They should account for "economic and social considerations."

■ They should have the "potential to result in lasting change."

The agency claims its standards are tough and that products are reviewed in depth before being allowed to add a DfE label to their packaging. However, its definition of the "safest possible ingredients" is as follows: "The product contains only those ingredients that provide the least concern among chemicals in their class."

In short, they may still contain chemicals. How green is that? The EPA maintains an extensive online list of accepted products in the DfE program, if you want to check it out. The nonprofit watchdog organizations Green Seal and the Environmental Working Group also publish their own lists and rankings of cleaning products based on their chemical content.

In the meantime, we'll share a list of tips originally compiled years ago, before "green cleaning" was a catchphrase. These first appeared in a 1998 edition of *Lodging*, a publication of the American Hotel and Lodging Association. Today, each point still applies for business owners who are interested in minimizing chemical use in their cleaning products.

1. Is it biodegradable? Whenever possible, biodegradable products are preferred for their minimal environmental impact.

2. Is it nontoxic? Choose products that can't harm workers, guests, or food products.

3. Does it contain phosphates? These encourage the growth of algae, which depletes oxygen and kills aquatic life in waterways. (You will learn more about these in Chapter 16's discussion of dishwashing soaps.)

4. Does it contain chlorine-based bleach? In wastewater, chlorine may combine with other chemicals to form toxic compounds.

5. Is it concentrated? Products that can be diluted in cold water save money and energy.

6. Does it contain a high concentration of volatile organic compounds (VOCs)? In sunlight, VOCs react with nitrous oxides to produce ozone and photoelectric smog.

7. Is it made from a petroleum derivative? Look instead for a purely oil-based product, with a pH level as close to 7 as possible to avoid irritation if the product comes into skin contact. (A pH level of 1 is the most acidic; 14 is the most alkaline; 7 is neutral.)

8-4 SANITATION

In foodservice, cleaning is just not clean enough. *Sanitizing* is the goal, which means the additional treatment of equipment and utensils *after* basic cleaning to kill germs and bacteria. It is a job that is never really finished, and the industry must do a better job when it comes to sanitation. In commercial kitchens, the U.S. Centers for Disease Control and Prevention report food-borne illnesses caused by cross-contamination and unsanitary conditions strike nearly 76 million people and result in 5000 deaths per year.

NSF International (the initials stand for the group's original name, National Sanitation Foundation) sets voluntary standards for the hospitality industry and acts as a national clearinghouse and arbiter of sanitation-related design ideas for equipment and facilities. As its mission statement reads, NSFI is "dedicated to the prevention of illness, the promotion of health and the enrichment of the quality of American living."

Although your goals don't have to be quite that lofty, following the NSFI guidelines ensures you will meet a thorough and rigorous list of standards and be able to pass your health inspections. Also, the NSFI allows its insignia (like a seal of approval) on equipment that meets its standards. You may see NSFI Standard 2 on foodservice equipment, NSFI Standard 3 on spray-type dishwashing machines, and NSFI Standard 4 on commercial cooking and hot-food storage equipment.

From NSFI's publication *Food Service Equipment Standards*, items that meet these standards "shall be designed and constructed in a way to exclude vermin, dust, dirt, splash or spillage from the food zone, and be easily cleaned, maintained and serviced." The standards also specify that the materials used in the construction of equipment withstand normal wear and tear; rodent penetration; and any corrosive action of food, beverage, or cleaning compounds. This includes the food contact surfaces as well as safely rounded corners and exposed parts or edges. NSFI also makes many suggestions for equipment installation with ease of maintenance and cleaning in mind. Among the terms defined in the NSFI standards are a few you should probably be familiar with, listed here alphabetically:

Accessible. Readily exposed for proper and thorough cleaning and inspection.

Cleaning. The physical removal of residues of dirt, dust, foreign matter, or other soiling ingredients or materials.

Closed. Spaces that have no opening large enough for the entrance of insects or rodents; an opening $1/32$ inch or less shall be considered "closed."

Corrosion resistant. Materials that maintain their original surface characteristics under prolonged contact with foods and normal exposure to cleaning agents and sanitizing solutions.

Food zone. Also called *food contact surface*, this includes the surfaces of equipment with which food or beverage normally comes into contact and those surfaces with which food or beverage is likely, in normal operation, to come into contact with and drain back onto surfaces normally in contact with food or into food. (Yes, read that one twice!)

Nonfood zone. Also called *nonfood contact surface*, this includes all exposed surfaces not in the food or splash zones.

Removable. Capable of being taken from the main unit with the use of simple tools such as pliers, a screwdriver, or a wrench.

Sanitizing. Antibacterial treatment of clean surfaces of equipment and utensils by a process that has been proven effective.

Sealed. Spaces that have no openings that would permit the entry of insects, rodents, dust, or moisture seepage.

Smooth. Surfaces free of pits or occlusions and having a cleanable (for food zones) Number 3 finish on stainless steel; or (for splash and nonfood zones) commercial grade hot-rolled steel, free of visible scale.

Splash zone. Surfaces other than food zones that are subject to routine splash, spillage, and contamination during normal use, also called *splash contact surface*.

Toxic. A food-related condition that has adverse physiological effects on humans.

The NSFI standards also require that all utility service lines and openings through floors and walls be properly sealed and fitted with protective shields or guards. When horizontal pipes and lines are required, they must be kept at least 6 inches above the floor and 1 inch away from walls and other pipes.

Sink interiors are also considered food contact zones and should meet the minimum standards; wooden bakers' tables, synthetic cutting boards, and shelves all have their own standards.

Interestingly, a great deal of thought is put into how the legs and feet of equipment are built. Here you will encounter terms such as *bullet leg* and *gusset*, which you probably never thought you'd need to know when you decided on a foodservice career. It's handy to have an organization like the NSFI paying attention to this type of detail and sharing its expertise.

In the chapters that follow, we'll talk more about how to buy and install equipment—from the legs up.

HACCP STANDARDS

Most foodservice businesses are spot-checked by health inspectors for cleanliness and correct food temperatures and cited for code violations. Illustration 8-2 is a sample inspector's report form. Too many restaurants try to please the inspector when what they should be doing is striving for clean, safe conditions because it's the right thing to do, day in and day out. The food safety system that's been in use since the 1960s is *HACCP—Hazard Analysis of Critical Control Points*, which was developed by the National Aeronautics and Space Administration to evaluate its methods of assuring that all foods produced for U.S. astronauts were free of bacterial pathogens. Nothing could be worse than having gastrointestinal problems in space!

Even today, decades later, HACCP is considered the absolute standard for food safety, far more effective than simply spot-checking for violations. It combines up-to-date technical information with step-by-step procedures to help operators evaluate and monitor the flow of food through their facilities. The core objective of the process is to identify and control the three types of food safety threats in any commercial kitchen:

1. *Biological contaminants or microorganisms.* These include bacteria, viruses, and parasites, which already exist in and on many raw food products and can be passed on by unknowing employees or customers.

2. *Chemical contaminants.* These can come from improper storage or handling of cleaning products or pesticides, from cross-contamination, or from substitutions of certain recipe ingredients.

3. *Physical contaminants.* These are the most common cause of food contamination— foreign objects in the food, including hair, bits of plastic or glass, metal slivers, and the like—which can be deadly if choked on.

Here are the seven basic HACCP steps:

1. Identify hazards and assess their severity and risks.
2. Determine critical control points (CCPs) in food preparation.
3. Determine critical control limits (CCLs) for each CCP identified.
4. Monitor critical control points and record data.
5. Take corrective action whenever monitoring indicates a critical limit is exceeded.
6. Establish an effective recordkeeping system to document the HACCP system.
7. Establish procedures to verify the HACCP system is working as intended.

The first step is to decide what hazards exist at each stage of a food's journey through your kitchen and how serious each is in terms of your overall safety priorities. On your own checklist, this may include these items:

- Reviewing recipes, paying careful attention to times for thawing, cooking, cooling, reheating, and handling leftovers.
- Giving employees thermometers or temperature probes and teaching them how to use them. Correctly calibrating these devices.
- Inspecting all fresh and frozen produce on delivery.
- Requiring handwashing at certain points in the food preparation process and showing employees the correct way to wash for maximum sanitation.
- Adding quick-chill capability to cool foods more quickly in amounts over 1 gallon or 4 pounds.

**FOOD ESTABLISHMENT
INSPECTION REPORT**

Business name: _____

Date: _____

Operator: _____

Address: _____

County: _____

City: _____

State: _____ Zip: _____

Inspection Type: _____

Inspector: _____

(Regular, Enforcement, Epidemiology, Follow-Up, HACCP, Investigation, Pre-Opening):

CRITICAL ITEMS – These are items that relate directly to factors which lead to foodborne illness They require IMMEDIATE attention!

Violation Code	DESCRIPTION	Correction Date

Inspection Time: _____ (minutes)

HACCP Time: _____ (minutes)

Risk Type   L   M   H

On-site Follow-Up Report Date:

_____

Follow-Up Report Date:

_____

Enforcement Inspection Required?   YES   NO

_____

_____

NONCRITICAL ITEMS – These violations relate to overall cleanliness and/or maintenance of food operations.

Violation Code	DESCRIPTION	Correction Date

SCORE
(Number of Critical Violation)

COMMENTS: _____

_____

_____

_____

_____

(Continue on reverse if needed.)

The items recorded above must be corrected according to the specified dates on this form. Failure to comply will result in an enforcement inspection, the cost of which will be charged to the business operator; or in denial, suspension revocation, or failure to renew your license. A hearing on any disputed charges may be scheduled at your request.

_____

Business Owner, Operator, or Mgr. on Duty

_____

County Health Inspector

ILLUSTRATION 8-2 A typical report form a health inspector might use when he or she tours a foodservice operation. Code violations would be noted at the bottom of the form.

IN THE KITCHEN

Critical Control Points for Protecting Food

1. Raw food is cooked to correct minimum internal temperature.

2. Hot food must be maintained at a minimum temperature of 140°F (60°C) or above while holding and serving.

3. Leftover hot food is cooled quickly and safely to 70°F (21°C) within 2 hours and from 70°F (21°C) to 40°F (4°C) or below within an additional 4 hours.

4. Leftover hot food is reheated quickly (within 2 hours) to 165°F (74°C).

5. Cold food must be maintained at 40°F (4°C) or below while holding and serving.

6. Refrigeration equipment should be maintained at 36°F to 38°F (2°C to 3°C).

Source: The International Food Safety Council, Chicago, Illinois.

There are as many of these possibilities as there are restaurants.

The second step is to identify *critical control points (CCPs)*. This means any point or procedure in your system where loss of control may result in a health risk. If workers use the same cutting boards to dice vegetables and debone chickens without washing them between uses, *that* is a CCP in need of improvement. Vendor delivery vehicles should be inspected for cleanliness; product temperatures must be kept within 5°F of optimum; expiration dates on food items must be clearly marked; utensils must be sanitized—and the list goes on and on.

The third step is to determine the standards and limits for what is acceptable and what is not in each of the CCP areas for your kitchen.

The fourth step in the HACCP system is to monitor all the steps you pinpointed in Step 2 for a specific period to be sure each area of concern is taken care of correctly. Some CCPs may remain on the list indefinitely, for constant monitoring; others, once you get the procedure correct, may be removed from the list after several months. Still others may be added to the monitoring list as needed.

Step 5 kicks in whenever you see that one of your critical limits (set in Step 3) has been exceeded and that corrective action must be taken.

Step 6 requires that you document this whole process. Without documentation, it is difficult to chart whatever progress your facility might be making. If a problem affects customer health or safety, having written records is also very important.

Finally, Step 7 requires that you establish a procedure to verify whether the HACCP system is working for you. This may mean a committee that meets regularly to discuss health and safety issues and to go over the documentation required in Step 6.

Despite its thorough and science-based approach to food safety, some operators are reluctant to use the HACCP program because of its technical language and the fair amount of procedural discipline and documentation it involves. For this reason, in June 2005 the U.S. Department of Agriculture released a new, simpler set of implementation guidelines aimed at school foodservice. Unfortunately, the name is as unwieldy as some believe is the HACCP program itself: "Guidelines for School Food Authorities Developing a School Food Safety Program Based on the Process Approach to HACCP Principles." Download it at www.schoolnutrition.org and take a closer look.

FIGHTING BACTERIA AND MOLD

Now that you know some of the temperature and cleanliness requirements for food safety, exactly how do you go about achieving them? Part of this critical task is selecting the best equipment for the job. The foodservice industry has been transformed to meet the challenges of HACCP. Perhaps the most stringent requirements have been noted in the area of refrigeration, as HACCP standards emphasize prompt refrigeration of cooked foods and keeping cold foods at a constant temperature. In the past decade, blast chillers have become standard in most large kitchens. (You'll learn more about this quick-chilling technique and its many applications in Chapter 15.) Display cases and salad bars have improved to maintain constant 41°F temperatures without freezing foods.

In other areas of the kitchen, warewashing equipment has been upgraded. Pot sinks feature more powerful sprayers and separate water heaters, helping ensure pots and pans are sanitized after use. We discussed mechanized handwashing systems in Chapter 6. Temperature and humidity monitoring equipment has also greatly improved. It is more affordable, more portable, and easier to use. A single temperature probe doesn't fit all situations, so thermometers and probes are designed expressly for fryers, freezers, walk-in coolers, storage areas, and so on. Many of them are color-coded so it's easy to see when a HACCP step is being met or violated. Others have remote recording devices to track temperature fluctuations and pinpoint potential problems.

Antimicrobials are making their mark in foodservice operations because of their ability to retard the growth of dangerous microorganisms. (These are not to be confused with *promicrobials*, the beneficial enzymes used in some cleaning products.) The healthcare industry has used them since the early 1990s, and it makes sense they would also have food-related applications. An antimicrobial is a substance that naturally retards bacteria growth. The companies that make them are somewhat secretive about the contents, but a few of the substances are silver ions, ozone injected into water, chlorine dioxide, and a compound called *nano-antibiotic mother granule*, or NAMG for short. Some antimicrobials can be mixed with plastic, rubber, and other materials in the manufacturing process to form a long-lasting safety barrier on cutting boards, knives and slicing blades, ice machines, wire shelving, appliance legs, and protective gloves. They become part of the molecular structure of the item, so they are effective for its useful life.

If there's a downside to the popularity of antimicrobials, it is the false sense of security they can impart in a busy kitchen. Employees assume a quick rinse-off is sufficient to clean a cutting board or that they can dip a utensil in an antimicrobial solution and the sanitizing work is done. In fact, antimicrobials are just one more level of defense against food-borne illness. They cannot prevent cross-contamination. They are not disinfectants, and there's no claim by manufacturers that they will kill microbes. They simply retard their growth by creating a surface that is unfriendly to microorganisms. The EPA regulates the use of antimicrobials (and their safety claims), as does the FDA. NSF International has a certification process for them if they come into contact with food during their common use.

Another top health hazard in commercial buildings today is mold. While fatal to only a small percentage of the population, mold spores trigger allergies and asthma and can prove debilitating to people with respiratory problems or compromised immune systems. Even people without these problems can suffer mold's irritating effects: a burning sensation in the eyes and nose, coughing, and skin irritation. Mold grows where moisture is persistent, so it is a particular problem in humid climates. If contained within walls, all affected portions must be dried, stripped, disinfected (to prevent more mold from growing), and dried again—or perhaps replaced. Kitchens are full of mold growth opportunities, which make it important to:

- Repair leaking plumbing fixtures promptly.
- Increase ventilation of damp spaces in cold weather.
- Dehumidify in warm weather.
- Install inexpensive meters that monitor both humidity and temperature.

The more absorbent a material is, the more likely it must be replaced if it becomes moldy. Bleach is typically used to kill mold spores in dish rooms and restrooms, but it is not effective on drywall or other thick materials because it doesn't penetrate deeply enough to do the job. In these cases, experts recommend a spray-on borate solution that crystallizes (and has the side benefit of repelling cockroaches, another humid-climate problem). Green-board can be used in new construction; this is drywall with fiberglass backing rather than paper. It is more expensive, but it prevents moisture buildup.

SAFE FOOD HANDLING

We've mentioned protective gloves, and, as in so many other industries, foodservice workers are required to wear them in many municipalities to prevent the spread of germs. Interestingly, however, the use of gloves in commercial kitchens remains a topic of debate. A University of Oklahoma study (reported in the May 2005 issue of *Food Safety Insights* magazine) revealed "no substantial difference" in the microbial counts of food samples handled with clean, bare hands versus gloved hands. Others complain that wearing gloves leads to laziness about handwashing practices; that the gloves themselves are awkward and can contribute to accidents; and that gloves pass along contamination because most workers don't change them frequently enough, or change them correctly—that is, by pulling at the wrist area to turn the glove inside out as it comes off the hand. In short, like handwashing practices, gloves are only as effective as the worker's resolve to use them correctly and the management's rules about doing so.

Gloves are easier to become accustomed to when they fit properly and are task-specific. The kitchen should stock gloves in a number of sizes and thicknesses. There are now cut-resistant gloves for workers who use knives and other sharp-edged tools. People with latex allergies can choose nonlatex gloves. Another potential breakthrough being studied at this time is a glove that, when exposed to light or moisture, emits chlorine dioxide, a water-soluble gas that can be used as a disinfectant.

Another key to using gloves correctly in the commercial kitchen is that they be changed frequently. We have observed gloved workers moving about the kitchen area, touching anything and everything in the process and wearing the same gloves for hours at a time. Remember, the gloves don't protect the customers if they are not kept clean, and they are to be used only for the task for which they are put on in the first place.

We discussed the importance of the handwashing station in Chapter 6, and the natural follow-up here is its correct use. To effectively manage sanitation, restaurant managers must set standards for methods and frequency of handwashing. Basic handwashing procedures are covered in virtually every food safety training program, so the culprit isn't usually *how* people wash their hands but *how often*. For some groups of customers—the elderly, children, and hospital patients, to name a few—requirements may be more stringent than the usual.

This type of management is a matter of process control, and good process control requires good documentation. Experts suggest determining a baseline by documenting handwashing practices meticulously for 6 to 8 weeks, whether you use a sign-in sheet or a wireless electronic device that logs who washed when. (There are meters or counters available nowadays for soap dispensers, towel holders, and water faucets.) The baseline is then communicated to the workforce, and ideas are solicited to improve on it, both individually and as a team. We're told that the use of hands-free or so-called touchless sink technology is a hit with younger foodservice workers, who seem to like its ease of use and think the new gadgetry is cool.

Concern is growing that, with or without exposure to human hands, cold foods are most likely to contain food-borne pathogens because they are not exposed to the heat of the cooking process. Let's take a fresh green salad as an example, and use commonsense disease-controlling steps. First, use vegetable brushes on nonleafy vegetables and wash them under running (not stagnant) water. A water rinse removes most fertilizer and pesticide residue, but it's no longer considered sufficient for most produce. Instead, fruit and vegetable cleaning solutions do a more thorough job. All too often, people use their prep sinks to wash greens—but how well have the sinks been cleaned? Do you also use the sink to defrost chicken, risking cross-contamination? Some foodservice kitchens purchase a semiautomatic greens washer. This machine agitates leafy greens (and most can do other types of produce too) to remove embedded dirt, then tumbles them into a mechanized salad spinner for drying. We've seen machines that inject ozone into wash water for sanitizing, or you can add a produce cleaning solution to the wash.

After washing nonleafy produce, you can steam most other fruits and vegetables for 15 seconds, then dunk them into ice water or place them in a blast chiller to quick-chill them. The wash-and-steam process helps sanitize them. If you use a blast chiller frequently for chilling instead of freezing foods, you should order one with a "gentle chill" or "soft chill" cycle. Too many people assume refrigerators are designed to cool foods down when, in fact, they are intended to keep cold foods cold. A blast chiller is a faster-working and therefore safer alternative.

As salad ingredients are cleaned and chopped, they may be sitting in the prep area at temperatures of 50°F or more. This is not a problem if they are promptly refrigerated to a cooler temperature, 34°F to 38°F. (This temperature seems to be optimum for commercial refrigerators, which are opened and closed frequently.) Avoid the tendency to store freshly made salads in large, deep containers—such containers don't allow quick enough chilling because of the depth and the density of the salad. Instead, spread the salad into containers no more than 2 or 3 inches deep to allow faster cooling, and don't put more than 10 pounds of product at a time in each shallow container. Cover with film or foil before chilling, making sure there's no layer of dead air between the top of the salad and the film or foil.

When kept in the 34°F to 38°F range, most uncooked foods have a safe-to-use shelf life of 7 days. Once they reach temperatures of 41°F to 45°F, their shelf life drops to only 4 days.

Hot foods also present risks for food-borne illness. As with cold foods, time and temperature are the keys to safety. Cooking (for the correct time and at the correct temperature) *pasteurizes* a food or liquid, sterilizing it by killing harmful microorganisms without chemically altering the food or liquid itself. This is why it's so important to track temperatures through the cooking process and to cook foods thoroughly.

Once a product is cooked and pasteurized, its shelf life begins again. At least one-third of all food-borne illness stems from improper cooling of foods placed into refrigeration. The accepted standards are:

■ If a food comes off the range or out of the oven at 140°F or hotter, it must be cooled to 70°F or below in less than two hours.

■ If food is received at a temperature of 41°F, it cannot sit at room temperature for very long. It must be completely chilled (to below 41°F) within four hours.

In either of these situations, a blast chiller or ice bath is a practical necessity to get the cooling job done quickly enough to keep the food safe. Putting hot food in a freezer is possibly the worst decision you could make. Instead of cooling the entire product quickly and uniformly, a thin layer of ice begins to form almost immediately on the surface of the food. The center or core of the food remains warm and is, in fact, insulated by the ice. This igloo effect is dangerous.

8-5 WASTE MANAGEMENT

Foodservice businesses of every size and type contribute to the growing solid waste problems in the United States, but most are taking at least a few steps to minimize what is sent to the landfill and down the sewer system. Becoming ecology-minded isn't always easy or inexpensive, but the alternatives are worse: a planet cluttered with junk, and higher fees and taxes to try to make headway in cleaning it up. It is estimated that Americans throw away almost 3 pounds of waste per day per person.

For the restaurateur, there are several good reasons to reduce waste:

■ It saves money. Decreasing your waste output decreases the size of the dumpsters you need and lengthens the time between trash pickups. You can also use composted materials on your outdoor landscaping, which saves fertilizing costs.

■ It protects the environment. Food waste that ends up in garbage disposals or landfills increases levels of odor and dangerous methane gas. Oil and grease from cooking and frying must be dealt with by your community's wastewater treatment plant.

■ It complies with state laws. Increasingly, more stringent legislation is being passed to manage solid waste. Waste-intensive businesses, including restaurants, are being watched more closely with these new laws in mind.

■ It has public relations benefits. Taking responsibility for doing your part to preserve the environment by becoming a so-called green business is impressive to your customers, many of whom are taking similar steps at home by recycling and reusing items.

Both the EPA and the NRA have worked hard to come up with an integrated waste management system that can be adapted for different communities and different types of foodservice facilities. The system has several components, the combination of which varies depending on the area, the type of business, and so on. The components are:

- Source reduction
- Reuse and recycling
- Composting
- Combustion or incineration
- Landfill use

We covered maintenance of the grease trap and management of oily waste in Chapter 6. Let's discuss each of the other waste management topics here.

SOURCE REDUCTION, REUSE, AND RECYCLING

The catchphrase in this area is "P2"—pollution prevention. *Source reduction* means making less waste in the first place, with a secondary goal of reducing the toxicity of the waste that still exists. In today's throwaway society, it's a long-term solution, but more and more products are being designed and used in ways that reduce their chances of ending up in the trash, and market research indicates consumers think positively about companies making these extra efforts. Some examples of source reduction include:

- Decreased use of mercury in dry-cell batteries and the introduction of many types of rechargeable batteries
- Advanced resin technology, which means milk cartons and plastic bags made with fewer materials
- Advances in tire design, which have increased the useful life of automotive tires by 45% since the 1980s

But food is a different story. The most recent USDA figures estimate Americans throw away 27% of all the edible food available in the nation. If just 5% of this discarded food (not even the rest of the trash) in the nation were recovered, it would provide one day's meals for 4 million people and save $50 million per year in solid waste disposal costs. In the same report, foodservice waste was blamed mostly on overpreparation, expanded menu choices, plate waste, and sales fluctuations that were beyond the operator's control—things like sudden weather changes that prompted fewer customers to show up on a particular day.

One method of source reduction is to buy less food, using up what you've got before you replace it. Computerizing your inventory will assist greatly in this type of planning. Labeling inventory and using the first-in, first-out method ensures nothing sits too long on storeroom shelves and goes bad. Ask your suppliers about products that meet your specifications but are minimally packaged. Ask them to take back and reuse their shipping boxes and pallets. Use and wash linens, kitchen towels, dishes, and silverware instead of disposable paper products.

These are just a few factors to consider when beginning a waste reduction program. Yours will be far more effective if you know exactly what you classify as waste. Environmental agencies, and possibly your own trash pickup company, can supply you with a *waste audit form* so you can start keeping track of what (and how much) you throw away. You might start by, literally, taking one day's trash and separating it into categories. How much is food waste? How much is cardboard? How much is recyclable?

Using the waste audit information, you can work with your trash pickup service to find out about the resources in your area for recycling. *Recycling* is the collection and separation of specific refuse materials that can be processed and marketed as raw materials to manufacture new products. Even as far back as World War II, there was a strong push in the United States to recover paper, steel cans, and other items, but when the war ended, the movement—billed then as a form of patriotism—lost its steam. By 1960 Americans were recycling only 7% of their solid waste; by 1986 the number had crept up to 11%. But during the 1990s the national recycling rate climbed from 12% to 27%, according to the Institute for Local Self-Reliance, a nonprofit group with offices in Washington, D.C., and Minneapolis, Minnesota. At this writing, dozens of cities and counties—including Ann Arbor, Michigan; Bellevue, Washington; Crockett, Texas; and Visalia, California—have reduced their municipal solid waste (MSW) to record-setting levels. In these cities, from 40% to 65% of what would otherwise be waste is being recycled.

You'd assume just about anyone would think recycling was a wise thing to do, but the entire waste industry butts heads with the EPA and some other industries. The EPA believes the waste industry's estimates of how much solid waste can be recovered, collected, and processed are unrealistically high, while industries that create new products claim jobs are threatened when people reuse instead of buying new. If there is a happy medium, we haven't yet reached it. The Institute for Local Self-Reliance website (www.ilsr.org) does a good job of keeping up on the ongoing debates.

The National Restaurant Association says about 7 out of 10 restaurants have recycling programs. Of these, 84% recycle paper and cardboard, 79% recycle glass, 74% recycle aluminum and tin, and 57% recycle plastic. Restaurants do make money from recycling—partly because they pay a little less for trash pickup and landfill fees and partly because waste-processing companies pay (usually by the pound) for the materials they receive. It is not a major source of revenue, however, so most restaurateurs view recycling as a break-even situation and a good, community-minded thing to do. Some even donate recycling proceeds to charity.

Another type of charitable effort is to participate in a food reuse program by donating unused produce and leftovers to the needy. There are several ways to accomplish this:

▪ Most states and sizable cities have a food bank that distributes large volumes of nonperishable goods (dried, canned, and prepackaged foods) to other groups that help low-income families. One of the best-known national networks of food banks is America's Second Harvest.

▪ Prepared and Perishable Food Programs (PPFPs) redistribute surplus prepared foods and perishables, usually for use at local homeless shelters. They are sometimes called *food rescue* programs. Most of them offer free pickup of these items. Each has its own guidelines for what it will accept and how to store the food before it is collected. The nonprofit group Foodchain has a list of all organizations that accept prepared, perishable food in most areas. Some communities have specific produce distribution programs for fresh vegetables and fruit.

▪ You can also contact homeless shelters, battered women's shelters, and similar organizations in your area and offer to cook for them periodically. Many charitable organizations depend on regular participation from restaurants, church groups, and the like, for their mealtime needs. Of course, you are donating your time as well as the ingredients in these cases, but it is well worth the effort.

Concerned about your legal liability in these situations? In the United States, a federal law—the Bill Emerson Good Samaritan Food Donation Act of 1996—and individual state laws address this issue. Known as a group as Good Samaritan laws, they protect food donors from most civil or criminal liability except in cases of gross negligence, recklessness, or intentional misconduct of the donor. Your food bank or PPFP may have an agreement for you to sign or may provide what is known as a *letter of indemnification* that spells out the rules and your legal protection.

COMPOSTING

In some cases, food scraps, soiled papers, and packaging make up 90% of a foodservice operation's waste. They can't all be recycled, but the food scraps can be saved and processed into hog feed, or the whole lot can be composted.

BUILDING AND GROUNDS

Source Reduction and Recycling Ideas

- Order condiments in bulk containers instead of single-serving packets.
- Use refillable containers for sugar, salt, pepper, and cream.
- Buy straw and toothpick dispensers instead of individually wrapped items.
- Use permanent coasters instead of cocktail napkins.
- Order items to be shipped in reusable containers such as tubs.
- Use linens for aprons, napkins, and restroom and kitchen towels.
- Use hand dryers in restrooms or cloth towels instead of paper.
- Design your menu cycles to improve the secondary use of leftover food (yesterday's chicken for today's sandwiches, salads, soups).
- Use reusable plastic pails or covered plastic containers to store foods.
- Switch to draft beer and fountain soft drinks over bottles and cans.
- Use chalk or marker boards instead of menu inserts to post specials.
- Purchase concentrated, multipurpose cleaning supplies.
- Use dishes and silverware instead of disposables, or use biodegradable plasticware.
- Don't store tomatoes and lettuce in the same container. The tomatoes naturally emit a gas that turns lettuce brown!
- Revive some types of wilting fresh vegetables by trimming off the bottom inch or so of their stalk or core and standing them in warm (not hot) water for 15 to 20 minutes. This works with celery, broccoli, lettuce, and parsley.
- Offer discounts to customers who bring their own mugs for beverages.
- Minimize the use of take-out containers, or use ones made from recycled paper.
- Use heat from refrigeration lines to preheat water for the water heater. It takes a load off of the condenser too.
- Check out the EPA's Green Lights program for a 30% lower electric bill.
- Set up an *ask-for-water policy*. Put a timer on the food disposal to save water.
- Turn organic waste into a rich soil supplement by composting.
- Use a polystyrene densifier, which crushes thousands of foam containers into a dry, dense 40-pound cylinder.
- Make sure you're paying your hauler for full loads: Compact trash to maximize bin space.
- Cut trash-hauling fees in half by recycling cardboard. Baling eases labor and storage.
- Make sure recycling bins are well marked and paired with regular trash bins.

Sources: Restaurants and Institutions, the British Columbia Waste Exchange, and the North Carolina Cooperative Extension Service.

Composting is a bacterial digestion process that naturally breaks down organic materials over time, turning them into soil enhancers that are useful in horticulture and agriculture. One of the most popular types of composting today is to set aside summer's grass clippings and fall's leaves from your lawn to rot over the winter and use as mulch to fertilize that same lawn in the spring. Cities and towns push this effort because up to 20% of the waste they deal with is leaves and grass clippings.

Composting can be rather smelly and unsightly, and, until a few years ago, it was unheard of in foodservice. Skyrocketing waste removal costs and vanishing landfill space, however, have revived interest in this back-to-nature option.

Foodservice operations within larger businesses that have landscaping needs—say, on a college campus or at a resort hotel—or even those with their own gardens might benefit from an on-site composting program. Smaller operations may allow their employees to take home scraps for their own compost piles. Still others hire a commercial composting company to dispose of food waste.

The first step is to ask about city and state composting regulations. Not every area has them, but some restrict composting to pre-consumer waste—that is, no leftovers allowed—as a health precaution. Increasingly, however, they allow food scraps, napkins, paper towels, and garden, yard, and tree debris.

The next step is a waste assessment, which should include sufficient information about the types and amounts of organic material your operation produces. You may be surprised at how much of what is typically thrown away can be composted: eggshells and paper egg cartons, coffee filters and grounds, teabags, waxed paper and most paper products, and even plasticware that is labeled biodegradable. However, some companies opt not to compost meat, meat byproducts (bones, fat), and dairy items because these decompose slowly, creating odor problems and attracting insects and rodents.

Staff training is another critical component of a successful composting project. Making the commitment to composting requires that workers understand how to carefully and thoroughly separate materials that can be composted from those that cannot.

A system must be devised to make this separation as easy as possible by placing clearly marked collection bins in the most logical, convenient locations for the kitchen workstations and dish room and with the back-of-house traffic patterns in mind. Depending on the final destination of the compostable material, there may be separate bins for paper products and/or food scraps. The bins should be lined with plastic bags, Yes, it's true that plastic bags probably are not the most environmentally conscious option, but they keep the bins clean and prevent odors and rodent infestation, and there is no biodegradable option at this point.

You also need places to store extra and full containers, either a compost bin on-site or a secure spot where they will be picked up for use off-site.

Most foodservice businesses do not reuse their own compost but rather have it picked up for use by their municipality or a third-party contractor. A commercial kitchen owner may be able to negotiate with other local businesses that could put waste materials to use—from hog farmers (who will take food scraps, including meat products), to landscaping companies. The Biodegradable Products Institute is a nonprofit organization that maintains an online database of commercial composting facilities in the United States and Canada.

Commercial compost businesses may be strict about what they accept so as to meet the health and safety laws they are required to uphold. And you'll need to be strict too, making sure the contractor sticks to the pickup schedule. Particularly in the summer months, frequent pick-ups are a necessity to minimize odor and to keep from attracting insects and rodents.

Cleanliness of the waste collection area helps alleviate this problem. When doing your own composting, it is important to mix kitchen scraps (high in nitrogen) with leaves and grass (high in carbon dioxide) to help control unpleasant odors. If your compost is not bagged, turn and mix the pile regularly to speed the composting process.

Yes, composting can be a hassle in an already busy foodservice kitchen. But it is also an earth-friendly step that can be used to your benefit, helping create a reputation as a green business that cares about the environment and takes responsibility for mindful waste disposal. If you take the time to do it, make sure your community knows about it.

COMBUSTION

Combustion (or *incineration*) is the burning of trash. Since the 1960s, researchers have experimented with ways to collect and reuse the heat generated by trash burning. By the 1990s, we burned as much as 9% of all solid waste; 6% of the time, we recovered the energy from this combustion process. However, the incinerators used to do the burning have built-in problems: They create ash and smoke. Two culprits are of particular concern: dioxins and furans, members of a highly toxic family of gases. Acids and metallic gases are also by-products of burning, so it is important that materials be carefully sorted before burning to remove anything that creates these chemicals or retards the burning process, including yard waste, glass, and metal.

The good news is, the EPA reports that more than 99% of pollutants can be filtered out of combustion emissions before they are released into the atmosphere. Ashes are typically sent to landfills, and the agency has regulations for their safe handling and proper disposal.

The United States is far behind on the international scene when it comes to use of combustion as a waste disposal method, burning only 14% to 20% of post-recycled waste. In Japan, 68% of waste is incinerated; in Sweden, 56% is incinerated.

LANDFILL USE

The landfill is the destination of most solid waste, but this luxury is in jeopardy. The number and capacity of landfills dwindle as human health and environmental concerns such as gas emissions and water contamination make headlines. Siting a new landfill—the technical term for starting one—has become more difficult as local, state, and federal governments adopt more stringent policies to help restore public confidence in their waste disposal decisions. There seems to be a nationwide drive to close municipal solid waste landfills (MSWLFs, in EPA terminology). The number was down from about 8000 in 1998 to 1800 in 2005. Fewer than half of the landfill sites that closed are being replaced. The EPA estimates that by the year 2013, only 1000 landfills will be in operation.

The remaining landfills are being better managed with a system called *landfill reclamation*. Landfill managers remove whatever recoverable materials they can (and sell them if there's a market), then combust or compact much of the remaining waste. This extends the life of the landfill site by using the space more intelligently, but it is also a more costly and equipment-intensive plan for a city or region to undertake. Of course, the fees paid by businesses and homeowners pay for this.

Reality dictates that landfills will always be needed because there will always be items to discard that cannot be reused or recycled. An estimated 55% of the 220 million tons of trash generated in the United States every year still ends up in landfills. That's much less waste than the nation generated even a decade ago—but it isn't good enough when so much could be reused. In managing your business, you must weigh the options carefully to ensure the landfill isn't just a last stop but a last resort for your throwaways.

■ SUMMARY

Your business' environment should make guests feel welcome, comfortable, and secure. It should also provide a safe, productive work setting for employees. Safety and cleanliness are top priorities in any facility. In the kitchen, sufficient exhaust capacity for hoods (canopies) over cooking equipment is not only necessary for a comfortable work environment; it is also regulated (and monitored) by health and insurance laws and inspectors.

Human engineering, or *ergonomics*, should be part of the design throughout your facility. This includes everything from nonslip flooring and comfortable heights for work surfaces and stairs to placement of exits and safe storage of cleaning chemicals. When choosing materials for flooring, select slip-resistant and highly durable material that is resistant to temperature, moisture, and stains and can hold up under heavy equipment and a lot of foot traffic. Carpet is suitable only in dining areas, never kitchens, and it must be kept clean and in good repair.

NSF International sets voluntary standards for equipment sanitation, but everyone in the foodservice industry recognizes these standards as critical. They cover the design and construction of equipment so it is easy to clean and sanitize and readily accessible when maintenance is required.

Most municipalities have adopted the NSFI standards as well as a set of rules for keeping food clean and at safe temperatures known as the HACCP system. The Hazard Analysis of Critical Control Points is just what its name indicates: a system to decide what possible hazards exist in each step of the process, from receiving to storage and preparation to serving; and doing whatever can be done to minimize these hazards. Documenting and monitoring these steps is a major part of HACCP compliance, and the USDA has simplified the rules in recent years to encourage greater participation.

Commercial cleaning and sanitizing often involves chemicals, and this chapter contains information about properly documenting each potentially hazardous substance in the form of Material Safety Data Sheets (MSDS). There's also a growing market for alternative "green cleaning" products, and numerous sources of information are available about how to choose such products.

Technology is making it easier for foodservice workers to minimize the chances of passing on dangerous viruses and bacteria with the development of antimicrobial surfaces, automated handwashing stations, and special wash solutions for produce. But these advances have one drawback: They can prompt workers to slack off on the basics of proper handwashing and protective glove use. This chapter indicates the need to think of hand sanitation as a process that requires regular monitoring to result in improvements.

Foodservice businesses have a civic duty to minimize the waste they generate. Do a thorough waste assessment to see what it is you're really throwing away, and determine what else might be done with it. Consider every option for recycling, composting, and combusting to lessen your business's impact on landfills. You may also be able to make arrangements with a local food bank or homeless shelter to donate usable food products you would otherwise discard.

An option like composting can be challenging but rewarding, and foodservice businesses have plenty of waste that can decompose to create rich soil amendments. The process takes time and staff training, however, and there are space and sanitation concerns about how and where to collect, sort, and store what amounts to rotting food and paper products, to avoid problems with odor, insects, and rodents.

■ STUDY QUESTIONS

1. What types of fire protection are required in a restaurant kitchen?
2. Why does the height of worktables matter in a kitchen?
3. Where should small tools and hand appliances be stored in a prep area, and why?
4. What is an aggregate floor, and is it a positive or negative factor in worker safety?
5. What are the main problems with these types of flooring?
   a. Quarry tile
   b. Composite sheet vinyl
   c. Vinyl tile
6. How do you choose a vacuum cleaner for commercial use?
7. When NSF International examines foodservice equipment, what is it looking for?
8. What is HACCP, and what is its role in the commercial kitchen?
9. How would you wash fresh fruits or vegetables by hand to sanitize them as much as possible?
10. Why is it not smart to put hot foods immediately into a refrigerator or freezer to cool them?
11. What should your first step be when creating a source reduction or recycling program in an existing restaurant, and why?
12. List two of the recycling tips you think would make the biggest overall difference in the amount of waste output of a foodservice facility, and explain why you chose them.

Klein Forest High School

Klein, Texas

Key Equipment List

1. Cold storage assembly
2. Reach-in heated cabinet
2a. Mobile heated cabinet
3. Exhaust hood
4. Convection oven
5. Fill faucet
6. 60-quart mixer
7. Mobile utensil rack
8. Baker's table w/bins
9. Dough divider
10. S/s wall cap
11. Fire suppression system
12. Gas valve housing
13. Tilting kettle
14. Trench liner w/grate
15. Tilting braising pan
16. Worktable w/utensil rack
17. Steamer
18. 2-burner range
19. Kettle
20. Soak sink
21. Disposer
22. Worktable w/ overshelf
23. Food processor
24. 20-quart mixer w/stand
25. Slicer w/stand
26. Preparation sink
27. Mobile pan rack
28. Snack bar counter
29. Cash register
30. Sandwich merchandiser
31. Chili/cheese dispenser

32. Drawer warmer
33. Ice cream dispenser
34. Beverage cooler
35. Slushie machine
36. Pass-thru heated cabinet
36a. Reach-in heated cabinet
37. Reach-in refrigerator
37a. Pass-thru refrigerator
38. Set-up table
39. Clean dishtable
40. Booster heater
41. Vent cowls
42. Dishmachine
43. Soiled dishtable
44. Hand sink
45. Ice machine
46. Serving counter #1
46a. Serving counter #2
47. Heat lamps
48. Guide rails
49. Air screen
50. Ice dispenser
51. Tea dispenser
52. Pizza cutting table
53. Pizza oven
54. Backcounter

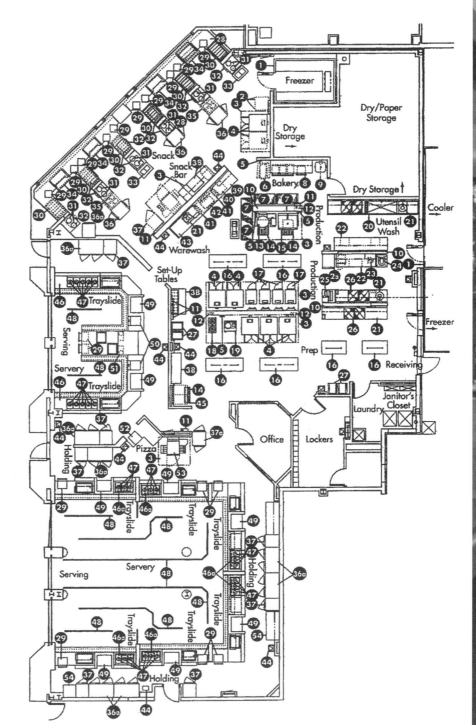

241

SITUATION

Predicting an increase in enrollment and a need for summer programs and other activities, the Klein Independent School District authorized this remodel of its high school cafeteria. As the district's director of foodservice put it, "The kitchen was stripped down to dirt floors and rebuilt from scratch." School cafeterias don't get makeover opportunities very often, so the team had to build flexibility into a design expected to last for 20 years or more. The 8900-square-foot layout includes a 4300-square-foot servery.

CHALLENGES

- Adding enough storage to handle the increased capacity.
- Getting most of the work done during the summer months when school was not in session (timetable: May to mid-September).
- Allowing for how tough the Texas heat and humidity is on the cooling system, which is located outdoors.
- Moving large groups of students through the cafeteria quickly and efficiently during short meal periods (up to 40,000 meals daily, with 30 minutes for breakfast and 1 hour for lunch).
- Maximizing food choices to keep student participation high.
- Cooking without fryers. Since 2009, the Texas Department of Agriculture prohibits schools from serving fried foods, so no fryers could be part of the menu planning. It also limits the use of sugar and fats, which reduces bakery and dessert selections.
- Maintaining sanitation and serving temperatures, which are absolutely critical.

COMMENTS FROM THE DESIGN TEAM

- "We added 1000 square feet of cold storage, which includes a walk-in cooler and freezer, into a large service yard outside the building. We married the storage with the building by putting in a door that leads directly into the kitchen."—Bob Millunzi, principal, H.G. Rice/Millunzi and Associates
- "Serving areas take up a lot of room in the front. So, we folded up the lines by creating a W shape and running them deep into the servery."—Bob Millunzi
- "All serving line modules, including their equipment, can be taken out easily and replaced if needed because trends change."—Jamal Hazzan, KISD Director of Foodservice

HIGHLIGHTS OF THE DESIGN

- 18 point-of-sale stations for quicker service.
- Bright, lively colors and bold graphics, reminiscent of mall food courts that are popular with teens.
- Dual-shelf merchandisers, both hot and cold, to allow pre-portioning of foods. Rear-loading capacity means staff doesn't get in students' way to replenish them.
- Hands-free sinks for handwashing, with electronic sensors.
- Electronic monitoring of all refrigeration temperatures.
- Two traditional service lines; a food court–style area with six lines; and a snack bar with à la carte items prepared to order.
- A management office placed in the center of the kitchen allowing a full view of all food production lines.

RESULTS

Student participation has increased from 75% to 80%, and the cafeteria's total revenue increased in the first year from $900,000 to $1 million.

TEAM MEMBERS

- *School principal:* William Lakin
- *Director of foodservice:* Jamal Hazzan
- *Foodservice managers:* Diane Evans and Pat Owens
- *Foodservice dietician:* Teresa Lane
- *Director, Capital Projects:* Don Blue
- *Architect:* Cheryl Lawrence, principal, RWS, Houston, Texas
- *Kitchen consultant:* Robert Millunzi, H.G. Rice/Millunzi and Associates, Spring and Dallas, Texas
- *Signage and décor:* Frank Medina, The Edu-Source Corp., San Antonio, Texas
- *General contractor:* Jeff Mathel, Marshall Construction, Houston, Texas
- *Kitchen contractor:* Glenn Redman, Custom Kitchens, Humble, Texas

Source: The full article about this project first appeared in the July 2007 issue of *Foodservice Equipment & Supplies* magazine, a Reed Business Information publication. The original article was written by Donna Boss; our encapsulation and the layout appear with permission.

9

BUYING AND INSTALLING FOODSERVICE EQUIPMENT

■ INTRODUCTION AND LEARNING OBJECTIVES

Equipment purchasing is not the world's most glamorous job, but foodservice businesses in the United States spend more than $7 billion a year on it, and no one can deny its importance. Making the wrong decision—by choosing equipment that is too small or not quite right because it's such a good deal; or a vendor that falls short of your expectations for training, installation, and maintenance—can become a major headache in terms of dollars spent and compromised service.

Equipment purchasing is not just for newly opening businesses. Every year, you'll find yourself replacing old or worn-out appliances, partially redesigning spaces, or just becoming aware of new items that do certain things better than the equipment you already have. Whatever your situation, there's so much to know about equipment that it will take the next half-dozen chapters to go into the specifics.

After reading this chapter, you will be able to:

■ Choose the appropriate gas or electric equipment.

■ Differentiate between buying and leasing new or used equipment.

■ Know how to have equipment custom built.

■ Describe new trends and technology in equipment design.

■ Write equipment specifications to ensure you'll get exactly what you want.

■ Understand proper equipment installation, service, and maintenance.

Today, most organizations use a team approach to make equipment purchase decisions. Often the end-user works with consultants and financial advisers to identify high-cost areas and seek lower-cost alternatives to equip the kitchen properly. The team prioritizes each piece of equipment by comparing costs of annual operation, costs of regular maintenance, and projected costs over the expected *life cycle*. Together, they develop a plan to achieve the original objectives of the business. Later in this chapter, we'll give specific details on how to analyze a purchase.

The restaurant's concept can be reinforced by equipment choices—a wood-burning oven for a pizza restaurant or an open-hearth broiler in a display kitchen, for instance. Remember the menu has a major impact on equipment selection, as the first part of a much larger picture. What you decide to cook and serve determines equipment needs . . . which dictates the layout of the kitchen . . . which establishes your labor needs . . . which sets your price points . . . which helps to configure the seating and choose the décor appropriate for those price points. Your goal is to make all the pieces fit together.

9-1 BASIC DECISIONS

The industry publication *Foodservice Equipment and Supplies* surveys its subscribers every couple of years to see exactly what their feelings are about the business and what prompts them to make big-ticket purchases. For 2013, almost one-third of the respondents (29%) said the economy is their biggest challenge—but that's down from 32% the previous year.

Nearly 24% of respondents planned to increase their equipment budgets, and 52% said their budgets would hold steady. This is encouraging news after multiple years of a sluggish economy that kept many foodservice businesses from launching big or expensive changes. Another 24% said their equipment budgets would decrease. Sales were forecast to be up slightly—but so were food prices.

The top equipment categories on their 2013 wish lists were "primary cooking equipment" (13.37%); refrigerators and ice machines (12.13%); and "food preparation equipment" (12.47%). We must add, however, that by far, the category cited by the highest percentage of business owners who planned to make purchases was "paper goods and disposables," at just over 17%.

What will drive those purchases? Equipment dealers who responded to the survey believe the reasons will be renovations of existing spaces (40%); designing and building new projects (31%); and replacement of older equipment (29%).

The survey doesn't include one interesting question: Is anyone buying used equipment rather than new? If you decide to buy new, you can opt to have appliances fabricated or accessorized to meet the specific needs and dimensions of your facility. This can be expensive, involving hiring an equipment consultant to write specifications for each piece of equipment and then paying a premium price to manufacturers to custom-make them. That isn't always the case, however, as manufacturers often allow minor changes during the fabrication process at little or no extra cost. In short—ask for what you want.

You can get almost the same custom service for much less money by selecting standard equipment from manufacturers' catalogs, which list and show many color choices and dimensions. To get an idea of what is available, visit manufacturers' showrooms or warehouses or attend a foodservice equipment trade show. At trade shows, special discounts usually are offered; sometimes even the display models are sold at deep discounts so the manufacturers don't have to ship them back to the factory. (However, you have to figure out how to get the display model from the show to your business.)

Manufacturers' and local equipment dealers' sales representatives have lots of experience and often are willing to help, but they're also trying to sell you their particular wares. If you're a beginner, or if you have an unusual kitchen space you're not quite sure how to cram everything into, consider hiring a foodservice facilities consultant. This person designs kitchens for a living. The term *consultant* may conjure up nightmares of budget-busting fees, but if you are honest with your consultant about what you can spend—and if you ask about fees up front—you can receive objective and invaluable advice. When buying equipment, remember that a mistake will affect your operation for a long time.

How do you decide the tough questions: lease or buy? new or used? Prior to making any large equipment acquisition, ask yourself these questions:

- Do I need this piece of equipment now?
- How will I pay for it?
- What capacity or size do I need?
- Should I estimate future capacity?
- Do I have enough space for it?
- Will the kitchen staff use it, and how hard will it be to train them?
- How useful are the available options and accessories?
- Are the written materials that come with it (equipment manual, service instructions) easy to read and understand? If necessary, are they available in languages other than English?

■ Is the equipment I need available locally?

■ Is quick, competent service and maintenance available locally?

■ Do local ordinances affect my use of this equipment?

In short, what makes a piece of equipment essential? If it is the most practical and least expensive means of getting the quality and quantity you need at the right time and place, your kitchen probably can't function without it. If it can be used to accomplish several tasks, it's probably worth having. Think about just how essential it is before you buy it.

9-2 ANALYZING EQUIPMENT PURCHASES

Mark Godward, president of Strategic Restaurant Engineering in Miami, Florida, was discussing kitchen redesign when he wrote a column for the December 2005 issue of *Foodservice Equipment and Supplies* magazine, but his three "Simple Rules for Improving Profitability" are good starting points for any discussion of what, and how much, equipment to buy. What are you cooking? Who are your employees, in terms of numbers and skill level? Godward's succinct rules follow:

■ *Only have the equipment that you really need.* Godward gives the example of using a contact or radiant conveyor toaster oven to toast sandwich buns. It's the piece of equipment that springs to mind for most such jobs; however, he points out that if your volume is not high, as he puts it, "It can be perfectly acceptable to toast buns on a flat grill and not pay for and house an additional piece of equipment." Items that normally might be sautéed also can be cooked on a grill, eliminating the need for purchasing, storing, and washing sauté pans and reducing the hood capacity required on a hot line.

■ *Ensure you have the best equipment available.* "The worst thing about making decisions solely based on cost," writes Godward, "is that opportunities to be more productive and effective will be missed." Here the example is the clamshell griddle: A two-sided clamshell is not inexpensive, but compared to heating sandwiches on a griddle, it cooks faster; the employee doesn't have to stand there and flip the sandwich over (saving on labor costs); and the clamshell ultimately reduces the need for griddle space and hood capacity. Similarly, the least expensive equipment to bake pizza is a deck oven, but an impingement conveyor oven provides a balance between cost and versatility.

■ *Ensure you have only as many as you need.* If one fryer vat is good, surely two are better and three will cover every possible need. Right? Not necessarily. Godward suggests making a capacity model that takes sales volume into account during peak hours and determining exactly what processes must be undertaken at each workstation. He cites the tendency for operators to add equipment as concepts and menus evolve, only to be reluctant to get rid of equipment, even when it is used less and less as time goes on. He also points out it's just as bad to have not enough equipment, which creates bottlenecks at workstations during busy times.

TOTAL COST OF OWNERSHIP

When determining whether a piece of equipment is essential to your operation, you must estimate two important figures:

1. How much money will it enable you to bring into the operation?
2. How much money will it cost during its useful life?

The first figure can be determined using information about the machine's output and efficiency, the costs of raw ingredients, and the sale price(s) of the items it allows you to produce. The second figure involves a series of calculations known as *life-cycle costing* (you'll see the term spelled this way or as two words: *life cycle*) or, the more current term, *total cost of ownership (TCO)*. The North American Association of Food Equipment Manufacturers (NAFEM) coined the latter phrase because the term *life cycle* can be so broadly interpreted.

The idea has been likened to the cost of owning another major piece of equipment with which everyone is familiar: a vehicle. You know, for instance, you'll need fuel, oil changes and other periodic maintenance, new tires, insurance, and so on. You know you can purchase a variety of nice options (air conditioning, upgraded sound system, etc.) or skip those. You also know some vehicles have better reputations than others for safety and durability, and that, in most cases, vehicles lose some of their value the moment they are driven off the dealer's lot.

It can be difficult to think about all of these long-term factors when you can see yourself in that shiny new car and you know you can qualify for the loan to drive it home. And so it is with foodservice equipment. Determining TCO is a way to analyze all of the factors, tangible and intangible, to make prepurchase comparisons. In 2007, NAFEM introduced an equipment life cycle cost tool that can be downloaded from its website (www.nafem.org) to guide prospective buyers through the process of determining these costs and making comparisons. The tool walks the user through a thorough set of "detail forms" to fill out. The forms contain all the costs associated with the purchase, service and repair, preventive maintenance, the actual operation of the equipment, and its eventual disposal. The "Operating Detail" form is reprinted as Illustration 9-1. It contains variables to factor in, in light, moderate, and heavy use.

What are some of the components of TCO? In addition to its list price, you'll consider:

- Freight costs to get the equipment to your location.
- Installation costs.
- Utility prices and energy efficiency are especially important when determining whether to select gas or electric equipment, although these costs also are among the least predictable. The NAFEM committee that developed its life cycle cost tool claims it can pinpoint power and water consumption within 90% based on laboratory tests.
- The costs of supplies, such as chemicals and filters, necessary for daily operation.
- The costs of accessories.
- Additional costs of having to ventilate or install plumbing in order to use the equipment.
- Labor costs to operate the equipment, including training costs.
- Insurance costs.
- Preventive maintenance. What does it take to clean it? oil it? calibrate its thermostats? program it?
- Repair costs, including parts and labor. Many feel a higher initial cost, if it includes local, dependable service, is usually worth the price.
- Trade-in or salvage value, and disposal costs, if necessary. Some types of appliances contain mercury, refrigerants, and other substances and must be disposed of as hazardous waste, for which there are additional fees.

After filling out the detail forms, the tool creates a spreadsheet that summarizes each category of expenses and allows the user to consider the total cost of a piece of equipment over its lifetime.

As you see, purchasing equipment is not as simple as choosing the most reasonably priced item you can find. A lower-priced choice may not be worth the money if it only lasts one-half

NAFEM Life Cycle Tool

Project Title	0
Supplier	0
Equipment Name	0
Model Number	0

Annual Utility Cost Summary

1A Electric Utility Costs
Electricity Rate ($/kWh)	$ -
Annual Electricity Consumption (kWh)	0
Total Annual Electricity Cost	$ -

1B Gas Utility Costs
Gas Rate ($/therm)	$ -
Annual Gas Consumption (Btu)	0
Total Annual Gas Cost	$ -

1C Water/Sewer Costs
Water Rate ($/ccf)	$ -
Sewer Rate ($/ccf)	$ -
Annual Water Consumption (gal)	0
Total Annual Water/Sewer Cost	$ -
Utility Inflation Rate	0%
1 *Total Annual Utility Cost*	$ -

Additional Annual Operating Costs

2 Consumables Cost
Supplies Cost	$ -
Supplies Inflation Rate	0%

3 Labor (operating/cleaning)
Hourly Labor Cost	$ -
Total Labor Hours	0.00
Labor Inflation Rate	0%
Total Annual Labor Cost	$ -

4 Other Costs | $ - |

Additional Comments:

Note:

Utility Cost Summary—Assumes utility expenses to operate the product during a fiscal year, including electric, gas, water and sewer connection costs.

Operating Labor Cost—Labor costs includes annual labor to operate and clean the appliance on a daily basis (excluding preventative maintenance) and may include incremental gains due to reduced labor required for production. This is useful for comparing similar but different processes.

Additional Operating Costs—Assumes miscellaneous operating expenses required for operation, including consumable supplies, e.g., air or water filters.

ILLUSTRATION 9-1   One of the worksheets that is part of the NAFEM Life Cycle Tool for determining TCO.

Source: Courtesy of North American Food Association of Food Equipment Manufacturers, Chicago, Illinois.

or two-thirds as long as a higher-priced model. Ice makers, for example, can be self-contained or installed with a remote compressor. The installation for the remote compressor is more expensive, but it may extend the useful life of the ice maker or reduce maintenance costs, as it is more easily accessible.

Less tangible factors to add to the mix are discussed in the next paragraphs.

Ease of use is a cost-related consideration because the more difficult it is to operate and maintain equipment, the longer it takes employees to learn the task and to use it properly. This increases labor costs, which affects your bottom line. Always think about *labor savings* as an equipment advantage. An example: By using a cook-chill system (described in greater detail in Chapter 15), a chef can prepare his or her specialties and quick-chill them to near-frozen temperatures, where they will keep safely for several days. Then, even if the chef is not there, a less skilled line cook is perfectly capable of reheating it for serving. Self-cleaning appliances may be more expensive initially, but they may also pay for themselves faster by reducing the labor it would take to clean them manually. With these types of options, you can schedule employees' time more productively at the least cost to the operation.

Projected use is also a consideration. The tilting skillet, the convection steamer, and the rangetop with oven are all examples of multiple-use equipment. In a busy kitchen, versatility is key. Combining several functions in a single piece of equipment is another way to increase workers' productivity. Whenever possible, purchase equipment that can serve more than one purpose.

Brand names mean a lot in the foodservice business. Any chef who has been around a long time has some marked preferences for certain types of equipment. A newcomer to the industry may be swayed by advertising or the recommendations of dealers, but the seasoned operator asks kitchen personnel what they like, and why.

New equipment should also come with *warranties,* which cover parts and workmanship for some period (usually not more than one year) and then "parts only" for another specified length of time. Generally, a warranty indicates the manufacturer will replace or repair, free of charge, any part that proves not to work properly due to "defects in materials and/or workmanship." Most warranties state they are valid only if no one has altered the equipment and if it has been correctly installed and maintained. In fact, most warranty hassles result because the equipment was not properly installed. You'll read more about types of warranties later in this chapter.

Payment terms for the equipment are important when money is tight. When you're spending a minimum of $2,000 for a commercial mixer, expect some strings attached unless you can afford to pay cash. If you borrow money from a bank to purchase equipment, the bank technically owns the equipment until the loan is repaid in full. If you're leasing your space, this gets a little sticky. The lease must include information about what to do if you fall behind on your payments, and the landlord must agree to grant the bank a first lien on all financed equipment. This means if the rent is not paid, the bank can repossess the equipment before the landlord can.

If the local bank is reluctant to lend money for equipment, ask the equipment dealer. Many dealers finance or lease entire restaurant installations. It is useful to have an attorney look over the legal aspects of these arrangements.

ALTERNATIVE CALCULATIONS

The National Restaurant Association suggests two other helpful calculations: simple payback (SB) and return on investment (ROI).

SIMPLE PAYBACK. This is the amount of time it takes for an appliance to pay for itself—not just its cost but also any savings you realize by using it. Let's assume we are evaluating the purchase of a commercial dishwasher and have received price quotes from two vendors. Vendor A's machine has a price tag of $7,500, with a life expectancy of 10 years and "annual savings" (features like lower utility costs and less detergent use) of $1,500, according to the manufacturer. Vendor B's machine costs $9,000, has a life expectancy of 10 years, and offers "annual savings" of $2,000.

To calculate simple payback (SP), divide the price of the appliance by the annual savings figure. The result is the number of years simple payback requires:

$$\text{Vendor A:} \quad SP = \frac{\$7,500}{\$1,500} = 5 \text{ years}$$

$$\text{Vendor B:} \quad SP = \frac{\$9,000}{\$2,000} = 4.5 \text{ years}$$

In SP terms, it will take 6 months longer to recover the initial purchase price for Vendor A's dish machine than for Vendor B's. However, this method does not take extra features of the machines into consideration—only their cost.

RETURN ON INVESTMENT. You can factor in extra considerations using the return on investment (ROI) method. To calculate ROI, subtract annual depreciation from annual savings and then divide that figure by the purchase price. The result is a percentage figure, your return on investment. The equation looks like this:

$$\frac{(\text{Annual Savings} - \text{Annual Depreciation})}{\text{Purchase Price}} = \% \text{ (ROI)}$$

Using the same dishwasher quotes, let's determine ROI for the machines from Vendors A and B:

$$\text{Vendor A:} \quad \frac{(\$1,500 - 750)}{\$7,500} = \frac{\$750}{\$7,500} = 0.10 \text{ (10\%)}$$

$$\text{Vendor B:} \quad \frac{(\$2,000 - 900)}{\$9,000} = \frac{\$1,100}{\$9,000} = 0.12 \text{ (12\%)}$$

The higher the percentage of ROI, the better. Again, Vendor B offers the better buy. Vendor B's machine returns 2% more of the original cost throughout its usable life than Vendor A's machine.

INTRODUCTION TO DEPRECIATION

A common method of determining a commercial appliance's useful life in the United States is to look at the depreciation schedule used by the Internal Revenue Service (IRS). U.S. tax laws permit a depreciation deduction for the "exhaust, wear and tear of tangible property" used in the normal course of business. Any property held by the restaurant for the production of income qualifies, as long as it is considered income-producing tangible property that has a useful life of more than one year. More specifically, what does *not* qualify is property used for personal purposes, such as a residence or vehicle not used in the business. Your flatware, glassware, plateware, linens, and uniforms are not depreciated because they are considered operating expenses, not tangible property. (Their costs can be written off in full in the year they are purchased.) Depreciation is also not allowed for food, beverage, or other inventories, land (apart from its improvements such as buildings), or any natural resource.

Most equipment can't be written off your income taxes in one lump sum; instead, the depreciation schedule allows you to write off a percentage of equipment costs every year during the standard life of the equipment. The figures are fairly arbitrary, but because so many people must comply with them, they have become the norm. Table 9-1 is an abbreviated list of useful life estimates for kitchen equipment and dining room furniture.

There are several methods of depreciation. Your accountant or tax adviser can make recommendations and should most definitely be consulted, but here are some of the details stipulated in the IRS's yearly publication #534, called *Depreciation*:

TABLE 9-1

Useful Life of Kitchen Equipment

EQUIPMENT	PROJECTED YEARS OF USE
Broilers	9
Dish and tray dispensers	9
Dishwashers	10
Food slicers	9
Food warmers	10
Freezers	9
Deep-fat fryers	10
Ice-marking machines	7
Milk dispensers	8
Ovens and ranges	10
Patty-making machines	10
Pressure cookers	12
Range hoods	15
Scales	9
Scraping and prewash machines	9
Serving carts	9
Service stands	12
Sinks	14
Steam-jacketed kettles	13
Steam tables	12
Storage refrigerators	10
Vegetable peelers	9
Worktables	13

Source: Arthur C. Avery, *A Modern Guide to Foodservice Equipment*, rev. ed. (Long Grove, Illinois: Waveland Press, 1991).

■ All restaurant equipment is assigned a 7-year useful life. This means that, over a 7-year time period, you can write off the cost of the equipment (one-seventh of the cost each of those years). Any automobiles, trucks, office equipment, and computers that qualify are assigned a 7-year useful life.

■ The method of depreciation depends on when the piece of equipment was put into use and how long its life cycle is estimated to be. Again, there are many methods of depreciation, so your accountant is the best source of advice. It is important to note, however, that once you choose a given method, you cannot change it without IRS approval.

To see a few examples, let's look at a chart showing how two methods would treat the same piece of equipment costing $10,000 (see Table 9-2).

Do not neglect items other than actual appliances that are also eligible for the 7-year depreciation. These include the gas, electric, and plumbing lines and connections necessary to operate all appliances, including computerized point-of-sale systems, exhaust hoods, and fire protection systems. As these items are purchased or replaced, keep and organize all receipts to be able to prove their monetary value for depreciation purposes.

Smallwares have their own set of rules about how their costs can be deducted. These are discussed in greater detail in Chapter 18.

TABLE 9-2

The Way to Depreciate Equipment Value
Depends on the Method Used

PURCHASE PRICE: $10,000

USEFUL LIFE: 7 YEARS

METHOD: STRAIGHT-LINE METHOD			METHOD: 200% DECLINING BALANCE		
YEAR	7 YEARS	DOLLARS	YEAR	7 YEARS	DOLLARS
1	7.14%	714.00	1	14.29%	1,429.00
2	14.29%	1,429.00	2	24.49%	2,449.00
3	14.20%	1,429.00	3	17.49%	1,749.00
4	14.20%	1,429.00	4	12.49%	1,249.00
5	14.29%	1,429.00	5	8.93%	893.00
6	14.29%	1,429.00	6	8.92%	892.00
7	14.29%	1,429.00	7	8.93%	893.00
8	7.14%	714.00	8	4.96%	446.00
	100%	10,002.00		100%	10,000.00

Source: Uniform Systems of Accounts for Restaurants (Washington, D.C.: National Restaurant Association, 1996).

9-3 RESEARCHING EQUIPMENT PURCHASES

After you know what piece of equipment you want, it's time to research its construction quality, practicality, and ease of use. The best place for observation is probably the equipment dealer's showroom or a food equipment exhibition, such as the giant trade shows held regularly by the NAFEM (every other year) or the NRA (every year).

Observe carefully when a manufacturer's representative demonstrates how to operate and clean the equipment. Notice where the controls are and try the procedures yourself. Controls should be accessible; moving parts should operate easily. Ask about safety features such as guards and shields, and check for hazards: sharp blades, hot surfaces, or open flames workers may have to touch or reach across, protruding or moving parts that may snag clothing or hands.

Think about ergonomic concerns, such as whether staff members have the physical ability to operate the equipment; some pieces require significant strength. Are the surfaces too high or too far away for a comfortable reach? Repeated bending, for example, can quickly cause fatigue and soreness.

Elsewhere, we've gone into detail about cleaning, but we can't stress enough that the cleaning process should be easy and quick to encourage employees to comply with hygiene procedures. Make sure the food contact surfaces can be wiped down easily. Be wary of equipment that requires a multitude of small pieces or fasteners that can be easily lost. At the same time, be sure removable pieces are not too large to be properly cleaned and sanitized with your present warewashing equipment. If additional cleaning equipment is part of your plan, make sure it is purchased by the time the rest of your equipment is installed.

The size of the equipment is an important consideration. European-made models are becoming more popular in U.S. kitchens, as they often are designed to fit in tighter spaces. Most major commercial appliances need a few inches, or as much as 1 or 2 feet, of clearance around them for ventilation, utility hookups, cleaning, and repair. Bring accurate measurements with you when you shop to save yourself frustrating experiences. Front access is the easiest, but several other types of easy access are employed to reduce repair and maintenance time and minimize equipment downtime. Another size-related caution: Don't buy *more* than you need. If you're ordering a pot sink and you know your largest piece of equipment is an 18 × 26-inch sheet pan, why buy a custom-made 30-inch square sink bowl when a standard 28-inch model will do?

But don't automatically assume you can't afford options and accessories. Some can improve equipment performance, save labor, and add versatility to a new unit. You owe it to yourself to find out what's available before you buy.

COOKING EQUIPMENT: GAS OR ELECTRIC?

In brand-new foodservice facilities, one of the earliest decisions to be made in the planning stages is the energy source used to cook the food: gas or electricity? Because this is often a matter of the chef's personal preference, the question will continue to be debated for decades. Without taking sides, the August 2002 edition of *Foodservice Equipment and Supplies* magazine did a good job of summarizing the advantages of each energy source.

GAS

1. Overall, natural gas is less expensive than electricity because it contains a higher cumulative amount of Btus (British thermal units) delivered from the point of extraction to the point of use. An example: A supply of 100,000 Btus at the well head, which is then converted to electricity, will have "lost" 73% of its original power by the time it is transferred through power lines to the restaurant, delivering only 27,000 Btus for actual use. Keep the same 100,000 Btus in natural gas form and deliver it through a series of gas pipelines to the same restaurant, and the restaurant receives 91,000 Btus, a net loss of only 7% (7,000 Btus).

2. As you learned in Chapter 5 of this text, electricity has an additional cost, known as the *demand factor*, which gas bills do not include.

3. Natural gas does not make additional demands on kitchen ventilation systems, which are determined by the cooking process, not the energy source.

4. Technological improvements in gas appliances include infrared fryers with 80% fuel use efficiency and griddles with consistent temperatures on their entire surface. Boilerless combi-ovens that use gas have almost eliminated most costly combi-oven maintenance problems.

5. Gas-fired bakery ovens produce moister products with longer shelf lives.

ELECTRICITY

1. Electric equipment is more fuel-efficient overall because more of the energy it uses goes directly into cooking the food.

2. Electric fryers are more efficient because the heating element (heat source) is located directly in the frying oil, which results in better heat transfer.

3. By design, electric ovens are better insulated, and the way their heating elements are placed gives them more uniform internal temperatures, which results in improved food quality and better product yield.

4. Induction rangetops, which use electricity, provide faster heat, instant response, and easier cleanup, and they contribute to a much cooler environment. (You'll learn more about how induction cooking works in Chapter 11.)

5. Electric equipment is more energy efficient because the way the thermostat controls the temperature, cycling on and off only as needed, means the appliance's actual power use is only a portion of its nameplate rating.

6. Electric utility providers often offer so-called step-rate purchasing for commercial customers, meaning a lower cost per kilowatt-hour as consumption increases.

There are as many details about equipment construction as there are pieces of equipment. Be aware the quality and workmanship you choose helps determine the life of your equipment.

Before you shop, make a complete list of attributes you are looking for. You will also need this information if you order custom-fabricated equipment.

The first question to consider is: What is it made of? The substances used to construct most foodservice equipment are stainless steel, galvanized steel, and aluminum.

STAINLESS STEEL. Stainless steel is the costliest and most commonly used material, and for good reason—if cleaned correctly, it is the most resistant to corrosion, pitting, and discoloration. In the case of cookware, stainless steel also does not impart flavors or odors to the foods being cooked in it. Stainless steel begins as iron, but chromium and nickel are added to form a tough, invisible outer layer that gives it its durability. (You learned how to clean it correctly in Chapter 7.) The most corrosion-resistant is 18/8 stainless steel, meaning it contains 18% chromium and 8% nickel. Chromium combines with oxygen to form a strong, corrosion-resistant film on the steel; nickel gives the finished product its flexibility, allowing it to be shaped into many forms. An important note: In order for manufacturers to meet NSF International sanitation standards, stainless steel that comes into contact with food must contain at least 16% chromium. The term *austenitic steel* means it is nonmagnetic steel made with 16% to 26% chromium and 6% to 22% nickel. 18/8 is a type of austenitic steel.

FOODSERVICE EQUIPMENT

Selecting a Range: Gas or Electric?

Mustard or mayonnaise? Regular or decaf? Rare or well done? Some foodservice dilemmas are easily addressed by examining personal preference. Others require a bit more practical thought and research, especially when deciding on something you'll have to live with and use for a long time to come. So, what's it going to be: a gas or an electric range? *R&I* asked experts on both sides for the relative benefits.

ELECTRIC RANGES

Kaye Hatch, executive director for the Electric Cooking Council, lists the benefits of electric-range cooking:

- Greater heat retention that keeps your kitchen cooler.
- No open flame, so fire hazards are lessened.
- New technology offers better burners. Halogen, infrared, and traditional coil burners are available now.
- Even heat in the oven chamber makes more consistent products.
- No gas flue to worry about, so the range takes up less space.
- No gas by-products to vent out of the kitchen.
- Not dependent on a finite fossil fuel. Electric ranges are ideal for places where natural gas is not readily available.

GAS RANGES

Tom Moskitis, managing director for external relations for the American Gas Association, explains the advantages of gas-range cooking:

- Better control of burner temperature. Whether up or down, gas burners respond instantaneously.
- Instant on/instant off burner power. There's no heat-up time and no cool-down time.
- Cheap operating costs. Typically, gas is 7 to 10 times cheaper than electric power.
- Features that let you control burner power. High-input and low-input burners are available for specific applications.
- Greater moisture content. Gas heat is moist, which is ideal for baking—it keeps foods moist while cooking in the oven chamber.

Source: Restaurants and Institutions, a division of Reed Business Information, Oak Brook, Illinois.

The American Iron and Steel Institute ranks stainless steel in five classifications, called *grades* or *types*, according to its chemical composition. Each grade is identified with a three-digit number; the ones you'll find most often in foodservice are Grade (or Type) 304, 301, 420, and, to a lesser extent, Grade 403. Grade 420 is used for cutlery, cooking utensils, and some cookware.

In recent years, nickel prices have been volatile, prompting steel manufacturers to experiment with alternatives that contain less nickel to keep costs down. They may replace some of the nickel with manganese or nitrogen; they may reduce both nickel and chromium content and add a bit of copper. These new alloys are acceptable alternatives, and some of the resulting equipment and cookware has the additional benefit of reduced weight.

You may be asked what **finish** you desire for your equipment, meaning its degree of polish or shine. Finishes are given numbers on a scale of 1 to 7: 1 is very rough; 7 is an almost mirrorlike shine. For most work surfaces, 3 or 4 (brushed or matte finish) is preferable because the metal can otherwise reflect glare from lights. The higher the finish number, the more expensive, so even choosing a 3 instead of a 4 can save 10% or so on the equipment cost.

GALVANIZED STEEL. *Galvanized* means the iron or steel is coated with zinc. It has the strength of stainless steel, but the galvanized coating or baked-on enamel used to prevent corrosion eventually chips and cracks, thus leaving the underlying steel to rust. Galvanized steel is still a good choice where appearance is not very important, as for equipment legs or the bracing that strengthens them. It is not recommended for areas of a kitchen that are usually damp or wet.

Another very important note: Galvanized steel emits toxic fumes in the case of fire and can endanger firefighters who might be exposed to them. Therefore, it is never recommended for exhaust hoods and, in our view, should be avoided in the kitchen due to this inherent property.

ALUMINUM. Aluminum is a soft, white element found in nature that must be converted to a metal of the same name. It is tempered (mixed with other substances) to improve its density, conductivity, strength, and corrosion resistance before being used in hundreds of manufacturing

---

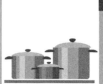

FOODSERVICE EQUIPMENT

Choosing Food Preparation and Storage Equipment

This advice was originally published at the turn of the twenty-first century, but it still applies to equipment purchases.

Choose only equipment that meets industry and regulatory standards. Check equipment evaluations published by NSF International (formerly the National Sanitation Foundation) and by Underwriters Laboratories (UL). Never use equipment intended for the home.

NSF International standards require:

1. Food-contact and food-splash surfaces that are:

   ■ Easy to reach
   ■ Easily cleanable by normal methods
   ■ Nontoxic, nonabsorbent, corrosion resistant, nonreactive to food or cleaning products, and that do not leave a color, odor, or taste with food
   ■ Smooth and free of pits, crevices, inside threads and shoulders, ledges, and rivet heads

2. Nontoxic lubricants

3. Rounded, tightly sealed corners and edges

4. Solid and liquid waste traps that are easy to remove

Source: Reprinted with permission from "Foodservice Equipment 2000," a supplement to *Restaurant Business 1995*.

applications. Tempered or alloyed aluminum can be almost as strong as stainless steel but not nearly as heavy. It can be sanitized, is rust-resistant, reflects heat and light, does not ignite or burn, can be polished to an attractive finish, and doesn't get brittle under cold conditions—making it a good choice for refrigeration units. Its thermal (heat) conductivity makes it useful for water heaters, condenser coils, and heating, ventilation, and air-conditioning (HVAC) system parts. One of the environmental advantages of using aluminum to make appliances is that it is fully recyclable.

WOOD. Everyone loves the look of wood, but few people realize the challenges it must survive in a busy foodservice setting. Wood countertops or wall paneling should never be used around waitstations, coffee makers, or anywhere that receives a lot of traffic or moisture. Never use particleboard in foodservice fabrication because it loses its shape and consistency when it gets wet. For countertops, plywood is acceptable if it is covered with plastic laminate or wood-look veneer, which should be glued on with an exterior-rated glue typically meant for outdoor use. Again, moisture is the issue, and you want your countertops as moisture-resistant as possible. The best plywood is graded with a three-letter code; if the last letter is X, that means the glue is exterior-rated. If money isn't an object, request marine-grade plywood, which is heavier (and more expensive) than regular.

SOLID-SURFACE MATERIALS. Attractive countertop options have been formulated from granite, marble, concrete, and man-made materials such as Corian and Formica. For long-term quality, many recommend granite because it is not affected by intense heat, as some other materials are. High-grade granites are quite expensive, but lower grades are available that are durable and won't break the budget.

OTHER CONSTRUCTION DETAILS. The *gauge* of a metal (abbreviated GA) means its thickness: the lower the number, the thicker the metal. Pots and pans are usually 18- to 20-gauge steel, as they need to be light enough to conduct heat well. Low-impact surfaces such as counter aprons or exhaust hoods are usually 18 or 20 gauge. Heavy-use and load-bearing surfaces, such as worktables and counters in food prep and delivery areas, should be 14 gauge. For surfaces in serving areas, 16 gauge is sufficient.

In terms of cost, the thicker the metal (or lower the GA number), the more expensive it is. This is why you use it sparingly, only in the areas where it is truly needed for safety and sturdiness.

It is important to reinforce equipment that holds a lot of weight or might be hit by heavy objects. A countertop that is not correctly reinforced can bow noticeably from the weight of equipment; a storage shelf can even crease or buckle if overloaded. Sturdy leg structure requires horizontal support to prevent wobbling or buckling. Tie rods can accomplish this on mobile racks, cross-rails, or worktables.

The least effective reinforcement method is simply to hem (turn under) the edges of the enclosing metal sheet frame, which doubles the edges and makes them somewhat stronger. This kind of reinforcement may be sufficient for cabinets, but if they are located in heavy-traffic aisles where mobile carts can hit them, more substantial framework is needed.

Equipment is most often held together by welds. *Welding,* the joining together of two pieces of metal by heating them, is by far the sturdiest and most permanent method but also the most expensive. A fully welded piece of equipment outlasts one that has been fastened by other means. It also costs more to be shipped, if it must come from elsewhere, because it is fully assembled.

Manufacturers like pop rivets because they're quick and comparatively inexpensive, but each pop rivet has a hole in the middle where debris can collect. They cannot be replaced if they snap off. Screws are also less desirable; they tend to vibrate loose from the metal when equipment is in use. The screw can fall out entirely, or the screw hole can become stripped beyond repair. Choose these less expensive options only for light-duty equipment.

9-4 BUYING USED EQUIPMENT

The used restaurant equipment market always seems to be glutted, and for budget-minded entrepreneurs, the lure can be tempting. However, you must think of used equipment like any other type of used goods. The seller usually accepts cash only and will not finance the purchase. The buyer typically accepts the merchandise as is, with no warranty or possibility of a refund if it breaks down 2 weeks after purchase.

Used equipment comes from several sources. It may be part of an existing business, sold as part of the overall ownership change. An owner may be closing a business and selling off individual pieces. A foodservice equipment dealer may buy the whole lot from such an owner, refurbish each appliance, and resell it piecemeal. An auction house may hold a public auction of equipment acquired from several foodservice businesses.

You can get some great deals—savings of up to 80% over purchase of new equipment—but buying used is a true bargain only if you are buying what you really need *and* it is in good condition. The piece may be dented or scratched; will that eventually mean leaks or rust problems? If current owners are getting rid of it because they've replaced it with something more energy efficient, will it become *your* energy drain next?

To determine how good a deal you're getting, you must first do your homework about new equipment. Find out exactly what the item would cost brand-new, fully installed, and ready to use. If you are purchasing a similar piece from a used-equipment dealer, pay no more than 50% of this brand-new price and get at least a 30-day warranty.

Restaurants USA, a publication of the National Restaurant Association, offers these additional guidelines for used-equipment purchase:

- Anything that needs repairs is a risk to the buyer. The availability of service for used merchandise is sporadic and should be determined first.

- Unlike new equipment that can be ordered to exact specifications, used pieces may not fit correctly into a kitchen. One-stop shopping to outfit an entire kitchen with used merchandise is almost impossible.

- You have no idea how much the item was used or how hard it worked. This can cause unforeseen problems.

- Service warranties, if any, are usually short: 30 to 90 days.

- Used equipment may be up for sale simply because it has been replaced by newer, more energy- and labor-saving models.

- Large equipment can be jarred and possibly damaged in moving. Does the sale price include professional delivery to your location?

Equipment resellers may do nothing more than clean a used item before putting it up for sale; if they have already refurbished and serviced it, plan to pay more for it. Do business with a reputable dealer who will allow you to have the equipment inspected before the sale by a repairperson you know and trust. If you can, ask the piece's former owner, or the local supplier of that brand of equipment, for the written service record on each piece you are thinking about buying. Ask other restaurateurs about their experience with this brand. Is it reliable, or is it a headache? You can also check the age of the piece by jotting down its serial number, then contacting the manufacturer or local supplier. Look for evidence of oil leaks.

Some types of older equipment fail to meet current health codes, which is no bargain if they cause you problems with the local health inspector. Others, such as exhaust hoods, may not meet the National Fire Protection Association's current standards for design and fire safety. We like the stance of Dupage County, Illinois (near Chicago), where local regulations require that any foodservice facility that is remodeled or reopened must meet all current and applicable codes, which settles the debate about grandfathering existing, older equipment into a newer facility rather than replacing it with more modern appliances.

You might shop only for used equipment that does not have moving parts or electrical components: sinks, tables, shelves, and stainless steel pans. Finally, here are a few hints from the experts on what *not* to buy used: ice machines, commercial dishwashers, and refrigerators. These items are most likely to have problems that crop up after being relocated. Also, gas ranges and ovens are a better used buy than electrical ones.

Sometimes equipment is sold at auction. In this case, plan to pay no more than 20% of brand-new value; be prepared to pay cash; and bring your truck and enough people to carry whatever you buy, because you will be expected to take it with you.

E-COMMERCE

A source of both new and used equipment, the Internet auction giant eBay is fast becoming a virtual superstore for foodservice equipment and supplies. eBay contains more than 10,000 restaurant-related listings for both new and used equipment and claims to sell 3300 of them per week. Traditional equipment dealers are using eBay as a way to expand their market share. The site's Smart Search feature allows prospective buyers to narrow the wares based on several criteria, including price range, geographic location, item category, and others.

Business courses held in many communities teach people how to buy and sell on eBay, and it's a handy skill to learn. The process is fairly simple. A seller registers and places goods for sale in an auction, with detailed digital photographs and a thorough description of each item, along with a minimum acceptable bid and a "Buy It Now" price—the latter is for someone serious enough to bypass the bidding process and pay top dollar. Potential buyers place bids on the item they want, and it is sold to the highest bidder when the auction closes in 3 to 10 days. Would-be buyers monitor their auctions daily, and even hourly, to see if they've been outbid. There are several payment options, and the buyer and seller work out delivery terms if the seller has not already made them clear on the auction site. Both buyer and seller can give written feedback to eBay about the other, which helps keep the transactions hassle-free. However, they are not always, especially when buyer and seller are individuals (not companies), located far from each other, or when the online description of the item sold was, shall we say, not quite accurate.

Many companies and websites offer foodservice equipment for sale, both new and used. Be careful to specify exactly what you need in terms of a base unit and any options or accessories. Many times, the options and accessories that appear to be included (based on posted photographs, for instance) are actually not.

Because e-commerce is crucial to businesses of the future, manufacturers are using an automatic identification system for their products, a sort of bar code as is now commonly used on many types of products sold at retail, to ensure exact identification of what is being ordered and prevent misbranding or mislabeling of lesser-quality merchandise as "brand name" when it isn't. NAFEM has a Foodservice E-Commerce Group (FEG) working to develop ethics, standards, and a workable infrastructure for this fast-moving part of the industry.

9-5 LEASING EQUIPMENT

Leasing equipment is an attractive option for first-timers in foodservice. However, as for many other types of leases, you end up paying more over the 3- to 5-year lease than the equipment would have cost to purchase outright. Roughly, the equipment is paid for by the 24th or 30th payment, but you've got 36 to 60 payments total. When you look at it that way, you're paying for a mighty expensive maintenance policy. And if your business doesn't make it, you may still be stuck with the remaining lease payments.

However, leasing does offer advantages. Generally, it's a way to get the equipment financed 100%, so to speak, whereas a bank lends only 85% to 90% for equipment you purchase and requires you to come up with the rest as a down payment. Seen this way, leasing leaves you with some cash on hand. And, as the payments are spread over several years or tailored to your cash flow, they may be less expensive month to month than a bank loan. Some lessors allow lease-purchase arrangements; others give you the option of purchasing the equipment at the end of a lease or turning it in on a newer model and a new lease. Leasing has tax benefits too: Lease payments typically are deductible expenses, with no interest or depreciation calculations to consider.

Some items are commonly leased because of the accompanying support and service from the company. Warewashing machines can be leased, and the dealer provides the maintenance and cleaning chemicals. Coffee-making equipment can be leased, with the coffee and all related service items supplied for a fee. A representative of the coffee company replenishes the items and provides regular machine maintenance.

The most prudent course of action is to do a cost analysis on a couple of specific pieces of equipment.

■ *To buy.* Consider total price, interest rate, deposit required, monthly payments, depreciation, estimated maintenance costs, and what the equipment will be worth at the time you make your final payment.

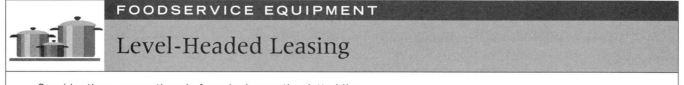

FOODSERVICE EQUIPMENT

Level-Headed Leasing

Consider these suggestions before signing on the dotted line:

■ *Determine what you hope to accomplish by leasing.* "Leasing should be a well-thought-out, well-planned financial business decision rather than something you have to fall back on as your only way to get funds for equipment," says Ron Sciortino, president of American Specialty Coffee and Culinary in Atlanta. "You want to know your leasing plan up front."

Sciortino cautions operators against leasing equipment if they're undercapitalized. "I've seen restaurants lease $10,000 to $20,000 worth of equipment because they wanted to go all out and do it right," says Sciortino. "Then 6 months to a year later, they're out of business. Then we have legal problems and court battles."

The biggest danger in embarking on a leasing program in an attempt to compensate for inadequate funds occurs when restaurateurs sign personal guarantees on the equipment they lease. "If you sign a personal guarantee, you're personally liable," notes Sciortino. "You could lose your house and a lot more."

■ *Explore all your options.* This detail includes approaching the leasing department at your bank as well as considering a conventional bank loan so you can purchase the equipment instead. "Always go to your bank first," says Kay Stephens, owner of Capital Financial Corporation in Cordova, Tennessee. "Depending on your locality and relationship with your bank, they may give you a lease through their own leasing department, and the rates might be lower. And you might even be preapproved."

■ *Protect your credit by making an informed but fairly rapid decision.* "Don't shop a leasing agreement around," warns Stephens. "Each time you shop around, a credit report is pulled on you and your restaurant. To the leasing companies, it either looks like you're shopping around or you've been turned down. If you've been turned down, why would they want to take a chance on you?"

■ *Deal with reputable suppliers.* "In the lease atmosphere, you're at the mercy of the supplier," remarks Sciortino. "If you deal with a shady individual, it's impossible for you to be safe, so look for those who have a reputation for honesty and for treating their customers with respect."

■ *Read the lease thoroughly before signing.* "You'd be amazed at how many people don't read the lease," says Stephens. "Leases are legally binding contracts drawn up by attorneys."

Source: Reprinted from *Restaurants USA* with permission of the National Restaurant Association.

■ *To lease.* Consider total price, interest rate, deposit required, monthly payments, and estimated maintenance costs not covered by the leasing company.

As you read the lease, be sure the document contains: a specific length of time the lease will be in effect; the dollar amount being financed as well as the total dollar amount of all the payments and interest (two very different totals, as you will see); and the amount of deposit required.

You must also be clear about the difference between a maintenance contract and a service agreement. The leasing company is technically not responsible for making sure the equipment functions properly. The manufacturer should provide the same warranty or service agreement you would receive if you purchase the equipment. If the leasing company offers a maintenance contract, that agreement is over and above the manufacturer's service agreement; it guarantees maintenance will be provided as needed.

9-6 TRENDS IN THE EQUIPMENT FIELD

In 1999, the North American Association of Food Equipment Manufacturers introduced a concept known as the **NAFEM Data Protocol (NDP)**, a standard for linking foodservice equipment to a central computer that collects helpful data. The technical definition of this system used by the NDP Steering Committee is: "a closed loop, back-of-the-house network connecting the production and support equipment to a central computer in a restaurant manager's office. It facilitates bi-directional communications between the kitchen equipment and the manager's PC-based workstation."

To be part of the same network, all the equipment must speak the same computer language, and NAFEM has gathered experts to create this language and develop software and a converter that can translate between equipment and computers. It has successfully tested NDP in real-life situations.

A technologically smart kitchen that functions as a network can convey multiple advantages to the business owner. Such a system can program and monitor equipment temperatures, remind employees about maintenance duties, and even call a manager at home when the walk-in refrigerator quits running after hours or a vendor when service is required. It can collect data about whatever you want to track: consumption rates for certain food items, minimum inventory levels (to determine when to reorder), HACCP safe holding times and temperatures of foods. You can see which work areas of the kitchen carry what power loads at what times of day. You can program appliances to heat up or shut off on cue, and you can track their energy consumption. You can keep maintenance records for appliances and be automatically updated when they need servicing.

Restaurants have been slow to embrace the technology because it is new, and cost is surely a factor. However, this is the type of trend that must begin at the manufacturing level. Some major companies now offer NDP-enabled equipment. Manufacturers say they've had more initial interest from schools and hospitals than from restaurants, possibly because of the opportunity to monitor HACCP-related food safety data in high-volume situations. NAFEM is convinced that NDP will "revolutionize" commercial kitchens, but this revolution is happening more slowly than the organization had hoped.

In the meantime, several manufacturers maintain their own proprietary software and hardware. Be sure to ask them to demonstrate that their products do what you intend or that they can "talk with" NDP products if that is necessary for your operation.

EQUIPMENT BREAKDOWN INSURANCE

In today's highly computerized and increasingly automated equipment world, all it would take is a power surge, brownout, or short circuit to put an appliance, or an entire restaurant, out of commission, suddenly and unexpectedly. The insurance industry has created a new type of insurance policy to mitigate these risks, which might not otherwise be covered under a standard commercial or business owner's policy (BOP). *Equipment breakdown insurance* can be purchased separately, although insurance companies increasingly are making it part of a wider commercial policy that includes other types of liability coverage.

When purchasing this type of insurance, it is important to understand what it actually covers. A policy that pays for your loss of business income during the shutdown and repair period is critical; it must also cover extra expenses such as rental of refrigerated storage to keep food from spoiling, purchase of an emergency power generator, the caterer you have to hire to finish a big order on a deadline, and so on. Some businesses pay more for so-called extended coverage (also known as *extended period of restoration*) in case they must work to get back up to previous levels of business after lengthy shutdowns. If your business is located in a hurricane- or flood-prone area, this idea is certainly worth considering. You'll want to be sure you're covered for perishable foods you lose as a result of a breakdown, for the costs of repairing or replacing the equipment itself, and for equipment you may use at an off-site event rather than in your primary location.

METRIC LABELING

We'll discuss CE marking, a quality control certification in Europe, later in this chapter. If the European Union (EU) has its way, equipment sold into its member countries will be required to bear metric labeling. This impacts spec sheets, technical manuals, product literature, and advertising as well as the nameplates on the equipment. The metric-only EU regulation went into effect in 2010.

The resurgence of the metric system is yet another sign of the globalization of our economy and, if the U.S. foodservice industry expects to remain competitive, its participants, from equipment service technicians to kitchen staff members, must learn to live with the needs of its second-largest group of customers in the world (at least, currently): in Europe. For those of you who can't recall the junior high school math classes in which this was surely covered, here is the basic conversion system from U.S. standard to metric measurement.

Area	One inch equals 2.54 centimeters (cm).
	To convert from inches to centimeters, multiply by 2.54.
	A 6-inch measurement would be: $6 \times 2.54 = 15.25$ cm
	One foot equals 30.54 centimeters (cm).
	To convert from feet to centimeters, multiply by 30.54.
	A 3-foot measurement would be: $3 \times 30.48 = 91.35$ cm
Volume	One ounce equals 29.57 milliliters (ml).
	To convert from ounces to milliliters, multiply by 29.57.
	An 8-ounce measurement would be: $8 \times 29.57 = 236.56$ ml
Weight	One ounce (of weight, not volume) equals 28.34 grams (gm).
	To convert from ounces to grams, multiply by 28.34.
	An 8-ounce measurement would be: $8 \times 28.34 = 226.72$ gm
Temperature	From Fahrenheit to Centigrade (C):
	From the Fahrenheit temperature, subtract 32 then divide the result by 1.8.
	A 98.6°F would be: $98.6 - 32 = 66.6$, then $66.6 \div 1.8 = 37.0$°C
	From Centigrade to Fahrenheit:
	Multiply the Centigrade temperature by 1.8 then add 32 to the result.
	A 100°C would be: $100 \times 1.8 = 180$, then $180 + 32 = 212$°F

9-7 WRITING EQUIPMENT SPECIFICATIONS

Equipment specifications are concise statements about a piece of equipment, written to explain exactly what is needed so potential sellers can supply exactly what you want. Commonly known as *specs* or *spec sheets*, they can consist of a few sentences, photographs or drawings with accompanying text, or several detailed pages. It all depends on who writes the specs, and why.

Restaurant owner/operators usually write specs using their own past experience or from reading and reviewing manufacturers' literature. A foodservice consultant often can improve the basic description, adding notes from his or her own experience. Manufacturers write equipment specs in their product catalogs, listing details on capacity, dimensions, utility requirements, and more. The longest and most detailed specs are written by people who must purchase equipment for tax-supported facilities, such as schools and prisons, because they are likely to shop several manufacturers and must make sure they can justify their final decision.

If you've never written equipment specifications, you can get ideas from manufacturers' catalogs, from trade journals, by consulting with local equipment dealers, or by attending a foodservice equipment trade show. However, in this chapter we'll give all the information you need.

LEGAL CHALLENGES IN THE SPECIFICATION PROCESS

How specific should a specification be? A growing number of experts are suggesting adding no-substitution language—that is, stating in writing that the equipment contractor must stick to the brand or model originally requested by the buyer or consultant when making the bid for the job. Equipment contractors say they are only trying to save their customers money by offering several alternatives, at different price points and from different manufacturers, but the other reason is that compiling such a bid may make their prices appear more competitive. Critics of this process (sometimes known as *value engineering*) say they aren't taking a crucial next step, though, which is checking out whether the alternatives they're suggesting are as good as the model originally specified by the consultant or restaurateur. Instead, the tendency is to recommend brands that offer the most profit to the equipment contractor. This debate is not likely to die down anytime soon but, increasingly, foodservice consultants are asking for ***proprietary specifications*** (also called *single-source specifications*) for all equipment in order to bypass it.

Of course, whenever one manufacturer's product is specified, other products are left out. Some unhappy manufacturers have taken their complaints to court, arguing that this amounts to an unfair trade practice or illegal restraint of trade, under federal antitrust laws. These laws have been broadly written and, so far, no court has declared that writing or using brand-specific spec sheets is illegal or unethical.

In some interesting cases, architects and consultants have been taken to court for requesting certain brands of building materials in their construction and design specifications. The rulings to date indicate these do not violate the antitrust laws as long as the manufacturers compete with each other freely and fairly to influence the architects' opinions prior to the specs being written. In one case, the judge stated that architects' general knowledge of the construction industry permits them to make "informed judgments that are in the client's best interest" and that architects are presumed to react to a healthy competitive environment and to specify what is most appropriate for a project or client.

In much the same way, foodservice consultants make hundreds of decisions on a project. They shop around, test products, learn from past experience and recommendations of others, and meet with manufacturers and equipment suppliers. The whole idea behind specs is to select the brand or model that meets all, or most, of the client's requirements.

STANDARDS FOR SPECIFICATIONS

A consultant, architect, dealer, manufacturer, or end user can be a *specifier*—that is, write specs. If you write your own, you can pretty much determine your own format. The ideas on these pages can be adapted to fit your needs.

The Foodservice Consultants Society International (FCSI) and NAFEM have developed recommendations for how specification sheets should be written and illustrated. Manufacturers' spec sheets have a standard format. Their most recent collaboration is known as the Specifier Identification System (SIS). It assigns an identification code to any new piece of equipment. The code is a series of letters and numbers, which appear immediately following the model number plus an asterisk (*). The SIS code stays with a model from the time the specs are written until the time it reaches the manufacturer's warehouse or showroom floor. Along the way, if a manufacturer has questions for the specifier, it can look on the NAFEM website for that SIS code, find the specifier, and contact them personally. There is no cost for being listed on the site as a specifier.

The terminology in spec sheets is precise, and for good reason. Next we've excerpted a few examples and suggestions written by Justin H. Canfield, author of *A Glossary of Equipment Terminology*, for the SECO Company, Inc.

INSTALLATION. Sometimes referred to as *erection*. It is most important that your specifications, whether they are for a complete installation or a requisition for a single piece of heavy-duty equipment, clearly specify the work to be done by the supplier when the equipment is delivered to you. Unless you state in your specifications, "Set in position designated on plan and anchor to floor," the supplier will probably dump it on your shipping platform and *you* will have to figure it out from there. Your specifications should also state, "After proper installation has been made as called for and mechanical connections have been completed by others, the supplier will start up and adjust this equipment, including the initial oiling and greasing if necessary, and demonstrate the use of it to any person or persons the owner might designate." Unless you spell it out in this or some similar manner, the cheap low bidder will not do it because he doesn't know how. At best, he will say: "Next time a factory man is in town, I'll send him around."

It is also important, on a complete installation, that you stipulate that a competent foreman be provided for the erection and placement of the equipment by an equipment contractor, a foreman who is able to counsel with other contractors in regard to connections required at the time of the installation of the mechanical connections.

In short, spell out the work you expect the supplier to do; do not leave anything to chance.

CLEANUP. Everyone knows what this means, but it is a provision frequently omitted through oversight when specifications are written. Specifications should clearly state, "Equipment contractor will clean up all debris made by his workmen immediately upon completion of installation and remove same from the premises." If this provision is not included in the specifications, the purchaser of the equipment will be left with quite an unsightly mess to clean up—at his own expense.

DETAIL DRAWINGS. These describe the drawings that must be submitted by the equipment contractor to the consultant, owner, and/or architect for approval. The drawings, usually done on computer nowadays, relate to specially built equipment to be supplied by the equipment contractor. These are submitted to the consultant, architect, and/or owner for approval before work is started so corrections, if required, can be made in advance. Whereas floor plans are usually submitted at a scale of ¾ inch per foot, these detail drawings show the plan elevation and certain cross sections of special fixtures prepared at a scale of no less than ¼ inch per foot so all of the details of construction can be clearly indicated.

GUARANTY OR GUARANTEE. Everyone knows what this word means, but it is so commonplace it is frequently overlooked. All kitchen equipment specifications written by reliable consultants contain a guaranty clause. However, people who are requisitioning individual pieces of equipment are cautioned to read the manufacturer's warranty or guaranty and, if it is not suitable for their requirements, stipulate what type of guaranty they want. If this exceeds the manufacturer's normal guaranty, there will, obviously, be an extra charge. This matter should be carefully weighed to determine if it is worth the extra expense or not.

QUALIFICATIONS. It is important in obtaining quotations for large quantities of equipment that you state in your specifications the qualifications you require of a bidder: financial ability; the fact that all of the equipment will be manufactured in one shop; that the bidder has the personnel and the engineering facilities to properly design, manufacture, and install the equipment; and that it be of uniform design and finish. Your consultant can help you in the proper wording of this all-important clause.

INSTALLATION INSTRUCTIONS. As Canfield also points out, you must be specific in your descriptions of how each item is to be installed. Here's a good example:

> *Shelves.* It is important in writing specifications that the shelves within a cabinet-type fixture be specified as "fixed," "removable," or "removable and perforated," whichever is desired. Insofar as overhead shelves are concerned, it should be clearly stated how they are to be mounted. If they are to be placed along a wall, specify the type of bracket (stainless steel, band iron, etc.). If the shelves are over a fixture in the center of the room, your specifications should state whether they are to be mounted on tubular uprights on all corners and in the center, if necessary, and of what material the tubular uprights are to be made, or if they are to be cantilever-type shelves.

BEGINNING TO WRITE SPECS

Now that you've seen the standard format, the job may seem even more challenging. When jotting down the first notes for your specs, be as practical as possible. Think about who will use the piece of equipment and what they'll be expected to do with it. Will it be used to prepare particular menu items? What capacity do you need? What type of power source will be used? Where will it sit in the kitchen? Does it need to be mobile to serve more than one area? As Canfield hinted, delivery, setup, and installation costs should also be included in the specs. These are often overlooked, resulting in unexpected additional costs.

Arthur Avery, in his *Modern Guide to Foodservice Equipment*, offers a good, thorough outline of general requirements to be included in written specs.

- *The common, easily recognized name of the piece of equipment.* For example, reach-in refrigerator, one-door.
- *A general statement of what the buyer wants.* A one-door reach-in refrigerator to be used by the hot-line cooks to store products prior to cooking.
- *Specific classification information.* This includes type, size, style or model, grade, type of mounting required, and so on. In some instances, drawings or diagrams will be helpful.
- *Proof of quality assurance.* Inspection reports or results of performance tests on the equipment.
- *Delivery and installation.* Who will do it, and when; how much you are willing to pay for it. Put your request in writing here.
- *Any specific requirements about construction.* This might include materials used to construct the equipment; utility details; performance parameters; certification by an agency, such as Underwriters Laboratories or the American Gas Association; warranty and/or maintenance requirements; and the need to be supplied with instructional materials about installation, use, or maintenance. (You'll learn more about certification agencies later in this section.)

Another suggested list of practical specifications comes from the Foodservice Information Library's *SPEC-RITE for Kitchen Equipment*. It includes 20 specific points to consider:

1. Who is the purchaser? (Who's paying the bill?)
2. Where should the equipment be shipped?

3. How is the equipment to be shipped, and who pays the freight costs?

4. What specific services are included in delivery: unloading, uncrating, setting in position, leveling, mechanical connections, start-up, use demonstrations?

5. If permits are required, who will secure them? If inspection is needed, who does it, and who pays for it?

6. List any appropriate standards of national agencies (electrical, mechanical) for this type of equipment.

7. Provide mechanical details, such as types of utility hookups, dimensions, and so on.

8. List the interior and exterior colors and finishes you want the piece of equipment to have.

9. Include any other options you would like.

10. Decide to include (or exclude) the "or equal" clause, which states you will accept something of equal value if your exact needs cannot be met.

11. Require in writing that all custom-fabricated equipment be of uniform design and finish.

12. Provide a deadline for delivery, including adequate time for production and shipping.

13. Outline warranty needs, including who will service the warranty, how long it will be, and what it should cover.

14. List installation responsibilities. Who pays for it? Is the cost included in delivery? Should installers be union or nonunion; at what rate of pay? Are there specific times of day the installation can (or cannot) be performed?

15. If equipment arrives early, before a new facility is completed, where will it be stored, and who pays for storage costs?

16. Taxes: What are they, and who pays them?

17. When and how will the equipment be paid for?

18. If changes are required, who pays for them, and how much?

19. If there are delays, who is responsible for additional costs (rush delivery, etc.) that may be incurred?

20. If the order is cancelled, what would be an acceptable reason? Will there be cancellation penalties, and how much? Who pays?

If you will be doing a lot of spec writing, we recommend the SPEC-RITE publication. It is a workbook to help equipment purchasers buy and sell intelligently. The book first guides the prospective buyer through a series of questions, offers generic guidelines for each equipment category, and then generates a specification template. The activity is structured to improve communications between the operator and the dealer or consultant.

As you can tell, this is much more complicated than a department store purchase. A few of the topics we've mentioned deserve further discussion: certification agencies, warranties, equipment start-up, installation, and maintenance.

CERTIFICATION AGENCIES

Certification means that equipment meets a set of minimum standards for safety and sanitation. In the United States, the Occupational Safety and Health Administration (OSHA) requires that foodservice equipment be certified. It is important, for obvious reasons, that this testing be done by independent third parties, not the manufacturers themselves.

Equipment is tested by several agencies in the United States and internationally. What are they looking for?

■ Materials used to make the equipment must be able to withstand normal wear, corrosive action of food, cleaning products, and even insect or rodent penetration. Nothing that comes into contact with food can impart to it any odor, color, taste, or harmful substance.

■ The equipment must be able to be installed, maintained, cleaned, and sanitized properly with reasonable effort.

■ The equipment must perform as expected, according to its purpose and the manufacturer's promises. A holding unit that has a temperature range of a certain number of degrees must indeed be able to hold food in good condition within that temperature range, and so on.

ILLUSTRATION 9-2 NSF International logo.

Courtesy of NSF International, Ann Arbor, Michigan.

Some groups not only do the testing; they also develop the standards based on their research. All of them also use the standards of the American National Standards Institute (ANSI) to bring consistency to the process. ANSI does not certify equipment; its job is to compile all the standards. Another organization, as its name implies, takes a more global view. The International Organization for Standardization (ISO) compiled the standards of 140 nations to develop a set of quality control documents, collectively called *ISO 9000*, to set guidelines for manufacturing, production, and management practices. ISO 9000 certification is difficult to attain and considered a real benchmark for doing international business. As our economy becomes more global, expect more companies to achieve it—and more major buyers to expect it. The McDonald's quick-service chain already does. Certification standards are reviewed by the groups that create them every few years and may be reaffirmed, revised, or withdrawn. An overview of the major certifying agencies follows.

NSF INTERNATIONAL. For our discussion, NSF International (formerly known as the National Sanitation Foundation) is the primary agency that ensures foodservice equipment is manufactured and installed in a safe manner for all concerned—guests as well as employees and managers. Founded in 1948, the organization is an authoritative and independent clearinghouse for users, manufacturers, and health authorities to solve sanitation problems together. The name was changed to NSF International in 1995 to reflect its worldwide influence.

The NSF International logo (see Illustration 9-2) is widely recognized and indicates that a particular piece of equipment complies with applicable food safety and sanitation standards. The standards are developed through research, testing, and equipment evaluation. Committees within the foundation, made up of government, user, and manufacturer representatives, review and determine the standards for equipment manufacturers, fabricators, and installers.

The NSF rule most applicable to foodservice facilities is Standard 2, but more than 20 detailed NSF standards deal with everything from commercial dishwashers to food carts.

UNDERWRITERS LABORATORIES. Underwriters Laboratories (UL) is the leading third-party product certification organization in the United States. Founded in 1894 to evaluate products for safe use at home and work, the UL tests more than 17,000 types, including all electric equipment and appliances used in foodservice settings. The UL standards are compatible with the National Electrical Code and other nationally recognized installation and safety codes.

A UL mark of approval (see Illustration 9-3) means representative samples of the product have met nationally recognized safety standards for fire, electric shock, and related safety hazards. To the end user, the UL Mark is an accepted symbol of safety certification.

The UL Classification Program evaluates and classifies industrial, commercial, and other products for more specific properties and hazards such as sanitation. Currently the UL publishes more than 700 standards for the benefit of the entire safety community.

ILLUSTRATION 9-3 Underwriters Laboratories logo.

Courtesy of Underwriters Laboratories, Northbrook, Illinois.

CSA INTERNATIONAL

ILLUSTRATION 9-4 Canadian Standards Association logo.

Courtesy of the Canadian Standards Association, Mississauga, Ontario, Canada.

ILLUSTRATION 9-5 ETL logo.

Courtesy of ETL, a division of Intertek Testing Services, Ltd., London, UK.

ILLUSTRATION 9-6 Conformité Européenne (European Conformity) logo.

Courtesy of CEMarking.net.

ILLUSTRATION 9-7 (a) A restaurant range and oven.

Courtesy of Southbend Corporation, Fuquay-Varina, North Carolina.

In 1998, the UL started a sanitation certification service as a complement to its safety testing services. Both electric and gas-fired appliance manufacturers can obtain a separate UL sanitation classification, which includes compliance with the appropriate ANSI/NSFI standards. This distinct marking, in addition to other UL marks, is widely recognized by regulatory authorities.

CSA INTERNATIONAL. Because Canada is the largest trading partner of the United States, it follows that U.S. manufacturers and others selling equipment in Canada should comply with Canadian standards. CSA International (formerly the Canadian Standards Association) is a nonprofit, membership-based organization that writes equipment standards and tests products for compliance with ANSI and international standards. Its primary goal is to ensure appliances and other equipment bearing its seal meet the minimum safety requirements it has established (see Illustration 9-4). If you're planning to do business in or import from Canada, find out about the applicable rules and regulations for all relevant equipment.

ETL SEMKO. ETL stands for Electric Testing Laboratories, a competitor and alternative to UL. The organization's SEMKO Division certifies both gas and electric equipment and works with NSF International to allow ETL clients to meet international certification requirements. Its logo is shown in Illustration 9-5. ETL and SEMKO were separate companies; both were acquired by a British firm, Inchcape, which became (and is now known as) Intertek Testing Services, Ltd.

CONFORMITÉ EUROPÉENNE. Not a French major? That means "European Conformity," the health, safety, and environmental protection standard for all products made or marketed in Europe. The ***CE marking***, as it is called (Illustration 9-6), is required for equipment, toys, and medical devices sold in the EU nations and in member nations of the European Free Trade Association (Iceland, Liechtenstein, and Norway). If you purchase equipment in Europe, look for this marking. If you sell equipment to a European customer, you'll need to know more. CEmarking.net is a good Internet website for basic background, or enter the phrase "CE Certification" in any Internet search engine to find plenty of companies ready to set you up for import/export business.

OTHER CERTIFICATION INFORMATION. About a half-dozen private laboratories, known as *third-party testing facilities*, assist the certification agencies. For instance, the Food Service Technology Center (FSTC) is operated by a foodservice engineering and consulting firm, Fisher-Nickel, Inc., to research manufacturers' competing claims. With the blessings of both natural gas and electric power industry groups, UL, and other professional organizations, the FSTC does extensive refrigeration and ventilation research as well as comparisons of equipment performance. FSTC is located in San Ramon, California, and has been operating since 1987.

If specifications seem overwhelming, here's a hint: For standard equipment such as a range, a manufacturer's catalog is the first point of reference and contains all kinds of information in a digestible format. Illustration 9-7 shows a picture of a range as well as its spec sheet containing technical information in great

Models: ☐ **436D-2G, 436D-2T**   ☐ **436C-2G, 436C-2T**

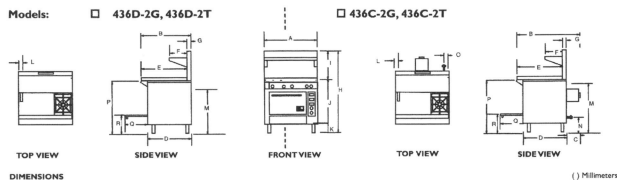

TOP VIEW   SIDE VIEW   FRONT VIEW   TOP VIEW   SIDE VIEW

DIMENSIONS

( ) Millimeters

MODEL	EXTERIOR											¾" GAS CONN.		ELECTRIC		Cook Top	Door Opening	Oven Bottom
	A Width	B Depth	C¹	D	E	F	G	H Height	I	J	K	L	M	N	O	P	Q	R
436D	36.5" (927)	32.5" (826)	—	29.75" (756)	29.625" (752)	12" (305)	2.75" (70)	56.5" (1435)	20.5" (521)	30" (41)	6" (152)	1.625" (41)	30.25" (768)	—	—	36" (914)	15.5" (394)	12.5" (318)
436C	36.5" (927)	40.5" (1029)	8" (203)	29.75" (756)	29.625" (752)	12" (305)	2.75" (70)	56.5" (1435)	20.5" (521)	30" (41)	6" (152)	1.625" (41)	30.25" (768)	10.25" (260)	3.75" (95)	36" (914)	15.5" (394)	12.5" (318)

MODEL	OVEN INTERIOR			BURNERS (BTU each)².			CRATE SIZE³			Cubic Volume	Crated Weight
	Width	Depth	Height	Open	Oven	Griddle	Width	Depth	Height		
436D	26" (660)	26.5" (673)	14" (356)	2 (26,000)	1 (32,000)	3 (16,000)	45" (1143)	39" (991)	58" (1473)	58.9 cu. ft. 1.67 cu. m.	535 lbs. 241 kg.
436C	26.125" (664)	21.75" (552)	14.25" (362)	2 (26,000)	1 (25,000)	3 (16,000)	55" (1397)	45.5" (1156)	81.5" (2070)	118 cu. ft. 3.33 cu. m	705 lbs 317kg.

NOTES:
1. Dimension C on Model 436C-2GL, 436C-2TL includes 2" minimum clearance between motor and wall.
2. OPTIONAL - Hot To Plate in lieu of 2 Open Burners at 12,000 BTU/Burner (24,000 BTU) Total
3. Units shipped with Splasher mounted, unless knockdown requested. Knockdown height 45" (1143mm).

UTILITY INFORMATION:

Gas– 436D-2GL, 436D-2TL total 132,000 BTU,
436C-2GL, 436C-2TL total 125,000 BTU
One ¾" male connection (for location, see drawing above).
☐ NATURAL
☐ PROPANE
Required operating pressure: Natural Gas 4" W.C., Propane Gas 10" W.C.

Electric–436C-2GL, 436C-2TL only
STANDARD: 115/60/1 Furnished with 6 ft. cord w/3-prong plug. Total max. amps 6.2.
OPTIONAL – 208/60/1 or 3 phase (190 to 219 volts) – supply must be wired to junction box with terminal block located at rear. Total max. amps 4.0.
OPTIONAL – 236/60/1 or 3 phase (220 to 240 volts) – supply must be wired to junction box with terminal block located at rear. Total max. amps 4.0.

CONSTRUCTION (BIDDING) SPECIFICATIONS:

☐ **ALL UNITS –**Commercial Gas Convection Oven - 36½" W x 36 "D (Including 6" high legs).

Exterior Finish: Stainless Steel front and shelf standard. Stainless Steel sides.

Range Top: Equipped with griddle and 2 each cast iron burners of round, non-clogging design. Center-to-center measurements between burners not less than 12" side-to-side, or front to back. Removable, one piece tray provided under burners to catch grease drippings.

Smooth Polished Griddle — ½" thick, hot rolled, Blanchard ground steel plate with high raised sides, 36½" wide X 29¾" deep.

Rigid Single Deck Stainless Steel Back Shelf. Sturdy Slip-On Design.

☐ **OPTION –**Thermostat griddle control with 2 throttling type thermostats. Temperature range of 100°F to 450°F (add prefix T-).

☐ **436D (STANDARD OVEN) – 32½" Deep**
Interior: Cavity sides top and back aluminized steel. Oven bottom and door lining porcelain enamel finish. Four sides and top of oven insulated with heavy, self-supporting, block type rock wool with oven baffle assembly constructed of Aluma-Ti steel.

Rack and Rack Guides: 2 position rack guides with one removable rack.

Door: Heavy Duty construction with extra heavy duty hinges and unbreakable quadrants. With cool tubular handle.

Controls: Oven thermostat low temperature type adjustable for 150°F to 500°F temperature range. Automatic safety pilot is hydraulic type.

Pressure Regulator: Supplied loose

☐ **436C (CONVECTION OVEN) – 40½" Deep**
Interior: Cavity, oven top, sides, bottom and door porcelain enamel finish. Back aluminized steel. Four sides and top of oven insulated with heavy, self-supporting, block type rock wool.

Rack and Rack Guides: Heavy-duty. 5 sets of rack guides on 2 ⅜" centers with 3 removable plated racks (For best results, no more than 3 racks should be used.).

Door: Heavy duty construction with extra heavy duty hinges. Revolutionary single spring and chain device for easy adjustment, with positive door catch, and stainless steel mesh silicone impregnated door seal.

Blower Fan and Motor: ¼ hp., 1725 rpm, 60 cycle, 115V AC, high-efficiency, permanent split phase motor with permanent lubricated ball bearings, overload protection and Class "B" insulation. Motor mounted to rear of oven. Motor serviceable from front of oven through oven cavity.

Electrical System: Wired for single phase, 115V AC with fused control circuit. 6 ft. cord and 3-prong plug supplied with each deck.

Gas Control System: Includes solenoids, pressure regulator, flame switch safety, pilot filter, pilot adjustment, and manual oven service shutoff valve.

Electronic Ignition: High voltage, spark-type igniter with flame switch safety device. Electronic ignition is activated by Oven Power On Switch, thus igniting standing pilot. With Oven Power switch in off position, the gas supply to the oven is completely shut off. Operates on 115V system (for 208/236 volt, system incorporates a reducing transformer).

Oven Heating: Aluminized steel bar burner. Dual flow fan recirculates heat directly from combustion area and within oven cavity.

Controls: Located in slide-out drawer, away from heat zone.
Controls include:
Power Switch: Controls power to oven.
Fan Switch: 2-position; on for normal operation, on for forced cool down.
Thermostat: Adjustable for 150°F to 500°F temperature range.
60-Minute Mechanical Timer.
Oven Ready Light: Cycles with oven burner.

MISCELLANEOUS INFORMATION

If using flex hose connector, the I.D. should **not** be smaller than ¾" and must comply with ANSI Z 21.69

If casters are used with flex hose, a restraining device should be used to eliminate undue strain on the flex hose.

For installation on combustible floors (with 6"-high legs) and adjacent to combustible walls, allow 6" (152mm) clearance.

Recommended – install under vented hood.

Check local codes for fire and sanitary regulations.

If the unit is connected directly to the outside flue, an A.G.A. approved down draft diverter must be installed at the flue outlet of the oven.

Two speed motors are not available on Restaurant Range Convection Ovens

Oven **cannot** be operated without fan in operation.

INTENDED FOR COMMERCIAL USE ONLY.
NOT FOR HOUSEHOLD USE.

First in Cooking, Built to Last.
southbend A MIDDLEBY COMPANY

1100 Old Honeycutt Rd. • Fuquay-Varina • NC • 27526
☎ (800) 348-2558 • (919) 552-9161
Fax (800) 625-6143 • (919) 552-9798

RR-04A   FORM RR03A/04A   PRINTED 07/96

ILLUSTRATION 9-7 (b) One page of its corresponding spec sheet.

Courtesy of Southbend Corporation, Fuquay-Varina, North Carolina.

detail: dimensions, diagrams, utilities, and an outline showing how to specify this particular information. Most manufacturers have similar sheets for each piece of kitchen equipment they manufacture.

WARRANTY SPECIFICS

Whether it is called a warranty or a guarantee, the major questions this document should answer are: Exactly what will (and will not) be covered, and for how long? Although some provisions are standard, each manufacturing company fashions its own warranty, and it generally costs about 4% of the cost of the product. Most warranties cover repair or replacement of defective parts due to faulty workmanship or materials and are generally in effect for one year. The warranty also covers labor necessary to make the repair or replacement, typically for a 90-day period. If the equipment is assembled away from the factory, or by nonfactory personnel, be sure responsibilities are clearly outlined if something goes wrong. The 90-day period usually begins the date the equipment is first put to use; sometimes it's a 120-day period that begins the date the equipment is shipped from the factory. It is wise to ask that the warranty begin on completion of the start-up demonstration, where all interested parties are present and the proper use of the appliance or equipment is explained. This will cover you if, for example, construction delays mean your kitchen is not up and running for quite a while after you've purchased the equipment.

The exact date the warranty goes into effect is something all parties should know and agree on: the owner/user, foodservice consultant, installer, equipment dealer, and manufacturer.

Warranties also have standard exemptions. Exemptions state that the manufacturer is not responsible for equipment problems that are the result of abuse or improper use. This means the piece must be correctly installed, not altered in any way, and maintained and operated according to the instructions in its service manual. You would be amazed at how often this crops up during warranty claims. So, to cover yourself, always make sure:

- The equipment is correctly leveled.
- It is connected to the correct power voltage.
- The motor is running in the right direction. (As unlikely as it may sound, if electric service is hooked up wrong, motors will sometimes run backward!)
- Gas, water, or steam pressure is at the manufacturer's suggested settings (not too high or too low).
- Adequate ventilation is provided.

Manufacturers also usually insist they are not liable for the cost of any lost product or workers' wages to produce that product if the equipment fails.

On heat-producing appliances, warranties typically do not cover basic adjustments of thermostats, Bunsen burners, and pilot lights; nor do they cover replacement of timers, light bulbs, indicator lights, or valve handles. Also, water-related problems (a major source of service calls) are not covered.

Of course, the warranty or guarantee won't do you any good unless you understand how to file a warranty claim if you need to. Make sure the procedure is clearly spelled out for you. Today, most manufacturers offer an online claim form that can be e-mailed to them.

In addition to the basic warranty, other limited warranties of up to 3 years may be available from the manufacturer at an additional cost. Under this type of warranty, the manufacturer agrees to pay for parts, labor, and "portal-to-portal" (round-trip) transportation to make repairs during the first 12 months; for the remaining 24 months, the owner/user is responsible only for the transportation costs. Generally, mileage and/or drive time limitations are specified for transport.

A warranty is not the same thing as a service agreement, in which the buyer pays an annual fee for regular maintenance of the equipment. There is no additional charge for parts, labor, or travel. If you can afford service agreements for your most valuable appliances and you trust the company that does the maintenance, such agreements do provide a certain amount of insurance against costly breakdowns and downtime.

Parts obsolescence is a real problem in the restaurant equipment industry. The best warranty allows a 5-year minimum period in which replacement parts and interchangeable assemblies will be stocked and readily available from the manufacturer. Also, look for the manufacturer's ability to provide instructions and service manuals in more than one language (Spanish, in particular) and duplicate manuals if the originals are lost or destroyed.

In looking through warranty materials, you may notice several kinds. Here is a list of the prevalent ones found in the hospitality industry:

Parts warranty. Covers repair and/or replacement of defective parts. May or may not cover the labor required to do the repair or replacement or any freight charges involved if parts must be ordered.

Labor warranty. Covers the labor costs involved in repair or replacement of defective parts. Total cost may be limited, and the repair person's travel time may not be included.

Refrigeration warranty. An extension of the standard parts-and-labor warranty for refrigeration units. Typically covers the compressor or other parts that might be damaged by compressor failure.

Service contract. Another name for *extended warranty.* Many manufacturers offer service contracts to cover repair and replacement costs for time limits beyond the standard warranty. The price of the contract may depend on its length (from 1 to 5 years), the type of equipment, and the standard policies of the manufacturer.

Carry-in warranty. Covers parts replacement but covers labor charges only if the equipment is brought to the repair facility for servicing. Labor costs are not covered if repairs are made at the owner/user's place of business.

More and more manufacturers are set up to accept customers' warranty claims online, which streamlines the hours they would normally spend handling claims by phone. Many, in fact, now outsource the entire warranty function. They're trying to reduce the considerable costs of warranty claims by handing them over to companies that specialize in warranty administration, which may be a good thing for equipment purchasers. There's nothing more frustrating than trying to negotiate with a manufacturer's customer service representative who can't answer your questions about what the warranty covers, or worse, can't even locate your warranty paperwork. Outsourcing has advantages for manufacturers too: Warranty tracking companies keep records of types of failures and can provide helpful feedback, allowing the manufacturers to improve their products based on that information.

9-8 START-UP, SERVICE, AND SAFETY

Your relationship with your new appliance depends a lot on whether you're happy with the seller, so part of your pre-purchase research should focus on the reputation and services of the dealer or manufacturer. An age-old gripe of restaurateurs who purchase new equipment is that dealers and manufacturers "sell 'em and forget 'em." However, the owner/user who makes an effort to keep in touch with these merchants—by sending in warranty cards, calling with questions, or dropping a note of thanks after a successful installation—gets more attention simply by realizing that communication is a two-way street.

Take the time to visit one or two sites where a particular piece of equipment is in use— and, if possible, take the service provider with you on these visits. No amount of sales literature can replace the experience of seeing equipment in operation and talking to the people

FOODSERVICE EQUIPMENT

Service and Support: How One Manufacturer Does It

Groen is a Dover Industries company that specializes in steam equipment—steamers, combi-ovens, steam-jacketed kettles, and so on. Thomas Phillips Jr. retired as president of Groen in 2001, but not before putting a multifaceted approach in place for the company's service and support functions:

1. *Start-up and application training.* As training programs evolved and grew more extensive, Groen saw the need for more user-friendly installation and operation manuals. The company met that need by incorporating more graphics—photos, diagrams—and nontechnical language into the resource materials to make them easier for employees to follow.

2. *Warranty and technical support.* One of Groen's most important long-term commitments is to provide customers with a quality warranty, a parts-stocking service network, and knowledgeable service technicians. With its network of authorized service agents already in place, Groen took the next step and implemented an aggressive "train the technicians" program. Phillips says service personnel are only as good as the training they receive. The training is supplemented by detailed diagnostic service manuals, which are updated regularly.

3. *Quick-ship inventory.* Groen recognizes not all users can afford to wait on special orders for equipment and created an "In-Stock" program designed to fill emergency quick-ship requests. It involved a multimillion-dollar inventory investment to have the most popular models and styles of all its equipment in stock so they'll always be available for shipment within 48 hours.

4. *Product training.* The Groen Training Academy is a factory-based school that includes a 3-day program of hands-on experience with all major equipment categories. It's most often attended by equipment dealers, but others are welcome.

5. *Accessibility.* Groen is committed to being a company that's "easy to find" and has extended its customer service function to 24 hours for emergency technical support. The company website (groen. difoodservice.com) allows customers to access complete operations manuals as .pdf files. Spec sheets, CAD drawings, price lists, and performance data are also available online.

6. *Layout and design services.* Foodservice consultants and dealers expect up-to-date equipment specs and resource material. When a unique situation calls for special assistance, a team of Groen engineers is available to make layout and equipment recommendations.

Source: Adapted from *Equipment Solutions*, a publication of Talcott Communications Corp., Chicago, Illinois (November 2000).

who use it the most. Your dealer should be willing to provide names of other customers. Be wary if they won't.

Ask about the types of in-house training the seller will provide for your employees. In response to industry growth and customer needs, many manufacturers have implemented standardized, well-organized start-up procedures. These demonstrations familiarize the owner/ user and employees with how to use and maintain the equipment and any of its accessories or attachments. It is important to have anyone who will use or clean the equipment—not just the owner—present at the demonstration and that it be conducted by a person who is fully qualified and authorized by the manufacturer to do so. Allow time for hands-on experience with the machine, and encourage your employees to ask questions: What kinds of simple, on-site service procedures can we perform ourselves? How often should this be cleaned (oiled, adjusted, etc.)? Have you noticed any common problems with this model in other kitchens? Where should we store the service manual? Whom do we call in case of problems if our boss isn't around?

A basic thumb-through of the manual should be part of the start-up demonstration, which should also cover warranty information and instructions on how to file a warranty claim. Schedule the demonstration at least 2 weeks in advance. Have more than one session if you need to so all employees who must use the equipment may attend. Ask for videos and a website address in case you need more information after the demo. Web-based customer service, in today's Internet-driven business world, is especially important. Can you get questions answered, and order parts, online?

Finally, it would certainly be an oversight (perhaps quite an amusing one) if you got everyone together for the demonstration and discovered the equipment was not ready to, shall we say, participate? So, prior to its big debut, make absolutely sure it is correctly installed, hooked up to all necessary utilities, equipped with all its accessories, and ready to do the job. Again, the manufacturer should be able to assist you with this.

EQUIPMENT INSTALLATION

Depending on the type of equipment, the people who install it may be electricians, plumbers, utility company representatives, or equipment dealers' sales or service personnel. A problem in the equipment industry has arisen because overall there is very little accountability—the lines begin to blur after the sales pitch and purchase. Who's responsible for setting up the equipment and making sure it works? If something goes wrong, who takes care of it? This problem is partly the fault of equipment manufacturers and dealers, who don't include installation in the cost of the bid, leaving the purchaser to find someone to do the job; and partly the fault of the purchaser, who typically opts for the least expensive installation option and then hopes for the best. Many of these headaches could be remedied with better communication and clear (preferably written) expectations on the part of purchaser and dealer.

Foodservice consultants generally include installation and connection requirements in their written specifications, making clear where the responsibility lies for actually setting the equipment in place and making the final connections correctly and safely. Because a major project includes many personnel from many trades, these instructions must be clear, and all the contractors must adhere to the written install-and-connect specs.

No matter who handles the job, however, both you and your local health department will want the equipment installed so the equipment and surrounding area are easy to clean. Not every piece of equipment can safely or conveniently be put on rollers or casters. Other pieces are too tall or too heavy to be wall mounted. Careful consideration should be given to alternative installation methods. Next, we illustrate and describe some of the most common options.

Floor mounting. Some equipment (e.g., a revolving-tray mechanical oven) is designed and built to be mounted directly on the floor or on a pedestal. It should be sealed to the floor around the entire base of the equipment.

Masonry base mounting. Reach-in refrigerators, heavy-duty rangetops, ovens, and broilers are sometimes mounted on concrete. The bases should be built at least 2 inches high and coved (rounded) where the platform meets the floor (see Illustration 9-8). The equipment should overhang the base by at least 1 inch but not more than 4 inches. The equipment must be sealed to the base around the perimeter, and all utility connections or service openings through the floor must be adequately sealed for sanitation and to prevent vermin from nesting beneath the equipment. Some designers have moved away from using **masonry bases** simply for greater flexibility in the future. Once concrete is placed, it is there for a very long time and may create challenges if different cooking equipment must be added.

Wall mounting. Mounting equipment on a wall is the most expensive installation option, but it is practical for sinks because it allows storage space beneath them. Wall mounting requires, of course, that the wall be reinforced well enough to hold the additional weight without damage to the building. To facilitate cleaning, a clearance of at least

ILLUSTRATION 9-8 When an appliance is mounted on concrete, the slab is called a *masonry base* and requires special installation.

Source: A Manual of Sanitation Aspects of Installation of Foodservice Equipment (Ann Arbor, Michigan: NSF International).

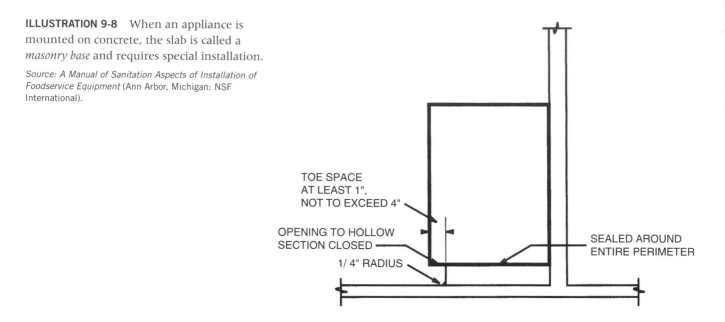

6 inches should exist between the lowest horizontal part of the equipment and the floor. The installer must also make sure that liquid waste, dust, or debris cannot collect between the equipment and the wall itself.

In any installation procedure, remember that all utility service lines and openings through walls or floors must be properly sealed to discourage insect and rodent infestation, an ugly reality when working around food.

When horizontal lines or pipes are required, they should be kept at least 6 inches above the floor and 1 inch away both from the wall and other pipes. This makes both walls and pipes easier to clean. If the pipes pass through the floor, they should be housed in protective sleeves or guards. Many health departments request that both horizontal and vertical exposed pipe runs be covered with a stainless-steel chase. Normally, the chase is C-shaped and affixed to the wall, then caulked to prevent insects or rodents from getting access to these spaces.

Drain connections should conform to the National Uniform Plumbing Code or to state or local plumbing ordinances. The diameter of the drain lines should be equal to or greater than the diameter of the equipment connection provided by the manufacturer. NSF International provides basic standards for all of these items.

EQUIPMENT MAINTENANCE

You've ordered it, installed it, and learned how to work it. Now you have to keep that piece of equipment in top shape, or your investment won't pay off over the long haul. Most restaurants are filled with so much equipment made by so many manufacturers that maintenance requirements are easy to neglect. Almost every seasoned foodservice operator can cite instances in which a well-paid technician was called out for nothing. The problem was a circuit breaker that was tripped and not reset; a pilot light that had blown out; an appliance that didn't work because it got too wet or too dirty; or (perhaps most embarrassing) an appliance that wasn't plugged in. How frustrating—and how expensive.

So let's introduce the fine art of preventive maintenance, or anticipating trouble. It is the best way to control service costs and to assure each piece of equipment a full, useful life. Think of it like servicing your car. If you spend all that money on the initial purchase but never change the oil or rotate the tires, you're asking for trouble.

The tasks vary according to the age, size, and type of equipment, but all facilities should have a preventive maintenance schedule. This includes a system for regular inspection and cleaning.

In Chapter 5 we discussed preparing a master checklist of every piece of electrical equipment in your building, including its location and model numbers or identification numbers. Expand the items on this list to include gas-, steam-, and battery-powered equipment, and you'll have a complete service list.

Make the list into a master maintenance schedule by looking through the service manuals and creating a simple document for yourself that includes the maintenance requirements of each item. You'll find most can be scheduled on a calendar, by day or week. Make sure to allot a certain number of hours for maintenance and specify who will do it.

Today's computer software makes it easy to keep online records, but it is helpful to have hardcopy backups. Some owner/operators keep a set of files in a single binder with one sheet for each piece of equipment. This equipment information sheet (such as the one in Illustration 9-9) can be in any format you choose, but it should contain all the technical details about the piece from serial number to date of purchase to manufacturer's website and phone and fax numbers. A separate

TYPE OF EQUIPMENT

Manufacturer	Serial No.	Date Acquired	1st-year Guarantee Expires	5-year Guarantee Expires
Service Contract	Cost	Weeks PM Service to be performed		

Critical Components	Part Number	Nearest Service	Service Company	Remarks

Recommended Spare Parts Inventory

Part Name	Quantity

Miscellaneous Data

_____
_____
_____
_____
_____
_____
_____
_____
_____
_____
_____

ILLUSTRATION 9-9 An equipment information sheet should be kept on computer, with a hardcopy backup, for every piece of kitchen equipment.

online record can list the itemized history of each piece of equipment: when it was serviced and by whom; for what problems or symptoms; what was done; and how much it cost. Once your maintenance records are organized, keeping them current will save you much time and hassle if anything goes wrong with equipment.

These records will also be invaluable for the other critical type of equipment maintenance, which is troubleshooting. Most types of equipment have a prescribed series of steps to follow to diagnose a problem. The dealer, manufacturer's representative, or service technician often is willing to show someone on your staff what to look for if there is an equipment malfunction. Troubleshooting helps the repair person arrive with the correct parts the first time out, saving you labor costs or the price of an additional service call.

Even the best-maintained kitchen equipment sometimes require service simply because parts don't last forever and eventually wear out. If you have the correct replacement parts on hand, you may be able to do the job yourself. One idea is to purchase replacement parts at the same time you purchase the piece of equipment. You're more likely to get a good deal on them because you are making the big-ticket purchase at the same time.

For many years, a debate has ranged among equipment suppliers regarding the use of original equipment manufacturers' (OEM) parts versus generic (and in most cases, less costly) alternatives. OEM parts are built to equipment manufacturers' specifications. Before you buy generic, consider these factors:

- Installing a generic replacement part probably will void the equipment warranty.
- The use of a generic replacement may modify the original design just enough to increase its fire risk when appliances have thermostats or electric elements.
- A generic part may not perform the same way as an OEM part does. For critical tasks like holding foods at certain temperatures, this could be a problem.
- The durability of the appliance may be compromised with the addition of a component not specified by the manufacturer, causing other parts to fail sooner.
- A generic part may require the equipment to use more energy.

Proponents of using generic parts say many OEM parts are made from the same components as generics, all purchased from the same original component manufacturers (OCMs),

FOODSERVICE EQUIPMENT

Seven Deadly Sins of Equipment Maintenance

1. Failure to keep condenser coils clean on all refrigeration equipment.
2. Neglect of cooling fans. In conveyor ovens, keep them clean and free of obstructions.
3. Nontreatment of water and/or failure to clean scale buildup from steam equipment, dishwashers, and ice makers.
4. Lax fire safety; failure to clean exhaust fan filters or have ducts inspected and cleaned regularly.
5. Dish machine abuse; failure to check temperatures and pressure requirements and to ensure daily that flatware isn't lodged in the wash pump motor.
6. Food mixer abuse, which usually is the result of overloading the machine or using the incorrect blade. This overloads the drive system and damages the motor or transmission.
7. Failure to maintain rotisseries. The drive system must be able to move freely or it will freeze. This means inspecting (and periodically changing) the pit bushings that turn as the drum wheel rotates.

Source: Frank Murphy, training director, GCS Service, Inc., an Ecolab service. Summarized from *Equipment Solutions* (September 1999).

and that the OEM seller is simply a middleman who marks up the cost of the part. In some cases, this probably is true. A good rule of thumb is a compromise: Purchase items that don't affect the inner workings of a piece of equipment (fryer baskets, oven racks, gaskets, springs, cast-iron grates, etc.) as generics, but purchase anything used in a motor, cooling, or heating device from an OEM source.

SAFETY AND TRAINING

The Golden Corral cafeteria chain, based in Raleigh, North Carolina, requires its unit managers to complete an intensive 1-week course on equipment use and maintenance. It's the best idea we've heard in a long time. Managers learn how to operate the equipment and study the instructional manuals, warranty documents, and service contracts. Corporate headquarters spot-checks units around the nation to ensure compliance with its maintenance policies.

Maintenance is being treated as more than an afterthought in today's restaurants, especially the big chains, and that's a good thing. It affects a variety of other functions: accounting (through TCO calculations); store operations (through training programs and scheduling of service); and capital spending. Maintenance also has a valuable marketing function: The well-maintained restaurant is seen by all as a cleaner, safer place to eat and work.

It is critical to get staff members involved in your company's efforts to care for equipment. Here are just a few ideas:

■ Reduce the damage caused by carelessness, abusive behavior, and vandalism by holding the staff accountable for the condition of the equipment when they complete a work shift.

■ Eliminate dents and gashes caused by carts and mobile equipment by providing adequate clearance around equipment. It's not smart design to have people transporting items through what seems like a maze.

FOODSERVICE EQUIPMENT

Maintenance Top 10 List

Much of this information can now be generated, stored, and updated online. However, it will be of no use to your kitchen staff unless they are able to access it, so it is important for them to have hardcopy backups.

1. Read the warranty information supplied with each piece of new equipment to learn what the manufacturer recommends.
2. Put all literature supplied in a three-ring binder.
3. Assign the binder to managers and require new managers to review it.
4. Prepare a one-page maintenance sheet and post it where it will be seen.
5. List appropriate clean dates and a box to initial and date when cleaning is completed.
6. Insist that managers pay attention to maintenance schedules.
7. Post a list of local service agencies, their phone numbers, and the brands they cover.
8. Upon start-up of a new facility, place a list in the front of the binder containing the make, model, serial number, date, and service and warranty period of each piece of equipment.
9. Set up service, cleaning, and inspection programs for all operational equipment with servicing agents.
10. Set up a regular exhaust hood and duct cleaning program to prevent fires.

Source: George Zawacki, senior associate, Cini-Little International, Inc. First appeared in *Equipment Solutions*, a publication of Talcott Communications Corp., Chicago, Illinois (March 2004).

- Protect both fixed and mobile appliances with rails, guards, and bumpers, which are offered as accessories.

- Catch little problems before they turn into big ones with a weekly or monthly check of all kitchen workstations. Look for missing screws, damaged or worn wires and cords, and bent panels or hinges. Get them corrected promptly.

- Make your staff aware of what maintenance costs the restaurant. Make maintenance the topic of some staff meetings in addition to training sessions. Solicit opinions from the staff about improvements that could be made.

- Make an effort to get "clean" utilities—that is, do everything you can to protect equipment from power spikes with surge suppressors, and treat or filter incoming water and air.

Performing your own maintenance and simple repairs may make you apprehensive, especially if you don't consider yourself handy. However, if you follow these safety lists and read the precautions in your appliance owner's manual or handbook, you can handle many minor items without paying for a service call.

In general:

- Be sure equipment has cooled down prior to attempting any repairs to it. The repair job will be done quicker, and there's no need to blister your hands.

- Water and hot oil do not mix. Don't use water near fryers.

- Keep equipment at least 6 inches away from walls.

- Clean equipment before starting any repairs or maintenance. This is critical but difficult to achieve in a busy kitchen.

- Know where the nearest fire extinguisher is and how to use it.

- Know where the nearest fire alarm is and how to shut it off.

- Know where the circuit breaker is for the appliance you're working on.

- Wear safety glasses.

The next two segments in this chapter contain specifics that could save your life.

GETTING GOOD SERVICE

No matter how handy you are or how well you've maintained your equipment, eventually the need will arise for a friendly visit by service technicians. The best time to select them is before you need them. In fact, the dealer or manufacturer's representative should be able to give you the names and phone numbers of reliable repair shops at the start-up demonstration. You should list these on your equipment information records. Another suggestion is to contact your local restaurant association, or even your competitors, to ask for their recommendations. A last resort is to flip open the telephone book and start calling.

How do you decide which service/repair firms are reliable? Here are some pointers from Nolan Marks's *On the Spot Repair Manual for Commercial Foodservice Equipment*:

- How long does it take the firm to respond to a service call? Then, when you've actually called them, clock them. Ask specifically about service and prices on nights and weekends.

- Once the service technicians arrive, how long does it take them to diagnose the problem? If it takes more than two trips for one piece of equipment, that's too long.

- Are you charged for more than one trip even though it's their fault they can't figure out what's wrong? A bad sign.

- After a visit, does the equipment work fine . . . for a day or two? This indicates they've fixed the symptom but not the overall problem.

IN THE KITCHEN

Safety Rules for Electricity

1. When you've turned off a circuit breaker to work on a piece of equipment, always put a piece of tape across it so no one else accidentally turns it on.

2. After the circuit breaker is turned off, always test the equipment with a voltmeter to make sure you turned off the correct breaker and no other circuit breakers need to be turned off.

3. Always check the voltmeter to make sure it works by testing it in a live outlet that you know works.

4. When doing a jump test: Turn off the power, place the jumpers across the switch to be jumped, turn the power back on to observe the results, turn the power off, and remove the jumper.

5. Never leave a switch jumped longer than the few seconds it takes to see the results of the test.

6. Never work on electrical equipment if the floor is wet.

7. If you must leave a piece of malfunctioning equipment, take precautions to ensure no one else will try to use it. Leave a big note taped to the front explaining the situation (e.g., "Out of Order. Will be back with parts this afternoon.").

8. When a piece of equipment is temporarily fixed, never let anyone use it until you are certain it is safe.

9. Never call a job done until you have thoroughly tested it and are 100% sure the equipment is safe and fully operational.

10. Always know where you can turn off the power quickly in case of a problem, whether at the plug or circuit breaker.

Source: Don Walker, *Manual of Gas/Electric Equipment* (Walker Publications, 1989).

■ Are two people sent on the job when it only takes one? If so, are you charged for both of them?

■ Does the repair person show up with few (or no) tools, diagnostic equipment, or replacement parts? This is a sure sign you'll need a return visit, which you'll probably be charged for.

■ Does the repair person spend a long time on the phone describing the problem to someone back at the office instead of dealing with it on site?

■ Does the repair person make several visits to replace one part at a time, hopeful at each trip that "This should do it"?

■ Ask about the availability and possible cost savings of rebuilt parts instead of new. Ask for the defective parts and inspect them for signs of wear or damage. Of course, if it's a warranty repair, the old part must be returned to the manufacturer.

To summarize what it takes to be a smart equipment owner, Marks also has a dozen handy tips he calls "Marks' Maxims." Follow them, and you'll minimize your facility's need for minor but costly equipment servicing and repairs.

Marks' Maxims

1. Check your power source.
2. Read the owner's manual.
3. Clean the equipment.
4. Educate your employees.

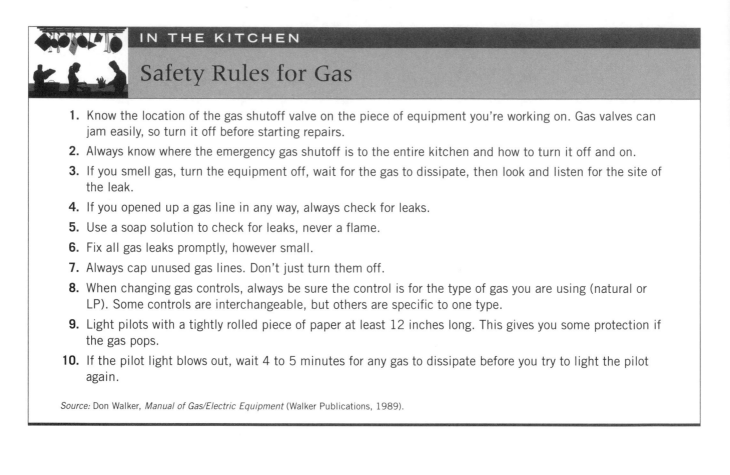

IN THE KITCHEN

Safety Rules for Gas

1. Know the location of the gas shutoff valve on the piece of equipment you're working on. Gas valves can jam easily, so turn it off before starting repairs.
2. Always know where the emergency gas shutoff is to the entire kitchen and how to turn it off and on.
3. If you smell gas, turn the equipment off, wait for the gas to dissipate, then look and listen for the site of the leak.
4. If you opened up a gas line in any way, always check for leaks.
5. Use a soap solution to check for leaks, never a flame.
6. Fix all gas leaks promptly, however small.
7. Always cap unused gas lines. Don't just turn them off.
8. When changing gas controls, always be sure the control is for the type of gas you are using (natural or LP). Some controls are interchangeable, but others are specific to one type.
9. Light pilots with a tightly rolled piece of paper at least 12 inches long. This gives you some protection if the gas pops.
10. If the pilot light blows out, wait 4 to 5 minutes for any gas to dissipate before you try to light the pilot again.

Source: Don Walker, *Manual of Gas/Electric Equipment* (Walker Publications, 1989).

5. Use the equipment properly.
6. Know your warranty coverage.
7. Don't let someone do for you what you can do for yourself.
8. Always disconnect the power source before working on equipment.
9. Choose a good service/repair business.
10. Check the obvious.
11. Ask for discounts.
12. Don't be a part-changer—changing one part per visit.

■ SUMMARY

Most foodservice operations use a team approach and do a lot of research to determine which pieces of equipment are best for their business. You can buy new or used equipment, lease equipment, or even have it custom-built. No matter what you decide, you must first choose the equipment specifications, or specs, that explain exactly what you want. You can write your own specs or select a specific manufacturer's spec sheet and use it as a guideline to shop around.

Cost, energy efficiency, projected use and useful life, size, brand name, type and length of warranty, availability of service and repairs in your area, and online support from the manufacturer—all are considerations as you shop for equipment and write your specs. This chapter includes a detailed list of items that can be part of written specifications, including who pays for what if the piece of equipment is late or if the order is cancelled altogether. Several certification agencies, such as Underwriters Laboratories, NSF International, and the Canadian Standards Association, create safety standards for equipment and test appliances to hold them to those standards. It's also important to keep up with trends and technology by reading industry equipment publications and attending an occasional foodservice equipment trade show.

This chapter introduced the concept of *total cost of ownership* (TCO), the foodservice industry's attempt to simplify the incredible amount of guesswork in determining the costs of owning an appliance. TCO includes many considerations other than initial price, from utility costs to maintenance and supplies to what it might cost to eventually dispose of the equipment. A program creates a spreadsheet to total these costs and allow accurate comparisons before purchase. Of course, in addition to cost factors, you must also consider how much money an appliance allows you to make as well as what it might save you—in terms of labor, utility costs, or even its trade-in value.

Trends in the foodservice equipment industry seem to address increasing the efficiency of equipment. The NAFEM Data Protocol allows much more automated and precise tracking of energy conservation and food safety efforts. Equipment insurance and specialized companies that handle installation and/or warranty claims are new attempts to take the hassle out of service-related problems for operators.

In equipment selection, there are three additional steps to getting your money's worth: correct installation that meets your building codes, training your employees to use the equipment correctly, and regular maintenance. Employees' knowledge and attitudes about keeping the equipment safe and sanitary, day after day, are keys to its longer useful life.

■ STUDY QUESTIONS

1. List three things to consider in deciding whether you need a particular piece of equipment for your kitchen.

2. What questions should you ask about the equipment seller's ability to service what you buy?

3. Generally, how much less should a piece of used equipment cost than it did when it was new? What should you think about (besides cost) in choosing used equipment?

4. What are equipment specifications, and why do you need them?

5. What kinds of information should be contained in equipment specifications?

6. What kinds of precautions must you take to make sure an appliance is adequately covered by its warranty?

7. What are the five types of warranties? Which ones would you say you definitely need, and which are optional?

8. What determines if a piece of kitchen equipment should be mounted on the floor, on a base, or on the wall?

9. What information should be listed on an equipment information sheet?

10. How would you check up on an equipment service/repair company? Name at least three things you would assess in deciding if it was a reliable firm.

David Yudkin

OWNER, HOT LIPS PIZZA, PORTLAND, OREGON

David Yudkin took over a struggling pizza business in Portland, Oregon, in 1989 and has transformed it—not only into a financial success but also a company with a national reputation for sustainable business practices. From its use of locally produced ingredients (complete with photos of the farmers on the company website) to deliveries by bicycle or tiny electric car, Hot Lips Pizza is the epitome of a "green" restaurant. The employees are just as passionate about it as Yudkin; the company has a turnover rate of only 14%.

Yudkin didn't plan to be a restaurateur. "My relationship to restaurants was working in them, from early on, to earn money," he explains. "When I got out of college, I said, 'Now it's time for a real job.' I got an office job, and it was the most boring, excruciating thing I ever did. So I got another restaurant job!"

Q: What prompted you to focus on sustainability?
A: Well, the initial goal was just to keep the doors open! When you're sinking or on fire, it's very clear what the mission is: You bail or put out the fire. We crossed that threshold about the same time several things happened. My wife and I consolidated ownership; we had several partners. We realized it was great pizza, and we had the goodwill of the community to help Hot Lips survive the hard start-up phase. And we started a family!

We figured the company had already survived the toughest years, but I saw the "big guys" looking at gourmet pizza as a sector, and I knew we couldn't compete head-on with highly capitalized chains. So there was a big question out there: What do we make of this company?

At that time, I heard a presentation from the Oregon Natural Steps Network about sustainability. We'd already taken eight stores, a commissary, and a management office—more locations than we had the infrastructure to support—and shrunk it down to two stores, self-contained. We were growing up again, literally; and for our third store, I tried to incorporate some of the theories I had about energy efficiency and design. We built a fourth store using what we learned, and then a fifth. . . . and I realized how many design questions there are, and that part of it is a throw in the dark!

Q: What kinds of things did you learn?
A: One of my assumptions is that people tend to look at the limitations of a facility. They say, "My customers go

here, and my cooks go here." That gets laid out, and after that comes the layout of the equipment—and if something has to give, it's usually the production area. Instead, we tried to give equal weight to the equipment and how it's laid out. For instance, our walk-in seems pretty low-tech, but it serves three purposes. There's a door on the pizza cook's side where the cook keeps the dough and backup materials. Then there's the retail side, which has the sodas handy and the beer taps sticking out of it. And finally, the man-door is on a third side of it, where the pantry person stocks his stuff. It is centrally located, right behind the counter, so instead of having three or four compressors for several units in different spots, we're down to one. That means reduced maintenance, materials, and heat production in the room, which affects the HVAC load.

Q: Do you use the heat from the compressor?
A: We do that with the pizza oven, which works theoretically, but not the way I had intended operationally. We had pizza ovens at 100,000+ Btu each, and then a hood over the ovens that had an electric motor, and a 5-ton air conditioner pushing cold air in, and, in the far corner of the room, we had another 80,000 Btu burner heating hot water. So my question

was, "Can we incorporate these?" I surmised that if I could get all my baking into one oven and channel all the excess exhaust into one channel, I could put in a heat recovery system and heat hot water. I spent $5,000, got the heat recovery unit, and, down in the basement, there's a 120-gallon tank with a pump, and the water continuously circulates, and it takes off about 100°F—that is, every gallon of water that goes in is coming out 100°F hotter.

In subsequent stores, we just took a coil of copper and rested it on top of the pizza ovens, so the water comes in from the city line and is preheated before it dumps into the water heater. It's not getting a 100°F jump—but I didn't put out the $5,000, it's much more low-tech, there are fewer moving parts, and I'm almost sure I'm ahead of the game in terms of return on investment.

We learned a similar lesson with lighting. In one store, instead of rewiring and putting in all new fixtures, I put in compact fluorescent lights. It clearly has shown a return on investment. It's amazing how much heat an incandescent bulb puts out, and if you have 72 of them at 100 watts each, that's a load of heat. The only downside to the CFL is the kind of light it puts out. We've also experimented with LED floodlights, which are supposed to last forever. They're blue, so they're not very attractive, but they're incredibly useful as night-lights, down-spots for security, exit lighting, and so forth.

In a new location, it's important to evaluate the existing light condi-tions before you do anything. In one store, we have great south-facing windows and good natural light. So, instead of putting in one circuit that controls all the lights in the dining area, we split it into three circuits and put rheostats on them. That way, we can light the store as needed—more light toward the inside of the store, where the food is displayed, and less lighting and energy use by the windows. It cost $100 per circuit and another $25 per rheostat. I had a higher-tech solution in mind, but the reality is very low-tech and a lot less expensive than spending $400 each for compact fluorescent fixtures.

Q: We've talked about saving energy; how about saving water?

A: You need to use water flow restric-tion fixtures. As part of our regular maintenance evaluations, we notice: Are the faucets dripping? Is a toilet running? Even with checks in place, I go into stores and find them dripping and running. We're training a mainte-nance person now to deal with all this.

It seemed to me that using dishwashers would save water, so I bought one. My expectation was that it would last 20 years, and it lasted 10. If one looks at the whole process of building, using, and then decon-structing a dishwasher, I'm not sure I could argue it was less resource-inten-sive than washing dishes by hand.

My point is, every one of these things involves a complex life cycle analysis and, when you're running a restaurant, it's hard to go that far with every single piece of equipment. But you can't just be interested in the cost—you must consider the impact of the equipment on the environment.

With flooring, for instance, you can look at vinyl and wood and concrete. We tried to select a surface so all we'd have to use on it is soap and water, trying to get away from the stripping and waxing and harsh chemicals. There's more and more pressure to report hazardous chemi-cals and have all the safety sheets on-site. Our rule now is, if you're a manager and you bring in a chemical product to use, it's your responsibility to put that sheet in the book and keep the book updated. [Laughs.] Now *that* discourages the use of chemicals!

Q: What are you seeing in terms of industry innovation for "green" equipment?

A: Not enough! When you look at the operation of a store, the cost of energy has not really caught up with the true impact of the energy. But in the coming years, as energy costs climb and become a larger percentage of the operating costs of a business, more attention will be paid to it.

I know there's a better pizza oven, but I buy a particular pizza oven because I know it performs well. It's not very energy efficient, though, and I've asked the manufacturers, "When are you going to redesign your oven?" Right now, I think the idea of replicat-ing the cool, wood-burning oven look drives them more than energy effi-ciency. I think the market is about to be pressured into it, but I don't think the manufacturers are especially proactive. When they see their sales slump, they'll become proactive.

Q: When you decided to do this for a living, did you have any idea you'd be involved in design and equipment issues to this extent?
A: Oh, no, I wanted to cook food! The first thing that really stunned me was learning how to manage the business—the food, the money, the balance sheet. What does the bank want? How do you deal with the personalities? It's really evolved into a conceptual thing. You're trying to envision spaces before they're built. Will customers walk into that space or not? What will they see? How do the salad and the pizza and the T-shirts the employees wear look together? It's been an incredible trajectory.

One of the things I've realized is that I'm really in the restaurant innovation business. I had to accept the fact that one day Hot Lips might not even be a pizza business anymore. The highly capitalized guys out there could easily displace what I'm doing. What I've realized is that we're just a little drip in a big pond and, if I don't share the knowledge I've gained, we're not truly sustainable. There's a social obligation to share the knowledge that transcends the competition. I think learning, applying, trying all lead to higher quality.

Q: And finally, David, what advice do you have for students in this field?
A: Being in the restaurant business takes a lot more capital than you ever imagined; and you can't do it all yourself. One of the hardest things, which I'm still learning, is to delegate to other people and accept the result that comes in, and understand that even if you did it yourself, there might have been shortcomings you don't see. So, surround yourself with people you think are smarter than you! Ultimately, it'll make your life easier.

10

STORAGE EQUIPMENT: DRY AND REFRIGERATED

■ INTRODUCTION AND LEARNING OBJECTIVES

The ability to stockpile some types of foods and other products is critical for a foodservice business. It allows you to take advantage of specials and discounts by purchasing in bulk and helps ensure sufficient product on hand for busy times. However, storage also comes with challenges: making the best use of often limited space, and storing foods in a manner that prevents them from becoming stale or contaminated, or contaminating other surfaces.

Food, beverages, and supplies must all be accounted for and properly stored until they are needed. The two basic types of storage are dry and refrigerated. *Dry storage* is for canned goods, paper products, and anything else that doesn't need to be kept cold. *Refrigerated storage* is for items that must be kept chilled or frozen until used.

In this chapter, you'll learn about both kinds of storage and about the types of equipment needed to outfit your receiving and storage areas for efficiency, safety, and conservation of space.

After reading this chapter, you will be able to:

■ Organize and properly equip a receiving area.

■ Identify the functions of scales, pallets, and carts in a receiving area.

■ Determine how to organize storage shelves and use shelving systems.

■ Explain how and why refrigeration works to keep foods fresh.

■ Identify the types of refrigerators used in foodservice.

■ Explain basic ice machine operation, maintenance, and sanitation.

You'll also learn how refrigeration systems work in refrigerators, coolers, freezers, ice makers, and specialty systems like beer kegs and soft-serve machines, and how to select them for your operation.

10-1 RECEIVING AND DRY STORAGE

You'd be amazed at how many deliveries arrive at the back door or loading dock of the average restaurant in a week. And they all have one thing in common: Someone on your staff must check them all for accuracy. Many owners and chefs go a step further, personally inspecting the quality of fresh items such as produce and seafood and rejecting on the spot those that do not meet their standards or expectations.

Having a well-organized receiving area, setting hours for deliveries, and designating employees who are responsible for accepting and storing incoming stock saves you time and money. Illustration 10-1 shows some of the smartest outdoor dock area features.

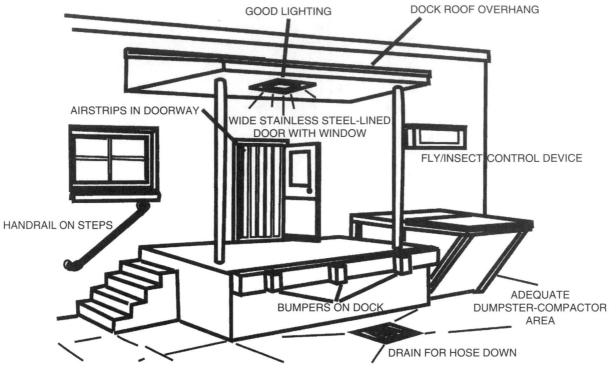

ILLUSTRATION 10-1   Smart features to include in a dock area.

Source: Carl R. Scriven and James W. Stevens, *Manual of Equipment and Design for the Foodservice Industry*, 2nd ed. (Weimar, Texas: Culinary and Hospitality Industry Publications Services, 2000).

Merchandise goes from the dock area into the receiving area, shown in Illustration 10-2. The well-equipped receiving area contains these basic items:

- Scales
- Pallets
- Carts and trucks
- Shelves

ILLUSTRATION 10-2   A receiving area that works well.

Source: Carl R. Scriven and James W. Stevens, *Manual of Equipment and Design for the Foodservice Industry*, 2nd ed. (Weimar, Texas: Culinary and Hospitality Industry Publications Services, 2000).

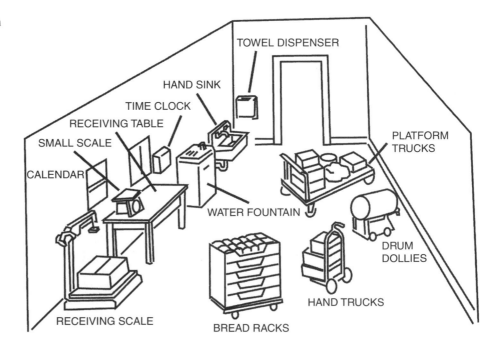

SCALES

Scales are used to weigh items received in bulk to make sure you're getting the amount you pay for. As food costs increase, checking the weights of items in the receiving area is considered a strong control point for a smart operation. If the scales aren't accurate, you're wasting time and money. Of course, scales continue to operate even when they're inaccurate, so the maintenance needed to keep them properly calibrated may be neglected. Dust, dirt, and food scraps can easily clog the sensitive inner workings of a scale and build up friction, which causes inaccuracy.

Scales really get a workout in most foodservice operations, so you want a good, sturdy model instead of the inexpensive, lightweight spring-operated ones, which quickly become inaccurate with heavy use. Scales can be mounted on a receiving table, or they can be floor-standing models. Make sure to provide enough space for easy access to the scales in your receiving area. Some of the things you'll be weighing are big and bulky.

The *digital electronic scale* has become the top choice in many foodservice receiving areas and kitchens too. Capacities vary, but these models have the greatest accuracy as well as being foolproof to read. Some units read out in both metric and U.S. units. Electronic scales also have automatic tare weight adjustment buttons (see Illustration 10-3). *Tare weight* is the weight of the container, which is used so you can subtract it to get an accurate *net weight* of whatever is inside the container.

Electronic or digital scales (they are called both) are all manufactured similarly (see Illustration 10-4). They contain a circuit board and a metal bar called a **load cell**. The bar has **strain gauges** attached to it that measure the way the bar bends when a product is placed on the weighing platform or tray. The strain gauges literally change electrical resistance when they are stressed by weight, and what is being measured is the voltage change that occurs as the bar is bent. The load cell then transfers this voltage change as an analog signal through a wire attached to the strain gauges. The analog signal is processed by an analog-to-digital converter (A/D converter) into a digital signal. A microprocessor reads the digital signal and displays the correct weight on an LCD screen for the user to see.

We share these details to illustrate how important it is to properly care for digital scales. If they are used to weigh items that are clearly too big for them, the internal bar can bend permanently, damaging the scale. Being dropped or shaken can also cause permanent damage. In addition, many scales use batteries. The most frequent cause of inaccurate readings and other digital scale malfunctions is a low or dead battery. Scales that are plugged into wall outlets should be kept plugged in and turned on.

Digital scales also do not perform well under extreme temperature conditions as their load cells are temperature-sensitive, especially to cold. They should be kept in moderately

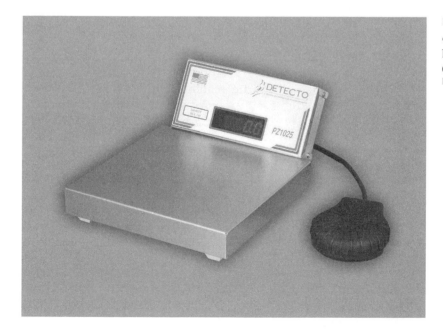

ILLUSTRATION 10-3 A digital electronic scale is compact and easy to read. There are models for both kitchen and receiving use.

Courtesy of Detecto Scale Division of Cardinal Scale Manufacturing Company, Webb City, Missouri.

ILLUSTRATION 10-4 The inner components of a digital scale.

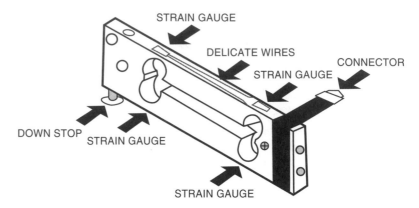

warm, dry conditions, from 65°F to 85°F. If they are moved from one place to another and the temperatures or humidity levels of the two places are quite different, manufacturers suggest letting the scale rest, unplugged and unused, for a few hours to adjust to the new surroundings. **Drift**—the scale industry term for a progressive change, up or down, in the numbers shown on the digital readout—may indicate a temperature problem or a source of static electricity that is interfering with the scale's balance. Repairs on most models would surpass the cost of a new unit.

Three other popular types of scales are used for receiving areas. The dial-type *counter scale* (see Illustration 10-5) rests on a table or bench, with a platform of 12 to 14 inches. It can weigh items from 50 to 200 pounds, in graduations of 1 ounce (for a 50-pound scale) to 4 ounces (for a 200-pound scale). The scale has a *tare knob*, which allows you to adjust for the weight of the container. Today's manufacturers generally supply weight information both in pounds and metric measurements. Often, when you buy a scale, manufacturers include the table, bench, or stand to hold it. A desirable characteristic is that the scale stand be on wheels. Most dial-type counter scales can be leveled for accuracy.

Another popular piece of equipment is the *beam scale*. Most models are portable, rolling on four platform casters, but there are also fixed, countertop models that take up less space. Beam scales come in capacities of 100 to 1000 pounds. The upright pillar on which the scale rests is offset to the left of the platform, which allows for a narrower scale. The *counterweights* (sometimes called *beam poises*) are open on one side to allow for easier reading. Counterweights usually are stored on a side bracket located at the top of the pillar; the biggest problem with counterweights is that they can be lost or misplaced if your staff is not careful. There are also double-beam scales; the second beam allows for tare weight adjustment of up to 100 pounds. Some manufacturers add a balance indicator at the top of the pillar.

A less popular model scale is the hanging dial scale, such as the ones you see in supermarket produce sections. It's handy because its double-faced dial allows people on both sides to see the reading, but it usually has just a 30-pound capacity, graduated in 1-ounce or ½-pound increments. The pan, which is often removable, is 18½ inches in diameter and 4½ inches deep.

The type of scale you choose for your operation is largely a matter of personal preference. Some say dial scales seem to wear out faster but are easy to use; others insist beam scales are more accurate but take longer to use. Electronic scales are the most expensive. No matter your choice, remember the scale will be a vital, hardworking part of your receiving area.

ILLUSTRATION 10-5 A counter scale is a necessity in a receiving area.
Courtesy of Hobart Corporation, Troy, Ohio.

PALLETS

A **pallet** is a low, raised platform on which boxes of products can be stacked and stored. They come in square or rectangular shapes, usually 36 × 48 inches or 48 × 48 inches, and can be made of steel or wood. In many cities, wood pallets are not acceptable for foodservice purposes because they are harder than steel to sanitize and could harbor insects or germs.

The typical pallet is 4 or 5 inches high, and inspectors (who usually insist that everything in a restaurant be stored at least 6 inches off the floor) exempt pallets from this rule. The advantage of storing things on pallets is that they can be lifted and moved by a small hydraulic lift or platform truck, which enables a single person to stack or transport several cases of product at a time. Pallets allow for storage flexibility because they can be moved around easily or upended and leaned against walls when not in use.

CARTS AND TRUCKS

The terms *cart* and *truck* sometimes are used interchangeably in foodservice, but they are actually different items used for different purposes. A cart is a **utility cart**, a rolling shelf unit most often used in restaurants to transport dishes and cleaning supplies. Utility carts can be made of light-gauge metal, usually stainless steel, or molded plastic. The right cart can make life a lot easier on employees by allowing them to organize and move things without wasted steps or mishaps. Recent innovations have produced carts that are ergonomically designed as well as carts that can fold up for more compact storage.

Depending on its size and construction, a utility cart can hold up to 1000 pounds of supplies or equipment. The shelves are usually solid but can also be made of wire. Space between shelves should be 18 to 20 inches high. Most carts have three shelves, with a middle shelf that can be adjusted at 1-inch increments. The bottom shelf is typically about 20 inches from the ground. Rolled or raised edges on the shelves prevent spillage of liquids onto the floor.

The correct name for a truck is **hand truck**. Also known as a **dolly**, it is a small, sturdy platform on wheels that is designed to support and move heavy items such as cases of canned goods. **Layer platforms** or **flat trucks** have four wheels and are strong enough to hold 10 to 15 cases. Flat trucks should have swivel casters to make them easier to steer. In special situations, hand trucks or flat trucks can be equipped with drum and barrel **cradles** to hold beer kegs or other drum-shaped objects steady as they roll.

SHELVES

Perhaps the top consideration for dry storage shelving is adjustability to accommodate the heights of various jars, cans, bins, and boxes. The shelves you choose should be adjustable in 1-inch increments.

The type of shelving you'll choose depends on two things: where it is to be used and what you plan to store on it. Some shelf types are best suited for refrigerated storage; others for ambient (room temperature) storage. Here are a few of the most popular choices:

■ Open-grid wire or mesh shelves allow for air circulation and visibility, but they are difficult and time-consuming to clean because of the many crevices where wires are joined.

■ Flat, solid shelves are easy to clean and extremely sturdy. The solid shelf also eliminates the possibility of food products spilling from one shelf to others and contaminating what is stored below. If you use these, order them with a lip around the edges to help contain spills. The downsides? Solid shelving is expensive, and it does restrict air movement around food.

■ Embossed shelves are alternatives to flat, solid ones. These are solid, but with ridges or slots on the surface to allow air circulation.

Shelves can be made of polymer (heavy-duty plastic); stainless steel; vinyl-coated steel; zinc-plated, chrome-plated, or galvanized steel; and anodized aluminum. Wire-style shelves can be ordered with hard, synthetic coatings (not the same as vinyl) designed for heavy use; some incorporate the antimicrobials we learned about in Chapter 8.

Shelves are coated not only to prevent rust and discoloration when exposed to humidity and food acids but also to provide friction so containers don't slide around easily. Again, your choice depends on their prospective location and function. Most manufacturers quote a weight load per shelf, and, when comparing the same finish and style of shelving, the cost increases as the load limit increases. Stainless steel is the most expensive material, costing at least twice as much as chrome-plated steel.

One shelf coating that definitely won't please the health inspector is paint. Special food-grade paint is required on any painted surface that comes into contact with food to ensure the food is not exposed to lead.

Most shelves attach to upright posts with plastic snap-on clips. Manufacturers make these easy to use and to adjust at 1-inch or 2-inch intervals. Polymer shelves sit inside a heavy-duty frame that attaches to the posts. The whole shelf can be removed from the frame and run through a dishwasher to sanitize it.

The width of a shelf should be 12 to 24 inches; any wider and it becomes more difficult to reach items at the backs of the shelves. Depending on what is being stored, height requirements between shelves vary from 12 inches (for gallon containers) to 18 inches (for #10 cans). They

ILLUSTRATION 10-6 High-density storage units consist of shelves that attach to a track. The track may be installed on floor, ceiling, or top-mount, and allows the shelves to be moved as needed.

Source: Carl R. Scriven and James W. Stevens, *Manual of Equipment and Design for the Foodservice Industry*, 2nd ed. (Weimar, Texas: Culinary and Hospitality Industry Publications Services, 2000).

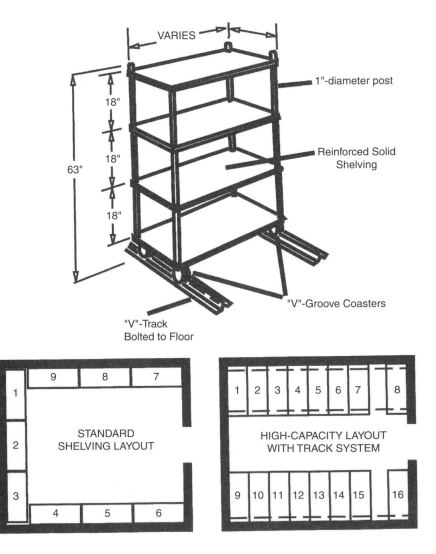

TRACK SHELVING, BOTTOM-MOUNTED

can be mounted on solid upright posts, on rolling racks, or on walls. Shelves come in standard lengths of 24 to 72 inches. Shorter lengths tend to be less efficient and more costly per foot of usable space; longer lengths have less weight-bearing capacity because of the increased span to be supported. Most storage areas utilize a combination, but you can't go wrong with a standard 4-foot shelf unit.

You must also match the upright posts to the width and height of the shelves you want. You may decide to trade the leveling foot at the bottom of each post for a caster, to be able to move whole shelf units as needed. Again, you're looking for maximum flexibility as your storage needs change. If you use casters, select the widest, thickest caster the post will accommodate. As you can imagine, a full shelf unit carries a lot of weight. Remember also that putting casters on anything raises its height by as much as 3 inches.

Height is a consideration because intelligent storage requires that you maximize available space. You can't store food items on the floor (or usually within 6 inches of it) for health and safety reasons, and you can't store it anywhere there is a risk of cross-contamination—near plumbing or sewage lines, under open stairs, and so on. Overhead storage is often an unexpected option. Aisles should be 30 to 36 inches wide if one person at a time will be working in the storage space; add 6 more inches if more than one person will be using the aisle at a time. If carts will be rolled in and out of the storage areas, leave a 10-inch clearance on each side of the cart.

Manufacturers of storage shelving are constantly coming up with new ideas to add space and efficiency. Shelf units with casters can be put on tracks so the user can roll individual shelf units as needed. The tracks can be installed on the floor or overhead; we recommend an overhead configuration because tracks on floors get clogged with debris. The layout of this type of storage area requires only a single central aisle because everything can be moved aside as needed to get to a particular shelf or product, as shown in Illustrations 10-6 and 10-7.

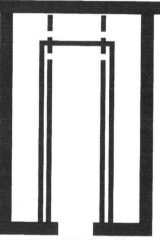

For 200 to 350 Meals / Day
8' 0" x 12' 6"
100 square feet

For 100 to 200 Meals / Day
6' 0" x 8' 0"
48 square feet

ILLUSTRATION 10-7 Suggested sizes for dry storage areas.

Source: Carl R. Scriven and James W. Stevens, *Manual of Equipment and Design for the Foodservice Industry*, 2nd ed. (Weimar, Texas: Culinary and Hospitality Industry Publications Services, 2000).

For 500 to 750 Meals / Day
12' 0" x 31' 0"
372 square feet

Finally, **dunnage racks** (*dunnage* is an old-fashioned word for baggage) are recommended for storage of bulky, unopened items such as bags and cases, which would otherwise take up valuable shelf space. Dunnage racks stand about 12 inches off the floor and come in dimensions from 18 × 24 inches to 24 × 48 inches. The newest models also have removable shelves for easy cleaning. Some establishments routinely designate the bottom shelves of their storage systems as dunnage racks for heavy or bulky items.

No matter what type of storage system you choose, as you shop for it, be sure it is quick and easy to adjust, to clean, and to customize as your menus and needs change.

10-2 REFRIGERATED STORAGE

Most people assume refrigerators and freezers are designed to make food cold and keep it cold. However, this is a common misconception. In fact, the basic principle of refrigeration is *the transfer of heat out of an enclosed space*. Cold is the absence of heat. The goal of any refrigerator is to remove heat. Removing heat extends the useful life of the refrigerated food by protecting it against decay and deterioration.

Now, let's take it one more step. Heat and heat energy are part of any food product, whether it is raw or cooked. Heat enters the refrigerator in three simple ways:

1. Through opening the refrigerator door (This is the major cause of heat transfer.)
2. Through products stored inside the refrigerator
3. Through the door edges and rubber grommets (tiny leaks, indeed, but significant because they are continuous)

The more heat that can be removed from a product (without the side effects that can be caused by too much cold), the longer the product can be held in usable condition. To achieve this ideal, three things occur simultaneously in the enclosed, conditioned environment you probably thought, until now, was just a fridge:

1. Temperature reduction
2. Air circulation
3. Humidity

Wide fluctuations in any of these conditions cause faster deterioration of the stored food; once deterioration begins, the process cannot be reversed. Cold can inhibit but not completely prevent the growth of most microorganisms associated with food poisoning. The temperature needed to accomplish this varies with the type of food, but remember the Hazard Analysis of Critical Control Points (HACCP) guidelines introduced in Chapter 8? They require refrigeration temperatures of at least 40°F (or for the food itself to have an internal temperature of 41°F).

Overall, the best refrigerator is one that can hold constant temperature and humidity levels and to circulate air at a steady rate. The other point to note is that no conditioned environment will be absolutely perfect. No matter what it's made of or how well it works, no freezer or refrigerator can fully halt deterioration, and it certainly can't improve the quality of the food over time.

Two terms you'll hear a lot in refrigeration: the **refrigeration cycle** and the **refrigeration circuit**. The cycle is the process of removing heat from the refrigerated space. The circuit is the physical machinery that makes the cycle possible. It's important to keep them straight as we delve into the inner workings of your refrigerator.

TEMPERATURE REDUCTION

Heat transfer from inside the refrigerator to the outside environment cannot take place without a temperature difference between the two. That different materials transfer heat at different rates also has an impact. There are four basic components of a refrigeration system. Their

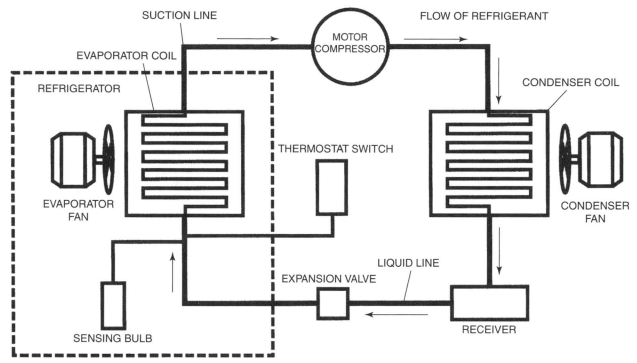

ILLUSTRATION 10-8 Components of a refrigeration system.

Source: Facilities Operations Manual (Washington, D.C.: National Restaurant Association).

names and functions are not unlike those you already learned about in Chapter 7, in our discussion of heating, ventilation, and air-conditioning (HVAC) systems. They are:

■ Evaporator

■ Refrigerant

■ Compressor

■ Condenser

These make up the major physical parts of the refrigeration circuit. The circuit is a closed system. See Illustration 10-8; notice how similar it is to the cooling cycle of an air-conditioning system shown in Chapter 7.

When you open the refrigerator door, warm air is introduced into the cooled, enclosed space. The warm air rises and is drawn into an ***evaporator***, a series of copper ***coils*** surrounded by metal plates called ***fins***. The fins conduct heat to and from the coils.

The evaporator coils hold liquid refrigerant, which becomes vapor (gas) as it winds through them. The vaporized gas is pumped by the ***compressor*** into the ***condenser*** (another series of coils surrounded by fins), where it turns back into liquid. As the gas is compressed, its temperature and pressure increase. The *expansion valve* is the small opening between the low-pressure (evaporator) side and the high-pressure (condenser) side of the system. This valve allows a little refrigerant or a lot to flow, depending on how much cooling is needed.

An alternative to the expansion valve is the *capillary tube*, or *cap tube* for short. Here's the difference: Expansion valves respond more quickly to temperature changes and are best suited to environments where doors are constantly being opened and closed. They are more expensive than cap-tube systems. The capillary tubes work well for storage-type refrigeration, when doors are not opened often.

In either system, the last step is the job of the refrigerator's ***thermostat***. When it indicates colder air is needed, the compressor turns on. Air circulates past the two sets of fins, removing heat from the circuit as the refrigerant flows.

There are top-mounted and bottom-mounted refrigeration systems—that is, the condenser coils can be located near the floor or at the top of the cabinet, above where the food is stored. In foodservice, top-mounted units are worth considering because the coils don't accumulate grease and dust as quickly as they do closer to the floor.

The goal in refrigeration, as with HVAC units, is a balanced system. The compressor, evaporator, and condenser functions must be sized to work well together.

THE CHEMISTRY OF REFRIGERATION

The environmental pressures of global warming have put increased emphasis on the need for alternative types of chemical refrigerants. The trend began in 1996, when the U.S. Environmental Protection Agency's (EPA) Clean Air Act banned the manufacture of the most common refrigerants because they contribute to the depletion of the ozone layer in the upper atmosphere that protects earth from the sun's ultraviolet radiation. These refrigerants were known by the trademark name of their manufacturer, DuPont, as Freon 12 and Freon 22, although we refer to them here by their generic names, CFC-12 and HCFC-22. In some circles, the fluorinated gases are simply referred to as *F-gases*.

CFC-12 is a chlorofluorocarbon (CFC), a chemical combination of carbon, chlorine, and fluorine. Chlorine is the culprit that zaps ozone. Under the EPA guidelines, CFC-12 was supposed to be phased completely out of use by the year 2000. In its place, related chemical compounds include hydrofluorocarbons (HFCs) and hydrochlorofluorocarbons (HCFCs). However, HCFCs (HCFC-22) are also slated for extinction by the EPA somewhere between 2015 and 2030. HFCs, without chlorine, have skirted the ban so far; in fact, the most widely used alternative refrigerant is now HFC-134a. In addition, glycol is now being used as a refrigerant. It works well in smaller systems but has proven problematic in larger freezer systems.

All refrigerants are hazardous when exposed to an open flame. Some of them contain butane or propane mixes blended into their formulas. If large quantities of refrigerant are released in a confined area, suffocation is a danger because refrigerant actually displaces oxygen. Breathing refrigerant can cause nausea, dizziness, shortness of breath, and even death. Thus, any type of refrigerant gas should be handled by a professionally trained technician.

Environmental risks and health warnings aside, the cooling power of the modern refrigerator comes from the repeated compression and expansion of a gas. As the gas expands, it cools and is cycled around an insulated compartment, chilling the contents inside. Ammonia, new chemical blends, and even space-age technology using sound waves to cool foods are other options that have been introduced recently with some success. In 2004, the ice cream maker Ben & Jerry's installed the first thermoacoustic freezer, developed by Penn State University researchers, in a retail location. *Thermoacoustics* is the premise that as sound travels through air, it alters the temperature of the air. A loudspeaker is used to create 170 to 195 decibels of sound (and yes, that's *very* loud) in a tube that contains inert, compressed gases (helium or argon), which are environmentally safer than CFCs. The sound causes the gas molecules to vibrate, expand, and contract. When they contract, they heat up; when they expand, they cool down. In refrigeration, the goal is to exhaust the heat generated as the soundwave is compressed and capture the chill as the soundwave expands. The loud screech emitted within the unit is muffled so it is heard as a quiet hum from the outside.

Thermoacoustics cools using a type of sealed motor developed in the 1800s by Robert Stirling of Scotland. You may hear the term **Stirling cycle technology** associated with CFC-free coolant ideas. A Stirling motor can be solar-powered, which is one of its potential "green" advantages.

Another promising cooling technology is *electromagnetism*. A magnetic refrigerator can cool by repeatedly switching a magnetic field on and off. The current prototype is made with gadolinium, a metal used in the recording heads of video recorders. Gadolinium and magnets are not cheap, but the technology shows great potential for two reasons: Electromagnetism is environmentally safe (no CFCs), and it does not require a compressor (no mechanical humming noise as the refrigerator cycles on and off). At this writing, Astronautics Corporation of America in Milwaukee, Wisconsin, (www.astronautics.com) is at the forefront of this field.

Expect more technological breakthroughs as climate-change headlines become more ominous. A group called Refrigerants, Naturally! (www.refrigerantsnaturally.com) was formed in

2004 by Coca-Cola, McDonalds, and Unilever to develop and test HFC-free refrigeration technologies, making commitments to eliminate HFCs in their point-of-sale cooling applications. The organization advocates the use of what it terms "natural" refrigerants, meaning substances that are found in nature—including air, ammonia, carbon dioxide, iso-butane, propane, and water.

How does the foodservice operator cope with the changes and the prospect of expensive new replacements for old workhorse refrigerators? Well, if your equipment is in good repair, you probably should do nothing as long as it lasts except keep it properly maintained. This especially means cleaning the unit's condensing coil once a month to prevent grease and dirt collection that block air circulation.

If your refrigerator needs repair, you have two choices: Voluntarily retrofit it to use an alternative refrigerant, or purchase a new unit that is already equipped to use the newer refrigerants. Retrofitting almost always requires more than one service call and includes these steps:

- Recovering the outdated refrigerant
- Changing the evaporation coil
- Replacing the filter or dryer, if necessary
- Recharging with the new refrigerant
- Checking performance for the first few weeks

The EPA now has a sophisticated set of rules for refrigerator repair. The EPA certifies repair technicians and their equipment, and requires that they recycle or safely dispose of refrigerant by sending it to a licensed reclaimer. The rules also state that "substantial leaks . . . in equipment with a charge greater than 50 pounds" be repaired. This means if the unit leaks 35% or more of its pressure per year, it needs fixing. As the owner of commercial refrigeration equipment, you are also required to keep records of the quantity of refrigerant added during any servicing or maintenance procedure.

The EPA website contains summaries of the rules as well as lists of acceptable alternative refrigerants that don't contain the ozone-depleting CFCs and HCFCs. They're identified with abbreviations and numbers, such as MP-39, HP-80, R-406a, and GHG-X4, which probably don't mean much to you as a foodservice professional. However, the important points to remember are:

- Use an EPA-certified technician, with certified equipment, to do your refrigeration repair work.
- Underwriters Laboratories (UL) is now authorized to test and approve alternative refrigerants, so look for the UL label on products.
- Keep your maintenance records updated. Violations of the Clean Air Act can result in fines of up to $25,000 per day.
- If you have more than one piece of older equipment, plan a gradual phase-out or retrofit program. Don't break your budget by trying to do it all at once.

This information should also serve as a caution when you are looking at purchasing used equipment. Is the owner getting rid of it because it no longer meets the environmental rules?

With that, we've discussed the first major process going on inside the refrigerator: temperature reduction. Now, let's examine the other two processes.

AIR CIRCULATION

The refrigerator relies on forced air to transfer heat. Fans inside the appliance move air around. The faster the air flows, the more quickly the heat is removed. For this reason, you don't want to do anything to block the airflow.

There are three basic types of forced-air systems in refrigerators. In *ceiling-type refrigeration*, a single fan is mounted on the ceiling of the appliance. This is adequate for small-volume interiors but is not used in larger refrigerators. Because it only has a single location, it might allow hot spots in the corners of the interior cabinet.

In *back-wall* or ***mullion-type refrigeration***, the airflow system takes in air above the top shelf and discharges it below the bottom shelf. The ***duct-type refrigeration*** system is a combination of the first two types. Here, the forced-air unit is located at or above ceiling level, and the air is circulated through a series of small air ducts vented to various spots on the back wall of the cabinet.

Just how important is air circulation? Well, the difference between safe and unsafe raw foods can be as little as 5°F to 7°F. Seafood, poultry, and red meats spoil within 18 to 24 hours if their refrigerated temperature rises above 42°F to 45°F, and you already know the HACCP guideline of temperatures no higher than 40°F. Would you rather risk a lawsuit and the resulting negative publicity from food-poisoning allegations or keep your refrigerator air circulating properly?

HUMIDITY

Humidity is the amount of moisture (or water vapor) in the air. At different temperatures, air can hold different amounts of water. In refrigeration, the type of humidity we are interested in is the ***relative humidity***, or how much of its maximum water-holding capacity the air contains at any given time, expressed as a percentage. For example, 85% humidity indicates the air is holding 85% as much water as it could hold at that temperature. Relative humidity greatly affects the appearance and rate of deterioration of many foods. If the air surrounding the stored foods has a very low relative humidity, for instance, the air naturally picks up moisture from the foods themselves, causing surface discoloration, cracking, and drying. If the air has a high relative humidity, some of the moisture will condense on food that is supposed to be kept dry, causing it to soften or grow mold or bacteria.

Fortunately, most foods do well in a relative humidity of 80% to 85%. To achieve this optimum level, manufacturers are concerned that the refrigerator's evaporator coils be large enough to operate at a temperature a few degrees lower than the desired temperature of the appliance. This differential reduces the amount of moisture that accumulates on the evaporator coils and keeps the moisture in the cabinet of the refrigerator instead. If the coils' temperature becomes too low, however, the moisture will turn to ice crystals and get stuck on them. In this case, airflow through the system is blocked and the moisture in the refrigerated space is depleted. As you can see, getting all the factors right is a delicate balance, with your food costs and food quality at stake.

In short, it is difficult to keep frost off the coils but necessary to keep them frost-free so they will operate properly. Adding heat to the area, to defrost the coils, can compromise the temperature of the food inside. A fairly new concept from Hussman Modular Defrost of Bridgeton, Missouri, does just what its name indicates: defrosts the coil in sections. The automated system defrosts coils at no more than 9 minutes per section and never defrosts adjacent sections at the same time, all programmed by an electronic controller capable of running up to six walk-ins. The idea works for walk-in and reach-in refrigerators but not freezers. It maintains food quality and saves energy by keeping the compressors from working overtime to compensate for frozen coils.

We mentioned most foods do well at 80% to 85% relative humidity, but fresh fruits and vegetables are exceptions. They require more humidity—up to 95%. To increase moisture content, you can slow the air circulation. This explains why there are separate closed produce bins in most refrigerators—to hold in natural moisture from the vegetables and to restrict airflow.

Freezers maintain an average relative humidity of only 30% to 35%. Any more moisture would automatically raise the temperature because it would hit the coils, freeze in place, and block the airflow, causing the freezer air to become warmer. The low humidity of freezers requires special food storage precautions. Use moisture- and vapor-proof wrapping to prevent the surface damage we know as freezer burn from occurring if any moisture condenses on the food.

10-3 SELECTING A REFRIGERATOR

Now that you know how a refrigerator works, how do you pick the ones that will work best for your operation? Manufacturers all have printed specifications, and your dealer or equipment consultant has recommendations. Some of the criteria you'll choose from follow.

Finishes. The surface of your refrigerator should be as sturdy as its interior components. Popular finishes include stainless steel, vinyl-coated steel, fiberglass, and coated aluminum; the latter comes in rolled, stucco, or anodized styles. In the commercial world, unlike the home, few models have porcelain or baked-enamel finishes. In fact, some health departments do not allow these finishes in commercial or institutional installations.

Construction. You can't kick the tires, but certain quality features are evident, such as overall sturdiness, door alignment, and how securely the handle is attached to the appliance. All-metal, welded construction is a plus, and having a seamless interior compartment is an NSF International requirement. In fact, the NSFI has a whole section on refrigeration, accepted by most cities as minimum standards. Also, look for ease of cleaning and self-defrost features.

Insulation. The most commonly used insulation is polyurethane, in sheets or foam, which has superior insulation qualities and even makes the cabinet a bit sturdier. Make sure it is non-CFC polyurethane foam, with at least an R-15 rating. Fiberglass is also acceptable, although it requires greater thickness to achieve the same results as polyurethane.

Doors. A small but critical detail is whether you want the door to open from the left or from the right side. Some models offer half-doors (you conserve cooling power by only opening half the refrigerator at a time) or full-length doors. Doors can be solid or made of shatterproof glass; they can have hinges or slide open and shut on a track. The way doors are opened can affect traffic patterns in the kitchen (see Illustration 10-9). Doors can also be self-closing, with magnetic hardware, to prevent being left ajar accidentally. The hinges should be stainless steel or, at least, chrome. Look for door gaskets that are easy to snap in place, not the old screw-in kind, as you will probably replace them during the life of the unit.

Handles. Stainless-steel or nickel-plated handles are best. You can select vertical or horizontal handles. They can protrude or be recessed. Be sure the handle is included in the warranty, as handles take a lot of abuse and may have to be replaced periodically.

Refrigeration system. It may be self-contained or, in the case of very large appliances, a separate unit. As we've mentioned, it may also be top-mounted or bottom-mounted. The accurate electrical current and capacity of the facility must be known so the manufacturer can supply the correct voltage and phase to meet the needs of the space. In some cases, additional expense may be involved to upgrade the electrical system. At any rate, look for the UL seal of approval, a sign the unit meets basic electrical safety standards. The system may be water cooled or air cooled. The most common in foodservice is the self-contained, air-cooled unit.

Remember, the capillary tube system is for refrigerators used for storage—not much door-opening action. The expansion valve system has quicker pull-down capacity—that is, it can pull the temperature down faster after the unit is opened. It is ideal for busy hot line situations where the refrigerator is constantly in use.

ILLUSTRATION 10-9 The way that refrigerator doors are opened affects workflow and traffic patterns in the kitchen.

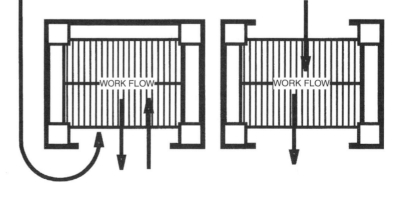

Drain Requirements. Most new refrigerators provide an automatic defrost system and automatic condensate disposal, which eliminates the need for a separate plumbing connection. Ask about it, however. A reminder: NSFI sanitation standards prohibit drains inside the refrigerator.

Accessory availability. You'll get shelves as standard equipment with a refrigerator purchase. Make sure they are adjustable. For foodservice, lots of additional items might improve efficiency: adjustable tray slides, drawers, special racks for serving pans (called **pan glides** or *pan slides*), and dollies or carts designed to convert a reach-in cabinet to a roll-in one. Think about these accessories when selecting the door, too. Certain doors seem to work better with some types of add-ons.

Warranty. Most manufacturers provide a 1-year warranty on parts in case of defective workmanship or materials; look for a separate 5-year warranty on the motor and compressor unit. Some manufacturers also offer extended service warranties.

Cabinet capacity. A properly designed refrigerator provides the maximum amount of usable refrigerated space per square foot of floor area and accommodates the sizes of pans you'll be using. There are plenty of complex guidelines for calculating capacity and needs, which will be covered elsewhere in this chapter.

Adaptability. Because today's foodservice operations have changing needs, manufacturers are building in features to maximize flexibility. One such offering is the convertible temperature option. With a flick of a toggle switch, a freezer can be converted to a refrigerator. This can be a pricey addition at the time of purchase, but if food storage requirements change, the option will pay for itself instantly. Another variation is the combination medium- and high-temperature cabinet, designed to thaw frozen products quickly and safely by introducing warmer air into the cabinet as needed. Additional fans and a temperature sensing device bring the unit back to its normal refrigeration level when the food is sufficiently thawed. In hybrid models, cabinets are separated into two or three sections, each with different cooling capacities.

More how-to-buy tips can be found in the next "Foodservice Equipment" box, reprinted with permission of *Restaurant Hospitality* magazine.

REACH-INS AND ROLL-INS

Most kitchens have too much or too little refrigerated space *at the proper locations* to meet their needs. In other words, the physical capacity may be adequate, but either it's not the correct type of refrigerator space or it's not flexible enough to be used to maximum efficiency. Having the right kinds of refrigeration actually can mean using fewer refrigerators and freezers, which saves energy and money. How do you accomplish this ideal? First, you must decide how much capacity you need. As we mentioned in our discussion of space allocation in Chapter 4, the norm in casual restaurants is to allow 1 to 1½ cubic feet of refrigerated storage space per meal served. In fine dining, this increases to 2 to 5 cubic feet of space per meal served. You use a refrigerator not only for storage but also to slowly, safely thaw frozen foods 24 to 48 hours before you need them.

Remember, roughly half of a refrigerator's total cubic footage is usable space. The rest is taken up by the unit's insulation and refrigeration system. (Walk-in coolers also contain aisles, which take up room. More about that later in the chapter.)

Another handy rule is that, in a reach-in refrigerator, 1 cubic foot of space holds 25 to 30 pounds of food. Divide the total weight of food you'll need to store by 25 or 30, and you'll have a good idea how much cubic footage you will need.

Only after you've determined your capacity needs can you take the next step: deciding how much floor space you have for refrigeration and what size unit will fit there. This way, you can calculate how many units you will need.

The third major step is to look at reach-in, roll-in, and walk-in options. A **reach-in refrigerator** is similar to the one you have at home: You pull open the door, reach in, and get what you want. In a commercial kitchen, the problem is that the refrigerator door is opened and

FOODSERVICE EQUIPMENT

Buying Refrigerators

Consider these factors when purchasing a refrigerator.

■ If you use a lot of sheet pans or steam-table pans, consider pan slides in lieu of wire shelves. Universal slides allow you to use either sheet pans or steam-table pans. A sheet pan on slides can also serve as a shelf when both are needed.

■ Many reach-in refrigerator manufacturers offer different lengths and widths for the same door configuration. For example, a manufacturer may make a two-door refrigerator in a 48-inch, 52-inch, and 58-inch width. Costs are all close. If you go with pan slides, then you want the narrowest unit that will fit the slides; anything wider is wasting area in your kitchen. If you will be storing larger items, such as case goods, a larger width may be the better buy.

■ To make the most of your worktable, consider undercounter refrigerators. Most can be bought with a worktable top as part of the unit. Garnish pans can be cut into the top for an added means of refrigeration.

■ Finish materials can often add to or reduce the cost of a refrigerator, but they also affect durability. An all-stainless-steel cabinet is the top of the line for most manufacturers. But if you forgo stainless steel inside the box, you can save 10% to 15% of the overall cost. If you can accept an aluminum finish on the refrigerator exterior except for the doors, an additional 20% or more savings is possible. The trade-offs are that aluminum is a soft metal and may be dented more easily than stainless.

■ Cool-down, rapid-chill, and blast chillers are becoming popular because health departments are more sensitive to proper food-handling procedures and storage temperatures. Any of these refrigerators cool foods quicker than a standard refrigerator and reduce chances of food-borne illness.

Source: Reprinted with permission of *Restaurant Hospitality* magazine, Penton Media, Inc., Cleveland, Ohio.

closed constantly, in heat that's a lot more intense than a home kitchen. A duct-type system, with louvered air ducts promoting airflow throughout the cabinet, seems to work best to counteract the inevitable blasts of warm air.

Inside the refrigerator cabinet, the wise use of space can increase your capacity by 30% to 35%. A simple, heavy-duty pull-out shelf system can allow full use of the bottom part of the unit without making employees stoop to retrieve things there.

Typical reach-in units range from a one-section, single-door unit with 22.7 cubic feet to a three-section, three-door unit with *more than* three times the capacity, at 74.7 cubic feet. Total storage capacity depends somewhat on the number of shelves in the unit—and, of course, the number of shelves depends on the heights of the products you'll store on them (see Illustration 10-10). They can be custom-sized to fit under counters or in small spaces.

Convertible reach-ins are basically freezers that can be converted to refrigerators with the flip of a switch located on the cabinet. Manufacturers offer these in one-, two-, and three-section units, so you can adjust for more refrigerator space or more freezer space, as needed. Adaptability is the key.

The bigger the foodservice operation, the greater the need for a ***roll-in refrigerator***. If your operation does a lot of batch cooking, for instance, you want to be able to move large numbers of meals on rolling carts in and out of refrigerated space. Carts mean less handling, which means less spillage, less heavy lifting, and so forth.

A reach-in unit can be converted to a roll-in by using a dolly on which a half or full rack of product rests. The rack has swivel casters and is latched onto the dolly. If the height of the dolly platform is compatible with the bottom of the refrigerator cabinet, the person holding the dolly can just position it correctly, tip it forward, and slide the rack of product into the refrigerator (see Illustration 10-11). This seems to work best when the reach-in refrigerator is equipped with 6-inch legs.

ILLUSTRATION 10-10 A commercial reach-in refrigerator allows its user a lot of storage flexibility.

Courtesy of Traulsen, a division of ITW Food Equipment Group LLC/Hobart.

The roll-in cabinet is similar to the reach-in except that, instead of 6-inch legs, it has a ramp at floor level so entire carts can be rolled inside. The floor of the roll-in is stainless steel. Capacities of roll-ins range from 35.3 cubic feet for one-section units to 113.2 cubic feet for three-section units.

A *pass-through refrigerator* is a variation of the standard reach-in. This cabinet has two sets of doors located opposite each other. Cafeterias and garde manger areas make good use of this special refrigerator when the kitchen staff places food in one side for servers to pick up at the other. On the service side, which gets opened more frequently, half-doors are recommended to help with temperature control. Another recommendation is to check the temperature of this unit more often for food safety reasons because it is likely to be warmer with all the activity. Choose glazed glass upper doors for easier product visibility, and consider using shelves in the upper portion of the cabinet and half-height carts in the lower section. This would allow a fully loaded tray of prepared salads, for example, to be transferred as needed from its storage in the lower section into the upper section to await serving. The workflow in your kitchen helps determine the suitability of a pass-through refrigerator.

ILLUSTRATION 10-11 A roll-in refrigerator is sized to accommodate rolling racks of food.

Source: Carl R. Scriven and James W. Stevens, *Manual Equipment and Design for the Foodservice Industry* (Albany, New York: Thomas Learning, 1989).

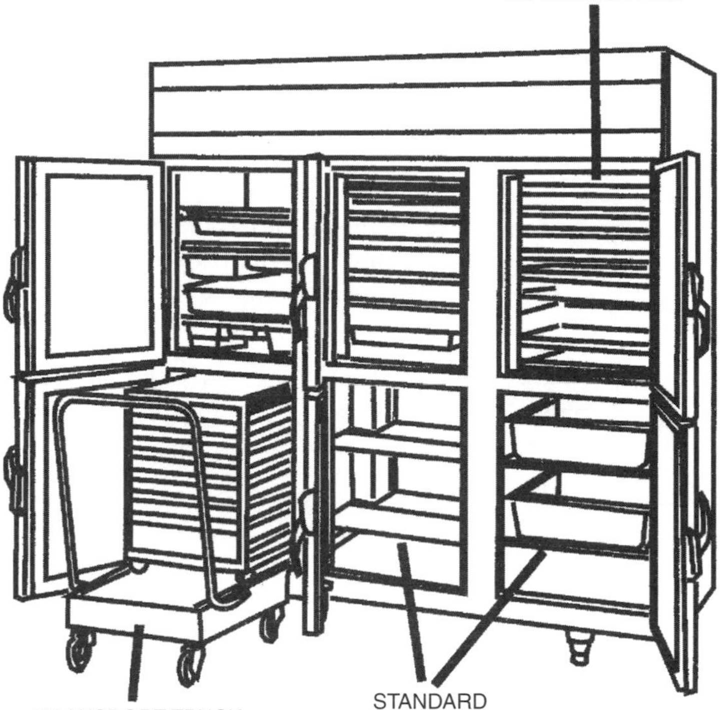

UNIVERSAL PAN GLIDES FIT MOST STANDARD PANS

TRANSPORT TRUCK MODULE FOR 18" x 26" PANS

STANDARD WIRE SHELVING

Other types of specialty refrigerators are made to fit under bars or counters. These are handy in confined areas, particularly around the hot or cold preparation lines. They range in storage capacity from 5.7 to 15.4 cubic feet, with one to three doors. There are also refrigerated prep tabletops, where 8 to 24 pans can sit on a chilled surface instead of using ice. Lately, these prep tops have come under the scrutiny of health departments for exceeding the maximum 40°F temperature to keep food safely chilled. One solution is to install the undercounter refrigerator and the prep top above with separate controls and separate refrigeration units even though they're run by the same compressor.

Reach-ins and roll-ins have lots of door styles to choose from. The basic decisions are whether you want the door to be full height or half height; on hinges or a sliding track; if it's hinged, what side it opens from; and if the door itself is solid or see-through glass. Some manufacturers are now building custom roll-ins for larger combi-oven trolleys.

Most equipment experts frankly prefer hinged doors over sliding ones. The door seal isn't as tight on sliding doors, and it can be hard to clean the sliding tracks. However, if aisle space is tight, a sliding door can be useful because it takes only half as much standing room to open as a hinged door. In a pass-through refrigerator, use sliding doors only on the servers' side; having them on both sides lets out too much cold air too fast.

The question of a full door or half-door is also one of temperature loss. The half-door is more energy efficient, especially if the unit is located close to the hot line, but it has the disadvantage of allowing staff to see only half of the contents at a time. Inside the refrigerator, you've got to organize the contents well to make half-doors work efficiently for you. When it comes to how the doors will open, give a lot of thought to the flow pattern of the workstations in the area. Open doors cannot help but obstruct the flow, and people should be able to see around them and maneuver around them safely. One manufacturer even boasts the ability to change door hinges on the spot within 30 minutes, a feature that might come in handy.

For door hardware, you'll have a choice of magnetic gasketing with self-closing hinges or positive-action fasteners with standard hinges. Eventually, both will have to be adjusted, although positive-action fasteners seem to require more frequent tweaking. Both are acceptable if the design and installation of the cabinet is sound.

A final word about reach-ins and roll-ins: Refrigeration systems can be either air cooled or water cooled, with a heavy-duty (often referred to as a *high-torque*) condenser, which must be kept clean for the system to work properly. In most foodservice settings, the air-cooled, self-contained system is used because a good kitchen ventilation system can remove the heated air given off by the refrigerator. The reason refrigeration experts don't recommend the water-cooled system for most kitchens is that it requires a water cooling tower. Although a tower is usually part of the building's regular air-conditioning system, it must be bigger (and take up more space) to accommodate the demand refrigerators place on the cooling system.

No matter which system you choose, correct voltage must be considered before you order or install your refrigerators.

10-4 WALK-IN COOLERS AND FREEZERS

A walk-in cooler is just what its name implies: a cooler big enough to walk into. It can be as small as a closet or as large as a good-size room, but its primary purpose is to provide refrigerated storage for large quantities of food in a central area. Experts suggest your operation needs a walk-in when its refrigeration needs exceed 80 cubic feet or if you serve more than 250 meals per day. Once again, you must determine how much you need to store, what sizes of containers the storage space must accommodate, and the maximum quantity of goods you'll want to have on hand.

The only way to use walk-in space wisely is to equip it with shelves, organized in sections. Exactly how much square footage do you need? The easiest formula is to calculate 1 to 1½ cubic feet of walk-in storage for every meal you serve per day.

Another basic calculation: Take the total number of linear feet of shelving you've decided you will need (A), and divide it by the number of shelves (B) you can put in each section. This will give you the number of linear feet per section (C). To this number (C), add 40% to 50% (× 1.40 or 1.50) to cover overflow—volume increases, wasted space, and bulky items or loose product. This gives you an estimate of the total linear footage (D) needed.

However, linear footage is not enough. Because shelves are three-dimensional, you must calculate square footage. So multiply (D) by the depth of each shelf (E) to obtain the total square footage amount (F).

Finally, double the (F) figure to compensate for aisle space. Roughly half of walk-in cooler space is aisle space. On paper, the calculations look like this:

$$\frac{\text{Linear Feet of Shelving (A)}}{\text{Number of Shelves (B)}} = \text{Linear Feet per Shelf Section (C)}$$

$$\text{(C)} \times 1.40 \text{ (or 1.50)} = \text{Total linear footage needed (D)}$$

$$\text{(D)} \times \text{Depth of each shelf (E)} = \text{Total square footage (F)}$$

$$\text{(F)} \times 2 = \text{Estimated square footage for total walk-in storage needs}$$

Another popular formula is to calculate that for every 28 to 30 pounds of food you'll store you will need 1 cubic foot of space. When you get that figure, multiply it by 2.5. (The factor 2.5 means only 40% of your walk-in will be used as storage space; the other 60% is aisles and space between products.) The result is the size of the refrigerated storage area you need. For a walk-in freezer, simply divide your walk-in refrigerator space by 2.

Larger kitchens serving more than 400 meals a day may need as many as three walk-in refrigerators: one for produce (41°F), one for meats and fish (33°F to 38°F), and one for dairy products (32°F to 41°F).

The walk-in is used most often to store bulk foods. Because this often means wheeling carts or dollies in and out, the floor should be level with the kitchen floor. This leveling is achieved by the use of strips (called **screeds**) that are applied to the floor.

Coolers don't come as a single unit; they are constructed on-site. The walls, ceilings, and floors are made of individual panels. Wall panels should be insulated to a rating of R-30, which means a 4-inch thickness. They come in various lengths and widths, with 12 × 12-inch corner panels at 90-degree angles. They can be as short as 7½ feet or as tall as 13½ feet. The most common type of insulation inside the panels is polyurethane, and the outside walls of the panels can be made of stainless steel, vinyl, or aluminum. Stainless steel is the most expensive, and aluminum—because it's the least expensive—is the most popular choice. If the walk-in is an outdoor installation, aluminum is the most weather resistant. Illustration 10-12 shows how walk-in panels fit together to form the structure. The installer ensures the unit has interior lighting.

The floor panels for walk-ins are similar to the wall panels. Load capacities of 600 pounds per square foot are the norm, but if you plan to store very heavy items (like beer kegs), a reinforced floor can be purchased with a load capacity of up to 1000 pounds per square foot.

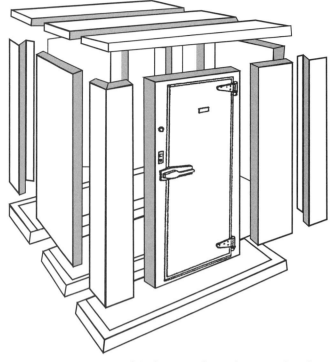

ILLUSTRATION 10-12 This diagram shows how insulated panels fit together to form a walk-in cooler.

Source: Arctic Industries, Miami, Florida.

The refrigeration system of a walk-in is a more complex installation than a standard refrigerator, primarily because it's so much bigger. Matching the system (and its power requirements) with the dimensions of the walk-in and its projected use is best left to professionals, but do note that a walk-in accessed frequently throughout the day requires a compressor with greater horsepower to maintain its interior temperature than one that is accessed seldom. A 9-foot-square walk-in needs at least a 2-horsepower compressor. The condenser unit is located either on top of the walk-in (directly above the evaporator) or up to 25 feet away, with lines connecting it to the walk-in. The latter, for obvious reasons, is known as a *remote* system, and is necessary for larger-than-normal condensing units with capacities of up to 7.5 horsepower.

Compressors are available in either piston or scroll types. Scroll compressors are normally available in larger-horsepower units, and they run more efficiently than piston compressors because the scroll is a turning screw that pushes the refrigerant through the system. A piston drive works more like an automobile engine; more energy is required to move the piston.

In a remote system, the refrigerant must be added at the time of installation. Smaller walk-ins have a plumbing configuration called a *quick-couple* system that is shipped from the factory fully charged with refrigerant; this definitely simplifies installation. However, you may need the added power of a remote system if your kitchen has any of these drains on the walk-in's cooling ability: frequent door opening, glass display doors, multiple doors per compartment, or an ambient kitchen temperature that's near 90°F.

Modern walk-ins sometimes offer a frozen-food section in addition to the regular cooler space. There are pros and cons to this concept. It may ease the load on the freezer because it's already located inside a chilled airspace; but it also can't help but reduce overall usable space because it requires a separate door.

You can also order your walk-in with a separate reach-in section that has its own door and shelves. Although this may save the cost of purchasing a separate reach-in, some critics claim that a walk-in is not designed to do a reach-in job, such as storing uncovered desserts. Do you really want them in the same environment as cartons of lettuce and other bulk storage items? Consider the possible cleanliness or food quality factors.

The doors should open out, not into the cooler itself. The standard door opening is 34 × 78 inches. Several door features are important for proper walk-in operation. These include:

- A heavy-duty door closer.
- Self-closing, cam-lift door hinges. If the door can be opened past a 90-degree angle, the cam will hold it open.
- A heavy-duty stainless-steel threshold. This is installed over the galvanized channel of the door frame.
- A pull-type door handle with both a cylinder door lock and room for a separate padlock if necessary.
- Pressure-sensitive vents, which prevent vacuum buildup when opening and closing the door.
- An interior safety release so no one can be (accidentally or otherwise) locked inside the cooler.

Other smart features that can be ordered for walk-ins are:

- A thermometer (designed for outdoor use, but mounted inside the cooler) with a range of −40°F to +60°F.
- A monitoring and recording system that keeps a printout of refrigeration temperature or downloads to a computer.
- Glass, full-length door panels (like those in supermarkets and convenience stores), sometimes called *merchandising doors*, either hinged or sliding.
- Heavy-duty plastic *strip curtains* inside the door. (One manufacturer claims a 40% energy savings with this feature.)
- A *foot treadle*, which enables you to open the door by pressing on a pedal or lever with your foot when both hands are full.
- Three-way interior lighting, which can be turned on from outside or inside the cooler, with a light-on indicator light outside. Inside, the light itself should be a vapor-proof bulb with an unbreakable globe and shield.

When space is at a premium, think about whether it is practical to install an outdoor walk-in unit. This is an economical way to add space without increasing the size of your kitchen, and you can purchase ready-to-use standalone structures with electricity and refrigeration systems in place. They come in standard sizes from 8 to 12 feet wide and up to 50 feet in length, in 1-foot increments. They range in height from 7.5 to 9.5 feet. Look for a unit with a slanted, weatherproof roof, a weather hood, and a fully insulated floor. An outdoor walk-in costs about half of the price of installing an indoor kitchen walk-in, so this is a moneysaving idea if it works in your location.

If your demands for walk-in space are seasonal, consider leasing a refrigerated trailer, available in most metropolitan areas on a weekly or monthly basis. They can provide an instant 2000 cubic feet of additional storage space that can be kept at any temperature from −20°F to +80°F. They use basic 60-amp, 230-volt, three-phase electricity. Ask if the lease agreement includes hookup at your site and service if anything goes wrong.

REFRIGERATION MAINTENANCE

Most refrigerators and walk-ins seem virtually indestructible and problem free, but you'll get longer life out of yours by following these safety and maintenance tips.

Clean the door gaskets and hinges regularly. The door gaskets, made of rubber, can rot more easily if they are caked with food or grime, which weakens their sealing properties. They can be safely cleaned with a solution of baking soda and warm water. Hinges can be rubbed with a bit of petroleum jelly to keep them working well.

Dirty coils force the refrigerator to run hotter, which shortens the life of the compressor motor. They should be cleaned every 90 days, preferably with an industrial-strength vacuum cleaner.

Walk-in floors can be damp-mopped but should never be hosed out. Too much water can get into the seals between the floor panels and damage the insulation.

A refrigerator works only as well as the air that's allowed to circulate around its contents. Cramming food containers together so there's not a spare inch of space around them doesn't help. Also try to keep containers (especially cardboard ones) from touching the walls of the cabinet. They may freeze and stick to the walls, damaging both product and wall.

Use a good rotation system: First in, first out (FIFO) is preferable. Or put colored dots on food packages, a different color for each day of the week, so everyone in your kitchen knows how long each item has been in the fridge.

10-5 SPECIALTY REFRIGERATION UNITS

Carbonated beverages and ice cream products often require their own storage in a busy restaurant. Because both must be refrigerated, we include them in this chapter.

Carbonated beverages are created on site at most restaurants by mixing syrup and carbonated water and propelling it through a system of chilled tubes (called *lines*) to the dispenser. There are two types of carbonated beverage machines, postmix and premix; and two types of dispensers, fountain and cobra gun.

The ***postmix system*** begins with the syrup concentrate moving through the chilled line and the carbonated water moving through another line to join at the dispensing head, where the newly mixed soft drink is poured into a cup or glass. A valve for each drink may be electric or manual. The syrup is purchased in 3- or 5-gallon tanks or boxes that sit under the dispenser. Valves on the syrup containers enable easy hookup to the system. Most systems use carbon dioxide as the gas that propels the liquids through the lines.

The ***premix system*** works much like the postmix, except the syrup and carbonated water are purchased already mixed and propelled up a single line into the dispensing head. Both premix and postmix dispensers usually contain room for ice storage and chutes to dispense ice into cups. Self-contained fountain units come in large standalone sizes with a dozen valves or more and in countertop sizes as small as 3 feet in width. They can be purchased from the same company that supplies the syrup.

Each type of beverage system has advantages and disadvantages. The concentrated postmix syrup is less expensive than the premixed product, but postmix requires more lines and is more complicated to maintain and hook up. In large operations where more than one dispensing site is needed (say, in a cafeteria), you can set up one central propulsion and cooling system and connect it to more than one row of dispensing valves.

In a bar, the dispensing unit is likely to be used only by the bartenders. It is called a ***gun*** (*bar gun, cobra gun, handgun,* or *speedgun*), and it's attached to a multiline hose called a ***cobra***. The top of the dispensing head features up to eight buttons you can push, each linked to a different line. The *cobra gun* can dispense as many as seven soft drinks and water from a single head (see Illustration 10-13).

ILLUSTRATION 10-13   The soft-drink system dispenser is known as a handgun, a cobra gun, or a speedgun.

Courtesy of Wunder-Bar™ Automatic Controls, Inc., Vacaville, California.

Soft-drink dispensers cool beverages one of two ways. One is a small electric refrigeration system that runs on a ¼- or ½-horsepower compressor and can chill up to 700 12-ounce drinks per hour. The other option is a *cold plate*, which is a metal sheet chilled by direct contact with ice in a bin. The beverage lines all run through the metal plate, cooling the syrup and water. The larger the plate and the more of its surface is ice covered, the colder the final product.

The temperature of your beverage dispenser is critical because carbonation tends to fizzle above 40°F, resulting in flat beverages.

DRAFT-BEER SERVICE

The popularity (and profitability) of draft beer requires another type of refrigeration: a system for storage of beer kegs and dispensing of beer. As we mention in *The Bar and Beverage Book*, 4th edition (published in 2007 by John Wiley & Sons, Inc.), the factors to consider when selecting a system are the numbers of beers you plan to serve; the potential beer volume needed at peak sales times; and the distance between the taps and the refrigerated storage where the kegs will be kept. From the book:

> A draft-beer system consists of a keg or half-keg of beer, the beer box where the keg is stored, the *standard* or tap (faucet), the line between the keg and the standard, and a CO_2 tank connected to the keg with another line. The *beer box*, also called a *tap box*, is a refrigerator designed especially to hold a keg or half-keg of beer at the proper serving temperature of 36°F to 38°F. Generally, it is located right below the standard, which is mounted on the bar top, so that the line between keg and standard is as short as possible. If more than one brand of draft beer is served, each brand has its own system—keg, line, and standard—either in its own beer box, or sharing a box with another brand.

A *direct draw* means the beer is served directly from the keg, traveling through no more than 6 feet of line. In a *long-draw system*, the beer passes through longer lines, and its quality depends in part on keeping these lines cold. The best long-draw line is heavily insulated and cooled with glycol. Small, electric pumps circulate the glycol and keep it cold (see Illustration 10-14).

In busy places where it is inconvenient to tap a new keg every time one is needed, the beer box can be a walk-in cooler, and kegs may be connected in series. Short lines are simply not feasible in this arrangement, so it is imperative that they be kept clean to avoid buildup. Beer lines are made of heavy-duty plastic or stainless steel. Some manufacturers run coolant lines alongside the beer lines to ensure the beer remains cold. A higher-pressure combination of carbon dioxide and nitrogen may be required to push the beer through a larger system.

The latest innovation is a glycol system that freezes the draft head, prompting some companies to guarantee cold beer at the tap as a selling point. Other glycol systems have discs or shelves that allow patrons to place their glass or mug on a chilled pad to keep their drink cold while they converse with friends.

Glycol can be used as a refrigerant in many appliances. The walk-in cooler shown in Illustration 10-15 uses a glycol system; it can also be used in walk-in freezers, cold-pan units, undercounter refrigerators, and the like. The advantage of a glycol system is that the glycol cools the walls of the refrigerated unit, eliminating the need for an *evaporator*.

This type of system requires a lot of piping, so it isn't practical for large installations, but it is proving to be well accepted by many foodservice operators.

SOFT-SERVE MACHINES

Another specialty refrigeration unit is the *soft-serve machine*, which produces frozen desserts from a liquid base mix. Mixes are available to produce ice cream, ice milk, frozen yogurt, gelato, and so on, in many flavors. Large machines are made for banquet-size crowds; smaller ones for individual customer servings are placed near the front counters of quick-service restaurants. When shopping for a machine, you'll want to know how much mix it will hold at a time. Capacities range from 8½ quarts to 72 quarts.

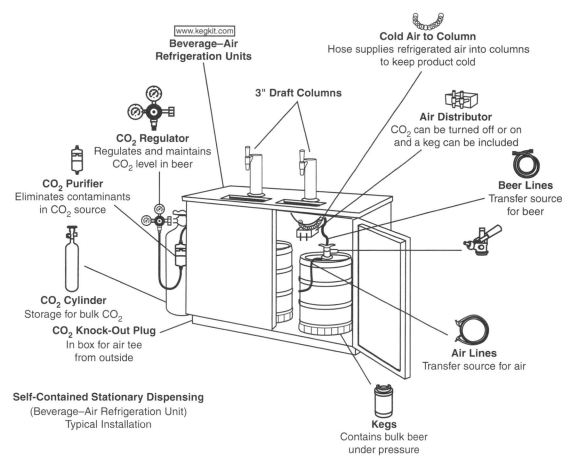

Beverage–Air Refrigeration Units
www.kegkit.com

3" Draft Columns

CO₂ Regulator
Regulates and maintains CO₂ level in beer

CO₂ Purifier
Eliminates contaminants in CO₂ source

CO₂ Cylinder
Storage for bulk CO₂

CO₂ Knock-Out Plug
In box for air tee from outside

Self-Contained Stationary Dispensing
(Beverage–Air Refrigeration Unit)
Typical Installation

Cold Air to Column
Hose supplies refrigerated air into columns to keep product cold

Air Distributor
CO₂ can be turned off or on and a keg can be included

Beer Lines
Transfer source for beer

Air Lines
Transfer source for air

Kegs
Contains bulk beer under pressure

ILLUSTRATION 10-14 A direct-draw setup installed under the bar.

Courtesy of Micro Matic USA, Inc., Northridge, California.

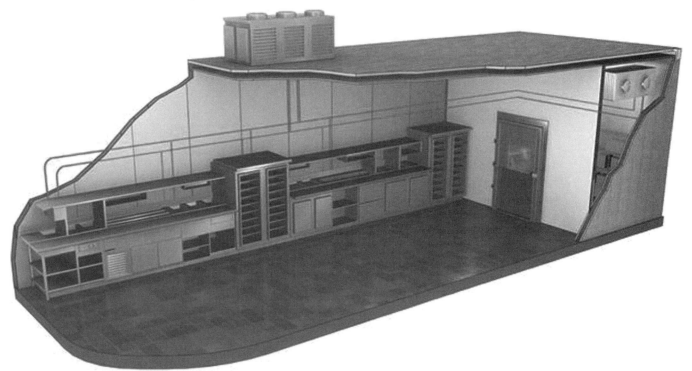

ILLUSTRATION 10-15 A glycol system.

Courtesy of Kairak, Anaheim, California, and Fortworth, Texas.

The soft-serve machine pumps air into the base mix, which gives it a soft consistency and increases the volume of the product. The percentage of air forced into the product is referred to as its ***overrun***; for instance, 16 ounces of mix with a 100% overrun produces 32 ounces of frozen product. A 50% overrun produces 24 ounces of product. Overrun accuracy is critical. Too much air causes a product that is thin and grainy; too little air causes it to freeze too hard and become difficult to dispense.

There are two types of soft-serve machines. In a ***gravity-feed machine***, the mix is loaded into a hopper. It flows as needed into a cylinder below, where it is frozen, scraped out of the cylinder, and dispensed. Gravity-feed machines are simple to operate and clean and are the least expensive.

The *pressurized* soft-serve machine uses an air pump to drive the mix into the freezer chamber, then forces it out through a spigot. Pressurized machines are more expensive, but they control the overrun air better than gravity-feed machines.

The refrigeration system is at the heart of the soft-serve machine. The simplest dispensers operate on 110-volt electricity, but machines with more features require 208-volt lines. Most units are air cooled; efficient water-cooled systems are also available. Air-cooled machines can be installed just about anywhere, but remember they vent hot air from the compressor into the environment—not efficient for an already hot kitchen area. Water-cooled machines dump the heat down the drain. There's no problem with ambient heat, but water use and sewage costs may increase slightly.

One of the key specifications for a soft-serve machine is its Btu rating. As you may recall from Chapter 5, the Btu (British thermal unit) is a measurement of how much energy a system or appliance uses. One kilowatt-hour is equal to 3413 Btus. The higher the Btu rating, the more mix can be frozen in a given period. Look for higher ratings in high-volume situations or where you want the product to be thicker than normal.

For any soft-serve machine, the most important consideration is cleanliness. It must be sanitized daily and serviced according to the manufacturer's maintenance schedule.

Cocktail freezers for making frozen drinks like margaritas and daiquiris are akin to the soft-serve machines. Match the machine to the type of cocktail being made. As the amount of sugar and alcohol in a drink increase, so does the time it takes to freeze the drink. A larger-capacity refrigeration system is needed for high-volume use, such as when you plan to serve frozen drinks by the pitcher, or for drinks that include high alcohol and sugar content.

Soft-serve desserts, shakes, and frozen drinks can be profitable, but the initial cost of the equipment is sometimes prohibitive. Most manufacturers offer a lease or a lease-to-own option. Calculate how many desserts or drinks you must sell per day to cover the combined costs of the equipment payment and ingredients.

ICE MAKERS AND DISPENSERS

All foodservice operations need ice, and the simplest way to meet that need is to have an *ice-making machine* that freezes, harvests, and stores ice automatically. Large standalone machines produce up to 3000 pounds of ice per day; medium-size, undercounter models make up to 200 pounds per day; small countertop ice makers deliver as little as 1 pound of ice per hour. You sometimes see ice makers referred to as ***ice cubers***.

An innovation in this area is the remote ice maker that delivers so-called tube ice—ice through a tubing system—to a dispenser or bin (see Illustration 10-16). This allows operators to direct ice from a single ice maker to two or more places simultaneously. The tube ice is considered a chewable ice, more similar to flaked ice than cube ice.

Before you buy an ice machine, you should determine not just how much ice you'll need but also *where* you will need it. If there are several sites for ice consumption—garde manger area, bar, waitstations—you might be better off with several smaller machines in various locations instead of everyone hauling ice from a single large unit. We'll talk in a moment about how to determine your ice needs.

Ice-making machines are refrigeration units. The ice is made when a pump circulates water from a tank. The water runs through tubing to a freezing assembly, which freezes it into a single sheet. The frozen sheet is then crushed or forced through a screen to produce ice

ILLUSTRATION 10-16 Tube ice machine with remote ice delivery equipment.

Courtesy of Follett Corporation, Easton, Pennsylvania.

cubes. Different types of screens produce different sizes and shapes of cubes. After the ice is crushed or cubed, it is automatically dumped into a storage bin. When the bin fills to capacity, a sensor inside the machine shuts it down until there is room to make and store more ice. Because most of the ice maker's parts come into direct contact with water, it is important that components be made of rustproof materials.

Ice maker capacity is determined by how many pounds of ice the unit can produce in a 24-hour period. However, any machine's output (and the quality of the ice itself) are affected by several factors:

Incoming water temperature. The ideal is 50°F; warmer water makes the machine work harder.

Room temperature. The ideal is about 70°F. If installed in an environment that has an ambient temperature of 80°F or higher, consider getting a unit with a water-cooled condenser to compensate for hot, humid, or grease-laden air.

Incoming water pressure. The minimum water pressure should be 20 pounds per square inch (psi); recommended pressure is between 45 and 55 psi. Anything higher than 80 psi causes malfunctions.

Water quality. Hard water cause the machine to work more slowly and almost always necessitates some kind of pretreatment before the water enters the machine. The fewer minerals and chemicals in the water, the more quickly and harder it will freeze and the more slowly it will melt. Filtration is almost always a good idea.

Read manufacturers' output claims carefully, and you'll find they are often based on ideal conditions: incoming water temperature of 50°F and ambient air temperature of 70°F. Generally, a 10°F increase in air temperature means daily ice output decreases by 10%. Also examine the water and energy usage figures provided by the manufacturer. You'll note a wide range: from 15 to 27 gallons of water to produce 100 pounds of ice, using from 5 to 10 kilowatts of electricity.

An additional resource to check is the Air Conditioning, Heating, and Refrigeration Institute (ARI), the national trade association that represents about 90% of manufacturers. Ironically, ARI data rates ice machine production capacities using more realistic conditions than the manufacturers' sales literature—with incoming water temperature at 70°F and ambient air at 90°F. ARI also rates machines by how many kilowatt-hours and how much water they need to produce 100 pounds of ice.

No matter where the ice maker is located, it needs a source of cold water and drainage. Particularly critical is a 1-inch air gap between the ice maker's drain line and the nearest floor drain. This necessary precaution prevents a backflow of soiled water into the ice bin.

Wherever you install the ice maker, proper plumbing is mandated by your local health department. A recessed floor beneath the unit is also recommended. Along with nearby drainage, this ensures that spilled ice does not melt on the floor and cause accidents. One smart option is to install an **inlet chiller** along with your ice machine. About the size of a household fire extinguisher, it collects the water that would normally be discharged from the ice maker into the drain. Instead, the water recirculates first through a series of copper coils in a chamber that contains fresh water on its way into the ice maker. The cold outgoing water chills the coils, which chill the fresh incoming water and allow it to freeze more quickly for faster ice production. The inlet chiller, which has no moving parts and uses no electricity, can save up to 30% on the electricity used to run the ice maker and boost its capacity by 50%.

Air circulation is also needed around the unit. An ice maker gives off warm air, like any refrigerator, and should be placed at least 4 inches from the wall to allow for ventilation.

Those are just a few of the factors to consider in your life-cycle calculation. Others are listed in the next "Foodservice Equipment" box.

Just as there are different machine capacities, there are also various sizes of storage bins. Most operators choose a combination ice maker and storage bin; by adding an extra 20% to the total capacity of each, you'll (theoretically) never run out. Ergonomics experts add that bins with a depth of more than 16 or 18 inches are hard to reach for employees that must scoop from the bottom. Look for storage compartments with volume sensors so production cuts off automatically when the bin is full. When ice tumbles out every time you open the bin, you're just wasting it.

A final important consideration is the length of time it takes the machine to complete one ice-making cycle. Under normal conditions, the whole freezing-harvesting-ejection period should take no longer than 15 to 20 minutes for the finished cubes to hit the storage bin.

Now that we've discussed the machine itself, what kinds of ice do you want it to make? The options are numerous.

■ Large cubes (*full cubes*) melt more slowly, so they're good for banquet situations where glasses must be set out early. They may be awkward in some glasses and may give the appearance of a less-than-full drink.

■ Smaller *half-cubes* stock better into most glasses and are preferred in bar settings.

■ Even smaller *cubelets* are suggested for soft drinks because they fill the glass or cup so well (and therefore make your soft drink supply go further).

■ *Nugget ice* is flaked ice that has been compacted into high-density, random-size chunks that melt slowly and are easier on blender blades than cubed ice. Nugget ice is a good choice for soft drinks and smoothies.

■ Round shapes fit a glass better at its edges.

■ Rectangular cubes stack better than round ones, leaving fewer voids in the glass.

Your clientele, your glassware, and your type of service determine the right shape for you.

In addition, you may have uses for crushed ice or flaked ice. You can buy an ice maker that produces crushed ice or one that makes ice cubes and then runs them through a special canister to crush them when needed. Crushed ice cools faster than cubed ice and is perfect for salad bars or fresh-seafood displays. A flaked ice machine, or **flaker**, produces soft, snow-like beads of ice, used mostly for keeping things cold, such as fresh fish, bottled wine, or salad bar foods.

How much ice does your operation require? *Foodservice Equipment and Supplies* magazine advises you to estimate depending on the type of business:

FOODSERVICE EQUIPMENT

Ice Maker Life-Cycle Factors

- Purchase price, tax, freight, start-up
- Installation (remote versus self-contained)
- Energy (at projected use rates, e.g., 300 pounds/day actual use from a 400-pound machine)
- Water
- Sewer cost (most machines use more than 12 gallons of water to produce 100 pounds of ice)
- Preventive maintenance cost
- Supplies (filters, cleaning agents, deliming chemicals)
- Labor (cleaning, changing filters, etc.)
- Service/repairs cost
- Footprint cost (e.g., by using remote, can you save space to devote to other equipment?)
- Annualized costs: all expenses amortized across projected life expectancy

Courtesy of *Foodservice Equipment Reports Magazine*, 2003, A Gill Ashton Publication, Skokie, Illinois.

- 0.9 pounds of ice per customer for quick-service restaurants
- 1.7 pounds of ice per customer for full-service restaurants
- 3.0 pounds of ice per customer for cocktail lounges

For most businesses, however, the figure changes daily and seasonally. You may need more ice during the summer than the winter, more during certain peak mealtimes, or more on weekends than on weekdays. The rule of thumb is to estimate your ice needs, then size your unit and storage bin to accommodate 20% to 25% more than your estimate. Running out of ice on a busy night is a restaurateur's nightmare, so most of them avoid it by buying the largest possible ice machine for the space they've got. However, it's probably smarter to calculate how much ice you'll need, and purchase the right machine for your needs. Here's how to do the calculation:

1. Estimate ice usage for one full week. You can figure every meal you serve will require 1 to 1½ pounds of ice.
2. Divide the figure by 7.
3. Multiply the result by 1.2.

The final total is your average daily usage. Also, remember it is two to three times more expensive to *make* ice than to *store* it. With this in mind, never scrimp on the size of your ice-holding bin. Get an ice machine with a bin that stores twice as much as the machine can produce in a 24-hour period. You may need this large bin capacity only 2 or 3 days a week, but it's more cost-effective than having your ice machine working all the time.

The other way to store ice is to stockpile it on low-volume days. Keep the machine running, bag the extra ice in your freezer, and bring it out as needed on busier days. It's a temporary solution, but if your operation is fairly small, it can work until you decide to upgrade to a larger-capacity machine.

We are noticing a trend toward centralized ice production for larger-volume businesses. This means a large machine with its compressor and condenser in a single location (usually outdoors) makes all the ice for a business instead of having several smaller, indoor ice makers in different spots. In the most sophisticated systems, ice is transported by pneumatic tube (think bank drive-up lanes!) to ice bins elsewhere on the property. The bins are automatically kept full without having to transport the ice manually.

THE DINING EXPERIENCE

Ice Service Guidelines

TYPE OF ESTABLISHMENT OR USE	ICE NEEDS PER DAY
Restaurant	1½ pounds per person
Bar/cocktail lounge	3 pounds per person (or per seat)
Water glass service	4 ounces ice per 10-ounce glass
Salad bar	30 pounds per cubic foot (multiply by the times you restock the ice during the day)
Quick-service restaurant	5 ounces per 7- to 10-ounce drink
	8 ounces per 12- to 16-ounce drink
	12 ounces per 18- to 24-ounce drink
Hotel—guest ice	5 pounds per room
Hotel—catering	1 pound per person
Health care—patient ice	10 pounds per bed
Health care—cafeteria	1 pound per person
Convenience store—self-serve drinks	6 ounces per 12-ounce drink
	10 ounces per 20-ounce drink
	16 ounces per 32-ounce drink
Convenience store—cold plate dispenser	50% more ice per day than listed above

From Ser-Vend International, a well-known ice maker manufacturer, come the handy Ice Service Guidelines listed in "The Dining Experience" box.

In many foodservice settings, ice is dispensed directly to the guest: quick-service restaurants, convenience stores, cafeterias, and institutional operations, to name a few. So, in addition to ice makers, *ice dispensers* are essential. The ice dispenser is also a refrigeration unit that freezes ice in a cylinder or *evaporator chamber*. As water enters the cylinder, it freezes against the wall and is chipped off as needed by a tool (called an ***augur***) that looks like a big corkscrew. The ice falls out of the cylinder into its storage bin as flakes or chunks. At the push of a button or lever, the ice is dispensed into a container: a paper cup or a hotel ice bucket.

Ice dispensers range in size from huge, freestanding floor models that hold 200 to 600 pounds, often found in hotels and healthcare settings, to countertop units that take up as little as 13 inches of counter space, produce 280 pounds of ice per day, and store up to 30 pounds. The largest countertop ice dispensers make 850 pounds of ice per day and store up to 200 pounds.

Manufacturers have come up with an ice dispenser to fit just about any foodservice application: commercial machines that can be programmed for hotels so that only guests' room keys open them; combination ice- and beverage-dispensing fountain units, the current standard in fast-food operations; ice dispensers with water-dispensing valves, popular in cafeteria settings. One of the biggest advantages of the dispenser is that it is more sanitary than reaching into an ice bin, even with scoops or tongs.

Finally, remember that an ice dispenser is close kin to the ice maker, with all the same considerations. Ambient temperature, water temperature and pressure, plumbing needs, water quality, cleanliness, and proper maintenance all figure prominently in ensuring its long, productive life.

ICE MACHINE MAINTENANCE AND SANITATION

You can order ice makers and dispensers with automatic cleaning features, such as air filters and self-rinsing ice contact surfaces, or you can do it the old-fashioned way. Either way, ice that is dirty, melts too quickly, or lacks uniformity usually signals a dirty or malfunctioning machine. Here are the most common problems and ways to prevent them:

Dirty ice. Clean and sanitize the machine regularly to remove algae, slime, and mineral deposits from the water. Bin walls can be washed with a neutral cleanser as long as you are sure to rinse them thoroughly. Always remove ice with clean hands and a sanitized plastic scoop, and never use the bin as a convenient place to stash foods or beverages. Check the air filter on air-cooled models at least twice a month, and wash it whenever it is dirty.

Small, cloudy, or broken cubes. This usually signals a problem with the water filtration system. Perhaps the water is not entering the machine at sufficient pounds per square inch because it's partially blocked. Try replacing the water filter by turning off the water first, then slowly opening the filtration cartridge or canister.

Blocks of ice stuck to bottom of bin. Keep the ice machine level and the drain unclogged to prevent melting ice from puddling at the bottom of the bin and refreezing to form blocks.

Lack of ice. First, confirm the electric circuit breakers (for both the machine and the condenser) are on. If it's a self-cleaning machine, be sure it is set to make ice instead of clean; the switch may have been tripped to the wrong position. Check the sensor inside the bin; tighten it if it seems loose. Finally, inspect the water supply. A partially closed valve or insufficient water pressure can interrupt or reduce the ice-making cycle.

Melting cubes. The fins on the condenser become clogged when they are dirty, which can interrupt the freezing process or make partial cubes instead of full ones. The fins should be clean enough to see through to allow the refrigerant to reach the right temperature to finish the freezing process. Inspect and clean fins at least every 3 months. If your machine has a condenser, it should also be cleaned regularly with a brush or vacuum cleaner.

These few problems are easily handled by your staff, but at times it's probably best to call the factory-authorized service person: when the ice-making cycle takes far too long, or when ice is not made even though the water supply seems fine.

By health code standards, ice is a food. A mandatory part of employee training should include handling ice so it remains safe and sanitary for human consumption. This means:

- Use dedicated containers for ice. Don't use containers that have also been used to store food.

- Store ice containers by hanging them upside down, far from the floor. Don't nest them in stacks, which is unsanitary.

- Clean and sanitize every ice scoop regularly. Have plenty of scoops, and store them by hanging them outside the machine—not sitting in the ice.

- Make employees wash their hands before they scoop or bag the ice; technically, they're not supposed to touch it at all without using clean disposable gloves.

- Do not allow anyone to eat, drink, or smoke around the ice machines.

Filtering water before it enters the machine improves ice quality and protects your machine from lime scaling, chlorine buildup, and slime. (Slime, incidentally, has a scientific name: ***biofilm***.) One popular manufacturer, Manitowoc Ice of Manitowoc, Wisconsin, uses an antimicrobial chemical to make all plastic parts that come into contact with water or ice. This retards but does not completely eliminate biofilm growth. Today's ice machines can be programmed to clean themselves automatically during the hours they are not in use, a process that takes about half an hour. Cleaning is recommended weekly.

■ SUMMARY

Having a well-organized receiving area for incoming goods involves selecting and maintaining accurate scales for weighing products; rolling carts and hand trucks for moving it from place to place; and pallets and shelves for storing it safely above ground. Your options for each are discussed in this chapter.

You'll need two types of storage: dry and refrigerated. Refrigerated storage requires air circulation, humidity (or lack of it) in the refrigerated space, and heavy-duty construction to withstand the rigors of the busy kitchen environment. Because most kitchens have too much or too little refrigerated space at the proper location to meet their needs, you must first decide how much capacity you need and then determine where to locate it. Often several smaller units, carefully placed, are preferable to one giant refrigerator.

Refrigeration technology has changed with the realization that many older refrigerants do environmental damage. Particularly when you are purchasing used equipment, it is important to make sure it can continue to be serviced as some of the older refrigerants (the F gases) are phased out and, in some cases, no longer legal to use. New cooling substances are being used, including glycol, and there is also a movement to advocate so-called natural refrigerants, including iso-butane and ammonia.

There are many options to consider when ordering a refrigerator, including sturdy thresholds, which prevent tripping; pressure-sensitive vents, which prevent vacuum buildup when opening and closing doors; safety door handles, which prevent accidental lock-ins or lock-outs; glass doors, which allow you to see inside; and temperature-monitoring systems, which promote food safety.

In addition to the standard models (reach-in, roll-in, and pass-through), there are a number of popular specialty refrigeration units: under-bar refrigerators, carbonated beverage machines, and soft-serve ice cream machines, to name a few. Serving draft beer from kegs requires a separate system with its own refrigeration capability for the kegs.

Walk-in coolers and freezers are excellent storage options. These are best used when equipped with shelves that promote air circulation and are configured for the types of containers you use. The refrigeration systems for these large units can be placed on their ceilings or even outside the building.

No foodservice business is complete without ice-making machines to dispense and store crushed or cubed ice. Your ice needs depend on how you use the ice. This chapter tells you how to determine the average daily ice usage in your facility. Because it is two to three times more expensive to make ice than it is to store ice, don't scrimp on the size of your ice-holding bins. You can also make more ice on slow days and stockpile it for busier ones.

Ice is considered a food, and all related safety and sanitation codes apply to its use in foodservice. It must be kept clean, and water coming into the ice machine should be filtered to keep out impurities that cause scaling and slime and impact the ice quality.

■ STUDY QUESTIONS

1. Create and describe a process for handling all incoming deliveries for a foodservice receiving area in a timely fashion.

2. What three processes inside a refrigerator work together to cool its contents? Are these processes the same no matter which refrigerant is used?

3. What is the formula for deciding how much refrigeration space you need for a walk-in cooler or freezer? Name three features you think are important for a walk-in, and explain why you think they are worth having.

4. How do you determine the capacity of an ice-making machine? How can you determine whether that capacity is sufficient for your business?

5. What is the difference between a premix and a postmix soft-drink machine? What are the advantages of each?

PREPARATION EQUIPMENT: RANGES AND OVENS

◼ INTRODUCTION AND LEARNING OBJECTIVES

The commercial kitchen **range** is a purchase with ramifications you will be living with for 7 to 10 years, the average life expectancy of this workhorse of an appliance.

There are actually two parts to most ranges: the *rangetop*, for surface cooking needs; and the *range oven*, for baking and other types of preparation that require the food be placed inside a heated cavity. Home models (the ones we often refer to as *stoves*) almost always locate both rangetop and oven on the same appliance. However, in foodservice, you can purchase separate, huge ovens without rangetops for volume baking or rangetop units without ovens. You can mix and match the pieces to fit your space and cooking needs. In fact, there are almost as many combinations of ranges and ovens as there are types of food.

In this area of equipment technology, professional cooks have reason to rejoice. Never before have science and industry collaborated to offer such a broad array of heat sources and equipment types, particularly ovens. The time when one range/oven combination was sufficient has passed. Today's foodservice kitchen is equipped with innovative, multifunctional integrated solutions for greater productivity and a higher food yield.

After reading this chapter, you will be able to:

- Identify the types of ranges and ovens, including custom models and options for certain types of cooking.
- Identify the sizes and utility requirements of ranges and ovens.
- Describe the basics of range and oven selection and purchasing.
- Explain cleaning and maintenance tips.

In this chapter, and in the other equipment-related chapters that follow, you will see that the growing use of multifunctional equipment has revolutionized working conditions. Most new appliances are easier to clean, and their designs minimize the need for heavy lifting. They are generally smaller, taking up about one-quarter less space than their traditional counterparts. They work more efficiently, which means lower cooking times and minimum weight loss of the foods being cooked. Cooking cycles and times can be electronically monitored, freeing up employees' time without a loss in food quality.

Before we look at these modern marvels, here are guidelines for selecting the right cooking appliances for your operation.

POWER REQUIREMENTS. What types of power are available, and what kind of load can your kitchen safely handle without shorts, brownouts, or other malfunctions? What installation costs are associated with this energy source? After you have your spec sheets in hand, ask an installer to visit your site and talk about how your existing power source fits your prospective purchase.

- *Gas.* Is the existing gas line large enough for another piece of equipment? Only so much gas will flow through a pipe at a given pressure. If existing equipment already draws most of the gas pipe's capacity, the addition of a new appliance could prompt all the gas equipment to function on less gas and, therefore, less power. Gas ranges that function in high-altitude locations (2000 feet or more above sea level) require adjustments to gas metering devices. If bottled gas is used, cook times will not be as fast, as bottled gas generates about 25% less heat than natural gas supplied directly from a gas line.
- *Electricity.* If your business operates in a high-rise building without a gas line to your floor, electric power is your only option. Should the fuse box be upgraded? Do you have sufficient wiring to service a 208- to 240-volt three-phase or single-phase piece of equipment? Bringing in a new line can be costly.
- *Propane.* If your needs include moving the equipment to set up in a remote location, such as offsite catering, portable appliances run on propane.

TYPE OF MENU. What are you cooking? How is it most efficiently prepared? Can some items be cooked in advance and reheated, or does the menu focus on individual, made-to-order foods?

QUANTITY. How much food do you need to meet peak demands? How much can be prepared in advance and held hot? Or, if the menu is mostly à la carte, cooked to order, how much rangetop space do you need at a time? Answers to quantity-related questions play out as you examine factors like how many racks can be held in an oven at one time and how many rack positions are available in the oven cavity.

SPEED. How fast must you be? Cooking at higher temperatures, using microwave ovens, setting the impinger/conveyor oven at its maximum speed—there are many ways to pick up the pace. Even with slow-cook methods, the smart operator comes in early and gets the rotisserie or smoker-cooker going so that, by mealtime, the food is perfectly cooked and ready to serve.

SPACE AVAILABLE. How much room do you have? Space affects not only the size of the appliance but also your ability to meet safety requirements for ventilation. Should you select a few multipurpose workhorses, or do you have room for specialized cooking appliances? Can you stack more than one appliance in a space and, if so, will people actually use it—or will they avoid it because reaching or bending to get to it is a hassle?

TYPE OF SPACE. If the kitchen is open for guests to watch as their food is prepared, take extra care to select appliances that are ready for public scrutiny. This may affect the finishes you select in terms of their ease of cleaning and maintenance. And what can the appliance do to help the kitchen staff function smoothly and efficiently? The last thing you want in the public eye is a close-up view of complete mayhem.

EMPLOYEES' SKILL LEVELS. Number of staff members, amount of turnover, amount of time for training, and complexity of recipes all figure into equipment decisions. Today's restaurateur can choose between the traditional kitchen equipment of past decades or highly computerized, programmable appliances that employees can control with the touch of a button. However, someone's got to be able to program these modern marvels as well as troubleshoot problems.

Making your checklist takes time, but don't let it discourage you. Choosing ranges and ovens, like so many of your appliance decisions, involves logistics, common sense, and budget considerations.

11-1 BASIC PRINCIPLES OF HEAT

Often, the physical dimensions or power requirements are listed for certain types of appliances. You certainly don't have to memorize these figures; rather, the idea is to acquaint you with the many options available from manufacturers and to prompt you to consider size and utility usage when outfitting your kitchen. Refer to Chapters 5 and 6 for more information about types of utilities.

Before you learn about the types of ranges and ovens available, it would be helpful to understand a few cooking-related terms. You may already have learned this in your culinary training; however, it is important enough to summarize briefly the cooking processes that are central to every kitchen.

Conduction is the simplest form of heat transfer. Heat moves directly from one item to another when they are brought into direct contact. Poaching, boiling, and simmering are all examples of conduction cooking.

Induction cooking creates heat when an electromagnetic field is created between stovetop and pan, creating currents that are converted to heat by the natural resistance of the metal pan. This electromagnetic field can be achieved only if the pan is made of ferrous metal, such as cast iron or stainless steel; nonmagnetic copper and aluminum pans will not work. Induction cooking surfaces do not become hot because the heat is transferred directly into the pan. When the pan is removed from the heat source, the induction process ends automatically, with no residual heat. Illustration 11-1 summarizes the process.

Convection occurs when heat is distributed by moving air, steam, or liquid. The simplest form of convection occurs when you stir a thick sauce, for instance, to circulate it and heat it thoroughly within the pan. Convection cooking was discovered when U.S. Navy researchers placed fans in standard oven compartments to see if food could be cooked more quickly when warm air was blown around it. They found that, indeed, the convection process cooked food faster and more evenly, no matter where in the oven the food was placed. Why? Because the hot air reduces the cool boundary layer at the surface of the food, allowing the heat to penetrate the food faster. More about convection ovens later in this chapter.

Radiation occurs when waves of energy hit the food and cause a heating reaction. Two types of radiation are used in the kitchen: infrared and microwave. In a broiler, the heating element becomes hot enough to release *infrared* radiation, which cooks food very quickly.

The other type, the *microwave* oven, used to be called *magnetron cooking* and was developed during World War II. U.S. airmen noticed they could keep their coffee hot by setting it near the plane's radar equipment box. The short electromagnetic waves emitted at the speed of light react to moisture, which is found in almost all foods. The moisture molecules in the coffee cup create friction as they try to align themselves with the microwaves, and friction creates heat. Today, this simple principle has found its way into almost every American home. More about microwave ovens later in this chapter.

How Induction Works

Although induction seems magical in how it works, there is a scientific explanation.

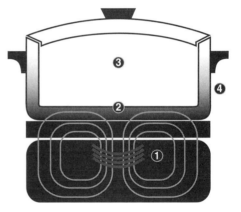

1. An alternating current in an induction coil produces an alternating magnetic field.

2. This magnetic field is instantly transferred and concentrated to the cooking vessel.

3. This concentrated magnetic energy in the cooking vessel causes it to heat up and start cooking.

4. When the vessel is removed from the heat source, the induction unit automatically shuts off.

Note: Induction cooking requires magnetic pots and pans to work effectively.

ILLUSTRATION 11-1   The induction process.

Courtesy of Garland Group, A Welbilt Company, Freeland, Pennsylvania.

11-2 THE RANGETOP

The most heavily used piece of equipment in any kitchen is the rangetop. It allows us to boil, sauté, simmer, braise, and deep-fry food, and hold it hot. Today, although new types of specialty equipment have been designed to perform these individual tasks, perhaps with greater efficiency, the rangetop is still the multipurpose king of the kitchen and the command post of most operations.

At first glance, the casual observer would say this workhorse has changed little over the past 20 years or so. Yes, the basic operating functions are intact. But special features set various manufacturers' models apart and may make a difference in your purchase decision: a narrower profile (24 inches instead of 36 inches) to fit smaller spaces, more energy-efficient burners, additional insulation, and so on.

In determining rangetop requirements, your selection begins with the menu and your full understanding of how the range will work in relation to other equipment. Consider the answers to these questions:

- How much *à la carte* (also called *to-order*) work is required by the menu? (This means items cooked individually instead of in batches.) How much will the range be used? Perhaps it will be required for selected meal periods only, or for different functions at breakfast, say, than at dinner.

- Are alternative appliances available for certain cooking duties: steam kettles, steamers, griddles, broilers, fryers?

- Where will you locate the range? In a display kitchen, enameled finishes and brass fixtures are appropriate and would be worth the extra money, compared to a typical kitchen hot line that is not seen by the public.

- How will the range fit into the layout, either existing or planned? The layout, of course, should be driven by how the hot line can work most efficiently. In most instances, the hot line will run parallel to a plating and pickup area, but in some cases, the two lines are perpendicular.

- How much space is needed during peak demand periods to hold food hot in pots and pans?

Total daily meal output is often used as a guideline for rangetop space. The simplest calculation is one restaurant range for every 50 seats, but other formulas exist. The requirements listed in Table 11-1 are a bit simplistic because they assume the presence of griddles and fryers but not steam equipment, which can be used instead of the rangetop to make stocks, soups, and sauces.

Another set of requirements, from J. Wilkensen's *Complete Book of Cooking Equipment*, is shown in Table 11-2. As you can see, determining how many ranges you'll need is not an exact science.

Commercial ranges offer a number of options not found in home models. For instance, they may be installed on steel or chrome legs of various heights over 6 inches. They may be equipped with casters for mobility. Stainless-steel shelves may be mounted above the back of the range for convenient storage space, and stainless-steel *spreader plates* can be placed between ranges and the appliances next to them to increase countertop work space (see Illustration 11-2).

Several manufacturers now offer options for refrigerated bases for the range units, which means you can tailor your equipment to the menu requirements. This adds versatility to the range lineup, as you can have some units supplied with ovens or cabinets and others with refrigeration. The cooks can now avoid added steps to a cooler or freezer or turning around to access cold product stored on the chef's table. These options are commonly referred to as *modular cooktops* or *fire and ice units*.

TABLE 11-1

Restaurant Range Needs

NUMBER OF RANGES/SIZE OF SPACE	MEALS PER DAY
2 ranges—6 feet in length	300
3 ranges—9 feet in length	500
4 ranges—12 feet in length	1000

TABLE 11-2

Range Requirements by Type of Cook-Serve Operation

TYPE OF OPERATION	MEALS PER WEEK	RANGE AREA (SQUARE FEET)	STEAM KETTLES OR COOKERS
Restaurant	1,000	24	No
Cafeteria	2,000	10	Yes
Hospital	2,100	24	Yes
Cafeteria	4,000	10	Yes
Cafeteria	8,500	14	Yes
Student union	14,000	36	Yes
Restaurant	14,000	24	Yes

Source: J. Wilkensen, *The Complete Book of Cooking Equipment* (Cbi Publishing, Workingham, United Kingdom).

When you look at ranges, always consider how you will clean them. Most of them have a drip pan under the burners to catch spills and grease. How easy is it to remove and clean this pan? Be sure the controls are easily accessible and protected from spills. Inspect the construction of each burner and order heavy-duty burners if you'll be using large, heavy pots on them. The legs of the range should be attached to heavy steel plates, which are welded to the body.

There are three basic types of ranges, each suited to a different type of kitchen operation: the medium-duty restaurant or café range, the heavy-duty range, and the specialty range. The latter has a number of configurations, several of which are mentioned in this chapter. Let's look at each type individually, as well as the burners that can be ordered for them.

MEDIUM-DUTY RANGE

This multipurpose appliance is also known as a *restaurant range* or *café range*. It generally measures from 36 to 60 inches in width. Its rangetop contains 6 to 10 open burners, which are 12 inches square and arranged two deep across the top of the range. As its name suggests, this type of range is suggested for smaller establishments with short-order menus or for settings where there is no need for constant, continuous use, such as church or nursing home kitchens (see Illustration 11-3).

Most often, though, the chef organizes the rangetop so that certain burners are reserved for certain duties. The front row may be used for sautéing, for instance, while finished products are held and kept warm on the back row (hence the popular saying that when something is delayed, it's been put on the back burner). There are dozens of handy

ILLUSTRATION 11-2 Spreader plates fit between appliances to give you more room to work. Many are also cabinets that provide extra storage space.

Courtesy of Vulcan-Hart Company, a division of ITW Food Equipment Group, LLC, Louisville, Kentucky.

ILLUSTRATION 11-3   A medium-duty or café range.

Courtesy of Vulcan-Hart Company, a division of ITW Food Equipment Group, LLC, Louisville, Kentucky.

rangetop configurations, some of which you can see in Illustration 11-4. A six-burner range can be used for quick sauté work in several small pans; a four-burner unit is well suited for large simmering stockpots. If the facility makes a lot of stock as a base for soups or gravies, a **uniform heat top** (usually called a *hot top* or an **even-heat plate**) should be considered. This flat surface distributes the heat over a larger area than the burner itself and makes more efficient use of the rangetop. A *step range* has its back burners raised above the front ones, making access to the back pans easier. You can also order three burners up front and two even-heat plates in the rear for keeping foods warm.

Ranges can be ordered with no oven, one oven, two ovens, or even refrigerated space underneath. Increased insulation provides sufficient temperature separation for units with refrigerated bases, a popular choice in smaller kitchens. Oven sizes are based on the dimensions of a typical baking sheet: 18 × 26 inches. If a facility has minimal baking or roasting needs, one oven is sufficient. Many cooks like the convenience of having an oven below their rangetop but feel the performance of the oven is not equal to that of a standalone oven. Further, placing an oven below the rangetop limits its size, whereas standalone ovens can be ordered in larger dimensions.

Most manufacturers offer three sizes of rangetops: 24-, 48-, and 60-inch widths, which may be added to in 12-inch modular dimensions to increase capacity. The back panel of the range may be either a small shelf located above the cooking surface or a simple 5-inch stub back that helps prevent food from spattering onto the wall. Many ranges also have 6-inch adjustable legs so the appliance can be leveled. As mentioned earlier, a smart option is to replace the standard legs with 6-inch swivel casters, at least 1½ inches thick, so the range can be moved for cleaning and servicing. Two of the casters should be lockable so the appliance won't move when in use.

HEAVY-DUTY RANGE

This model is similar to the medium-duty range but is made of heavier materials to withstand the rigors of high demand, and large, heavy pots and pans (see Illustration 11-5). The **heavy-duty range** is best suited for long hours and high-volume cooking. Its four open burners are rated up to 30,000 Btus per hour per burner for hotter and faster cooking than the 20,000-Btu output of café range burners. Heavy-duty rangetops fit together to form a solid line, or **battery**, in almost any combination of grate tops, hot tops, and griddles. Its modular design makes it easy to customize. Burners may be 12 or 18 inches wide. Where space is tight, half-width ranges (12 to 17 inches wide) with only two burners may be specified. A front rail with a **belly bar** keeps cooks from accidentally leaning on the hot surface when they are working.

An additional feature of the heavy-duty rangetop is a 7-inch stainless-steel shelf located directly in front of the burners. The heavy-duty range has its manifold (the pipe in which gas is delivered to the range) on the front of the range. The café range has its manifold at the back, so it can't sit as close to the back wall as a heavy-duty range. For gas-powered ranges, another handy option is a high flue riser made of stainless steel, which helps vent the flue gases to the exhaust hood. It minimizes spills inside the range as well as protects the back wall from spatters. One or two storage shelves may be mounted on the flue.

Electric—Six French Plates
24" x 24" Griddle

Electric—High-Speed Units
and French Plates

ILLUSTRATION 11-4 Rangetops can be configured in many ways, as seen in these sketches.

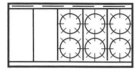

Gas—Two Even-Heat Plates
and Six Open Burners

Electric—24" x 24" Griddle Plate
and Two French Plates

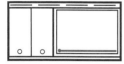

Gas—Two Even-Heat Plates
and One 36" Griddle Plate

Gas—One Even-Heat Plate
and Four Open Burners

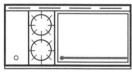

Gas—One Even-Heat Plate,
Two Open Burners, and
One 36" Griddle Plate

Gas—Two Even-Heat Plates
and Two Open Burners

Gas—Two Open Burners
and One Griddle Plate

SPECIALTY RANGES

If a popular type of food or method of cooking exists, you can be sure some manufacturer will create a custom range to facilitate it. A few current options in this ever-expanding market include:

Stockpot range. A short range with a large open burner used to heat stockpots; also found in bakeries, where it is handy for melting chocolate. The rangetop burner is a series of

ILLUSTRATION 11-5 A heavy-duty range is part of the hot line, or cooking center, which includes several appliances that must all be placed under a commercial exhaust hood.

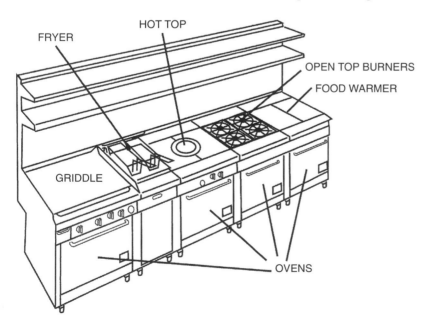

FRYER

HOT TOP

OPEN TOP BURNERS

FOOD WARMER

GRIDDLE

OVENS

concentric rings, concentrating the heat at the center of the burner with a gradual heat decrease toward the perimeter. Its heating capacity is impressive—55,000 Btus per hour—and the rangetop is usually an 18-inch square. However, double- and triple-size units are available.

Taco range. Designed for Mexican restaurants. Pans and pressure cookers fit into its recessed burners. On a single gas range, you can accomplish the multiple duties of preparing meats, refried beans, and rice.

Wok range. As you probably know, a *wok* is a bowl-shaped cooking pot used to cook foods quickly in Asian cuisine. This range features recessed, circular burners with rings that can be adjusted to accommodate large woks for stir-frying under high heat conditions, up to 106,000 Btus per hour. Step-range models set the back burners slightly higher than the front ones, allowing cooks to control six woks at a time easily. Removable trays, shelves for spices, or even a built-in water faucet may be available. Also called a **Chinese range**.

Any type of food that's cut into small pieces can be cooked in a wok with a minimum of fat, which makes it practical for glazing and sautéing, not just Asian cooking. In commercial kitchens, woks usually have two handles, but you can find them with a single handle, more like a big frying pan, which takes less cooktop space. Either type of wok fits on this unique burner.

Tabletop range. A partial rangetop that consists of two burners, this is used where space is at a premium. The burners can be situated side by side, for a depth of 16 inches, or front and back, for a depth of 28 inches. The 12-inch burners have a normal heating capacity of 20,000 Btus per hour.

TYPES OF BURNERS

There are also several types of burners. You will hear some referred to as **drop-in burners**, which means they are ordered in modules of one or two burners that drop in to the rangetop and are connected there. This makes it easy to mix and match types of burners depending on what you plan to cook. There are freestanding burner units too.

The most common *open burner* is covered by a one-piece, heavy-duty metal grate on which to set pots and pans. These burners are also referred to as *atmospheric burners*, as they use air from the surrounding atmosphere to mix with the gas and ignite the flame. Typically, the grate is 12 inches wide and 25 inches long. An open gas burner is sometimes called a **grate top** or *graduated heat top*. Some grate tops are made in concentric circles called *rings*; these can be removed individually, enabling the chef to set a pan closer to the flame. The heat may emerge from the burner through *jets* or rings. Jets concentrate the gas flame and aim it directly at a spot, allowing food to cook quickly in highly concentrated heat; rings disperse the heat over a wider area and are better for melting, sautéing, and cooking duties that require steady, even heat.

We've already mentioned hot tops, which can be installed instead of open burners. **Griddles** are flat, like hot tops, but made of a thicker steel plate. A griddle has the major advantage of a trough for convenient removal of grease from its surface as well as a removable drip pan to catch the grease and short walls on three sides that help prevent spattering.

A combination **broiler/griddle** may also come in handy. This hybrid includes tubular burners for the griddle and infrared radiant bulbs for the broiler.

Perhaps the biggest breakthrough in recent years is in induction cooking, with the ability to better control heat. The latest technology includes a current sensor on the coil, which detects whether a pan is on the cooktop and precisely controls the power applied (and, therefore, the heat delivered) to the pan. Most induction cooktops offer a temperature range of 90°F to 440°F in 5°F increments. If a pan's material or dimensions are not suitable for the range—too small, or not magnetic—the burner will not heat.

In general, the larger the range, the greater the number of burner options and combinations available.

ELECTRIC RANGETOPS

Thus far, we've discussed gas-powered rangetops because gas is by far the most widely used heat source in the commercial kitchen. However, in some sites the only available source of energy is electricity, and plenty of ranges fit this situation. Instead of Btus per hour, electric burner heat input is described in kilowatts (kW).

An electric range requires 208 to 240 volts, either one-phase or three-phase. This is an important detail because compatibility of voltage and phase must be determined when the appliance is ordered. Otherwise, you may find yourself making expensive modifications—or returning your shiny new range to the manufacturer.

Hot tops and griddles are also available for use on electric ranges. Like their gas counterparts, the rangetop components are modular and easy to mix and match.

The proper installation of an electric range determines its efficiency. It must be carefully unpacked and assembled by an experienced professional. First, the nameplate should be checked to be sure the voltage, ampere rating, and number of phases are all correct and match the available power supply. The electrician should install a ground wire and connect the equipment for safe use. The appliance should be plugged into a convenient outlet so it can be disconnected for cleaning behind and beneath.

Before placing the new range in service, the instructional manual should be read by anyone who will be using it. A quick run-through of switches, thermostats, and other parts saves time and trouble in the future.

An oily protective coating is put on some appliances at the factory to prevent them from getting scratched in transit. This film can be softened by rubbing the appliance exterior with a soft cloth saturated with cooking oil. After the coating has been rubbed off in this way, the range can be washed with a clean cloth and mild soapy water and rinsed clean with another damp cloth. Harsh detergents and steel wool will damage the surface. Electric appliances should not be hosed down or steam cleaned because the excess moisture may damage their electrical components.

After the range body has been cleaned, it's time to turn on the rangetops and ovens, using a procedure known as ***burn-in***. Turning on the burners and ovens and letting them heat up before you need them burns off any protective coating, smoke, or odors and readies them for regular use. Tubular high-speed heating elements don't need this process, but other heating surfaces do. The hot top can be set at its highest heat for 30 minutes or so, the oven at 400°F for 3 to 4 hours.

Trying out the ovens and burners is a good idea anyway because you can also check the performance of the appliance against its specification sheet. The manufacturer claims the rangetop preheats from room temperature to 350°F in 4½ minutes. Does it? Check it! When all burners are turned on to the same temperature at once, do they heat evenly, or is one sluggish? It may be a loose wire or a defective control knob or heating element. Whatever the case, find out early—before you're in the middle of a rush and discover the problem the hard way.

Heavy-duty electric ranges are 36 inches square and are found in high-volume foodservice operations. Medium-duty ranges measure 30 inches square and are usually seen in smaller establishments. Each size is available in several combinations of surface units.

RECTANGULAR HOTPLATES. These are 12 × 24 inches wide and capable of temperatures ranging from 250°F to 850°F. Their heat controls can be either so-called ***infinite-heat knobs***, which allow for small adjustments, or standard low-medium-high knobs. Because they are made of solid, 1-inch-thick cast iron, rectangular hotplates respond slowly to temperature changes. They use only 5 kilowatts of electricity per hour, but they take as long as 15 minutes to heat and longer to cool down. If a kitchen requires a lot of griddling, a griddle is still a better way to go than these hotplates.

FRENCH HOTPLATES. These are smaller (10 inches in diameter) and lighter in weight than rectangular hotplates, but the burners are solid, not coils. Made of iron, they are typically used for medium-volume cooking and à la carte cooking. Their power use is not even a whole kilowatt per hour,

(a)

(b)

ILLUSTRATION 11-6 These two drop-in, two-burner hot plates show the difference between (a) a coil-type and (b) a French hot plate.

Courtesy of Star Manufacturing International, St. Louis, Missouri.

so they provide an economical method of cooking small quantities of food. However, they can require up to 15 minutes to heat to a maximum temperature of 500°F. When cool, the round French hotplates can be scoured with a damp cloth and mild abrasive. Illustration 11-6 shows the difference between a standard, coil-type hotplate and a French one.

HIGH-SPEED SURFACE UNITS. These are a type of French hotplate with a kick, so to speak. They're designed for practically instant heat (within 2 minutes) and quick response to temperature adjustments, yet their power usage is still a modest 2100 to 2600 kilowatts per hour. High-speed surface units are tubular in shape, 8 inches in diameter, and supported by a rugged 10-inch ring. The trade-off when either type of French plate is used is that manufacturers suggest pots and pans be chosen correctly to fit the smaller diameter and output of the burner. As an example, one manufacturer specifies that the bottom surface of the pot or pan should not exceed 10 inches and its overall capacity should not be larger than 16 quarts.

ELECTRIC INDUCTION RANGETOPS. The induction cooktop has become a symbol, of sorts, of the modern kitchen. Trends in meal consumption away from home continue to move toward fast, light, tasty cooking that uses a minimal amount of fats and oils. As you read earlier in this chapter, induction cooking does not require an external heat source. The smooth, solid surface of the induction rangetop is usually made of tempered ceramic or glass, with circular markings to indicate where to place the pans (see Illustration 11-7). There are even concave induction burners for woks nowadays. The evenly distributed heat makes induction cooking ideal for even the most fragile ingredients.

This type of range is also highly efficient, using less energy than either its gas or traditional electric counterparts. Both the range and its burner units are sometimes referred to as **hobs**. Some hobs automatically revert to a power-saving standby mode when not in use. However, they work only when the right kinds of cookware are used. In a word, it must be magnetic. Most steel and cast-iron pans are fine, but those made of aluminum, copper, and some types of stainless steel are not magnetic and, therefore, will not work on an induction rangetop. Pans that have a nonstick surface can be heated, but the nonstick coating usually cannot withstand the intense heat of induction and will become unusable in a matter of days. The best pan for induction cooking is completely flat on the bottom, maximizing its surface contact with the rangetop. An empty pan left on an induction surface too long or too often will transfer heat back into the hob, which can damage its interior circuits.

Because it is not porous, an induction appliance is ready to cook very different foods after a quick cleanup without the risk of taste transfer. It works quickly, does not heat up the kitchen, and requires less ventilation (exhaust and supply air) than more traditional rangetops. Notice we said *less* ventilation, not *no* ventilation! The earliest induction cooktops were not required to be placed under an exhaust system, but this is no longer the case in some jurisdictions. Ventilation is most definitely needed to discharge the intense internal heat these units generate and prevent the circuits in the unit from overloading. Induction ranges should not be used near other heat-generating equipment—fryers, griddles, broilers—where circulating hot air might be picked up by the range's intake fan, in which case the range will not be able to cool itself down.

The heaviest-duty induction rangetop requires a 208- to 240-volt electrical outlet. It generates from 3500 to 5000 kilowatts of power and has a hefty price tag ($3,500 or so). Portable one- or two-burner hobs are available for off-site catering. They can be plugged into standard 120-volt electrical outlets, use from 1200 to 1500 kilowatts of power per hour, and can be heated from 160°F to 440°F.

Factors to consider when purchasing induction cooking equipment include the power requirements, approval of reliable safety and sanitation authorities (like UL and NSF International), overall costs, availability of qualified repair service, and whether the cookware you plan to use is suitable.

ILLUSTRATION 11-7   An electric induction burner.

Courtesy of The Middleby Corporation, Elgin, Illinois.

11-3 THE RANGE OVEN

The range oven, located beneath the rangetop, is the principal method of large-volume, dry-heat cooking in most restaurant and commercial settings. Like rangetops, range ovens can be powered by gas or electricity. Just remember, even if the appliance cooks with gas, the oven still has electrical needs: for the timer, lights, and, in the case of convection ovens, to run the fans inside the oven.

These ovens are as hardworking as any commercial kitchen appliance, and they are usually on all day long.

The basic range has a single oven below the rangetop. It may be used for roasting, baking, braising, smoking; for finishing sautéed and grilled items; or simply for storing hot food until needed. Even its swing-down doors are useful as a holding platform to support heavy pans. Most commercial gas range ovens have a 40,000-Btu capacity, with heavy-gauge, double-wall construction (known as a *flame spreader*) that distributes heat evenly throughout the cavity. Electric range ovens require from 1250 to 5000 kilowatts of power per hour.

The oven cavity is large enough to hold a standard sheet pan, which is 18 × 26 inches. Like home kitchen ovens, they come with at least one chrome-plated rack on which to place pans, and the position of the rack can be adjusted up or down inside the oven cavity. There are also a few half-size or space-saver ovens, suitable for small roasting pans or half-sheet (18 × 13 inch) pans.

In commercial use, there are quite a few smart options to consider when shopping for a range oven. Many of them evolved to help busy chefs use both oven and rangetop in tandem with minimal hassle. They include:

- Oven controls mounted on the side instead of at the front of the appliance. The knobs are temperature sensitive and can be damaged by repeated blasts of heat. Also, side-mounted controls minimize the chance that a cook standing at the rangetop will lean on the oven control and accidentally adjust it.

- Infinite-heat controls that allow the most precise adjustments are preferable to knobs with only a few heat settings (the old low-medium-high controls).

- The oven door should be hinged on the bottom and should open flush with the deck so food can be slid in and out easily on sheet pans.

- The handle should be smooth and well insulated so it stays cool and can be opened safely without a potholder.

- Fully open, the door should be able to support full pans of food, as much as 200 to 250 pounds. The door should also be counterbalanced to allow it to stay partially or totally open without being held in place manually.

- Heat-treated glass oven doors are preferable because you can see the food inside without having to open the door. However, in our experience, most operators fail to keep the glass clean. Instead of being useful, the result is an eyesore.

- Inside the oven cavity, the floor or bottom of the oven is called the **deck**. The deck should be made of at least 14-gauge steel, with raised sides and back to help catch spills. Try to choose a deck that is removable for cleaning.

- Ask about insulation, which is crucial to the oven's ability to hold heat. A minimum of 2 inches of rock-wool insulation is recommended.

- Easy-to-clean surfaces and a self-cleaning cycle for the oven cavity are recommended.

- Ovens must be leveled for some products to bake correctly; cakes and cheesecakes are among the most sensitive items. Leveling requires not only that the floor be perfectly flat but that the oven legs themselves be height adjustable. We've already mentioned the wisdom of mounting the oven on adjustable casters to permit rolling when necessary.

11-4 CONVECTION OVENS

Convection, as you'll recall, uses fans to circulate heated air around the oven cavity, reducing cooking times by 25% to 35%. Because the heat transfer is so much more efficient, foods can be baked or roasted at lower temperatures, which minimizes shrinkage and maximizes yield per pound. Standard recipes may have to be altered for best results in a convection oven.

Manufacturers have engineered the airflow in these ovens for better uniformity, giving operators an even quicker finished product with better results. So, in recent years, convection ovens have all but replaced conventional ovens everywhere—except under the rangetop.

Convection ovens come in three basic sizes:

1. Full size, which accommodates standard 18 × 26-inch sheet pans.
2. Bakery depth, which accommodates standard sheet pans placed either lengthwise or widthwise in the oven.
3. Half size, which holds the smaller, 18 × 13-inch half-sheet pan.

Because of their precise airflow patterns, convection ovens don't do as well with a variety of pan sizes. Therefore, it makes sense to determine the pan sizes you plan to use before you buy the oven. The key to successful convection oven use is proper air circulation around the food. It is therefore critical that the oven not be overloaded or improperly loaded. The food is placed on pans that are loaded onto shelves (racks) inside the oven cavity. The number of racks is determined by the height of the food being cooked. Like cook-and-hold ovens, many convection ovens automatically hold food hot after it's been cooked.

The type and sophistication of controls are other critical decisions. Top-of-the-line models can run preset programs with variables of time, temperature, and internal fan speeds, but some operators feel overly complex controls limit who is able to use the oven correctly. Think about the skill levels of the employees who will be doing the cooking before selecting the most highly technical option.

One option that can be useful is a two-speed fan. Convection ovens with slow-roasting capabilities use a lower fan speed for low-temperature cooking. A lower speed or pulse option is also handy for delicate products, such as muffins and cakes, when a higher-velocity fan might botch the results.

The typical oven is about 6 feet tall, 3 feet wide, and 3 to 4 feet deep. There are also half-size ovens, some made to fit on countertops. Special models can be ordered for baking, wide enough to accommodate baking sheets by length or width. The airflow design is critical in these ovens to achieve balanced heating and browning. Computerized controls allow the oven to be programmed by recipe. Electronic sensors inside the oven slightly lengthen cooking time to compensate for temperature drops caused if the door is opened too long or too frequently.

Some models feature electric meat probes that allow three products to be cooked at different times and temperatures.

When ordering a convection oven, pay special attention to its doors. Full-sized ovens have double doors that open simultaneously when one or the other is pulled open. They can open from each side or top and bottom. A single, counterbalanced door—more like a traditional range oven—is also available, hinged at the bottom or on either side. Another option is a single or double pane of glass on the door.

Convection ovens can be powered by gas, propane, or electricity (110, 240, and 280 volts). There is a general feeling among chefs who do a lot of baking that electric ovens yield a moister product. Gas models are required to have a venting system; check your local ordinance about electric models.

The popularity and versatility of convection ovens have prompted many manufacturers to create interesting hybrid forms. There are now double-oven models, with two separate cavities and control panels. They don't necessarily take up more space or use more energy either. Bakeries may choose the combination proofer and convection oven, with separate cavities for letting bread rise and then baking it. The proofer has special humidity controls; the oven has a built-in steam generator. Dual-compartment steamers and convection ovens work independently of each other and allow the same floor space to do double duty.

Another interesting hybrid is the steam convection oven. Although these are not combiovens (shown in Chapter 14), they can be programmed to add steam to the cooking chamber. This helps increase the speed of cooking and avoid the drying that may happen to some foods in a pure convection oven environment.

11-5 OTHER OVEN TYPES

DECK OR STACK OVENS

An oven manufactured with more than one cavity and set of controls is called a *deck oven* or *stack oven* because the heated, insulated boxes are stacked on top of each other in double-deck or even triple-deck configurations. Deck ovens are needed when production is high and space is limited. Depending on the needs of the kitchen, ovens can be stacked in any configuration—one regular oven and one convection oven, for instance (see Illustration 11-8).

The term *deck oven* comes from the way the oven is used: Food is set directly onto the deck or bottom of the oven cavity to cook (although some deck ovens also have interior racks or multiple decks to pack more products into the compact space). The deck itself is made of either stainless steel or ceramic; bakers prefer ceramic decks or stone hearth decks for more even distribution of heat. At least one manufacturer offers an air deck oven that uses impinged hot air to eliminate hot and cold spots in the traditional deck.

ILLUSTRATION 11-8 Deck ovens (or stack ovens) are so named because their multiple compartments are stacked atop one another.
Courtesy of Vulcan-Hart Company, a division of ITW Food Equipment Group, LLC, Louisville, Kentucky.

Contemporary gas deck ovens are generally classified in four broad categories:

1. *Traditional-style deck oven.* Each individual oven is either 8 inches high (for baking) or 12 inches high (for roasting) and, as we mentioned, can be stacked. The smallest ones hold two half-sheet pans (each 13 × 18 inches); the largest hold eight full-size sheet pans (each 18 × 26 inches).

2. *Motorized, convective deck oven.* A single baking cavity is equipped with three horizontal baking hearths made of perforated nickel-plated steel. This oven has a reversing fan system that circulates air evenly and enhances its heat transfer capabilities.

3. *Vaulted deck oven.* A single baking cavity has a larger, arched opening that provides easy access. Some have a secondary burner under the oven cavity to increase baking speed.

4. *Turntable deck oven.* The largest of the deck oven family stands more than 6 feet tall, with three or four horizontal, rotating, circular baking decks with diameters of 48 to 56 inches, perhaps made of ceramic (stone hearth). Multiple access doors maximize its efficiency.

The same basic guidelines for purchasing a single range oven also apply to deck ovens: doors that open flush with the deck, insulated handles, and so on. Insulation requirements are greater for multiple ovens: 4 inches of rock wool or fiberglass are recommended.

Another important recommendation is to order individual control panels for each deck, enabling them to be used simultaneously for a variety of duties. Temperatures for each oven range from 175°F to 550°F. Deck ovens may be ordered with or without steam capability.

Control compartments for gas-fired ovens are located below each deck; electric ovens may have controls either directly below or at the side of each deck. Electric ovens also have two sets of heating elements, just like a home oven—one on the top (for broiling) and one on the bottom (for baking). Also located in the control compartment are three-position heat switches for each element. An observation: Side controls are easier to access than bottom controls, especially when the oven doors are opened frequently.

Both gas and electric deck ovens have flue vents at the back of the appliance, which may be controlled by a hand-operated lever, also found in the control compartment.

Each deck holds two 18 × 26-inch sheet pans. Electric and gas requirements for each model are listed in manufacturers' catalogs and, as you've already learned, should be checked and rechecked to ensure compatibility with the kitchen before purchase.

At least one manufacturer has introduced a blower-and-duct system within each deck so a regular oven can function as a convection oven with the flick of a switch.

The chief disadvantage to deck ovens is that they are large and heavy. If you are pressed for space in your kitchen, you may wish to consider alternatives. Your designer/consultant can help you determine what equipment best suits your needs and the space you have to work with.

IMPINGER/CONVEYOR OVENS

As its name suggests, a *conveyor oven* allows the operator to place uncooked food on a moving surface (the conveyor belt) and program the appliance to move the food through the heated oven cavity at a certain speed; then the food emerges ready to serve. Conveyor ovens can be used to cook everything from meats and sandwiches to pastries and breads. Their advantage is product uniformity. At this writing, Burger King is testing a conveyor broiler system that promises 25% to 50% less energy use than broiling food conventionally, which you'll learn more about in Chapter 13.

Impinger/conveyor ovens are standard equipment in many busy foodservice facilities. One meaning of the word *impingement* is "to strike sharply," and that is the basic premise of the air impingement process in cooking: to blast high-velocity, heated air into the oven cavity. The air is aimed to concentrate on food that travels horizontally on a moving conveyor belt made of stainless steel or wire mesh (see Illustration 11-9). The belt can be adjusted to move at different speeds for different lengths of cooking time. The air moves with enough force to displace the natural layer of colder air that directly surrounds the piece of food being cooked, resulting in a shorter cooking time than traditional ovens.

Because impinger/conveyor ovens are automatic, high-quality output is consistent with minimal staff supervision, and little training is required to operate the ovens. The hot-air jets are located above and below the conveyor. Some ovens even have separate controls for different zones within the oven, allowing the air temperature and pressure to be set independently. This so-called *zoned cooking* is another feature of the most technically advanced ovens.

The air is forced through *finger panels*, which look almost like screens laced with small holes. Different types of panels are used to cook different types of food, but the basic premise is that items that cook quickly (e.g., pizza) need panels with fewer holes to restrict airflow, while thicker items (e.g., lasagne) requiring longer cooking time need more holes in the panels to allow more hot air to hit them. Food, depending on type, can be placed in pans, from thin aluminum for fish or frozen french fries to thicker stainless steel for pork chops. Porcelain cookware is also acceptable. If you'll be baking protein products—burgers, sausage, chicken, and the like—look for ovens with built-in grease and smoke controls.

Impinger/conveyor oven models feature four heat sources: Infrared and quartz models are electric, while natural-convection and forced-convection models may be gas powered or electric. No matter the heat source, most models need a small electric motor to move the conveyor belt. The gas-fired ovens operate from 39,000 to 180,000 Btus; most are in the

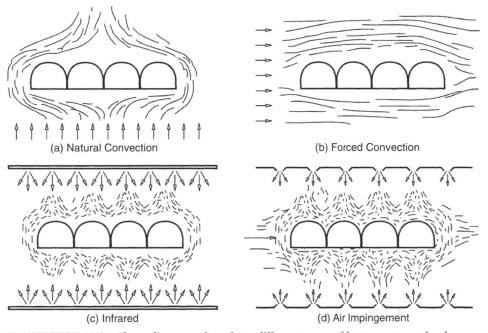

(a) Natural Convection

(b) Forced Convection

(c) Infrared

(d) Air Impingement

ILLUSTRATION 11-9 These diagrams show how different types of heat penetrate food differently. (a) Natural convection is the least efficient method of penetrating the coldest air layer, directly around the food. (b) Forced convection is a little more efficient than natural convection, because heated air is blown horizontally across the food, displacing the cold air around it. (c) Infrared, radiant heat cooks the surface of the food but does not impact the cold layer of air around it, which can affect cooking time. (d) Air impingement aims air at the food from several directions, forcing aside the cold boundary layer and prompting fast, uniform surface cooking.

Source: Based on figures from Lincoln Foodservice Products, Inc., A Welbilt Company, Fort Wayne, Indiana.

70,000- to 75,000-Btu range. The electric models require 10 to 27 kilowatts per hour and three-phase power. All of them require exhaust canopies. Temperatures in impinger/conveyor ovens range from 300°F to 600°F, and the conveyor speeds are adjustable. The conveyor belt is usually about 7 feet long, although only about 3 feet of it is located within the baking chamber. The oven does not have doors on the sides, but they usually have a small viewing door near the middle so you can check the progress of the cooking line or place items halfway down the belt if they don't require the full conveyor length to cook (see Illustration 11-10).

The most enthusiastic users of this oven are carry-out pizza restaurants, because of the speed and ease with which pizza can be cooked, but their manufacturers say conveyor ovens are versatile enough to be used a hundred different ways. Impinger/conveyor ovens can be double or triple stacked and/or placed on stands with casters. Options include twin belts, which run side by side at different speeds to cook different foods simultaneously.

PIZZA OVENS

The pizza oven is a special high-temperature application of the deck or stack oven, which can be set from 300°F to 700°F. The deck (also called a *hearth* in this case) is made of two pieces of 1-inch-thick ceramic or a single piece of steel. As pizzas are so flat, the height of the oven cavity can be as low as 7 or 8 inches. Multi-pizza tabletop models are available for places where space is tight.

The power requirements for pizza ovens are 82,000 Btus per hour per deck for gas-fired models and 7.2 kilowatts for electric ovens. One manufacturer has produced a model with a removable center divider; each half of the oven has its own burner and thermostat. Other manufacturers have introduced ovens with an **air door** or *air deck*. These pizza ovens have no solid door; instead, a blower fan circulates a curtain of air across the opening to prevent heat from escaping. To further increase energy efficiency, at least one manufacturer has doubled the thickness of insulation in its pizza oven walls, claiming a 50% energy savings over conventional ovens. Another twist is a gas-powered pizza oven that, for display purposes, has a visible flame inside that simulates the look of wood-burning. Its exterior can be ordered with an attractive brick or tile face. All of the atmosphere and none of the smoke!

ILLUSTRATION 11-10 Using a conveyor oven, you can control cooking speed two ways: the temperature inside the oven and the speed of the conveyor belt.

Courtesy of Star Manufacturing International, St. Louis, Missouri.

As you can see, there is no shortage of creativity when it comes to getting something cooked. Attend any restaurant equipment show to catch up on the latest trends.

MECHANICAL OVENS

The main characteristic of the **mechanical oven** is that the food is in motion while inside the oven compartment. Mechanical ovens are used in large-volume operations, including schools, hospitals, and other group feeding situations. Bakeries are also frequent users, particularly when they produce a single item in volume, such as loaves of bread or rolls.

The two basic mechanical oven designs are the *revolving oven* and the *rotary oven*. Their operation is easier to understand when you think of their respective nicknames, which are the Ferris wheel and the merry-go-round.

The revolving oven does operate much like a Ferris wheel. Flat trays are loaded between two rotating wheels inside the heated oven chamber. The wheels turn slowly, and the food on the trays cooks as it rides around and around. The oven door is small, to restrict the escape of hot air, and a control knob is located near the door that lets a person stop the rotating wheels and retrieve any tray when it comes parallel to the door.

The rotary oven operates on a similar principle, except the trays rotate around a vertical axis like a merry-go-round.

Mechanical ovens are usually assembled on site. The most common problem with them is that it can be difficult to level the trays, either end to end or side to side. Trays with burned or built-up food, or trays that have warped because they were not allowed to sit in the oven during warm-up periods before baking, are also hard to level.

Because they require so much space, mechanical ovens are rarely found in restaurants. One exception is in the Southwest, where the oven is modified to smoke meats with the addition of a firebox (a compartment in which wood is burned to produce smoke) and an exhaust fan (to circulate the smoke through the oven).

RACK OVENS

The **rack oven** has gained popularity in restaurants because of its efficient use of space. When floor space is at a minimum, you can purchase a tall, thin oven chamber into which racks of product can be rolled (primarily for baking or roasting). The smallest rack oven is 4 feet square and 6 feet tall. It may take up a bit more space than a full-size convection oven, but it can accommodate as many as 20 sheet pans simultaneously; double rack ovens hold 40 sheet pans. Some racks lock into their oven slots and then rotate during cooking for more even heat distribution.

Add the options of a self-contained steam system and built-in hood, and you have a baking and roasting workhorse! Of course, you'll also need proper-size bakers' racks for this oven (see Illustration 11-11). The racks are loaded with food, then wheeled into the

ILLUSTRATION 11-11 A rack oven is tall and just wide enough for rolling racks of product inside it.

Courtesy of Lang Manufacturing Company, Everett, Washington.

baking chamber on heat-resistant casters. Purchase racks that allow you to space the individual pans from 2 to 6 inches apart.

The best rack ovens are super-insulated so they require minimal clearance around their sides and back, and they are flush with the floor so the racks can be wheeled in and out easily.

Remember, if your rack oven makes its own steam for baking, you must be able to provide water connections and drainage.

COOK-AND-HOLD OVENS

The **cook-and-hold oven**, or *retherm oven*, as it is sometimes nicknamed, is an excellent choice for preparation of convenience foods (see Illustration 11-12). Schools and hospitals use it to reheat large quantities of frozen prepared dishes—ravioli, macaroni and cheese, dishes made with sauces or gravies—and keep them hot until they are served. The cook-and-hold advantage is its low, steady temperature, which heats a product without drying it out or burning it.

Where large quantities of roasted meats are consumed, cook-and-hold capability is a must. Radiant and convection heat combine to cook foods slowly while minimizing shrinkage. When the cooking cycle is finished, the oven's preset timer switches it to a holding mode that keeps the food warm. Holding is actually recommended—for beef roasts, for instance—because enzymes within the meat provide their own natural tenderizing process as the finished roast sits for at least 90 minutes.

There are three basic types of cook-and-hold food production:

1. One type of oven uses natural convection (no air movement), cooks at a slightly higher temperature, and maintains a humidity level of 90% to 95%.

2. Another roasts at slightly lower temperatures with a slow-moving air current and humidity levels of 30% to 60%.

ILLUSTRATION 11-12 Cook-and-hold ovens allow you to prepare foods in advance and hold them up to 24 hours.

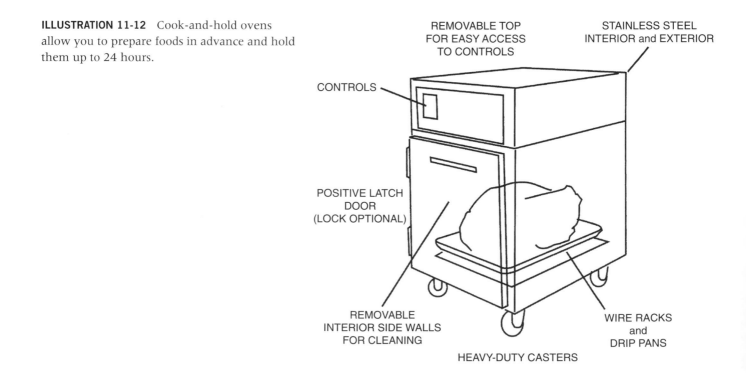

3. A third option is the smoke-and-hold oven, which can cold-smoke (i.e., smoke at low temperatures), then slowly roast anything from salmon to tomatoes to hazelnuts. A regular cook-and-hold can be converted to a smoke-and-hold with a firebox attachment (which allows the oven to burn wood and charcoal) and an ash collection tray. Smoke-and-hold ovens have slightly higher electric power requirements. They also need exhaust hoods, which are not required otherwise.

In addition to being well suited to high-volume cooking, cook-and-hold ovens are energy efficient because, although cooking times are longer, temperatures are lower and require less total energy than conventional ovens. Some foodservice businesses do their slow cooking overnight because these ovens are safe and low-temperature enough to be left unattended.

The temperature range for a cook-and-hold oven is from 140°F to 245°F; 140°F is the lowest safe holding temperature that prevents harmful bacteria from multiplying. Some ovens have timers (up to 100 hours of roasting time can be programmed) or use electric meat probes to determine when products are done. Either way, the oven automatically returns to its holding temperature when prompted. If a programmed oven experiences a large temperature change, this triggers a warning beeper or caution light. Tabletop ovens can be ordered with a carving station on top for use in buffet lines.

Because they cook so slowly, meats in a cook-and-hold oven should be placed on wire racks so they will brown uniformly; a sheet pan can be placed on the oven deck to catch any drippings.

The power requirements of cook-and-hold ovens range from 120-volt single-phase to 380-volt three-phase. No special ventilation is required, so the ovens can be rolled on 5-inch casters. The capacity for a single cook-and-hold oven is about 90 pounds of product; stacked double ovens can hold up to 180 pounds.

SMOKER-COOKERS

Commercial meat smokers are often called **smoker-cookers**. The upsurge in popularity of barbecued meats, particularly pork ribs, has created a cottage industry of sorts as barbecue restaurants and civic clubs sponsor cooking teams that enter outdoor cook-off competitions with their secret recipes and traveling meat smokers in tow on wheeled trailers.

The smoker-cooker is a type of oven, made to create smoke. There are two types of smokers:

1. One popular style of smoker looks like a cylindrical metal container with grills suspended inside on which food is placed. In fact, this smoker-cooker doubles as a grill if it's kept open while the cooking is done. Attached to one end of the cylinder is a firebox, in which charcoal or wood chips are placed to make a fire. When wood is used, it is often dampened with water or beer to make it smoke more as it burns. The use of liquid also helps the meat stay moist as it cooks. Portable models are mounted on trailers or equipped with casters or wheels (see Illustration 11-13).

2. The cabinet-style *oven smoker* uses gas or electricity to start the fire instead of matches. Wood or charcoal is added to the firebox. A draft door on smaller, home-use models—or damper on the smokestack of larger units—controls the even draw of smoke and heat through the cavity. The food inside cooks slowly, taking on a subtle, smoky flavor.

A smoker-cooker can be as simple as a stovetop unit that smokes a few portions of fish at a time or as complex as a conveyor-driven machine that can be loaded with hundreds of

ILLUSTRATION 11-13 The smoker-cooker uses wet wood or charcoal to create a moist, smoky, slow-cooking environment.

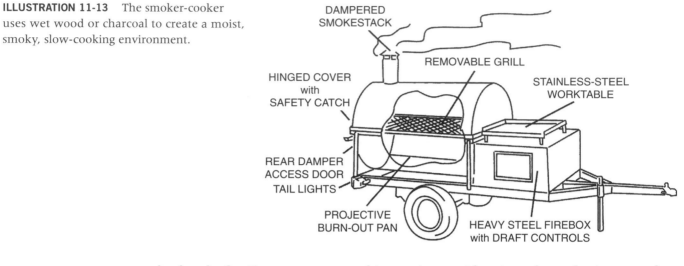

DAMPERED SMOKESTACK

REMOVABLE GRILL

HINGED COVER with SAFETY CATCH

STAINLESS-STEEL WORKTABLE

REAR DAMPER ACCESS DOOR TAIL LIGHTS

PROJECTIVE BURN-OUT PAN

HEAVY STEEL FIREBOX with DRAFT CONTROLS

pounds of pork ribs. Temperature control is a major consideration when selecting a smoker. Look for a unit that can produce the correct heat and smoke levels for the foods you intend to smoke; seafood, poultry, and beef each have different requirements. Foods are either cold-smoked (at temperatures below 100°F) or hot-smoked (at temperatures from 165°F to 185°F). Chefs will tell you a side of salmon, intended to be smoked at a low temperature, can be ruined within minutes if the smoker temperature is too high.

Barbecue aficionados have strong feelings about which types of wood are best; the choices include oak, hickory, and mesquite. Some people also add nutshells, stalks of herbs, or grapevines to the fire to produce flavor variations.

Another factor is the heat source. For long-term smoking (more than just a few hours), an electric heating element is best for consistent temperature control; it is also far less labor-intensive than a fire that must be constantly tended. Shelves or hanging hooks for meats should be adjustable in larger models. Ventilation is an issue in urban settings. To address environmental concerns, an oven smoker may be equipped with a precipitator unit to filter and cleanse discharged smoke before it is exhausted to the outdoors. Cleanup is also a challenging chore unless you have a unit that can be hosed out with a high-pressure spray, a drip pan for easier grease cleanup, and removable grills or racks. Thermometers should be well sealed so smoke doesn't condense inside and make them hard to read. Temperatures should be checked near the top of the smoker, where they are generally the highest.

The dimensions of full-size smoker-cookers vary from 9 to 12 feet in length and from 5 to 6 feet in height. More important is the grill space, which varies on commercial models from 18 to 30 square feet. For smoking, about 10 pounds of charcoal and up to four pieces of wood are needed; for grilling, a full-size smoker-cooker may need as much as 40 pounds of charcoal.

An alternative to the standard smoker-cooker is the *pit smoker*, a combination meat smoker and pressure cooker. Food is placed in a cooking chamber and the doors are shut. As the temperature inside rises, pressure (15 psi) builds up along with the smoke, forcing a faster cooking process as the food is smoked with a handful of wood chips. None of the smoke escapes from the cabinet, so no exhaust hood is required. Perhaps purists don't consider this "real" smoked barbecue, but it turns out 45 pounds of ribs in 90 minutes, or 40 pounds of beef brisket in 2½ hours.

ROTISSERIE OVENS

Food trends come and go, but the universal appeal of rotisserie cooking endures. *Rotisserie* comes from an old French word meaning "to roast." The sensory appeal of skewered meat turning over and over as it cooks above dancing flames evokes mouthwatering in most of us.

The modern-day rotisserie oven contains rows of metal spikes, called *spits*, on which meat is placed. One (or more) small electric motors rotate the spits as warm, moist air circulates through the oven cavity, slowly roasting the meat while the moisture minimizes shrinkage (see Illustration 11-14). Other rotisserie models are equipped with either hanging baskets (for placing food that won't stand up to skewering) that rotate carousel-style or a vertical, moving ladder with product on each rung. The vertical rotisserie is often called a **continuous cooker**. An advantage of a vertical machine is that it allows the operator to add more products during the day, always at the bottom position on the ladder. Programmable models allow presetting of cook times and temperatures for different types of food.

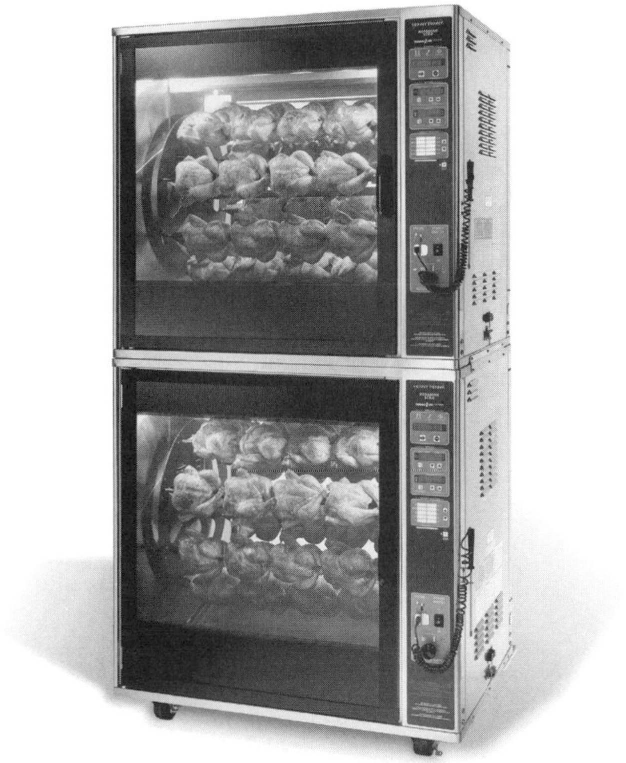

Rotisserie ovens are usually used for baking whole chickens, and capacities range from 6 whole (3-pound) chickens to 70 of them. The small countertop models make an attractive display with glass doors and halogen lights. Occasionally rotisseries are used for barbecued items, such as sausage or ribs. Accessories include heavy-

ILLUSTRATION 11-14   Rotisseries make attractive display ovens as well as cooking slowly and minimizing meat shrinkage.

Courtesy of Henny Penny Corporation, Eaton, Ohio.

duty angled spits large enough to hold a slab of prime rib or a whole turkey and baskets for fish or vegetables. Multiple products can be cooked simultaneously in a rotisserie oven, as long as care is taken not to cross-contaminate. For example, raw chicken juices shouldn't drip onto vegetables. Better yet are models with separate cooking chambers, either side by side or stackable.

Cooking power can be provided by gas or electricity. In gas models, an electric power source is also needed to run the motor that turns the spits. A single motor can turn all the spits or, in larger units, multiple small motors turn sets of spits. The heat source is located at the top or bottom of the oven cavity; some have a second heating element in the center of the rotating shaft. A few rotisseries burn wood or charcoal to impart a smoky flavor. (Not all jurisdictions allow wood-burning rotisseries, citing air quality problems or insect problems associated with wood storage.) As an alternative, gas models may be equipped with ceramic logs at the base of the unit to simulate the open-fire look.

Some chefs prefer gas-fired infrared burners, which produce fast, high-intensity heat that melts away the layer of fat just beneath the skin of the chicken. Inside the oven, a water pan provides moisture and a drip pan catches grease.

Restaurants with outdoor seating or extensive off-premise catering business sometimes buy portable rotisserie ovens. Although the spits are rotated electrically, the cooking is done with liquefied petroleum (LP) gas (also known as *propane*) or a meat smoker–style combination of wood and charcoal. The spits are often adjustable from 13 to 24 inches from the charcoal surface. Before buying, always find out how much weight the spit can support so you'll know not to overload it.

A clean, well-stocked rotisserie oven can be a wonderful sales tool. Customers can see the food as it cooks, and, because they can be made without added fat, rotisserie meats are considered healthful. Take-home food chains like Boston Market have capitalized on this trend. One result of good rotisserie cooking is a moist, tender chicken that has been basted by juices from surrounding chickens as they revolve in the oven.

As attractive as they are, one complaint about rotisseries is that they generate a lot of heat and require good ventilation. Ensure at least 36 inches of clear space around a rotisserie so employees can work nearby without discomfort. Heatproof gloves should be used to avoid burns while removing and reloading the spits. The visual appeal of this oven also makes regular cleaning absolutely necessary. All removable parts should be cleaned daily. Spattered grease is a major problem, so the drip pans, spits, and drains should be easy to remove and replace without tools. Overcooking and undercooking are concerns, and both are often the result of poor maintenance. The fans that circulate the air, for instance, must be cleaned regularly.

One more rotisserie decision to make is whether you need your meat in one batch at a time or in a continuous flow. Batch production is ideal for institutional settings that require a high volume of product all at the same time. Continuous production works for restaurants that need smaller amounts of cooked product at staggered times. Rotisseries can also be ordered with a warming cabinet where finished product can sit while a new batch is being cooked.

ILLUSTRATION 11-15 A wood-fired or wood-burning oven.
Courtesy of Wood Stone Corporation, Bellingham, Washington.

WOOD-FIRED OVENS

A *wood-fired oven* contains a well-insulated cavity in which wood is burned. The heat generated by the burning wood is stored and then released slowly and evenly for a flavorful method of roasting and baking. The heat is retained within 4-inch-thick stone or brick blocks, which store enough to cook for long periods without replenishing the fuel. The wood-fired oven (see Illustration 11-15) cooks like a traditional oven, with heat generated at the bottom of the cavity that rises to the top, but it cooks much more quickly. It can retain temperatures from 350°F to 620°F, even if the fire is tended only once an hour.

The accepted name for this type of oven is evolving. Most manufacturers call it a *wood-fired* or *wood-burning oven*, but others favor **brick oven**—from the days when ovens were usually made of brick—or even **stone hearth oven**. No matter what you call it, the intense heat of this oven produces benefits found in no other type of baking or roasting equipment. Its fiercely hot, fast-cooking method seals in meat juices, caramelizes sugars, and produces full flavor profiles for a wide variety of menu items.

Most operators decide to use gas-powered models that offer efficiency, fresh-roasted flavor, and visual appeal. This is partly because in many urban areas and commercial buildings it is difficult to get permission to use a purely wood-burning oven without an existing fireplace flue or a special variance. In terms of ambience, it probably doesn't matter to the customer whether the open flames are created by wood or by gas jets. Wood/gas combinations (sometimes called *gas-assist ovens*), or all-gas ovens, are also easier for the staff to use. Infrared burners at the oven floor maintain a constant temperature, with natural wood or adjustable gas flames boosting it as needed. These ovens vent through a flue collar located above the door; venting must adhere to specific standards. A gas-powered oven can share hood space with other equipment under a standard exhaust hood. But if yours is an all-wood oven or a wood/gas combination, the situation changes rather drastically. National Fire Protection Association (NFPA) Standard 96 says the by-products of solid and nonsolid fuels cannot be mixed in the same ventilation system, meaning if the oven does burn wood, it must have a separate chimney that vents to the outdoors and a hood with specs to suit wood fuel.

Most manufacturers custom-make these heavy ovens, which weigh 5000 pounds or more. The domed interior walls are made of alumina (unrefined aluminum), high-temperature ceramic, or refractory cement. Typical dimensions of the oven exterior range from 4 to 8 feet, with an interior cavity size of 9 to 31 square feet. It is important to note the **thermal head space** of the oven. The more head space it has, the more heat stays inside the oven cavity instead of being vented away. The deck of this oven is called its *floor*, which can be made of individual bricks or tiles or cast as a solid piece of alumina, ceramic, cordite, clay-and-aluminum, and the like. Those who favor bricks or tiles say they are easier to replace when one cracks. One-piece floor proponents believe heat retention is better in a single piece without seams or gaps.

Insulation is another point of differentiation. Some manufacturers put it between the oven's steel outer shell and its domed roof in the form of spun ceramic fiber or a cast mixture of perlite and cement. Others insist it is smarter to wrap the whole oven in a blanket of insulation before putting on its outer cover.

No argument—a wood-fired oven is a hefty acquisition. The weight of the oven makes installation a tricky and technical job. Some floors must first be reinforced with structural steel; venting, ductwork, and air pollution prevention measures may be expensive.

An optional *mantle* is a sturdy ledge that extends the work space near the oven door; a wood storage box (located below the door) is also optional but recommended. A metal **ash dolly** on wheels is a safe, efficient means of storing and transporting wood ashes, which should be removed daily from the oven when it is cool. All ashes and coals should be doused with water to be sure they are fully extinguished before final disposal. After use, the oven cools—a relative term—to about 400°F. It takes at least an hour to bring it back to its peak operating temperature (550°F to 625°F).

Firewood is chosen for these ovens based on the smoky flavor it may impart, how easy it is to ignite, and how much moisture it contains. Properly dried firewood should not contain more than 20% moisture, or it produces too much smoke and not enough heat. You can buy a meter that reads the moisture content of wood, which you can use to check incoming loads before you pay for them. Pressed-wood products are never appropriate for a wood-burning oven; they contain chemicals that may damage its interior. It's a good idea to nurture a relationship with a reliable wood supplier because wood quality and delivery schedules are so important. You must also have a separate storage area for curing newly delivered wood so termites and other wood-borne pests are not introduced to the kitchen.

11-6 MICROWAVE OVENS

Demographers claim the microwave oven has had the most influence on our eating patterns since the development of the freezer. Most American households have at least one microwave in service. At home, they're used mostly for reheating.

In foodservice, today's professionals have been forced to rethink their views of microwave cooking. Once frowned on as a necessary evil for kitchens with limited space, they have come to be indispensable in the well-equipped kitchen—and not just for retherm (reheating). Commercial microwaves have more power and durability, and better warranties, than ever before. They're easy to install and operate. They don't take up much space. They work quickly without affecting taste or nutrient content, and some microwaved foods can be cooked and served in the same dish, minimizing cleanup. Another benefit: They help keep the kitchen cool and comfortable.

When you're shopping for microwaves, remember that home-use models are not at all appropriate for commercial use. A home model is not meant for continuous use and might burn out within a few hours under the rigors of restaurant cooking. Commercial doors are also much sturdier than home ones, made to be opened and closed hundreds of times a day. In fact, most health inspectors and fire insurance companies require commercial-grade microwaves, and the manufacturer will not honor a home microwave's warranty if it is being used in a commercial setting.

Here's how the cooking process works. Between one and four electronic magnetron tubes located in the back ceiling of the oven converts electricity into microwaves. A fan (sometimes called a **wave guide**) located in the top of the oven pushes the waves from the tube(s) into the oven cavity, where **wave stirrers** distribute them evenly to prevent hot spots in the appliance. The waves themselves are not hot, which keeps the inside of the oven cool (see Illustration 11-16).

Many people believe an item in a microwave heats from the inside out, which is not exactly true. In fact, the waves heat only the molecules of moisture (water or fat) in the food. Both the microwaves and the molecules have positive and negative electrical poles, so the molecules try to align themselves with the microwaves. These alignment attempts, at 2.5 million times per second, create intense friction, which makes the heat that cooks the food. Exactly where the waves heat depends on where the moisture is located within the food. For instance, because there is more moisture in the center of a baked potato, the highest temperature is attained there. If you have baked a potato in a microwave, you may have noticed that if you take the potato out immediately and cut it open, you are more likely to find a rather hard, lumpy consistency. After microwaving, the potato should be wrapped in a clean kitchen towel and allowed to sit for 5 to 10 minutes to fully distribute all the heat built up in the center. The result is a much better, more evenly cooked potato.

ILLUSTRATION 11-16 The major parts of a microwave oven.

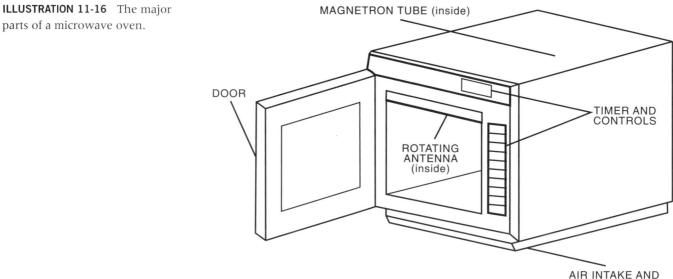

Microwaves can be used to safely defrost frozen foods too. Vegetables and seafood can be steamed or poached in the microwave with excellent results. High-powered countertop models cost the same as or less than most tabletop steamers, and require less power, no water supply or drain lines (and therefore no deliming process), and no preheating or standby mode.

You have surely heard that metal dishes or aluminum foil should never be used in a microwave oven. This is because the metal deflects the waves away from the food, preventing cooking and possibly aiming them back at the magnetron tube, overheating it. (Yes, you've probably noticed the inside walls of the oven are metal, but they deflect the rays back toward the food—a big difference!) Paper, plastic, glass, and ceramics are all good alternative materials for containers. If the containers are large, their contents probably must be stirred or rearranged partway through the heating process for best results.

The microwave oven should never be operated empty. It needs something inside to absorb the energy, even if it's only a cup of water. Most microwaves are equipped with an automatic shutoff if they are left on, to prevent them from shorting out. An oven spattered with food and not cleaned regularly strains the heating element, which can cause a breakdown or, at the least, shorten the life of the appliance.

So much for cooking tips (and with microwaves, there are hundreds of them). In the current generation of commercial-grade microwave ovens, power is more important than size. The more power you have, the faster foods cook. Power is measured in watts. For low-volume operators such as snack bars and service stations that use them only for warming, a 700-watt microwave oven is sufficient. Coffee shops, diners, and fast-food eateries need at least a 1000-watt microwave to heat precooked foods to serving temperature. High-volume restaurants, hospitals, and so on can get 1400- to 2700-watt microwave ovens, capable of defrosting and retherming in bulk quantities. Cavity size and cooking speed are the major differences between models. Large cavities accommodate two dinner plates or 10 × 12-inch plastic pans.

You'll notice the lighted digital display and control panels are similar to home-use microwave ovens, but commercial models have programmable cooking times, and some can do a computerized inventory of types of dishes cooked in them. A buzzer sounds at the end of each cooking cycle. Smaller units operate on 120 volts of electric power, while larger, heavy-duty ovens require 208 to 240 volts.

The cabinet sizes of microwaves range from 13 to 25 inches wide, 16½ to 25 inches deep, and 13 to 19 inches high. The microwave is generally a space-saver for a busy kitchen.

We just said size wasn't important for microwaves, but there are a few gigantic microwaves that blast up to 40,000 watts! A model that's popular on cruise ships has a cooking cavity 8 × 4 feet and is used primarily for speedy defrosting. It can thaw 85 pounds of chicken in 90 seconds. Even larger ones in food-processing operations are configured as conveyors that can safely thaw up to 12,000 pounds of frozen product in an hour. Like walk-in coolers with remote compressors, these huge microwaves can have their power generators installed in remote locations. They require 440- to 480-volt electrical hookups and plumbing for water-cooling the generator.

Microwaves, unlike X-rays, are not radioactive. However, the U.S. Food and Drug Administration limits the amount of waves that can leak from an oven to 5 milliwatts per square centimeter of oven cavity. Safety standards specify that the oven door have two independent but interlocking systems that automatically stop the oven when its door is open. Doors are also equipped with seals and absorbers to prevent radiation leakage.

You may still see "Microwave Oven in Use" notices posted at some restaurants, which were put up in past years to protect individuals with heart pacemakers. Today, however, the popular belief that microwaves disrupt pacemaker activity has been proven incorrect. It's still smart to check your local ordinances, as many of them continue to require the warning signs.

Because the microwave oven offers instant cooking on demand—no warm-up times, no heat loss when the door is opened—it is extremely energy efficient. Smart restaurateurs use the microwave in tandem with other cooking appliances to save energy and keep the kitchen cool. For instance, you can partially cook a steak in the microwave, then sear it to a quick finish on the broiler, reducing broiling time. Or an Italian restaurant may precook its pasta and meat sauce and simply retherm an individual serving in the microwave just as a pizza for the same table is about to come out of the oven. The microwave oven can give you the ability to schedule cooking activity to meet the demands of guests.

A few manufacturers offer combination microwave and convection ovens. This single unit can heat and brown pastries, poultry, and other foods that require both heating and careful control of their exterior color. These combination ovens heat and brown quickly, with a 1000-watt microwave and convection cooking of up to 475°F. Half a chicken in this combination oven for just 4 minutes looks like it has been roasting on a spit for 30 minutes! It's a practical, versatile combination you may want to consider.

As you make your purchase decisions, be honest with yourself about exactly what you will use the microwave for. Melting, cooking, and defrosting are distinct tasks requiring different wattages and speeds. You may need several microwave ovens, each for a different task.

11-7 OVENS FOR BAKERIES

The quality of bread served by a restaurant is, to some customers, a measure of the overall quality of the establishment. Chains and independent outlets now offer an ever-growing variety of artisan-style breads, rolls, croissants, muffins, and pastries. Including freshly baked goods on a menu increases profit and value in the eyes of the diner, and the sights and scents of baking are enticing enough to create a market niche. But what types of appliances do you need to achieve this? Whether bakery items are produced on site or prepared in commissaries and finished on location, their production relies on durable and (in most cases) high-capacity ovens and proofers as well as holding equipment that presents the products attractively to customers.

Perhaps the most basic question for a bakery operation is whether to use rack ovens or deck ovens, both of which have already been described in this chapter. The rack oven offers high volume for items such as bagels, producing them quickly, but the deck oven, which hearth-bakes breads directly on a ceramic deck, produces a better crust and is more appropriate for artisan breads. Deck ovens are also adaptable enough to use for entrées, casseroles, pizzas, soft pretzels, and cookies. Not everything must be placed directly on the deck; items can be placed in pans for baking too. The disadvantage is that a deck oven usually requires more space than a rack oven to produce the same quantity of product. Deck ovens use more power than rack ovens and also require more skilled labor to use them correctly. They are trickier to use with frozen or partly baked dough. So, in smaller spaces or for use by unskilled employees, a computerized, programmable rack oven is ideal.

Specialty ovens used for point-of-sale bakeries now allow such options as baking flatbreads (those that don't have to rise) in 2 to 3 minutes so customers can have fresh bread to order. Wood Stone, a manufacturer of wood-fired ovens, has introduced a line of gas-powered and gas-assist hearth ovens designed for high-volume bread production.

An alternative to the wood-fired oven for baking is the **steam tube oven**, which looks like a wood-fired oven with its high, domed cooking cavity and larger opening. This heavy oven is made of concrete, with hollow tubes placed above and below the cooking cavity and a gas-powered firing chamber at the bottom of the oven made of brick. The tubes are filled with water, which turns to steam in the intensity of the 580°F heat. The steam circulates through the tubes, then flows back to the firing chamber as water, to be reheated and recirculated. The high temperature and moisture combination is perfect for baking.

The oven tubes may also be filled with thermal oil, an excellent heat conductor, but this increases the cost and complexity of the system, requiring pumps, additional controls, and a knowledgeable staff member to maintain the oven. Using water in the tubes requires no maintenance and no moving parts.

PROOFERS AND RETARDERS

Proofing is the process of warming dough to allow the yeast to activate and cause the dough to rise before baking. The specialty appliances that both start and stop the proofing process are called *proofing cabinets* (**proofers**) and **retarders**. It's true that yeast products can proof just sitting on a shelf, but if you want them to reach their optimal size and texture consistently, you'll use a proofer, which maintains the correct temperature and humidity for the type of baked goods you are making. Incorrect humidity is a concern because it may cause dough to dry out, crust too soon, split open, or form irregularly.

After the dough has risen, you probably could just put it in the refrigerator to stop the yeast growth. But, again, there's an appliance to do the job more precisely. The retarder offers humidity control. For small spaces, you can order a combination proofer/retarder; this can be programmed to retard at night, then proof in the early morning hours so you can bake fresh bread for the breakfast rush. For larger kitchens, there are proofing cabinets that fit beneath an existing oven as well as roll-in proofers and even mobile proofers. Many of them do double duty as warmers for food cooked in other appliances.

A combination oven/proofer can turn any foodservice operation into a full-service bakery. Small enough to fit behind a counter but powerful enough to yield reasonable quantities, this machine consists of two chambers: the proofer (usually on the bottom), with temperature and humidity controls, and the oven on top, both with clear glass doors for easily checking on both processes. The oven is most often a convection oven and can be used for almost any cooking application; the proofer does double duty as a holding cabinet.

11-8 NEW OVEN TECHNOLOGY

You've probably learned by now that the science of cooking is constantly being toyed with in appliance manufacturers' laboratories, and the results are impressive. The breakthroughs in this field nowadays are products that combine several cooking technologies and computerized, highly programmable appliances. Panasonic, for example, launched a microwave/combi-oven/grill in 2007. It allowed the operator to program up to 99 recipes onto a computerized memory card, eliminating the need for manual setting of times and temperatures each time the appliance is used. The technology also means a restaurant or catering company with multiple outlets can e-mail recipes to each location; these can be saved on secure digital (SD) memory cards without having to program the individual appliances separately.

The primary drivers of innovation in today's market are improved energy efficiency, enhanced product quality, and speed of service. *Super-cookers* were developed to use two or three heat transfer methods simultaneously for incredible speed. For instance, a combination of microwave and convection cooking, sometimes called *combi-wave*, adds the speed of microwaving to the browning power and efficiency of convection cooking. Traditional microwaves have never been very good at browning, crisping, and broiling, but

(a)

(b)

ILLUSTRATION 11-17 (a) The TurboChef Encore rapid-cook oven uses radiant heat, high-speed air impingement, and side-launched microwave. (b) The TurboChef Turbo2 evenly and consistently toasts, browns, and crisps food up to 12 times faster than traditional cooking methods.

Courtesy of The Middleby Corporation, Elgin, Illinois.

add the forced air of convection fans at temperatures of up to 475°F, and the problem is solved. The downside is that the combi-wave unit is small, so its output is limited, and employees may have to experiment to find the right combination of ingredients, time, and temperature for the best results. The units are programmable, though, and easy to use.

Rapid-cook ovens (see Illustration 11-17) combine radiation with heat impingement, a technology pioneered by TurboChef, Inc., a company perhaps best known for supplying the ovens in Subway sandwich shops. Instead of the heated air hitting the food and bouncing off, as in typical impingement cooking, the rapid-cook unit contains a fan that sucks air all the way around the food and out the bottom of the oven cavity at a rate of up to 60 miles an hour. This means the heat transfer between the air and food is constant, resulting in uniform browning and crisping. At the same time, microwaves penetrate the food. Dishes cook 10 to 15 times faster than on a conventional griddle or broiler or in a conventional oven. A pizza in 70 seconds? A rack of lamb in 55 seconds? Cinnamon rolls in 8 seconds? No problem, says the manufacturer. And no exhaust hood is required, making these ideal for confined spaces. They're about triple the cost of a conventional oven, but TurboChef says the speed justifies the cost. Its countertop unit is no more than 3 feet square, uses 208- to 240-volt three-phase power, and is programmable for multiple types of menu items. (Note that whenever microwaving is a component of the cooking process, the types of pans that can be used are limited to non-metal, such as plastic or glass.) In 2006, TurboChef announced its tentative intention to develop the rapid-cook technology for home use.

Another innovation is the FlashBake oven, which uses a combination of intense light and infrared energy. The light waves (not microwaves) penetrate the food, cooking it quickly, while the infrared waves brown the food surface. The idea here is to control and balance the two energy sources by computer so as to cook both interior and exterior simultaneously in the shortest possible time. You can program up to 50 cooking cycles on the FlashBake oven control panel. Its footprint is less than 6 feet square, and it weighs only 121 pounds. It requires 208 to 240 voltage and one- or three-phase electricity, about a 60-amp circuit at maximum power. The door locks when it's cooking and unlocks automatically when the cycle is finished.

Speed is, once again, the driving force behind this type of oven technology. FlashBake sales literature boasts a 2-minute cooking time for a boneless chicken breast that would otherwise take about 7 minutes to broil, and 1 minute for french fries that would require 10 to 13 minutes in a fryer.

Super-cookers may not be miracle-working replacements for all other appliances, but they are worth a second look. Work with manufacturers, ask a lot of questions, and witness a couple of demonstrations to ensure any particular super-cooking appliance really meets your needs before you spend the money.

Another way to integrate two functions in a single appliance: the super-cooker with a base that is a refrigerator. Raw product is stored in the refrigerated compartment until the moment it is needed, saving space as well as a few steps for the employee doing the cooking.

It certainly is important to keep up to date on new technology. Just remember, if it doesn't improve your bottom-line results, there's no need to spend money on it.

11-9 CLEANING AND MAINTENANCE

In Chapter 6, we discussed basic troubleshooting for burners on gas rangetops. Here are additional tips for oven components and types.

The purpose of an oven is to put out a lot of heat, so it is important that the vents to dissipate that heat are always kept clean and free of debris. Overdone or underdone food often signals that airflow is somehow being blocked. The solution may be as simple as moving the appliance farther from the wall so airflow around it is not restricted.

Pilot lights and gas connections should be checked on gas-powered models, and wiring on electric models. Periodically check doors to make sure their seals are tight and they close correctly. Otherwise, you are probably wasting heat that escapes through a misaligned door.

Cleaning your commercial appliances is as easy as cleaning your dishes. Most equipment surfaces are made of stainless steel, which resists corrosion and is practically unharmed by moisture, detergents, food acids, salts, or anything else corrosive. These are solid sheets, not just a coating or surface that can be chipped off.

APPLIANCE SURFACES

The secret of maintaining stainless steel lies in cleaning it frequently to prevent buildup of surface deposits, which may be harmful over time. Ordinary food or grime can be removed with soap and water, applied with cloth, sponge, or fiber brush. Be careful with abrasive cleansers, though; use them gently and rub in the same direction as the polish marks on the steel surface so small scratches cannot be seen.

Baked-on food on rangetops and ovens requires a bit more effort. You can make a paste out of water and any of these substances:

- Ammonia
- Magnesium oxide
- Powdered pumice
- French chalk

Rub as gently as possible in the direction of the polish marks, or use a scouring sponge or stainless-steel wool for more resistant stains. Stainless-steel wool is different from plain steel wool, which is too abrasive and should be avoided. Also, avoid using steel scrapers, wire brushes, or files, all of which contain iron particles that may become embedded in and rust your stainless-steel surface.

The exterior appliance surfaces can be enhanced with lemon oil or a good grade of furniture polish wherever it is not in contact with food and when the polish odor is not objectionable. Polish removes grease, fingerprints, and smudges from stainless-steel finishes. Equipment manufacturers offer suggestions for cleaning products and methods, so be sure to ask when they demonstrate for your staff. You can also contact the American Iron and Steel Institute's Committee of Stainless Steel Producers in New York City for its recommendations.

MICROWAVE CARE

Microwave interior cooking cavities should be wiped out daily with a soft sponge or cloth, mild detergent, and warm water; do not use oven cleaner! If foods are cooked on and appear to be solidly stuck on walls or ceiling, boil a cup of water in the oven first, which will help loosen them. Remove the spatter shield on the oven ceiling weekly for cleaning. Also check the air intake and discharge areas of the microwave. Keep them free of dust and debris, and make sure they're not blocked or the oven will overheat. Some people say if an oven's controls are malfunctioning, unplug it, wait about one minute, plug it back into the wall, and try again.

Microwave ovens appear to be sturdy, but they are subjected to rigorous use in most restaurants. Don't slam their doors, and place dishes inside carefully. And of course, never put foil or metal objects in a microwave.

CONVEYOR OVEN CARE

Conveyor ovens put out a lot of heat, so it is especially important that their cooling fans and filters be checked and cleaned at least weekly. The oven exterior, interior chamber, conveyor belt, crumb pans, and inspection window should be wiped off daily, but only with whatever cleaning products are recommended by the manufacturer—and only when the oven is fully cooled. Some electric-powered ovens have their own computerized cleaning cycle but should still be wiped clean to remove debris.

DECK OVEN CARE

Exteriors and interiors of deck ovens can be cleaned daily with mild detergent and water on a cloth. It is not advisable to spray water directly into the oven cavity. In terms of routine maintenance, when deck ovens malfunction, it is often because they are not completely level or the vents that remove heat from the oven cavity are blocked. Manufacturers also recommend a break-in process for curing a new deck or hearth, especially a ceramic one. Periodically check the heat shields under the deck. If you notice the deck isn't heating evenly, these shields may need to be replaced.

ROTISSERIE CARE

Check burners and fans for debris. Any heat-transfer surface should be cleaned only with products recommended by the oven manufacturer. Some cleaners, while effective, are corrosive enough to damage ceramic or metal parts.

On gas units, the manifold valve should be open all the way, and the quick-disconnect on the gas line should be securely closed. Carbon and grease builds up on the hub assemblies and gears of the motor that turns the spits so, in addition to external cleaning, the drive system should be partially disassembled for a major cleaning at least twice a year. If the rotisserie is consistently overloaded, or loaded improperly, it may automatically shut itself off to protect the motor. Correct loading and cleaning prevent this.

WOOD-FIRED OVEN CARE

Like deck ovens, wood-fired ovens must be properly leveled to avoid uneven cooking results. A new oven should be fired up as hot as possible for an hour a day for about a week before you cook anything in it. This curing process tempers the floor and walls of the new oven. Damper controls must be adjusted manually. If the oven uses a gas pilot light, keep it clean.

A good way to wipe out the surface of the (completely cooled) oven is with a damp mop. After a full day's work, there is no need to wait around to empty the ashes from the wood fire. Just close the oven door, and do the cleaning in the morning before starting the fire again.

A word of caution: If your wood-fired oven is gas-assisted, be aware that crumbs in the oven should not be pushed into the burner chamber where the gas manifold is located. The crumbs will eventually clog the burners, which can cause the oven not to light.

■ SUMMARY

The most heavily used piece of kitchen equipment is the rangetop. Unlike your stove at home, in foodservice you can mix and match a rangetop with either an oven cavity or refrigeration cabinet beneath it to create the appliance that's right for your commercial kitchen. On the range surface itself, you then select the types of burners or cooking surfaces you want. These decisions are based on the types of cooking you will be doing and whether other appliances in the kitchen could perform some of the tasks you would otherwise do on a range. Also consider how easy it will be to clean your range, because you will be cleaning it often.

The three basic types of ranges are the medium-duty or café range; the heavy-duty range, which is made for higher volume and large, heavy pots and pans; and specialty ranges for certain types of cuisine. Whether to choose a gas or electric range is largely a personal preference.

Ovens present even more choices. You can order a single oven or a deck oven with more than one cavity and separate temperature controls for each. Large-volume operations may want to consider a mechanical oven, which moves the food through the cooking cycle on a rotating wheel or conveyor belt.

The old-fashioned flavor and ambience of cooking with wood is available with the use of smoker-cookers and wood-fired ovens. In many jurisdictions, however, wood burning is regulated for air quality reasons. Wood/gas combinations or gas-fired ovens with a simulated fireplace appearance are two ways to avoid these problems.

Microwave ovens, once frowned on by chefs who didn't consider them real ovens, have become kitchen staples for safe but speedy thawing of frozen products as well as retherming precooked dishes. Microwave technology has also been combined with convection and/or impingement to create so-called super-cookers that cook foods in a fraction of the time of conventional appliances.

This chapter has introduced and explained more than a dozen oven types—how they work and why they are used. Now, the purchase choices are yours!

■ STUDY QUESTIONS

1. What is a café range? How does it differ from a heavy-duty range?
2. What is the difference between conduction and induction?
3. Would there be any advantages to using a French hotplate instead of a rectangular hotplate on your electric rangetop?
4. Name three important considerations when you are ordering a door for your new oven.
5. What is the difference between a range oven and a deck oven?
6. Why would you choose a mechanical oven for your foodservice business?

7. Explain how an impinger/conveyor oven works.

8. What should you take into account when buying a smoker-cooker? List three important considerations.

9. Why should you choose a commercial instead of a home-use model of microwave oven for restaurant use?

10. What are three considerations you should make to avoid complications when installing a wood-fired oven?

Café 5555, Banner Thunderbird Medical Center

Glendale, Arizona

SITUATION

Since 1983, this hospital has grown from 75 beds to more than 500. It also provides extensive outpatient services on its 32-acre campus in northwest Phoenix, Arizona. As part of a 200-bed expansion project, a new 23,000-square-foot kitchen and cafeteria was planned. Café 5555—named for the address of the hospital—is the servery and indoor/outdoor dining area included in the expansion. Budget for the foodservice portion of the project: $10 million.

CHALLENGES

- Providing continuous foodservice during the 1-year construction period. Eleven mobile trailers were set up on what had been the hospital's helicopter pad, creating a 6000-square-foot interim food production space to serve two temporary cafeteria spaces and all patient room-service meals.
- Sufficient room for storing the large rolling carts used to deliver meal trays to hospital patients. Due to existing public corridors that could not be moved, only about one-quarter of the storage space desired was available.
- The corporation that owns this hospital and others changed its systemwide menu during the construction. Assumptions that made been made about equipment sizes and placement didn't fit the new menu requirements, necessitating adjustments.
- The need to keep kitchen labor costs down.
- Incorporating multiple service styles: self-service cafeteria, pre-packaged grab-and-go items, some display cooking, and room-service meals for patients.

COMMENTS FROM THE DESIGN TEAM

- "The logistics [during construction] created an intense situation. Even the temporary trailers housing the kitchen equipment had to be health-department approved and earthquake safe."—Richard Dieli, Dieli Murawka Howe
- "We used many different types of lighting, including LED and recessed and pendant lights, to keep the space interesting and vibrant. We also wanted to have a very open flow throughout the servery so people could move easily through the space."—Russell Combs, NTA Architecture
- "The same racks used in the combi oven are rolled into the blast chiller, which is extremely convenient, sanitary, and time-saving."—Jamie Palenque, executive chef

HIGHLIGHTS OF THE DESIGN

- Half the cafeteria seating is indoors; half is outdoors.
- Nearly half the 23,000-square-foot space consists of dry and cold storage, employee lounges and lockers, and administrative offices.
- A flight-type dish machine and tray accumulator system were chosen for their speed and durability. The window to get dishes, trays, glasses, and flatware clean between busy times is just 90 minutes.
- Flexibility was built into the cafeteria space with rolling, movable walls that can be used to open only a portion of the servery during the late evening hours when the full space is not needed.
- LED-lit television monitors serve as menu boards, making it easy to display and change the daily food offerings.
- Fun restaurant touches not often seen in a healthcare setting, such as a Mongolian grill and a brick oven, were added. The grill is used for stir-frying and also for cooking customers' make-your-own burritos; the wood-burning oven is where pizzas, pasta dishes, and casseroles are finished.
- Simple touches—like salad bar items presented in a variety of contoured shapes—add interest and visual appeal.

RESULTS

From lobby to cafeteria, the goal was to give the medical center's new public spaces the look and feel of a high-end, resort-like healing environment—with delicious food, lots of variety, and a lively atmosphere to attract local residents as well as patients' families and the hospital staff. Annual sales are targeted at $2.5 million, with 44,000 transactions per month. Three other hospitals in the company are undergoing similar conversions from contracted foodservice to operating their own kitchens.

TEAM MEMBERS

- *CEO, Banner Thunderbird Medical Center:* Tom Dickson, FACHE
- *Director, Culinary and Nutrition Services:* Julie Spelman, MBA, RD
- *Executive chef:* Jamie Palenque
- *Director of design and construction, Banner Thunderbird:* Jim Lucas
- *Senior project manager, Banner Thunderbird:* Pradeep Dugar
- *Architect:* Russell Combs, AIA, senior associate/senior project designer, NTD Architecture, Phoenix, Arizona
- *Interior designer:* NTD Architecture
- *Foodservice consultant:* Dieli Murawka Howe (DMH), Richard Dieli, FCSI
- *Equipment dealer:* R.W. Smith and Co., Orange County, California; Scott Roczey, director of sales

Source: The full article about this project first appeared in the March 2011 issue of *Foodservice Equipment and Supplies* magazine. Donna Boss wrote the original article; our encapsulation and the layout appear with permission.

12

PREPARATION EQUIPMENT: FRYERS AND FRY STATIONS

■ INTRODUCTION AND LEARNING OBJECTIVES

As much as most people love their fried foods, these have become a topic of concern in recent years. The reason is that *trans fatty acids* (TFAs) in the oils typically used for frying clog arteries, which can lead to heart disease, and the Food and Drug Administration (FDA) says they are unsafe to consume in any amount. (The American Heart Association recommends a person's trans fat intake be limited to less than 1% of his or her total calories.)

Therefore, most foodservice operators have moved to the use of TFA-free (or *zero trans fat*) cooking oils as the public and consumer advocacy groups demand healthier frying oil products.

In fact, in our experience, most school kitchens have moved completely away from having deep fryers in their equipment lineup. The combi-oven has given them an alternative preparation method for serving fried items. They simply use a light spray of cooking oil across the food products and then allow high-temperature convection to do the cooking. This alternative approach has worked well, although it does require that the operator buy a pre-blanched, oven-ready product for good results.

For those foodservice businesses that still want to incorporate traditional frying in their repertoire, this chapter is for you.

After reading this chapter, you will be able to:

■ Explain how fryers work.

■ Describe ways to maximize the life of fryers and frying oil.

■ Choose the appropriate fryer for your foodservice operation.

■ Identify the components of a fry station.

■ Describe the latest developments in fryer technology.

To fry means to cook in hot fat or oil, and frying methods include sautéing, stir-frying, pan-frying, and deep-fat frying. While the basic frying techniques remain the same, the equipment needed to accomplish them in a commercial kitchen has evolved to be safer, more energy efficient, and easier to clean. In past decades, it was generally accepted that you had to sacrifice either performance or energy efficiency. Today that is no longer the case.

The fryers we discuss in this chapter fall into several basic categories. The most common is the deep fryer (which used to be known as a deep-fat fryer), in which food is immersed in hot oil. We also cover pressure fryers, conveyor fryers, and air fryers. A host of options, including oil-less fryers, are now available to foodservice outlets.

12-1 DISSECTING THE FRYER

In foodservice, the fryer is a piece of equipment as standard as a rangetop. In only isolated exceptions—perhaps nursing homes and hospitals—fryers are not an integral part of the kitchen. They can operate on electricity, natural gas, and propane gas.

The receptacle in which the oil is placed is the **frying kettle**, also known as a *bin* or *vat*. It is usually made of 16-gauge stainless steel and should have rounded (coved) corners for easy cleaning. Kettles come in widths of 11 to 34 inches; their depth, from front to back, should not exceed 24 inches. This is primarily a safety feature, so the person using the fryer need not reach very far over the heated oil. The kettle has fill lines on its interior wall to indicate the proper oil level. The newest fryers feature insulated kettles, which manufacturers claim increases energy efficiency by about 10%.

Frying kettles are usually cube-shaped, but some are Y-shaped, widest at the top and tapering to a thin cone at the bottom. Heat is applied to the upper part of the Y, and the narrow bottom part functions as the **cold zone** where, as you'll learn in a moment, crumbs fall during the frying process. Kettles also vary depending on where their heat source is located. The heat conductors in a Y-shape fryer, called an *open-pot fryer*, are located outside the kettle, which allows the kettle to contain a deep, narrow cold zone. In a square or rectangular *tube-type fryer*, the conductors are tubes that run through the square or rectangular-shaped kettle, just a couple of inches above the bottom. The cold zone is located beneath these tubes.

Smart foodservice operators match the shape of the kettle or bin to the type of food they fry. For instance, to fry a large number of doughnuts simultaneously, you need a long, wide area. (A 34 × 24-inch frying bin can make up to six dozen doughnuts at once.) Doughnut frying bins also have a swing-up drainboard to allow the doughnuts to dry after frying.

The size of a kettle is a measurement of its capacity to hold cooking oil. An 11 × 11-inch kettle, for instance, is referred to as a 15-pound fryer because it can hold 15 pounds of oil. The largest kettle, 34 × 24 inches, can hold 210 pounds of oil. Some manufacturers rate the sizes of their fryers by the amount of french fries they produce in an hour. More about how to measure output later in this chapter.

FOODSERVICE EQUIPMENT
Safety Tips for Fryers

1. There is an open flame inside the fryer. The unit may get hot enough to set nearby materials on fire. Keep the area around the fryer free of combustibles.

2. *Do not* supply the fryer with a gas not indicated on the data plate. If you need to convert the fryer to another type of fuel, contact your dealer.

3. *Do not* use an open flame to check for gas leaks!

4. Wait 5 minutes before attempting to relight the pilot to allow any gas in the fryer to dissipate.

5. Never melt blocks of shortening on top of the burner tubes. This will (a) cause a fire, and (b) void your warranty.

6. Water and shortening *do not* mix. Keep liquids away from hot shortening. Dropping liquid frozen food into the hot shortening causes violent boiling.

7. At operating temperature, the shortening temperature is greater than 300°F. Extreme care should be used when filtering operating-temperature shortening to avoid personnel injury.

8. Ensure the fryer can get enough air to keep the flame burning correctly. If the flame is starved for air, it can give off a dangerous carbon monoxide gas. Carbon monoxide is a clear, odorless gas that can cause suffocation.

Source: Pitco Frialater, Concord, New Hampshire.

The food is lowered into the hot oil in a *fryer basket*. These are also made of wire mesh stainless steel or chrome-plated steel. The mesh allows oil to flow easily through the basket and surround the food. Baskets come in two depths: 4 inches and 6 inches. A fryer can have a single basket, double basket, or more, depending on how many fit side by side in the kettle. The basket, which often looks like a wire mesh saucepan, has a long handle on one side and a metal hook directly opposite the handle so it can be hooked in place on the kettle or on a *basket rack* for storage. Make sure the handle is insulated for comfortable use. Baskets often are made of copper or brass and coated with stainless steel. Be sure none of the interior metal is exposed because frying oil breaks down faster if exposed to it.

As you might imagine, a steel kettle full of hot oil can be a burn hazard if not enclosed somehow. Thus, the kettle fits into a fabricated metal *fryer cabinet*, also made of 16-gauge steel. The kettle, at least in smaller models, should be easily removable for cleaning.

The smallest fryers are called *drop-in fryers*. The kettle drops in to its metal cabinet, where it fits snugly. The controls for the fryer are located on the front vertical surface of the cabinet. The bottom of the cabinet is removable for cleaning. The drop-in fryer is mounted on a countertop. Most manufacturers provide sealing gaskets for mounting, which fit on the bottom of the kettle. Drop-in unit capacity ranges from 15 to 30 pounds.

The *countertop fryer* is a drop-in fryer that stands on 4-inch legs. The legs are either stainless steel or plastic.

The *freestanding* or *floor-model fryer* is the workhorse of any operation that mass-produces fried foods. It has a capacity of 28 pounds or more, and its cabinet stands on four adjustable 6-inch legs or rolling casters. If casters, the front wheels should lock and the rear wheels should be able to swivel. A freestanding fryer with two kettles and two sets of controls is called a *split-vat* floor model (see Illustration 12-1).

Freestanding models often have spreader plates like ranges or collapsible shelves to provide extra work space. If the bottom part of the cabinet is empty, it can be used for storage. However, in some large units, the **filtration system**, which keeps the frying oil clean by filtering particles out of it, is located in the cabinet.

Other handy options are automatic touch-time basket lifts (controlled by a timer, they lift the full basket out of the oil automatically) and a **crumb tray** at the bottom of the kettle, to remove and clean.

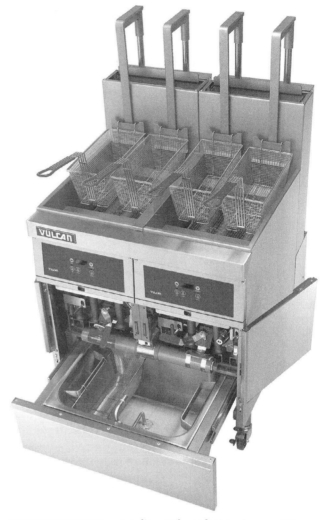

ILLUSTRATION 12-1 A split-vat fryer has two separate kettles and sets of controls.

Courtesy of Vulcan-Hart, a division of ITW Food Equipment Group, LLC, Louisville, Kentucky.

12-2 HOW FRYING WORKS

Let's take a closer look at what happens in the fryer kettle. Oil can be poured in as a liquid or as a solid (lard) that melts in the fryer. How the oil is heated depends on the type of fryer you purchase; we'll introduce the heating options and explain how they work in a moment.

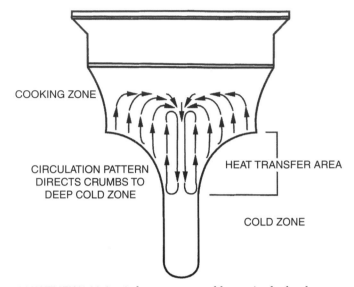

COOKING ZONE

CIRCULATION PATTERN DIRECTS CRUMBS TO DEEP COLD ZONE

HEAT TRANSFER AREA

COLD ZONE

ILLUSTRATION 12-2 A deep, narrow cold zone in the kettle helps ensure crumbs and debris don't mix with the hot oil above.

The oil reaches an optimum frying temperature, from 325°F to 375°F. Hotter temperatures than this—say, above 400°F—are not good for frying because the oil begins to decompose and burn, and the food absorbs more oil and loses nutrients. The oil also begins to break down if it is exposed to the following items, all of which are found routinely in the frying process:

- Water (foods with a high natural moisture content)
- Sediment (crumbs, flour, food particles)
- Salt
- Oxygen

This is why filtering and regularly changing fryer oil is so important. Over the years, equipment manufacturers have improved fryer heating techniques to help prolong the life of each batch of oil. Today's fryers have heating elements located about 2 inches from the bottom of the kettle. They divide the kettle into two zones: the **cooking zone**, which is the hot oil above the heating elements, and the *cold zone*, which is the 2-inch space between the heating elements and the bottom of the kettle (see Illustration 12-2). This cold zone is critical to both the frying process and the life of the oil. This is where all the crumbs and debris fall during frying, preventing them from mixing with the oil and damaging the fried foods. Because it is cooler than the rest of the oil, the cold zone prevents these crumbs from cooking or burning, which would cause the cooking oil to deteriorate. More about preserving the life of your oil later in this chapter. The only type of fryer that does not have a cold zone is the *flat-bottom fryer*, which is used for food items that float on top of the oil as they fry, such as doughnuts and fish.

An important term you'll hear in fryer comparisons is **heat recovery time**. This is the time it takes for the oil to return to optimum cooking temperature after cold (or frozen) food is dropped in to be cooked. For the most part, manufacturers have shaved heat recovery time to 2 minutes or less in commercial fryers. Why? If the food is placed in oil heated to less than 325°F for more than 2 minutes, it begins to absorb the oil. Instead of frying crisply, it becomes soggy and greasy and takes on the taste or smell of any other food the oil may have come into contact with. A speedy heat recovery time prevents this.

A fryer overloaded with thawing french fries has to work harder, cook longer, and compromise oil quality as the moisture on the fries drips into the hot oil. Water in oil also causes dangerous spattering.

Most fryers have control panels at the front of the unit, with an on/off switch and a melt/fry switch (the latter to distinguish melting oil from frying product). The melt cycle is used only to liquefy a solid block of new shortening placed in the kettle. This is not supposed to be done when the fryer is already at its optimum cooking temperature; the resulting **temperature shock** can scorch the shortening and overwork the heating elements of the fryer.

A red light signals that the fryer is on. On manually controlled fryers, a dial can be set to the desired oil temperature, but many are now computerized and can be preset for cooking different products. Most modern fryers have two thermostats. The **cycling thermostat** regulates frying temperatures up to 400°F. The **high-limit thermostat** is another thermostat installed as a safety feature to detect overheating. It turns the fryer off automatically if the oil temperature reaches 435°F. Some electric models, if attached to a vent hood, also shut off automatically when the hood's fire extinguishing system is activated.

When automatic basket lifts are used, additional push-button timers can be adjusted from 0 to 15 minutes' frying time and automatically reset. Each basket is governed by a separate timer. Other innovations include electronic sensors, which measure oil absorption, heat, and doneness, signaling when the food is ready.

GAS FRYERS

Gas-powered fryers have made major quality strides in recent years. Today they require less gas to operate, and their energy efficiency ratings have almost doubled from only a few years ago. Gas-fired fryers are heated in one of two ways: with an atmospheric burner (which mixes air with gas to ignite a flame) located under the frying kettle, or by injecting the gas flames through tubes located along the bottom or sides of the kettle. These tubes contain flame slots, or baffles, which aim the flames for maximum heat distribution and efficiency. Manufacturers have added larger tubes for faster heating and made improvements in the baffles that allow them to distribute heat more evenly, extract more heat from the energy source, and reduce wasted heat. Some gas models have infrared burners, which we'll discuss in greater detail in a moment. Others have instant-on electric igniters rather than gas pilot lights, or they use so-called pulse combustion to efficiently fire up when turned on.

Perhaps the most obvious difference between electric and gas fryers is the way they are cleaned. The kettles of most electric models are lifted out for cleaning, while the gas models are not. Also, gas fryers contain a **fryer screen** that separates the cold zone at the bottom of the kettle from the rest of the oil. The fryer baskets rest on the screen, above the burner.

The energy required for a gas-powered fryer ranges from 25,000 Btus (British thermal units) for a standard 11-inch-square kettle that holds 15 pounds of oil, all the way up to 260,000 Btus for the largest 210-pound kettle. Of course, gas models require a gas input line, which should be ¾ inch in diameter.

INFRARED AND INDUCTION FRYERS

One type of gas-powered fryer uses infrared burners to cook the food. Remember, infrared heat—like the sun's rays—transfers to objects only on direct contact. Because the heat is directed to the food itself, not to the surrounding area, it penetrates the food more quickly than other types of heat transfer (see Illustration 12-3). The result is faster cooking time with less energy use, and frying is no exception. Infrared fryers are 20% to 70% more energy efficient than their electric and standard gas-fired counterparts. It would take a conventional fryer 120,000 Btus of natural gas to do the job of an 80,000-Btu infrared fryer.

The infrared burner consists of ceramic plates or metal screens full of tiny holes (about 200 per square surface inch). A mixture of gas and air is forced through these holes by a blower or fan, burning at a surface temperature of 1600°F. The oil is heated so quickly that the frying process takes only a fraction of the time conventional fryers require, making infrared a good choice for high-volume operations. Infrared fryers have fast heat recovery times between batches and allow lower cooking temperatures, which means the oil lasts longer.

The latest electric model is an induction fryer, which cooks with electromagnetics; when the metal container comes into contact with the induction burner, heat is transferred immediately to the container (and thus, to the food). In this case, no heating elements are immersed in the frying medium, and frying can be accomplished at 600°F instead of the nearly 800°F required for conventional fryers. This results in a cooler cold zone and cooler flue temperatures—all making for a more comfortable kitchen.

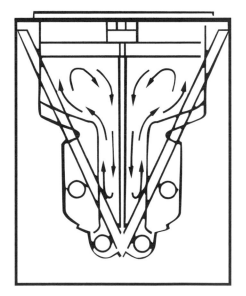

ILLUSTRATION 12-3 The cold zone of the infrared gas fryer is at the lower sides of its kettle. Circulation of the hot oil forces debris down and away from foods being fried.

ELECTRIC FRYERS

Electric fryers are usually tube-type fryers, with heating elements contained in stainless-steel tubes immersed directly in the oil. This makes them very energy efficient. The amount of the element that comes into contact with the oil is important; more surface area contact heats faster and minimizes heat recovery time. As the hot oil rises around the elements, it creates a rolling action that quickly heats the oil at the top part of the fryer, leaving a cool, quiet zone at the bottom of the kettle where sediment and food particles settle.

The elements swing up from near the bottom of the kettle for cleaning. Generally, you lift them out of the kettle, turn them on, and they'll burn themselves clean. If this isn't done regularly, burned food particles stick to the heating elements, compromising their effectiveness and using more energy.

From room temperature, it takes about 6 minutes for most electric fryers to preheat to 350°F using liquid oil. Solid shortening takes longer because it must first melt. The proper procedure is to set the thermostat to 250°F and pack the shortening tightly around the elements. The lower temperature permits the shortening to melt slowly and cover the elements without smoking. Many electric fryers have a melt setting that pulsates the temperature gently during the melting process. After melting, the thermostat can be readjusted to the desired frying temperature.

Electric fryers typically use less oil per unit of product than their gas counterparts. Their cold zone is not as deep, and the heat they generate throughout the kettle is uniform. Power requirements for electric fryers vary with their size and capacity, ranging from 5.7 to 36 kilowatts per hour and from 208 to 240 volts.

Several manufacturers have introduced electric induction fryers to the market. In these fryers, heat is created inside the tubes by induction coils. The metal tubes create a magnetic field inside that generates heat, even though the tube itself is not connected to any power source. An induction fryer can heat oil efficiently when its tubes are about 600°F, instead of the 750°F to 800°F needed for a conventional electric fryer. It's important to note the *oil* doesn't get that hot—the *tubes* get that hot, in order to transfer sufficient heat to the oil. The more moderate temperature prolongs the life of the frying oil, and the fryer itself gives off less heat. Induction fryers have large cold zones at the bottom of the kettle that are about 100°F lower in temperature than the cooking zone.

Electric induction fryers have many of the same advantages as infrared gas fryers. They are well suited to large-volume cooking, they extend oil life (by as much as 35%), they have quick heat recovery times, and their lower temperatures reduce oil spattering. No elements must be moved for cleaning. The most common induction kettle size is 15½ × 13¾ inches; it uses 14 kilowatts per hour to heat 50 pounds of frying oil. Most have internal filtration systems, with a dump station located to the right or left for handy disposal of crumbs and/or used oil.

VENTLESS FRYERS

When you can't make (or can't afford) expensive roof modifications, you can use self-contained fryers, often called **ventless fryers**. *Ventless* means the fryer has an internal air filtration system to capture grease and return cleaned air to the kitchen without venting it to the outside.

For mall food courts, sports arenas, sites with limited space, and in historic buildings where modifications are not permitted, ventless hood frying is perhaps the only choice. However, check first with local fire officials to be sure your building and fire codes permit their use.

These fryers began as countertop models, about 34 inches high, on 4-inch adjustable legs. The procedure is to place the food in a front hatch and push a button; the fryer automatically slides the food into a basket and sends it into the hot oil. When the preset cooking cycle is finished, the basket automatically tilts itself for 10 seconds to drain, then dumps the food into a chute, where it rolls into a catch pan located outside the unit. The self-contained fryer takes 6 minutes to heat 27 pounds of oil and can cook for up to 10 minutes per cycle. Much larger, floor-standing models are now available.

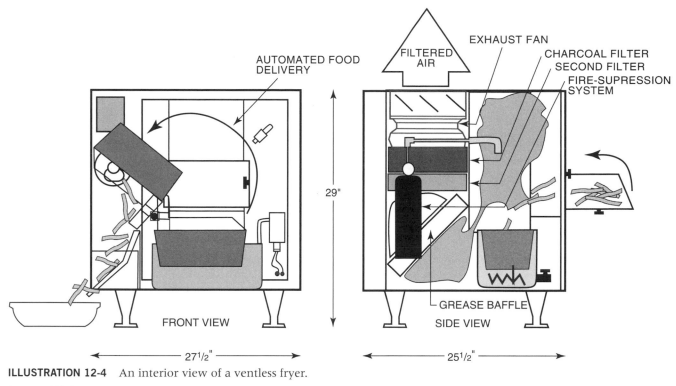

ILLUSTRATION 12-4 An interior view of a ventless fryer.

Courtesy of Motion Technology, Inc., Natick, Massachusetts.

One attractive feature of the self-contained fryer is its built-in systems: filtering, draining, and air purifying. The drain setup eliminates the danger of handling hot oil by draining the oil directly from the fryer into a sealed safety canister. And because the smoke and grease-laden moisture from the frying process pass through an air purifier, there is no need to vent this fryer under an exhaust hood. When the oil temperature reaches 413°F, or when the unit has operated for more than 16 consecutive minutes, it shuts itself off.

Another option is a standalone ventless hood unit with a built-in fire suppression system that can be purchased separately and used with a traditional fryer. See Illustration 12-4 for a closer look at the interior workings of a ventless fryer.

AUTOMATED FRYERS

Computer-controlled fryers can be programmed easily to turn out consistent product every time regardless of the size of the batch. They take a lot of judgment out of the frying process, which can mean better product consistency, less waste, and less time spent training employees to use them.

A computerized fryer can, for instance, fry 16 chicken quarters in 16 minutes at 325°F; a load of chicken tenders at 360°F is finished in 4 minutes; and full-capacity loads of french fries or small shrimp take only 90 seconds. Timers show how much longer the cooking cycle will last, and a bell or alarm signals the end of the cycle. The computerized fryer automatically raises and lowers baskets and shakes them for a preprogrammed time period to allow grease to drip off. Sensors inside the kettle instantly read the temperatures and, if overheating occurs, the unit can shut off, sound an alarm, flash a warning light—or all three! And, in a dual-kettle model, the computer keeps track of both cooking cycles at once. The advantages are obvious: Expert estimates indicate you can cut the labor costs associated with frying foods by up to 30% by automating the process as much as possible.

Cashing in on the computerization trend, a couple of manufacturers sell remote computer systems for regular frying kettles. These are basically electronic probes that can be installed inside the kettle, attached to the heating elements, or dropped into the kettle as the product cooks, and they keep track of oil temperature and signal doneness. They may work off the same power source as the fryer or plug into a standard wall outlet.

12-3 FRYER CAPACITY AND INSTALLATION

We've already mentioned that fryers come in many sizes and that they're often identified by the number of pounds of oil they hold. One way to gauge capacity is to know that for every 5 pounds of oil the kettle can hold, you can load and fry 1 pound of food. So, a fryer that holds 50 pounds of oil should be able to fry up to 10 pounds of food per batch.

Another way manufacturers rate a fryer is by how many pounds of french fries it can fry in 1 hour. The general rule is that a fryer produce product equal to 1½ to 2 times the weight of oil it holds. For example, a countertop fryer with an oil capacity of 15 pounds can produce about 25 pounds of french fries per hour. Remember, this rule does not necessarily apply to *all* fried foods, but it's helpful in determining basic fryer capacity.

Finally, consult the capacity-per-hour guidelines put out by the manufacturers, usually in chart form. They'll give you examples of product, frying temperature, and hourly output. Table 12-1 is an example.

Ask about the fryer's Btu input. The higher its Btu input, the better chance it will maintain an even temperature, and the faster its recovery time. For gas fryers, start with the basic premise that 600 Btus are required to fry 1 pound of french fries. Because most gas fryers operate at about 50% efficiency, you can estimate their production capacity by dividing the gas input (total number of Btus) by 1200 (600 Btu times 2). Example: A 60,000-Btu fryer divided by 1200 equals a 50-pound capacity of french fries.

The key is to buy a fryer with both the desired capacity and the shortest possible recovery time. Many professionals have found two smaller fryers do the job better than a single large one. Dual fryers give you the ability to cook two products at the same time or to turn off one fryer during slow times. Another note: Some foods (e.g., french fries) can be blanched prior to final frying, which decreases frying time and therefore increases overall fryer capacity.

Hot oil is a dangerous commodity not only because it can cause serious burns but also because it is highly flammable. The National Fire Protection Association requires that fryers be located at least 16 inches away from any piece of equipment that uses an open flame, such as broilers or rangetops. Unless it is oil-free, the fryer must also be located under a vent hood because it gives off grease-laden moisture (as well as gas fumes, in gas-powered models).

The sides and back of the fryer must be at least 6 inches from the walls to be ventilated properly, allowing an unobstructed flow of combustion air. If a gas fryer is installed in a spot where incoming airflow (to its burners and blower motor) is restricted, it will build up abnormally high temperatures and eventually short-circuit the electrical part of the motor—its controls.

Fryers must always be mounted on sturdy legs or heavy-gauge metal stands. Installing them flat restricts air circulation. Fryers should be correctly leveled from front to rear and from right to left. Before use, they must be calibrated.

For all of these reasons, fryer installation is not something most general electricians or plumbers are qualified to do. By all means, hire a professional installer. The manufacturer or equipment dealer can recommend one.

For both safety and convenience, it's smart to flank the fryer with counter space instead of other equipment. After all, you need a place to put incoming and outgoing food. This may not be possible because so much equipment must fit under the (usually limited) vent hood space, but it's worth considering. As an alternative, place the fryer at the end of the hot line and set a worktable or rolling cart beside it.

TABLE 12-1

Fryer Cooking Guide

FOOD	TEMPERATURE[1] (°F)	TIME[1] (MINUTES)[1]
POTATOES, FRENCH FRIES (³/₈-INCH CUT)		
Raw to done	350	6
Blanched, only	350	3
Browned, only	350	3
Commercially treated	350	6
Frozen, fat blanched	350	2
Potato chips	350	3–4
Potato puffs	360	1½
SEAFOOD		
Frozen breaded shrimp	350	4
Fresh breaded shrimp	350	3
Frozen fish fillets	350	4
Fresh fish fillets	350	3
Fresh breaded scallops	350	4
Breaded fried clams	350	1
Breaded fried oysters	350	5
Frozen fish sticks	350	4
CHICKEN		
Raw to done	325	12–15
Croquettes	350	3–4
Turnovers	350	5–7
Precooked, breaded	350	3–4
MISCELLANEOUS		
Breaded veal cutlets	350	3–4
Breaded onion rings	375	1½–2
Precooked broccoli	350	3
Precooked cauliflower	350	3
Precooked eggplant	360	3–4
Breaded tamale sticks	360	3
Fritters	375	4–5
French-toasted sandwiches	375	1
Yeast-raised doughnuts	375	1
Hand-cut cake doughnuts	375	1½
Doughnuts	375	2–3
Glazed cinnamon apples	300	3–5
Corn on the cob	300	3
Turnovers	375	4–5

[1]Allow for minor variations from these suggested times and temperatures according to the weight, texture, density, and other characteristics of the foods you use.

Source: Texas Utilities Electric Company, Dallas, Texas.

Also, remember to clean the fryer thoroughly before its first use. The manufacturer provides cleaning instructions, but it never hurts to break in a new kettle by boiling a solution of 1 part vinegar to 10 parts water in it; this removes the manufacturer's grease that is often used to shine and protect a new kettle before delivery. Drain and clean the kettle regularly. Always rinse with clear water until all vinegar odor is gone.

12-4 FRYING OIL CARE AND CONSERVATION

The flavor, aroma, and texture of fried food depend largely on the quality and condition of the frying oil (see Table 12-2). There are a number of ways foodservice professionals prolong the life of their oil, and there are plenty of reasons to do it. Other than the food itself, the oil to fry it in is your most expensive food-related cost. We've already mentioned that, in cooking, oil is affected greatly by temperature, moisture, food particles, salt, and more. In fact, as it is heated, oil changes its chemical properties even without food in it. So far, science is unable to provide an antioxidant or antifoaming agent that stops oil from naturally deteriorating with use. However, there are ways to minimize this deterioration.

First, the background about frying fats and oils. The oil used for frying is different from oil used for baking or making salad dressings. Butter and most animal fats are not suitable for frying because they contain high percentages of free fatty acids, which break down quickly. Frying fats and oils fall under the broad classification of *lipids*, which contain two other fat-related compounds: cholesterol and lecithin. The latter, in food, is an emulsifier, meaning it keeps something in suspension in a liquid. All of these products contain both "good" fats (polyunsaturated and monounsaturated), which prompt a human body to produce so-called good (HDL) cholesterol and remove so-called bad (LDL) cholesterol from the bloodstream to the liver for reuse or excretion; and "bad" fats (saturated), which can cause a buildup of LDL cholesterol in the bloodstream. The problem with *trans fat* is that it contributes to higher LDL levels.

How are trans fats created? The visible difference between fat and oil is that fats are solid at room temperature, while oils are liquid. In the past, semisolid fats were used for frying. They had such high melting points that some restaurants found it easier to melt them on the rangetop first, then pour them into the frying kettle for use. Technology solved that problem with the development of liquid frying fats with longer life expectancies and minimum decomposition, although the process of processing, so to speak, is a downright unappetizing series of steps that includes degumming, bleaching, deodorizing, and **hydrogenating**. The latter means the oil is chemically combined with hydrogen molecules, which increases its stability—and also creates trans fatty acids. The resulting product is referred to as *partially hydrogenated oil*. It has been widely used to increase the flavor and shelf life of many snack foods—until, in recent years, it landed at the center of the TFA controversy.

There are other ways to process cooking oil, creating *monounsaturated* or *polyunsaturated* products without the trans fats. However, they contain saturated fats instead—and, although these are not considered as harmful as trans fats, too much of either is a health risk associated with heart disease, stroke, diabetes, and more. It is perhaps ironic that Americans eat about five times more saturated fat than they do trans fat.

Food scientists are working hard to increase safer, more healthful cooking oil options. At least one alternative is made from a specially bred soybean. In 2 June 2005 article in *Food Processing*, Kantha Shelke, ingredients editor, explained some of the latest technology and techniques, which are summarized briefly below:

Blending fully hydrogenated fats with liquid oils. Examples include palm oil–based Sans Trans and a mixture of tallow and corn oil marketed under the brand name Nextra.

Enhanced stability through breeding. The Monsanto Company developed a low-linoleic soybean and made the seeds available to farmers starting in 2005. Among the first oils made from low-linoleic soybeans is Asoyia. Dow's entry into the canola oil market is called Natreon; Trisun is a sunflower seed–based oil.

Molecularly rearranged solutions. A process with the intimidating name of *enzymatic interestification* rearranges the fatty acids in the oil molecules. Food scientists say it is more environmentally friendly than traditional oil processing and also a cost-effective way to produce oils that are lower in fat.

Trans fat functionality from other ingredients. Some companies are trying to replace fats rather than manipulate their molecules for more healthful results. For example, Z-Trim is a corn-based product; another fat replacement that has worked well in baked products is a rice-based syrup developed by California Natural Products.

TABLE 12-2

Characteristics of Common Cooking Oils

OIL	USES	SMOKE POINT	FLAVOR CHARACTERISTICS	HEALTH/NUTRITION CHARACTERISTICS
Canola	Deep-frying, panfrying, sautéing, baking	High: 400°F	Mild flavor	Lowest in saturated fat of all oils; helps lower cholesterol levels
Corn	Deep-frying, panfrying	High: 450°F	Light taste; can be used in place of olive oil	Believed to reduce bad cholesterol in the arteries
Grape Seed	Deep-frying, panfrying, sautéing	High: 400°F	Light taste; can be used in place of olive oil	Believed to reduce bad cholesterol in the arteries
Nextra	Deep-frying (primary), panfrying, sautéing, grilling	High: 450°F	Tallow, meaty flavor; generally a flavor enhancer	No cholesterol and no trans fat; decreases LDL
Olive	Sautéing, stir-frying	Low to Med: Unrefined: 320°F; Extra Virgin: 406°F; Virgin: 420°F	Bland to very strong, depending on type	A monosaturated oil; the green/golden variety has more antioxidants
Peanut (refined)	Stir-frying, deep-frying, wok cooking, sautéing, grilling	High: about 450°F	Can add a rich, nutty taste, but does not absorb or transfer flavors	Contains resveratrol, which is associated with reduced cardiovascular risk and reduced cancer risk
Safflower (refined)	Deep-frying, panfrying, sautéing, baking	High: 450°F	Bland, flavorless	High in polyunsaturated fats; helps reduce total cholesterol and LDL
Sesame (refined)	Wok cooking, dressings, flavoring	Medium: 410°F	Pungent; used to flavor many Asian dishes	High in vitamin E (antioxidant), helps lower cholesterol
Sunflower (refined)	Deep-frying, panfrying, sautéing	High: 450°F	Generally bland	High in polyunsaturated fats; helps reduce total cholesterol and LDL
Vegetable (typically refined soy oil)	Deep-frying, panfrying, sautéing, baking	High: 450°F	Generally mild flavor	Soy oil is high in polyunsaturated fats; helps reduce total cholesterol and LDL; partially hydrogenated vegetable oils (PHVO) contain trans fats, which may increase risk of heart disease

Source: "Finding the Right Chemistry," *Restaurant Startup & Growth* (June 2005).

Advanced nutritional solutions. Some companies are trying other ingredients to make oils last longer without hydrogenation. These include the addition of omega-3 fatty acids found naturally in fish oil and antioxidants found naturally in palm oil.

All of these companies have websites and are doing good jobs touting their research. And, with about 8 billion gallons of frying oil in use per year in the United States alone, there is plenty of incentive for them to continue.

OIL QUALITY FACTORS

The best frying oil has enough stability to withstand the rigors of high frying temperatures, which is called *having a high* **smoke point**. If the oil produces thin, bluish-white smoke as it heats, you know it is overheating. Soon after, you'll notice a sharp, pungent odor reminiscent of a carnival midway (where food aplenty is being badly fried!). Turn down the kettle temperature immediately.

Another sign of oil deterioration is foaming, which happens as the fat breaks down into its oily components, becomes thick or gummy, and takes on a rancid odor. This sometimes occurs when the heating elements are not completely covered by the oil, causing them to overheat and smoke. The result is a layer of black carbon on the heating elements, which makes them less efficient. Electric fryers and their movable, swing-up elements are particularly prone to this; be sure they are not exposed to the air during frying.

When the kettle is not in use, one good oil-saving technique is to turn it down to a lower setting, between 190°F and 275°F, and then cover it. Certain manufacturers have added an automatic turn-down feature as an option. A kettle not in use should also be covered, but only if you first let the oil cool to its lower temperature. Covering a hot kettle accelerates the breakdown of the oil because moisture condenses on the hot lid.

Another way to increase the usefulness of your frying oil is to control the rate of **fat turnover**. This is the amount of fat used each day compared to the total capacity of the kettle. Experts recommend a total daily turnover of oil—that is, if the kettle's capacity is 30 pounds, a total of 30 pounds of fresh oil should be added in small amounts throughout the day. This won't cause the kettle to overflow because, as food is fried, it retains from 10% to 40% of the frying oil. The small amounts of new oil, added to maintain the kettle at its fill line, keep the overall batch of oil fresh. Another similar rule of thumb is that, at all times, fresh oil should account for one-third of the total oil in the kettle. One important caveat: Don't add fresh oil to oil that is already brown or foamy. You're just wasting it.

Fat turnover is determined in large part by how much oil is absorbed into the products you're frying. This is known as **oil take-up**. The oil take-up of a french fry, for instance, is 20% to 40%, which means the finished fry has that amount of frying fat in it. Undesirable oil take-up rates occur if the oil temperature is too low or if the food is allowed to stay in the oil too long.

FILTRATION

Research done by the PG&E Food Service Technology Center in San Ramon, California, has shown no difference in oil longevity used in gas or electric fryers, but the same tests showed that oil with a typical useful life of 10 days can extend that time to 15 days or more if it is filtered properly. In fact, the single most important way to protect your oil is to filter it regularly to remove crumbs. A little debris is a by-product of any fried food, and your job is to strain the debris out of the oil so it doesn't stay suspended in the oil and burn. Removable crumb trays and *skimmers*, to skim floating particles off the oil, are helpful. But nothing works quite as well as filtering. Most fryers today come with automatic filtering systems. However, for small kettles, you'll probably have to filter by hand.

What you'll need is a stockpot with a *larger* capacity than the fryer kettle, a metal filter holder, and a filter that fits the holder. Filters can be made of paper, cloth, or fine-mesh wire screen; the disposable paper ones are the handiest, but filter papers are messy. Some say the wire screen is the most eco-friendly, but it is also the most expensive. It is *vital* that the oil temperature is *below* 100°F before filtering, or you risk getting burned.

Most electric kettles can be lifted out and their contents poured through the filter into the stockpot. Gas fryers usually have a piece of pipe screwed into the bottom of the fryer that can be loosened, causing the oil to drain out through the filter and into the stockpot. This is the case with most countertop and drop-in and some of the smaller floor models. After filtering, the oil must be returned to the kettle. Some systems require a hose and nozzle, which is aimed manually into the kettle as oil is pumped back in. Others pump the oil directly back into the kettle through a hole in the kettle wall or a ring that encircles the kettle.

There are two types of automatic filters: The built-in system is stored beneath or beside the fryer; the portable system sits in its own cabinet and can be rolled and attached to fryers anywhere in the kitchen.

PORTABLE FILTRATION SYSTEM. For large floor-model fryers, a portable filter system is recommended. It is typically a four-wheeled cart, mounted on rolling casters, which can be stored underneath the fryer kettle. An electric pump pushes the oil through a filter at the rate of 5 to 7 gallons per minute. The pump has a ⅓-horsepower motor and operates when plugged into a standard 120-volt electrical outlet. Portable systems are also handy because they allow you to wheel the used oil away for disposal (see Illustration 12-5). Most have a safety feature: The filter will not turn on unless the fryer is turned off. This is critical. A fryer can be ruined if it is turned on when the vat is empty.

You probably will pay as much for a portable filtration system as you do for the fryer itself, but if you use a single filter for several fryers, it's money well spent. Before you buy, find out:

- ■ How efficient is the system at sucking all the oil from the filter pan? Some pans are constructed such that it is impossible to extract all the oil.
- ■ Can solid shortening be used with this system? Typically, the answer is no, because it can solidify and clog both filter and lines.
- ■ How is the filtered oil returned to the kettle? Each manufacturer seems to have a different type of line for accomplishing this.

BUILT-IN FILTRATION SYSTEM. This is contained in the cabinet below the fryer. A built-in system automatically transfers the oil from the kettle to a tank below the kettle, where filtering occurs. Fryers with built-in filtering systems are more expensive, but, where space is at a premium, they are especially handy. A single built-in system can also be set up to filter the oil from several side-by-side fryers (see Illustration 12-6).

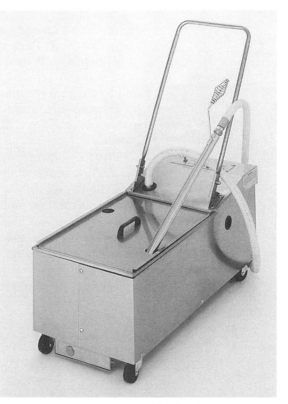

ILLUSTRATION 12-5 A portable filtering system can roll on a cart to serve more than one fryer.

Courtesy of Frymaster Corporation, A Welbilt Company, Shreveport, Louisiana.

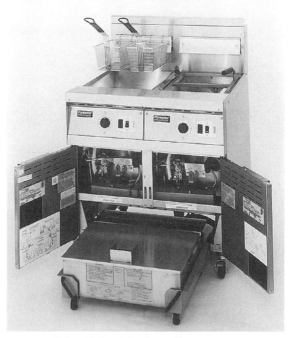

ILLUSTRATION 12-6 A built-in filtering system is contained in the cabinet below the fryer.

Courtesy of Frymaster Corporation, A Welbilt Company, Shreveport, Louisiana.

The built-in filter system automatically shuts off the fryer and opens the kettle drain, then turns on the pump and opens the return line (so the filtered oil can be returned to the kettle), in a sequence of steps. These steps require a person to pull a couple of handles on the machine and perhaps to manually turn on the pump.

Most manufacturers offer a product called *fry powder* to condition frying oil. They claim that adding the recommended amount of powder to the oil during the filtering process has an antioxidant effect, absorbing particulates and improving its life between 25% and 40%. But fry powder is controversial and probably not needed if sufficient care and cleaning routines are followed.

Filter manufacturers continue to make safety strides with the goal of keeping hot oil away from kitchen workers. No matter what type of system you select, the filtration schedule is critical. Frying oil should be filtered a minimum of once a day; where consumption is high or the menu includes a lot of breaded items, filter after each 8-hour work shift. Use the filtering time to continue maintenance of the fryer and filter, keeping them free of residue. The filtration system should also be checked periodically to make sure all the lines are clean and snugly connected, with O-rings properly seated and not worn out; and that elements, temperature controls, and the like are clean and secure. These types of maintenance, which should be outlined in your operator's manual from the manufacturer, save from $60 to $500 in service call costs per visit.

OIL DISPOSAL

A final consideration is to dispense with filtration altogether, and let an outside contractor handle it. Mahoney Environmental of Joliet, Illinois, sells a system to pump used oil through pipes to a remote tank where it can be hauled away regularly.

A national company called Restaurant Technologies, Inc. (headquartered in Eagan, Minnesota), offers an interesting agreement: Buy oil from them, and they install two tanks at your site, one for fresh oil and one for waste oil. Both tanks hook up to a port on an exterior wall, where tanker trucks regularly tap in, fill up the fresh oil tank, and drain the waste oil tank. The company says a large chain customer may pay 10% to 15% more overall for its "cooking oil management system" but save in other ways with its convenience.

12-5 PRESSURE FRYERS

Pressure fryers provide some of the most versatile and profitable cooking applications possible in small spaces. Today's pressure fryer can turn out 14 pounds of chicken in less than 10 minutes.

This specialty appliance cooks food with a combination of hot oil and steam. The steam may be generated from moisture in the food, or it may be added during the cooking process. Manufacturers claim their pressure fryers cook food more quickly than regular fryers and that the food is tenderized by the use of steam. Because we know moisture breaks down cooking oil, the system is designed to prevent condensation on the lid from falling back into the oil.

The pressure fryer is basically a fryer with a tightly sealed lid, usually a rubber gasket (see Illustration 12-7). The kettle is filled with cooking oil. When food is added to the hot oil, the lid is shut and the steam that naturally builds up inside raises the pressure in the kettle. The pressure causes the food inside the kettle to tumble and roll, like clothing in a washing machine. The frying time is shorter, and the oil temperature can be lower, which means less oil is absorbed into the food. This frying method saves energy and cooking oil in addition to being fast, making pressure frying a popular choice for quick-service restaurants. When done correctly, the fried food emerges crisp, golden, and moist.

Pressure fryers can be powered by gas or electricity. Gas models, depending on capacity, require between 40,000 and 80,000 Btus per hour. Electric models require from 120 to 208/240 volts and operate (again, depending on capacity) on 5.2 to 13 kilowatts per hour.

Like other types of fryers, pressure fryers come in countertop and floor models and should stand on legs or casters to facilitate ventilation. The smallest countertop models hold 10 pounds of oil and produce 6 pounds of food per 9-minute frying cycle; the largest floor models hold 43 pounds of oil and fry 14 pounds of food per frying cycle. Control panels and thermostats are similar to those on regular fryers, except that pressure fryers also have a pressure gauge. When shopping for this appliance, look for programmable, computerized models that adjust themselves for the load size, can be used with or without the steam pressure function, and have a built-in oil filtering system.

A cousin to the pressure fryer is the **controlled evaporation cooker**, also known as a *CVAP* or *Collectromatic Fryer System*. The CVAP technology involves holding food at peak quality in a combination of heat and steam. A heater and thermostat control the amount and temperature of the steam. CVAP fryers can be used with their lids open or closed and have built-in continuous filtration capability.

ILLUSTRATION 12-7   A pressure fryer.

Courtesy of Henny Penny Corporation, Eaton, Ohio.

FOODSERVICE EQUIPMENT

Hints for Successful Frying

1. Do not bring the fryer to full frying temperature without having food in the kettle.
2. Never salt food over the kettle.
3. Never fry bacon in the kettle.
4. Never load the fryer basket more than two-thirds full. Overloading results in soggy food that is not uniformly cooked.
5. Bite-size foods cook more uniformly and don't stick together if the basket is lifted and shaken now and then during the frying process.
6. Try to fry pieces of food that are similar in size so they'll all be done at about the same time.
7. Soap residue is an enemy of the frying oil. Use *only* the cleaning solutions recommended by the fryer manufacturer.
8. Clean the kettle by boiling water in it—not once but twice. After the second boil, rinse the kettle and wipe it out thoroughly.
9. Keep the exhaust hoods above the fryer clean. The last thing you want is grease from the vent hood dripping into your kettle.

ILLUSTRATION 12-8 The fry station contains everything needed to prepare and hold fried foods before serving.

Courtesy of Frymaster Corporation, A Welbilt Company, Shreveport, Louisiana.

ILLUSTRATION 12-9 Fry holding stations offer the ability to keep fried foods at optimum temperatures, ready to serve, without cooking or drying them out.

Courtesy of Hatco Corporation, Milwaukee, Wisconsin.

12-6 THE FRY STATION

When large volumes of food are being produced, the frying process requires an assembly-line setup. The complexity of the operation dictates the need for a *fry station* and whatever storage it requires (see Illustration 12-8).

Fry stations include the fryers themselves, landing space for food as it moves into and out of kettles, food warmers, basket racks, a filtering system, and a product storage area. In the typical fry station, you'll see:

Food warmers. These are fixtures that hold 250-watt infrared bulbs with heavy-duty chrome shades, in sizes ranging from two to eight bulbs. Many styles are available, including warmers that can be mounted on walls, counters, shelves, or their own rolling carts. A two-bulb unit installed under a shelf can warm an area the size of a standard hotel pan—about 12 × 20 inches. Most food warmers plug into standard electrical outlets (see Illustration 12-9).

Landing space. Counter space is absolutely necessary to a successful fry station, where the food can land as it goes in and out of the fryer. Collapsible shelving, either beside or in front of the fryer, is preferable. Rolling carts or tables, the same height as the fryer cabinet, are the next best option.

Filter system. We've already discussed the importance of filtration in the frying process. In a fry station where space is tight, the top of the filtering unit is often designed to work as counter space or as a place to mount food warmers. Portable filtering systems can be used where space is at a premium, so they can be stored elsewhere.

Dump pans. As their name suggests, *dump pans* are used to dump food as it moves into or out of the fryer. They are made of stainless steel. Dump pans for finished food are usually perforated to allow grease drainage.

Bagging area. This is where employees stand to put the finished product (usually french fries) into bags or other containers for sale. Doing this directly in front of the fryer only hinders the frying process. Think about the flow concept, and you'll see that the critical decisions in laying out a fry station are where to put pans and where to bag product. This is also the area where you can get creative with such popular items as french fries. Today, it's not uncommon to sprinkle them with herbs or spritz them with malt vinegar—and it all happens in the bagging area.

Basket rack. This metal rack is a necessary storage tool for busy fry stations. Full baskets of food can hang suspended on it, ready to be submerged into the kettle. Basket racks come in different sizes; the largest can hold up to 32 baskets.

Operations that require high-volume fried food output also should plan to save energy by using their fry station capacity wisely. Studies show that fryers actually are in use only about one-quarter of the time they are on, which wastes hundreds of dollars a year in energy as well as compromising oil quality. As most of them take less than 15 minutes to preheat, a start-up and shutdown schedule for fryers should be part of your kitchen logistics.

12-7 BUYING AND MAINTAINING FRYERS

Whatever type of fryer you select, examine it for five crucial characteristics over and above whether it fits your budget:

1. *Capacity.* Can it meet the volume you require for the types of foods you intend to fry, based on your peak demand amounts?
2. *Simplicity.* Is it easy to use and clean? Can you train new employees to use it (or, if it's computerized, to program it) correctly and safely? Does it have a full, clear set of instructions? Nowadays, most operators choose fryers with built-in oil filtration systems just because they are easier to clean.
3. *Reliability.* Does it require minimal maintenance? Is a local service person available for repairs or questions after the initial installation?
4. *Energy efficiency.* Look for the federal Energy Star rating as well as information about how well the cabinet is insulated. The most energy-efficient fryers have shorter recovery times and higher production rates.
5. *Space allocation.* Does it fit in the space you have in mind? Will it complement or conflict with other kitchen operations? If it must be vented, can the exhaust hood handle the additional load?

FOODSERVICE EQUIPMENT

Fryer Life Cycle Factors

1. Purchase price, tax, freight, installation, start-up
2. Energy
3. Hours of use per day, calculating percentage of use at full load and at idle
4. Average energy use per hour (Btus or kW)
5. Labor (e.g., dedicated employee versus using basket lifts)
6. Preventive maintenance (costs for labor and supplies to clean, filter, oil, etc.)
7. Shortening usage and disposal
8. Service repair costs
9. Higher component repair/replace rates for some types of units
10. Faster aging with high-volume usage
11. Level of staff training may affect how much abuse equipment receives
12. Additional energy costs (added HVAC load)
13. Annualized costs: all expenses amortized across projected life expectancy

Source: Foodservice Equipment Reports, a Gill Ashton publication (Skokie, Illinois: May 2003).

Additional considerations accrue primarily because fryer manufacturers now offer so many options:

- *Atmospheric or infrared burners.* Gas-powered fryers with traditional atmospheric burners (which mix gas with air from the surrounding atmosphere) are simpler overall than infrared burners, which also contain a blower or fan to force air through the system.

- *Nonbreaded or breaded products.* For nonbreaded products, a traditional kettle—wide at the top, with a narrow cold zone at the bottom—is preferable. For breaded products, an open fryer and wider cold zone at the bottom works well. Only a fryer with tubes (either gas or electric) works for products like fried chicken, in part because it is impossible to heat such a large amount of oil except with tubes immersed in the oil, and in part because the cold zone must be located below the heating tubes to catch debris.

- *Solid-state or computerized controls.* If you fry only a few items that work well in a similar temperature range, you'll do just fine with a fryer with basic, solid-state controls that can be adjusted by hand and a timer and beeper to alert you when the frying cycle is done. However, if you fry many types of products that require a variety of times, temperatures, and cooking cycles, computerized controls are preferable.

- *Gas pilot or electric ignition.* The electric ignition is preferred today, as it eliminates the need for a pilot light.

- *Automatic or manual basket lifts.* This requirement seems to depend on the menu type and the experience level of the staff. Quick-service restaurants with limited menu offerings and workers who specifically staff the fryers don't seem to mind the manual lifting of baskets in and out of the oil. Operations with more varied menus and multiple uses for the same fryers find auto-lift models a good solution.

When fryers must be replaced, it is most often because of a leak in the vat. You'll see congealed oil on the floor beneath the unit or somewhere else it's not supposed to be. Carbon buildup from the heating process can cause leaks. They also can be the result of grease buildup from lack of maintenance, or of rust, if the fryer cabinet has come into repeated contact with water. Leaks are not minor problems; they can be fire hazards, and they are expensive to fix.

Some parts of a fryer, such as the *cycling thermostat*, are simple and inexpensive to replace. You'll notice a thermostat needs attention when the fryer's *high-limit sensor* seems to be triggered too often.

Frymaster, a major fryer manufacturer, summarizes the necessary maintenance processes in three simple steps that it calls the "ABCs of Fryer Preventive Maintenance":

1. *Attend to filtration.* Frymaster recommends filtering twice daily.
2. *Boilout.* Draining and disposal of old oil and boiling a cleaning solution in the fryer is recommended every 3 to 6 months.
3. *Clean the blower.* For high-efficiency gas fryers only, Frymaster recommends removing the blower motor and housing from the fryer every 6 months and spraying the housing and blower wheel (*not* the motor) with a degreaser.

The company also suggests, at least once a year, checking wiring and connections, the calibration of the thermostat, and the condition of the filter pan as well as testing the recovery time of the fryer.

12-8 FRYERS OF THE FUTURE

As we've mentioned, frying is big business in most restaurants, so new types of fryers are always on the drawing board. Here we discuss a few of the latest options to consider when conventional fryers simply don't work in a particular space.

OIL-FREE (GREASELESS) FRYERS

And you thought this was a contradiction in terms! Known by a variety of names—including hot-air fryer, greaseless fryer, vacuum fryer, no-oil fryer, and forced-air food system—this type of machine is well suited for small jobs where speed is important in the frying process. It is in greater use in Japan than in the United States, but snack food manufacturers are taking a particular interest in it.

The greaseless fryer is often a countertop model, but higher-volume units are available up to 3 feet tall. Food is loaded into some type of basket and inserted into a heated chamber, where it is blasted with hot, fast-moving air; the result is a product that resembles fried food, without the frying. There are two general types of oil-free fryers: ones that use infrared or radiant heat to speed the cooking process and ones that flash-heat the oil found naturally in the food. Either way, the process results in a crispy, just-fried flavor. Frozen foods can be placed in these fryers, and a microprocessor gauges how large the load is and how frozen it is, then heats accordingly.

As the hot air turns the frozen moisture into steam, the steam is exhausted away. However, because no oily vapors are released, the machine does not need to fit under an exhaust hood. Instead, it contains a built-in grease trap and air-filtering system that must be cleaned regularly. A greaseless fryer might be appropriate for a low-volume operation, such as a convenience store, where foodservice is not the major thrust of the business. In delis and bistros, bars with appetizer menus, mall and airport quick-service eateries, it can be used to prepare onion rings and french fries, pizza, battered fish, some types of sandwiches, steamed vegetables—all oil-free.

Another type of greaseless fryer uses electricity to steam frozen or blanched french fries and quickly air-dry them. The countertop model cooks about 3½ pounds of fries in 4 minutes.

Current greaseless fryers on the market need from 2100 to 6000 kilowatts of power to operate. They run on 110- or 220-volt electricity and can heat up to 550°F, with cooking times of up to 6 minutes.

The obvious advantage of no-oil frying is that you don't have to continually purchase and store oil or shortening. These fryers are easy to use and clean, and they eliminate costly hood and venting requirements. They may also lower insurance costs. An interesting disadvantage is that many of these fryers look like microwave ovens—your employees must remember not to reach into the oven cavity, or use plastic in it. Most no-oil fryers cook smaller-volume batches of food, not sufficient for a busy operation. Finally, the use of cooking oil imparts a distinctive flavor to foods, part of the reason people love them so much! Air-fried foods do taste slightly different from their oil-fried counterparts.

SPIN-FRY TECHNOLOGY

Pitco is now testing a new fryer with a fry basket that goes through a spin cycle at the end of the cooking cycle. The basket automatically rises out of the oil and spins for a preset amount of time to eliminate as much oil from the product as possible. This new technology boasts less absorption of cooking oil into the product; the product holds longer as a result.

This technology holds promise for operators to reduce oil usage, minimize oil absorption, and offer a crisper product that remains fresh and ready to sell for a longer period.

PASTA COOKERS

The nearest cousin to the fryer is the pasta cooker, growing in popularity for small-batch cooking. It's a kettle seated in a cabinet that is similar in construction to the fryer kettle. It comes in both electric and gas-powered models and is designed to cook, rinse, chill, and reheat—a system tailor-made for perfect pasta. The two-tank cooker looks like a two-kettle fryer. On one side, pasta is cooked in bulk in stainless-steel baskets; on the other, it is reheated in convenient portion-size plastic baskets (see Illustration 12-10).

Using a pasta cooker has two advantages. One is speed; gas models with more than 150,000 Btus of heat input bring a full tank of water to boil within 10 minutes. The other advantage is having an automatic timer, which ensures the pasta is not overcooked and no one has to stand there and test it for just the right *al dente* doneness.

The pasta cooker's capacity is rated by how much water it can hold. It takes 1 gallon of water to cook 1 pound of dried pasta, and the pasta will absorb 1½ times its weight in water. This means the smallest gas-powered cooker, a 6-gallon pot that holds 6 pounds of dry pasta, yields 15 pounds of cooked pasta. Gas-powered kettles range in size from 14 to 24 inches. The largest holds about 17 gallons of water. Electric pasta cooker capacities are slightly larger, ranging from 7 to 19 gallons of water.

Manual models require someone to maintain the proper water level and raise and lower the cooking baskets into the water. With automatic models, a push of a button begins the preset cooking cycle. Sensors control the water level in the kettle, filling it automatically and ensuring the heating elements don't turn on until they are fully submerged. This prevents accidental damage to the kettle from elements that might overheat and scorch if the water level is too low. Automated arms lower and raise the baskets full of pasta, which still must be moved and rinsed by hand after cooking.

Because cooking pasta produces excess starch, a pasta cooker also has an overflow valve and drain mechanism at the front of the unit, allowing starch and foam to be drained off regularly. The largest floor units may have an optional pasta-rinsing sink beside the cooking and reheating kettles. The baskets that fit into the pasta cooker can be full-size or half-size (sometimes called a **split basket**). For individual orders of different types of pasta, a carousel is a convenient option. It holds up to nine plastic perforated cups that can be cooked simultaneously.

A sauce warmer is another option on some models, capable of warming a standard-size steam table pan.

Thermostats on the front control panel gauge water temperatures from 60°F to 250°F. A series of indicator lights alert you when the kettle is being filled or heated, or simply when the power is on.

Pasta cookers have basic plumbing requirements because they must be hooked up to a water source and have the capability to be filled and drained automatically. A 1-inch (or 1¼-inch) drainpipe is located at the front of the cabinet and extends 6 inches up from the floor. Your local plumbing code specifies whether this pipe should be open and whether it should contain a grease trap.

ILLUSTRATION 12-10 A pasta cooker.

Courtesy of Vulcan-Hart, a division of ITW Food Equipment Group, LLC, Louisville, Kentucky.

■ SUMMARY

Frying is the process of cooking food in hot fat or oil. It creates delicious results, but it has received quite a bit of negative publicity as a result of the harmful components of frying oils that are imparted to the fried foods. Luckily, that has prompted a new range of innovations, including greaseless frying options that use very hot, fast-moving air to cook the natural oil already found in some foods. Another promising technique is a fryer with a spin cycle to remove excess oil from foods after frying.

Done properly, frying can produce food that is as tasty as it is attractive. In today's health-conscious climate, this means paying special attention to the use of frying oils that do not impart trans fatty acids (TFAs) to food. Trans fats have been found to clog arteries and lead to a variety of serious health problems. For the same reason, it also is important to reduce the amount of saturated fat in foods. Correct frying times and temperatures can help operators achieve this.

Traditional fryers are made up of several basic parts: the kettle (also called a *bin* or *vat*), where the oil is placed and heated; the basket, where food is placed to be lowered into the kettle; and the cabinet, which houses the whole process, with room for the kettle and equipment to clean and filter the system. The kettle contains an all-important cold zone located below the heating elements. In *open-pot fryers*, it is a deep well that is colder than the rest of the kettle, where crumbs and sediment can fall; in *tube-type fryers*, where heating elements are located in tubes near the bottom of the kettle, it is the couple of inches between the tubes and the kettle floor.

To prolong the useful life of your frying oil, it is absolutely necessary to filter and change it regularly. Frying oil can break down when exposed to water, sediment, salt, and oxygen, all of which are common in the frying process. It can also burn if heated too much, especially if it contains debris that hasn't been filtered out. Signs of oil deterioration include foaming, smoking, and a pungent odor. Some fryers are drained and cleaned by hand; others contain their own internal filtering system. Portable filtration systems also can be purchased and hooked up to the fryers as needed.

When shopping for a fryer, ask about its heat recovery time. This is the time it takes for hot oil to return to its optimum cooking temperature after you drop in a batch of cold food. The heat recovery time should be no more than 2 minutes to prevent the food from absorbing too much oil and becoming soggy.

Fryers are powered with electricity or gas. Gas infrared fryers work more quickly than traditional fryers, and pressure fryers use a combination of heat and low-temperature steam to cook food. A fryer is chosen based on its output capacity, usually listed in pounds per hour by manufacturers.

Many foodservice establishments serve enough french fries to warrant the setup of a fry station, an area that includes one or more fryers and filtering systems, landing space for the food, warmers, racks to hang fryer baskets, a bagging area just before serving, and storage space for products.

Fryers are a specialized type of equipment and should always be installed by a professional who is familiar with them. Traditional fryers turn out enough heat that they must be installed under hood ventilation; greaseless fryers are the only options in spaces where hood ventilation is either impractical or impossible.

A cousin to the fryer is the pasta cooker. Like a fryer, it contains a kettle seated in a cabinet, and it is designed to cook pasta, then reheat it as needed for individual portions. Because the kettles hold water, they require a plumbing source in addition to gas or electricity.

■ STUDY QUESTIONS

1. What factors contribute to the breakdown of frying oil over time? Can any of them be prevented to lengthen the life of the oil?

2. What can a foodservice business do make its fried foods as healthful as possible?

3. In determining a fryer's output, which is more important: how much oil it holds at one time, or how much product it fries in a certain period?

4. What kinds of precautions are taken to make sure a fryer is safely installed?

5. Why would an automatic turndown feature be beneficial on a fryer?

6. What is a pressure fryer, and why would you use one?

7. In addition to the fryer itself, what else do you find in a fry station? List at least four of the six components.

8. What are the signs that your frying oil is deteriorating and should be changed?

9. How does a pasta cooker work?

10. What are the four things you should find out about a fryer once you have decided it's in your price range?

Jim Petersen

FOODSERVICE FACILITIES DESIGNER, C.I.I. FOOD
SERVICE DESIGN

Jim Petersen lives in Lapeer, Michigan, but his work as a kitchen designer has taken him to such far-flung spots as Saudi Arabia and the island of Diego Garcia in the Indian Ocean. He graduated from the Cornell School of Hotel Administration—as did his father, his uncle, and a cousin—and has been a kitchen designer for more than 35 years.

Jim is a regular lecturer at colleges and universities and has also spoken to the American Society for Hospital Engineering and the American Correctional Food Service Association. He is a member of the Foodservice Consultants Society International (FCSI), the Michigan Society of Architects, and the American Correctional Association.

Q: Why go to the Cornell School of Hotel Administration if you were going to be a designer?
A: You don't really declare majors there, but because I was interested in design, I selected my courses based on that interest. I knew I didn't want to be married to the type of work schedule that comes with operating hospitality facilities.

I had always been interested in architecture and was a good drafts-man—in those days when you still did drafting by hand. It wasn't until my junior year that I started thinking about kitchens specifically. So it was a marriage of my interests.

Q: What was your first job in the industry?
A: When I graduated, I took a job at Cini-Grissom Associates, which has since morphed into Cini-Little. I was in their Rockville, Maryland, office, just outside Washington, D.C., for almost 5 years. I didn't know that much about them when I interviewed but came to find out that John Cini is also a Cornell graduate and many of their employees at the time were also hotel school graduates. They are among the largest foodservice design companies in the world. A number of people who have passed through their offices are very well known in the industry.

Q: Did you feel like you were well prepared for the job—or was it more like, "Wow, what have I gotten myself into?"
A: The hotel school provided a good base of knowledge to operate from, but in many respects, learning how to design something requires on-the-job training. People become foodservice consultants from a lot of backgrounds, not necessarily academic. They may start as equipment dealers, manufac-turer's representatives, operators, or other industry segments and then become consultants. The best first step to take is to consider becoming a member of FCSI (Foodservice Consultants Society International).

Q: You've said you very rarely design restaurants. What kinds of projects do you take on?
A: I've designed a couple of private kitchens, but they've been in places like Saudi Arabia, where they have the kind of budget and culture that

includes such unique items as an integral slaughterhouse.

Most of the projects I do are funded by tax money—for instance, correctional facilities, schools, and military bases. In some respects, they're almost always big projects with more money to invest in the kitchen. For example, I designed a cook-chill operation for a correctional facility that prepares 30,000 meals a day. This was back in the 1980s, and the value of the kitchen equipment alone was over $2 million. Depending on how you calculate a fee, that's pretty significant.

Q: Are you a fan of cook-chill systems for large facilities?
A: Not necessarily. Cook-chill has its own set of demands. It's great when it is used properly, but cook-chill equip-ment is expensive, and the system requires a lot of discipline in regard to documentation and labeling and food safety. It probably wouldn't be a viable solution for a single high school, but if you have a large district that prepares food for 20 schools in a central location, it might be a good option.

Q: What are the challenges of designing a kitchen for a correctional facility?
A: Depending on the type of facility, you may have significant

challenges, both in terms of design and operations, with your workforce. If it's a county jail where the sentences are typically a year or less, there are probably a number of civilian employees. But if you're in a prison, there are a handful of supervisors and a lot of inmates.

From an inmate's perspective, working in the kitchen is a pretty good job. You may get a little bit of extra food, a little more sense of freedom. You're not stuck in your cell, you may get to wear a special uniform instead of the regular prison duds, and you're also gaining some job skills for when you are released someday. In fact, I've designed some prisons that include well-equipped culinary arts kitchens.

There are service-system questions. Are people going to eat in a cafeteria, day rooms in their housing pods, or in their cells? How much space is available for a dining area—do they have to eat in shifts, or all at once? With a prison, it's literally a captive audience. They have to be fed, and that may be one of the few things they have to look forward to. If the food isn't good, it can become an issue.

Prison designs also present interesting demands in terms of supervision. One of my clients likes to have elevated offices, although they have to get a waiver of ADA accessibility requirements to do it. In the few high walls you have, there is a lot of glass. In maximum-security prisons, you may have to create small rooms where a person is locked in to do their work; they have to pass the knives or other dangerous tools out through a slot before they can come out of the room.

As a designer, you have to think about every piece of equipment and every tool as a potential weapon—nails, pieces of glass, slicer blades. If you have a big kettle, the paddle used to stir it can be a weapon. There are problems with vandalism sometimes. The inmates don't necessarily have much to lose, other than forfeiting a pretty nice job or getting some time tacked onto their sentences, so they may get into trouble. I've heard of a case where they jammed an old-fashioned wire case for milk cartons into a flight-type dishwasher and in less than 5 minutes trashed the single most expensive piece of equipment in the facility.

Q: The receiving aspects of it must be difficult, too. How do they get supplies?
A: Clients frequently don't want any more than 3 days' worth of storage inside the secure perimeter because if there's what they euphemistically call "a disturbance," they don't want the inmates to be have access to food. One of the good guys' bargaining chips is food: "You're going to starve if you don't surrender." The other consideration is that prisoners sometimes vandalize, and if you have a lot of food stored and they decide to leave the refrigerator and freezer doors open, or throw knives into the fan coil blades, the losses can be major.

Some correctional facility designs include ingredient rooms, like big hospitals or hotels, to control theft. You have to sign out every ingredient, and very few people have access to the room.

To be able to take advantage of volume buying opportunities, some prisons locate a large warehouse outside the secure perimeter. That way, foodservice vendors can deliver to the warehouse, which avoids many security issues related to making deliveries inside the secure perimeter. Of course, if you're in a big city, you probably don't have the room to do that. There also are projects where a single kitchen is built off-site to serve four or five institutions in the area.

Q: What appeals to you about this type of design work?
A: I enjoy working with the other design disciplines—architects and engineers, foodservice operators—and coming up with answers to these challenges. I designed some restaurants early in my career, and one frustration was often that restaurant owners didn't have much money to invest and it was sometimes difficult to get paid. With these large, publicly funded projects, it may take a while, but you know you will be rewarded for your efforts.

I'm also not an interior designer, and in restaurants, they're very concerned with décor decisions. That's not my thing. With institutional facilities, you're really getting involved with the details of the equipment—things like electrical requirements, moving a certain volume of air, the security challenges.

I actually do more school facilities than I do prisons. School projects are smaller, but both are important segments of my business because, fee-wise, they're similar. In both of them, I enjoy the coordination of a large project without the egos you sometimes find in a restaurant. Of course, in any project you can have

one player—an engineer, a warden—who makes the whole thing tedious. I started a correctional project in 2004 that wasn't completed until 2012. During its course, they had a change in governors, a change in wardens; it started and stopped when they decided at one point they couldn't afford it; there were political issues with some people in state government who didn't like certain contractors. By the time it was done, I was the only person who had been involved in the project from the beginning.

Q: Do you think it's better to specialize in a particular type of design work or to be a generalist who can do whatever types of jobs come in the door?
A: I think if someone decided, "I'm only going to design four-star restaurants," or "I'm only going to design prison facilities," they might be surprised at how the market changes over time and how it will affect their business.

Some bigger firms have minimum requirements for the size of the fee or the project. But I'm a small office and am not averse to doing small projects. I also do some pro bono work, like soup kitchens or homeless shelters, once in a while. I'm prepared for any kind of work I am qualified to do.

One FCSI standard is that you don't accept engagements for any kind of work you aren't truly qualified to do. In those cases, you could partner with someone who has had that type of experience on a project. You just have to be adaptable.

Q: Are you asked to "go green" in your design projects in terms of energy and water conservation, waste reduction and so on?
A: Yes. Water conservation never used to be an issue in Michigan, but it is now. A lot of equipment manufacturers are jumping on the green bandwagon, and that's good, but some of the stuff that is called "green" really isn't, necessarily. If a restaurant purchases a fryer that's green because it uses only half the amount of BTUs as they used to, they think, "Great!" But the problem is, the recovery time might be twice as long—so yes, you save energy, but you can't make fries as fast. You need to think about things like that.

A conscientious designer, even before green was popular, would always choose the least expensive viable alternative but also consider the operating cost and the moral or ethical cost of what's best for the operator and best for the area. If a given facility's infrastructure includes such systems as steam or chilled water, you should take advantage of those opportunities.

Part of the problem with green is that it still very often means "expensive." If a well-known restaurateur has the money and wants to make a statement and be at the leading edge in the community, that's great. But if you're working with a school board that has to pass a bond in order to be able to finance a project, it's a different story. I may suggest a pulper-extractor because it would be a good way to minimize the volume of waste sent to a landfill, but if it's going to add $8,000 or $10,000 to a $250,000 budget—they'll say, "We just don't have that."

Q: And finally, do you have any advice for prospective designers?
A: Be prepared to start at the bottom, and don't get fixed on a particular direction you want your career to go. Don't necessarily count on your college experience and training to give you all the nuances—working with engineers, knowing the applicable codes. There are currently no nationally accepted health codes, for instance. And local interpretation of codes that *are* recognized nationally can change from person to person, sometimes even in the same jurisdiction. So much of it is regionalized.

An internship is a good way to become familiar with AutoCAD programs and AutoQuotes [an online foodservice catalog program]. Work for an equipment dealer and see how that part of the industry works. When someone says, "Find me a 40-gallon tilting kettle that uses gas and also has an automatic mixer available," you need to be able to do that.

At Cornell, we had to get summer credits, and in my case, they were not design-oriented. I got some real-world experience working in hotels. Or work in a kitchen. You'll soon see the design flaws and realize how important good design is when you see how far you have to walk with those heavy bus tubs to and from the dishroom.

13

PREPARATION EQUIPMENT: BROILERS, GRIDDLES, AND TILTING BRAISING PANS

■ INTRODUCTION AND LEARNING OBJECTIVES

At its simplest, broiling is an encounter between food and fire, where intense, direct heat creates a beautifully browned exterior and moist, flavorful interior. However, several terms seem to be used interchangeably to describe this cooking method: Broiling, grilling, and griddling are all types of dry, radiant-heat cooking. In many cookbooks, the words *broiler*, *grill*, and *griddle* are often used as if they refer to the same piece of equipment. So, for the purposes of this text, let us clarify these terms.

A *broiler* is a piece of surface cooking equipment with its heat source located above the food (an **overbroiler**) or below the food (an **underbroiler**). The food rests on a *grate*, and marks are left on the food from its contact with the hot grate. The act of creating these cross-hatched marks is called **scoring** or *branding*.

A **griddle** is also a piece of surface cooking equipment. Its heat source is located below the food, and the griddle is usually a smooth, solid surface. However, a griddle can be ordered with grooves in it, for other types of cooking.

Technically, a **grill** is not a cooking appliance. Rather, it is a grid of metal bars set over a heat source, on which food is placed to cook. However, outdoor barbecues for home use are often called *grills*, and the word is also used commonly in restaurant names, as in Clancy's Bar and Grill.

You will see the words *broiling* and *grilling* used interchangeably, but *griddling* refers specifically to cooking something on a solid-top griddle. The point of all this wordplay is to help you notice all three terms and to clarify in your own mind exactly what is meant when they appear in recipes, sales literature, or any other text.

To further confuse the novice, each of the appliances has several common names.

After reading this chapter, you will be able to:

■ Identify and explain how broilers and griddles operate.

■ Identify and describe the types of broilers and griddles on the market, and their uses.

■ Describe installation, utility, and maintenance requirements.

Finally, we'll introduce you to the **tilting braising pan**, another versatile piece of equipment that complements (and even replaces) the griddle in some kitchens.

13-1 BROILERS

Professional chefs say the pure, true flavor of food is best showcased when it is broiled. But this preparation method has become more precise and complex with the assimilation of international cuisine and flavors into U.S. kitchens, creating the need for bolder flavor profiles using spices, herbs, chiles, and other gourmet ingredients. Today's chef soaks bundles of herbs, grapevines, or mesquite chips in water and throws them onto the fire to add aroma and flavor, then finishes the broiled food with a dollop of herbed butter.

Broiling is a dry-heat method of cooking, which means little or no liquid is used in the process. Instead, the food is cooked by its proximity to direct radiant heat. Broiling gives foods an appetizing appearance complemented by an inviting aroma—the result of juices dripping onto the radiant heat source to produce a distinctive smoky flavor. It's a popular cooking method because it is convenient and quick, but most consumers don't realize what an art form it is, or the value of the skilled and experienced "broiler man" in a kitchen—who, nowadays, can just as easily be a "broiler woman."

The ideal in broiling is to cook the surface of the food perfectly while keeping the interior of the product tender and juicy. Because direct heat performs the former function much more quickly than the latter, broiling requires close attention and a true sense of timing. To get it right, there is no substitute for experience.

The chef must also take care to select only top-quality meats and vegetables, as broiling will not mask the imperfections of tough cuts of meat or inferior produce. Broiling should not be restricted to steaks, because this cooking method is easy and popular among health-conscious diners who want fresh, simply prepared foods.

To achieve these ideals, each operation requires something different from its broiler(s). Some need large volume; others want a broiler that is attractive as well as functional, for front-of-the-house exhibition cooking. Broilers are generally lumped into three broad categories:

1. Overhead broiler (also known as *overfired*, or an *overbroiler*, or a *hotel broiler*)
2. Charbroiler (also known as *underfired*, or an *open-hearth broiler*)
3. Specialty broiler (their names vary; you'll learn about them later in this chapter)

Let's take a closer look at each of these categories.

OVERHEAD BROILER

The term *overhead broiler* refers to the location of the heating element above the food being broiled. It is a heavy-duty piece of equipment designed for high-volume output. In the typical battery of commercial kitchen appliances, it is installed beside the range. It measures the same width and height as the range.

A big advantage of this type of broiler over other types is the separation of the radiant heat source from the drip pan at the bottom of the broiler. This eliminates most of the smoking and danger of grease fires.

Modern broilers usually preheat in less than 2 minutes. The broiler compartment is heated by burners, either ceramic or infrared. The burners may be located at the top and center of the broiler, spaced evenly across the top width of the broiler, or placed on the sides of the unit, aimed inward at the food. Gas burners use between 65,000 and 100,000 Btus (British thermal units) per hour; electric burners use from 12 to 16 kilowatts per hour. Gas infrared broilers can reach full operating temperatures within 90 seconds, with a blower that mixes air and gas and can cut conventional broiling time in half (see Illustration 13-1).

There's no wasted space on a commercial broiler. You'll find a high shelf directly above the broiling area, an overhead oven instead of the shelf (heated by the same burners as the broiler, used primarily as a warming or holding space), or a standard range-type oven

ILLUSTRATION 13-1 An overfired gas broiler with a conventional oven compartment below.

Courtesy of Montague Company, Hayward, California.

underneath the broiler. When an oven is part of the appliance, it has its own separate control panel.

If there's no need for an extra oven, you can install a second broiler unit. The **double broiler** needs a space at least 76 inches tall, including room for its 6-inch legs or casters. One criticism of the double broiler is that it can be uncomfortable to use. When the top broiler is working, the chef's face is parallel to the hot broiler; when the bottom broiler is working, the chef's body gets the heat. Experts suggest using a double-broiler arrangement only when space is at a premium and production needs are high. In any case, to reduce energy consumption and give the chef a break, make sure the broiler cavity is well insulated, and turn it off whenever it's not in use.

Note that combi-ovens are capable of scoring with special grids to mark the meat. Consider your menu to determine if most of your needs are for banqueting or short-order. A combi-oven with a small charbroiler may suffice if you do a lot of banquets. In a short-order envi ron ment, however, the combi-oven is most likely not your best option because of the need to score quickly and finish the cooking process by order.

Every broiler has a *flue*, which vents smoke and fumes out the rear of the unit. Most installation guidelines suggest at least an 8-inch clearance at the top of the flue. Local ordinances usually require broilers to be located under the exhaust hood—and their exhaust requirements are substantial, from 900 to 1200 cubic feet per minute (cfm).

The **grid assembly** is the metal grill on which food is placed to be broiled. It should roll in and out smoothly on ball bearings and be easily adjustable to place it closer to (or farther from) the heat source, depending on what's being broiled. Today's manufacturers offer up to a dozen stop-lock grid adjustments, between 1½ and 8 inches away from the burners. To score or brand steaks for a charcoal-fired look, the broiler operator preheats the grid by raising it closest to the burners for 5 to 10 minutes before loading it. Sometimes—in a high-volume banquet situation, for instance—the meat is scored hours, or even days, ahead of time, then placed in sheet pans, refrigerated, and finished in convection ovens prior to serving.

Cooking speed is determined by the distance between the food and the heating elements, plus the size and/or thickness of the product being broiled. A chart of suggested broiling times for meats is reproduced in Table 13-1, courtesy of Texas Utilities, an electric company.

At its topmost position, the grid is capable of reaching temperatures near 600°F, so it's obvious the grid should have a pistol-grip handle that stays cool to the touch and is easy to hold. A sloping **grease tray** located under the grid reflects heat up and aims grease run off into a **drip pan** at the bottom of the broiler cavity. The drip pan can be removed for cleaning.

Using an overhead broiler requires practice. Broilers have a series of switches or gas adjustment knobs that allow you to set the heat level in the broiler by section. There is a separate switch for specific zones (areas) of the broiler, and each switch can be adjusted to low, medium, or high heat. The three heat settings can be misleading to a newcomer. "Low" produces a heat setting one-quarter that of "high" over the entire section of the broiler controlled by that switch. "Low" is not for broiling, only for holding foods. "Medium" provides broiling heat, and "high" produces cherry-red radiant heat. The broiler should be preheated for 5 to 10 minutes before use. During idle periods, the broiler can be used for browning casseroles, making croutons, or keeping food warm.

Consider these points when planning overhead broiler installation. If the broiler is located beside a fryer, open-hearth broiler, or griddle, consider ordering it with stainless-steel sides, which make cleaning the inevitable spatters much easier. Install it well away from side and back walls for adequate ventilation space.

TABLE 13-1

Overhead Broiler Cooking Guide

MEAT	THICKNESS (INCHES)	WEIGHT RANGE (POUNDS)	TOTAL TIME[1] (MINUTES) (ONE-HALF TIME ON EACH SIDE)		
			RARE	MEDIUM	WELL DONE
Beef					
Club and rib steak	1	1–1½	6–8	8–10	10–12
Porterhouse	1	1½–2	6–8	8–10	10–12
Porterhouse	1½	2½–3	8–10	11–13	14–16
Porterhouse	2	3–3½	10–12	13–15	16–18
Sirloin	1	2½–3½	6–8	8–10	10–12
Sirloin	1½	3½–4½	8–10	11–13	14–16
Sirloin	2	5–5½	10–12	13–15	16–18
Ground beef patties	¾	6–8 ounces each	3–4	4–6	6–8
Tenderloin	1		6–8	8–10	10–12
Lamb					
Rib or loin chops	¾	2–3 ounces each	8–10	10–12	
Double rib	1½	4–5 ounces each	11–13	14–16	
Lamb shoulder chops	¾	3–4 ounces each	8–10	10–12	
Lamb shoulder chops	1½	5–6 ounces each	11–13	14–16	
Lamb patties	½	6–8 ounces each	4–6	6–8	
Ham and sausage					
Ham slices	½	9–12 ounces each	6–8		
Ham slices	¾	1–1¼	8–10		
Ham slices	1	1¼–1¾	10–12		
Pork sausage links		12–16 to the pound	4–5		
Broiling chickens (drawn) halves	1–1½		18–22		

[1]It is recognized that broiling times may be varied by chefs to suit individual tastes and preferences.

Note: The suggested broiling time is based on a thoroughly preheated broiler with units switched on high heat. The distance of the grid from the heating units will vary from 1 to 2 inches for thin and rare steaks to 3 to 4 inches for thick, medium to well-done steaks.

Source: Texas Utilities Electric Company, Dallas, Texas.

Avoid locating a refrigerator next to the broiler. You'll waste a lot of energy as the two run constantly to compensate for the other's cold and heat. However, you do need to have a refrigerator near the broiler area so the broiler operator has easy access to it.

Finally, consider casters and so-called quick disconnects to move the broiler more conveniently for cleaning. Ask the manufacturer or installer about this so it can be done without disrupting the gas or electric supply or the fire suppression system.

CHARBROILER

Charbroiling is the term for broiling food with flames, smoke, and radiant heat. It is popular in display kitchens because it's showy and fun to watch, but charbroiling also imparts a nice charcoal flavor to the food. The *underfired* broiler has its heating elements located beneath the food. The flames and smoke result when the cooking juices drip onto the heating elements. The open flame look has also prompted the name *open-hearth broiler* (see Illustration 13-2).

ILLUSTRATION 13-2 A gas charbroiler.

Courtesy of Wolf Range Company, Compton, California.

Unlike the overfired broiler with its movable grate and grid assembly, the underfired broiler cooks food at a fixed distance from the heat source. Its grate may have only two or three positions. Instead, it is the cooking temperature that varies. The chef can broil at anywhere from 300°F to 700°F, but charbroilers are at their best when they are between 550°F and 625°F. Heat is transferred in three ways: conduction (from the heat of the broiler grate); convection (as heat transfers from grease and vapors produced by burner combustion); and radiation (as heat reflects upward from the radiants).

The underfired broiler is much more compact than the overfired broiler. It preheats faster, and its infinite-heat controls allow precise temperature control that takes much of the guesswork out of producing uniformly high-quality broiled foods. For comparison with overhead broilers, we've reproduced a chart of suggested cooking times for charbroiled foods in Table 13-2.

The four basic fuels used to create heat for charbroilers are charcoal, wood, gas, and electricity. They vary in their efficiency, ease of use, and ability to impart flavor to food. Charbroilers are the modern, high-volume answer to Dad's backyard charcoal barbecue, and the ones that use charcoal are simply larger versions of the backyard model. Gas charbroilers use ceramic, stainless steel, cast iron, or igneous rock ("lava rock") to form a radiant bed above the gas burners that heat it, and so these items are known as *radiants*. When highly

---

TABLE 13-2

Charbroiler Cooking Guide

STEAKS	DONENESS DESIRED	INFINITE-HEAT CONTROL SETTING[1]	SUGGESTED COOKING TIME (MINUTES)
Frozen			
½ inch thick	Rare to medium	8	4–7
¾ inch thick	Rare to medium	7	7–8
1 inch thick	Rare to medium	7	12–14
1¼ inch thick	Rare to medium	6	23–25
Rerigerated			
¾ inch thick	Rare to medium	High	4–5
1 inch thick	Rare to medium	High	7–8
1¼ inch thick	Rare to medium	High	12–13
1½ inch thick	Rare to medium	High	14–18
Canadian bacon	Medium	400	2–3
Minute steaks and boiled ham	Medium	500	2–3
Hamburgers	Medium to well	550	2–4
Sausage patties	Medium to well	550	3–4
Ham steaks and beef tenderloin	Medium	600	5–7
Club steaks	Rare to medium	700	4–7
New York strip steaks	Rare to medium	700	5–10

[1]Infinite-heat controls are adjustable from 1 to 8 and to "high." Thermostat is adjustable from 300°F to 700°F.

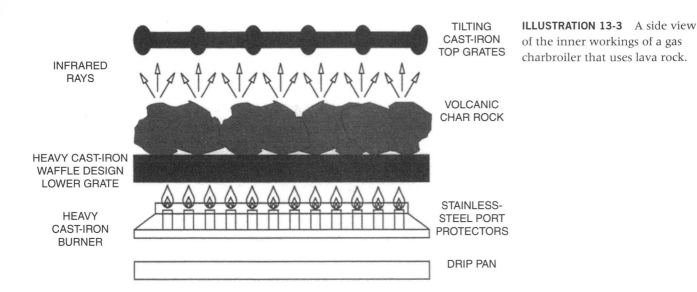

INFRARED
RAYS

TILTING
CAST-IRON
TOP GRATES

VOLCANIC
CHAR ROCK

HEAVY CAST-IRON
WAFFLE DESIGN
LOWER GRATE

HEAVY
CAST-IRON
BURNER

STAINLESS-
STEEL PORT
PROTECTORS

DRIP PAN

ILLUSTRATION 13-3 A side view of the inner workings of a gas charbroiler that uses lava rock.

porous lava rock or ceramic is used, it must be replaced periodically to avoid grease or carbon buildup. Electric models may have long, cylindrical heating rods, a cast-iron grate located above electric burners, or heating units embedded in the underside of the broiler grates (see Illustration 13-3).

Burners underneath the radiants may be made of steel alloy or cast iron. Steel alloy is the least expensive. One thing to look for in selecting a broiler is how the burners are shaped and spaced. There are straight tubes, U-shaped, H-shaped, and S-shaped configurations. A burner produces a row of flames for every 6 inches or so of its width, although some models stretch that to 9 inches. Just remember, the closer the rows of flames are together, the more even the bed of heat will be. If the rows of flames are far apart, they create hot and cold spots on the broiler surface. This is not necessarily a bad thing if you're trying to cook two steaks simultaneously, one medium and one well done, and know where to put them on the broiler to achieve these results. But if the temperature variance is too extreme, or if you are not familiar enough with the appliance, you may have problems.

The burners beneath some radiant beds include *reflectors*—also made of stainless steel or cast iron—that absorb heat and aim it toward the grate in a more concentrated fashion than a simple open flame. Stainless-steel radiants are less expensive, preheat quickly, and are easier to clean than their cast iron-counterparts. Cast iron may be slow to heat but holds its heat longer and is therefore handy for long shifts of high-volume production. Both steel and iron radiants are popular among high-volume operations because they provide even heat and require minimal handling—no turning stones, replacing lava rock, and so on. However, rocks and briquettes look great for display cooking, and some broiler operators like moving them around to get exactly the results they want. (Incidentally, kits are available to convert a radiant broiler to a briquette one.)

Manufacturers also offer a variety of grates. Spacing of the rods on the grate is important. The optimum is probably 5/16 inch, and no more than ½-inch spacing; otherwise you will drop pieces of foods with light or flaky surfaces. Here are your "really grate" choices:

Floating rods. Usually chrome plated, **floating rods** are easy to clean and prevent sticking. They are pretty much standard on most charbroilers today. They're also called *free-floating grates* because they allow for repeated expansion and contraction without warping.

Cast-iron flat. These grates are excellent for heat retention but awful for foods that tend to stick. They are also prone to warping.

Cast-iron wavy. These also retain heat and are made with the score-markings that have come to identify broiled foods.

Meat grates. These feature fine marks to make the broiled meats look like they've been scored. They are designed to deliver maximum heat for thick cuts of meat.

No matter what type of charbroiler you choose, the combination of heat, smoke, and flame raises safety concerns. Grease drips onto the heat source during cooking, although most grates are tilted slightly to allow at least some of it to flow away. An accumulation of grease in the drip pan can catch fire. Some manufacturers add baffles to separate the drip pan from the heat source, but the simplest precaution is to keep an adequate water level in the drip pan, check it, and clean it frequently.

Charbroiling produces a lot of smoke as well as grease. Most city ordinances require the broiler to be under an exhaust canopy and an air filter located no more than 48 inches from the surface of the broiler. A separate flue and/or exhaust system may be required for wood or charcoal broiling. In short, investigate local fire and health codes before you purchase anything that broils using live flames.

Three types of charbroiler require special mention: the drop-in, countertop, and freestanding models. The drop-in charbroiler is electric, installed in a stainless-steel counter with a gasket separating the broiler from the countertop (see Illustration 13-4). The heating element doubles as the grid in this compact unit, which measures from 16 inches square to 16 × 32 inches and uses only 10 kilowatts of power per hour. The drip pan may be a drawer, equipped with a splash guard and a funnel-like receptacle, all removable for cleaning.

A simple front control panel includes a switch (with "off," "broil," and "clean" settings), a dial (with "off," "low," and "high" settings), and a signal light indicating whether the unit is on.

Despite their small size, drop-in charbroilers have specific installation requirements. Their exhaust requirements are higher, ranging from 900 to 1200 cubic feet per minute. The unit must be located at least 1 foot from any side wall, at least 5 inches from a back wall, and about 4 inches from any other piece of countertop equipment.

Drop-in charbroilers are meant for restaurants or other foodservice facilities with low production needs. However, they may be a valuable addition to any small operation strapped for space. They come in widths from 15 to 42 inches, with a standard depth of 24 inches.

ILLUSTRATION 13-4 A drop-in charbroiler for countertop use.

Courtesy of The Middleby Corporation, Elgin, Illinois.

The countertop charbroiler is a standalone unit. It may run on gas or electricity. The electric countertop model looks similar to the drop-in charbroiler. It has 4-inch plastic or stainless-steel legs, which are both removable and adjustable, and stands about 16 inches high. This broiler can fit snugly in a space no more than 2 feet square, and uses 10 kilowatts of power per hour.

The gas-powered units seem more popular than the electric ones, perhaps because the grates come in a variety of widths, from 24 to 72 inches. Gas countertop charbroilers are taller than electric ones, from 25 to 36 inches, and they generally hold more food. They can be mounted directly onto a counter or on a stainless-steel stand. The burners are arranged one for every 6 inches (or 12 inches) of broiler width. Each burner uses between 15,000 and 45,000 Btus per hour; each burner usually has its own heat control.

Like their full-size counterparts, these smaller units often have adjustable grids or grates on which food is placed for broiling. If these are not adjustable, they at least slant slightly forward to allow juices to flow into a built-in grease trough.

Freestanding or floor-model charbroilers are used in large facilities as part of a production line-type output or in restaurants with a brisk à la carte business. The broiler space itself is not much bigger than the largest countertop unit, but it is housed in a stainless-steel cabinet for storage beneath the broiling area. It's generally placed on a floor stand, on casters to facilitate cleaning. Again, the grates are positioned either flat or tilted; cast-iron radiants or ceramic "coals" are heated by individual burners; and there is one burner for each 6 inches of broiler width. Freestanding charbroilers also feature backsplashes of up to 10 inches to minimize spattering.

13-2 SPECIALTY BROILERS

SALAMANDER. Perhaps its unusual name derives from the fact that the *salamander* is so adaptable, fitting into its surroundings much like the lizard that is its namesake. A miniature version of heavy-duty broilers, it measures from 10 to 13 inches deep and from 23 to 28 inches wide. It's small enough to rest on a shelf or counter, or it can be mounted directly above a range or spreader plate. During slow times when the main broiler is turned off, it can be used as an energy-efficient backup broiler. When things are really hectic, the salamander makes a convenient plate warmer (see Illustration 13-5).

Gas-powered salamanders may have from one to six infrared burners. Depending on the number of burners, its gas input requirements are from 30,000 to 66,000 Btus per hour. Electric salamanders vary from 5.2 to 7 kilowatts per hour and require outlets of either 208 or 240 volts. Typically, control panels include a switch to set the broiler's temperature to "high," "medium," or "low," as well as controls that allow the operator to turn on the left- or right-side elements separately or all together.

Infrared salamanders offer several advantages. They preheat in about 90 seconds instead of the 12 to 15 minutes a radiant-heat model takes. When not in use, radiant-heat units are kept in standby mode, which is not necessary with infrared units. And infrared units keep the kitchen cooler because they heat only the food, not the surrounding airspace.

Grids or grates are removable for cleaning and can be positioned and locked in place between 1⅓ to 4 inches away from the heating units. Grease deflectors are attached to the underside of the grids, channeling hot drippings to a drain pan or drawer, also removable.

A salamander installed on a high shelf or above any combination of hot range equipment should sit at least 18 inches above the lower workspace; the shelf should be reinforced to prevent an unexpected tumble onto the hot top below. Some manufacturers charge extra for heavy-duty wall-mounting brackets, but they are worth the cost. If the salamander is going to sit on a counter, you might also consider more expensive stainless-steel legs, which are sturdier than plastic.

CHEESEMELTER. The *cheesemelter* is another type of specialty broiler. Its name explains its most common use: to melt cheese, primarily on Italian, Mexican, and Tex-Mex dishes. While it accomplishes that task quite nicely, it can also be used to brown, poach, and boil. Its primary function, however, is to finish food, not cook it, so cheesemelters often have lower overall heat output than salamanders. Some units are constructed

ILLUSTRATION 13-5 The salamander often is installed above the hot line so it can be used as a food warmer as well as a broiler. This model also has a fry-top range (griddle) and oven below.

Courtesy of Wolf Range Company, Compton, California.

ILLUSTRATION 13-6 The cheesemelter is used most often for its namesake task: melting cheese atop foods just before serving.

Courtesy of Vulcan-Hart, a division of ITW Food Equipment Group, LLC, Louisville, Kentucky.

to turn on only when food is placed on the grid or rack, thus saving energy (see Illustration 13-6).

Installation methods are similar to the salamander—as a countertop or wall-mounted unit. A few models have a pass-through feature similar to the pass-through refrigerator with doors on two sides.

Most cheesemelters, from 2- to 10-foot widths, have only a single adjustable (three-position) rack on which to place food. Additional racks often can be installed.

Cheesemelters can be powered by gas or electricity. The infrared units seem the most popular; they preheat within 1 or 2 minutes and do not need to be left warming indefinitely on a standby mode. As with other broiler types, the burners are individually controlled, and the drip pan is removable.

FOOD FINISHER. This innovative category of surface cooking equipment emerged in response to operators' demands for multifunctional kitchen machinery. The term *finishing* typically means putting the final touches on a dish. This may mean giving it a final sear to toast or brown, melt, or crisp something. The term is also used to indicate reheating previously or partially cooked foods. The *food finisher* does all of this, and quickly, requiring about half the time of heating in a conventional oven.

Whether at a food kiosk that needs fast cheese-melting over nachos or a hotel that provides limited-menu 24-hour foodservice, the simplicity of the machine facilitates easy operation. As shown in Illustration 13-7, it looks like a cross between a microwave oven and a toaster oven. Unlike a salamander, which stays at full power throughout the business day, the food finisher can switch itself to standby mode when not in use, remaining "on," but at 20% of its full power until it is needed again. Some restaurants use food finishers instead of cheesemelters and/or salamanders.

CONVEYOR BROILER. For maximum cooking speed, a conveyor broiler includes both underfired and overfired radiants. Food is placed on a stainless-steel conveyor belt; in some units, the food is loaded onto the belt at the front and emerges fully cooked at the back; others are front-load and front-return. These broiler types cook faster because foods are exposed to both top and bottom heat. A burger that would normally take 4 or 5 minutes to broil can emerge from a conveyor broiler in less than 2 minutes. If you must broil quickly to keep up with volume, or if you simply need a broiler that can react quickly to a surge in demand, this is it. Conveyor broilers offer other important advantages besides speed: reduced labor and consistent results.

ILLUSTRATION 13-7 The food finisher is countertop size and can do a number of last-minute duties quickly, including reheating, melting, and toasting.

Courtesy of Hatco Corporation, Milwaukee, Wisconsin.

The downside of the conveyor broiler is that the food items must be uniform in size and thickness to cook evenly. For hamburgers or frozen waffles, this is no problem; for steaks or chicken breasts, it might be. Conveyor broilers also can be difficult to clean. The largest conveyor broilers have as many as three separately controlled, programmable conveyor belts and heat chambers that can all work at once on different settings. Conveyor broilers can be all-gas, all-electric, or a combination of gas burners on the bottom and electric on top. The smallest countertop unit measures 15 × 23 inches, requires 208/220–240 volts of power, and can cook at least 100 hamburgers per hour.

WOOD-BURNING BROILER. The *wood-burning broiler* has made a comeback in recent years on the dining-out scene, although few manufacturers produce broilers that burn real wood or charcoal anymore (see Illustration 13-8).

Of course, the commercial model has to be larger than Grandma's old wood stove, and certainly more efficient. The wood or coal is contained in a firebox inside the broiler. A damper controls the flow of incoming air, which, in turn, controls the intensity of the fire. The fire heats the cooking grates on which food is placed.

The unit is made of double-wall stainless steel; between the two steel layers are fireproof insulation and a firebrick lining that holds the heat inside. Ashes are removed through two drawers at the bottom of the unit.

Wood-burning broilers come in sizes that give you cooking surfaces as small as 4 square feet or as large as 10 square feet. At least one manufacturer added an optional rotisserie unit to its wood burner, which may be removable or installed as a permanent fixture.

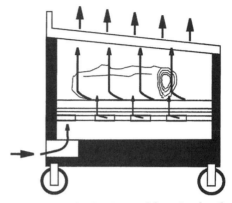

ILLUSTRATION 13-8 A wood-burning broiler creates heat in a deep firebox below the food.

ROTISSERIE. Some people consider the rotisserie a type of broiler. After all, it also cooks food over a flame or heat source. You already read about rotisseries in Chapter 11, and a couple of specialty appliances fall into this category. One is the *hot dog roller grill* so popular at movie theaters and in convenience stores. The other is the *gyros machine*, a countertop oven with a vertical rotating spit in a glass-front cabinet to permit public display. Gas jets or electric heating elements line the sides of its semicircular wall. A traditional Greek food, the gyro (pronounced "year-oh") is made of a chopped lamb or lamb-beef mixture and formed into a cone-shaped slab that fits onto the spit. The broiled layers are thickly sliced off the cylinder and made into delicious sandwiches.

The third and most recent import to the United States is the *churrasco*, an open-flame broiler from Brazil that cooks meats on long, thick spits made to resemble swords. It is the showpiece of South American *churrascarias*, barbecue restaurants that specialize in this type of cooking and casual, all-you-can-eat menus, called *rodizio* service. These restaurants have become popular in many U.S. metropolitan areas for their huge portions and reasonable prices.

PORTABLE BROILER. We just mentioned the Greek gyros machine, with its side-mounted broiler elements and display case. Another popular option seen in mall food courts and at street festivals and corporate picnics is the *portable broiler*, which offers up to 10 square feet of cooking area. Its heat source can be liquefied petroleum (LP) gas, wood, or charcoal. The wood- and charcoal-burning models have grates that can be adjusted to four positions above the flames. It is wise to order an optional hood or cover for your portable broiler; although it weighs more than 70 pounds, the hood can be removed if not needed, and keeps the appliance cleaner and in better shape when not in use. LP gas models use a 40-pound fuel cylinder that is mounted beneath the firebox area for easy portability. LP gas, also commonly known as *propane*, provides the advantage of lighting all four burners with the push of a button. Each burner has a separate control knob. One fuel cylinder produces a total of 130,000 Btus.

COMAL. The *comal* is a specialty broiler used in Mexican restaurants for display cooking of fajitas. The meat is positioned on a grill over gas-heated briquettes.

13-3 BUYING AND MAINTAINING BROILERS

Consider these points as you choose your broiler(s):

Your menu. If delicate items will be broiled—say, gratin-type casseroles—you'll need a broiler that heats from the top. For just about every other item, either top or bottom heat works equally well. Infrared broilers are capable of the highest temperatures, but they work best with foods of uniform thickness (like T-bone steaks). Radiant heat works better with foods that have highs and lows (like lobster tails, which are thicker in the middle than on the end). The few other exceptions are specialty items, like Greek gyros, for which a specialty (vertical) broiler is appropriate.

Heat source. Again, multiple choices: electricity, natural gas, LP gas, charcoal, or hardwood. Electric broilers offer a trouble-free, clean cooking source. The benefits of gas are high-heat cooking and glowing flames that are attractive to customers in exhibition cooking. Remember, the heat output of a gas broiler fueled with LP gas is as much as 25% lower compared to natural gas, so be sure to calculate this when determining the size of cooking surface you'll need.

Charcoal briquettes offer quick, long-lasting heat. An alternative, particularly for exhibition broiling, is the lump "charcoal" made from chunks of hickory, oak, or other hardwood. These have none of the petrochemicals found in traditional charcoal, and their random sizes make them especially handy; use the smallest pieces for a quick-starting fire and larger ones for a longer-lasting fire. If you select lump charcoal, you need a place to store it and a reliable supplier to keep the storage place filled. Fresh wood chips can be used, even in gas and electric models, to impart a smoky flavor.

Amount of food. This includes how much you must cook at one time as well as how much is cooked during a set period. If you're broiling chicken halves for an outdoor catered event and plan to serve them directly from the broiler, you'll need a bigger cooking surface than if you plan to batch-cook them early and hold them in a warmer. When broiling steaks to order, you need enough broiler surface to hold the maximum number of steaks that may be ordered at any point. Size requirements are determined by a few factors:

- The temperature of the food to be broiled
- The size of the pieces to be broiled (both thickness and surface area)
- The temperature to which most of the food will be cooked (medium, well done, etc.)
- Distance between the food surface and the heat source

Once you have determined those basics, you'll find differences between manufacturers in these areas:

Grate materials. Cast iron has long been the standard on underfired broilers, but a growing number are available with stainless-steel grates, which minimize food sticking to them. In either case, the grates must be examined closely to ensure they can stand up to the rigors of daily use.

Grate position. On some broilers, the grates lie flat, while on others, they are slightly sloped. The sloped grates help drain grease away, which is helpful in reducing flare-ups. This minimizes the amount of smoke and potential fire hazards and makes it a little safer to lean against the front of the appliance, which is the natural tendency as you broil. Some grates are adjustable—they lie flat to be used like a rangetop or slope for broiling. Grates should also be easily adjustable up and down, to move them closer to or farther from the heat source depending on what you're broiling.

Temperature zones. Distance from the heat source is key, but some broilers also offer separately controlled temperature zones. The overhead or hotel broiler has two burners with separate control knobs. Some charbroilers offer the highest heat at the back of the grate, medium heat in the middle, and lowest heat at the front. Portable underfired broilers can have up to six sections, each with its own controls.

Grease tray. The location and size of the tray or pan that collects drippings are important. It must be easy to remove and large enough to catch grease from the entire grate area. The pan must be able to hold water, as this minimizes smoking and is helpful in providing moisture as food is cooked. The pan should have baffles to help prevent spilling as it is removed and cleaned.

Preheat functions. Some broilers are ready to use within a few minutes; others require as much as 15 minutes to reach optimum broiling temperature. The preheating times are usually longer with charcoal- and wood-fired models.

Radiants. Depending on whether you want an overfired or underfired broiler, the heating elements of gas and electric models may be made of stainless steel or ceramic. Stainless-steel radiants heat quickly, but ceramic briquettes or lava rock can be attractive. Infrared models generate the highest heat, up to 1500 °F, far hotter than ceramic radiants can reach. The number and type of radiants in these broilers affect the consistency of the cooking. You may find hot spots or cool zones at the edges of the grate.

Portability and flexibility. You can purchase a griddle top for use with some broiler models, which allows you to cook a wider variety of foods. An optional rotisserie or clamshell (discussed elsewhere in this chapter) is available with some models. If you plan to use your broiler for off-premise catering, order it with heavy legs and large-diameter casters, bigger than the standard 4-inch casters.

Before we move on to our discussion of griddles, here are a few troubleshooting and maintenance hints that should make life with your broiler a lot easier.

First, the grates or grids must be kept clean for best performance. Food and the inevitable carbon buildup that is part of the broiling process must be scraped from the grates with a spatula or, better yet, a wire *broiler brush* designed for this purpose. Allow this gunk to accumulate, and you risk contaminating the food. Daily, or as often as necessary, lift the grates off the broiler and take them to the pot sink for a good, thorough scrub. The grid assembly can also be lifted from the broiler and put in the pot sink for cleaning with a soft wire brush. If your broiler is electric, never use water to clean the heating elements. Wipe surfaces of the unit with a damp cloth, but clean the tubular elements by turning them on high and allowing them to burn off any food residue.

Like many metal surfaces, grids and grates should be lightly greased before use (or after washing) to prevent food from sticking to them. This seasoning also helps prevent rust from forming.

Drip pans should be checked often and emptied and cleaned daily. Just because your unit has a big grease pan doesn't mean you should wait until it's full to dump and scour it. Drip pans can hold up to 1 quart of water, which prevents accumulated grease from catching fire. Check the water level frequently.

Consider replacing lava rocks or ceramic briquettes every 6 months. The intense heat of broiling causes the radiants, and even the lower grates, to warp or crack. Check them periodically; change them as needed.

Cooking tips: Preheat the grates before using them to mark or brand foods. Don't press the natural juices out of meat as it broils. Turn it only once on the grate, and don't salt it until it has been cooked and removed from the heat. Salt can damage heating elements.

Follow the manufacturer's recommended preheating instructions. There's no need to waste energy by preheating for too long or at unnecessarily high temperatures. Broilers also should be loaded to maximum capacity to make the best use of the energy. This may mean running half of the broiler on high power to sear rare meats and the other half on low power for the slower cooking of well-done meats.

Inspect and clean the ventilation system above your broiler often, both for safety and for health reasons. On gas-powered broilers, check the burner ports to make sure they are clear of debris. If they are blocked, you can clear them with a wire brush or a drill fitted with a proper-size bit. If the gas pilot light won't stay on, it may be due to clogged orifices or an air draft that is blowing them out. Check the kitchen for the source of the draft.

If an electric-powered broiler has an element that does not seem to be getting hot enough, it might not be getting enough voltage. If your unit needs a 240-volt outlet but is plugged into a 208-volt outlet, it may work—just not efficiently.

Finally, you may live in a region with strict air pollution guidelines that require you to control smoke or grease output from your broiler (especially underfired units) by installing expensive rooftop turbines and electronic precipitators. If this is the case, read on. The grooved griddle, which we discuss later in this chapter, may be an alternative for you.

13-4　GRIDDLES

As mentioned earlier in this chapter, the griddle is a piece of surface cooking equipment, an appliance on which food is heated from below by gas or electric heating elements located directly under the flat surface. In most instances, the food is cooked on one side at a time, and the metal griddle top is coated with oil to prevent it from sticking (see Illustration 13-9).

No matter the size of restaurant or type of cuisine, the griddle is practically indispensable. It is usually placed in the center of the hot line or à la carte cooking section of the kitchen, right in the middle of the action along with rangetops and fryers.

The benefits to cooking on a griddle are many. They include:

- The coagulation of proteins, such as those found in eggs, fish, and meat
- Flavor enhancements, such as those that occur during browning meats or toasting buns
- Tenderizing certain items, such as steaks and chicken breasts
- Development of a crispness or crust for potatoes or bacon and the like

However, a griddle's chief disadvantage is that it is not energy efficient. Unless its surface is entirely covered with burgers or pancakes to be flipped, the heat simply escapes into the exhaust hood under which the griddle must sit. Depending on the model and what's being cooked, griddle efficiency ranges from 30% to 70%. Even locating a griddle too close to an outside door can lessen its efficiency when cold drafts hit it in the winter.

The griddle's cooking surface is also called the **plate**. Surrounding the plate, you'll find the **trough** and **splash guard**. The trough (also called a *gutter*) is a recessed area that surrounds the griddle top on three sides. Grease buildup and food particles are scraped into the trough with a spatula to keep the surface clean and smooth. At the back or far side of the griddle, the splash guard (also called a *fence*) is a 3- to 6-inch-high piece of metal that prevents food from falling behind the griddle; it also controls grease spatters.

Whatever the size of griddle surface, you want one that heats evenly, without hot spots or cold spots. Modern research has enabled griddle manufacturers to even out these spots with strategic placement of heating elements and more sensitive thermostats. Most griddle plates are divided into sections with separate heat controls so different foods can be cooked simultaneously. These sections are called **heat zones**. As you face the griddle, its heat zones run horizontally—from left to right, not from front to back. Each heat zone is between 12 and 18 inches wide, so a large griddle may have up to six of them. Electric griddles seem to have more precise heat zone control than gas griddles; for the latter, a separate thermostat is recom mended for each burner to gauge the temperature of each 12-inch zone.

Zoning is a fundamental concept in griddle cooking. Theoretically, you can cook different foods at very different temperatures side by side on two separate zones, a huge advantage in fast-paced kitchens. However, extreme temperature differences may affect each zone to a certain extent, so it's best to cook foods that require widely differing

ILLUSTRATION 13-9　A countertop griddle.

Courtesy of Vulcan-Hart, a division of ITW Food Equipment Group, LLC, Louisville, Kentucky.

temperatures in nonadjacent zones. Each zone can be set for cooking temperatures from 150°F to 450°F.

Beneath the plate, the gas griddle burners are either *atmospheric* (conventional, individual gas flames) or *infrared* (combustion within a ceramic housing that heats up and glows instead of producing a flame). Infrared technology eliminates one maintenance problem: the buildup of carbon soot on the underside of the griddle plate from flames hitting it for hours at a time.

Griddle tops range in length from 18-inch countertop models to 72-inch floor models, with depths (from front to back) of between 20 and 32½ inches. If aisle space permits, you might consider adding additional depth, which is available in 6-inch increments. That doesn't sound like much, but on a 24-inch griddle, it adds another square foot of cooking space, and another 3 square feet on a 5-foot-long griddle. A 7-inch-wide *work board* can also be ordered to provide landing space at the front of the plate. However, before you start adding depth, think about the person who has to lean over the griddle to do the cooking. Add only as many extras as make sense ergonomically.

Griddles can be retrofitted with a heated top plate, which can be *contact* (the top plate actually touches the food) or *noncontact* (the top plate heats from above, but without touching the food). This also increases the power requirements for the griddle, so be sure your system can handle it. The top plate is typically ordered to fit on just one heat zone, either the far left or far right side, and it may be smooth or grooved.

Yes, most folks think of a griddle as a perfectly smooth surface, but some griddles have grooved plates to create the seared look for burgers or steaks. Grooved plates can cover the entire surface of a griddle or just a section of it. Generally, the grooves are ⅜ inch wide and spaced ³⁄₁₆ inch apart. Some appliances can be fitted with detachable grooves and griddles so they can work like griddles or broilers.

As with broilers, griddles also come in countertop and floor models; the countertop ones may be drop-in or freestanding units. Each can be ordered as a gas or electric appliance.

Drop-in countertop griddles vary in width from 2 to 6 feet and in depth (front to back) from 2 to 2½ feet, which gives you a cooking space of between 400 and 1663 square inches. (Or think of it this way: You can cook 32 three-inch burgers on the smallest griddle, up to 120 at a time on the largest.) Like other griddles, countertop models have thermostatically controlled heat zones for every 12 inches of surface width. They're on 4-inch adjustable legs made of plastic or stainless steel. The grease troughs run the full width of the griddle plate, at the front and rear. Incidentally, it's a handy feature to have a trough at both front and rear because some people prefer to scrape forward while others prefer the grease and food particles be scraped to the rear of the griddle and out of sight.

The power requirements for a single-sided electric griddle vary from 3 to 25 kilowatts per hour, and they're available in single-phase or three-phase installations, from 208 to 480 volts. Gas countertop griddles usually have a rating of 20,000 Btus per hour per burner; there's a minimum of two burners on the smallest unit. Some manufacturers boast 30,000-Btu-per-hour burners. The largest double-sided gas griddles may use 90,000 to 140,000 Btus. Again, the higher the Btu capacity, the faster the heat recovery time, and that's important when you're loading your griddle with frozen foods.

Countertop griddles may also be ordered with work boards on the front, which give the operator a 7-inch surface for resting plates and product. And you can choose two grease drawers, one at each side of the unit, instead of a single drawer beneath the plate. Griddles can be ordered in combination with open burners; or a special bar can be installed on one side of the griddle onto which a pan can be hooked. The pan can be filled with buns, for instance, to load with burgers or hot dogs as they come off the griddle.

You can make a countertop griddle into a floor model by mounting it on a stand that places the griddle surface no more than 36 to 38 inches from the floor. (That is the most comfortable work height for most people.) Or you can buy a floor-standing unit. Either way, the smartest choices for legs are 3½-inch swivel-type casters, with front casters that lock in place to prevent rolling when the equipment is in use and unlock to roll for cleaning or repositioning. In gas models, a quick-disconnect feature also enables easy movement.

However, the basic location of the griddle is usually determined by city ordinance: under the exhaust canopy.

Large standalone griddles often are sectioned so different areas can be used for different foods. A griddle in a breakfast operation, for example, can cook hotcakes, bacon, and hash-browned potatoes simultaneously. Before you buy, decide on the combinations you'll need and, of course, whether you'll require a two-sided model. Preheat times are similar (15 to 23 minutes) whether the power source is gas or electricity.

13-5 SPECIALTY GRIDDLES

HEAVY-DUTY GRIDDLE. A variation of the floor-model griddle is the *heavy-duty griddle* with an oven below. It's considered heavy duty because its gas burners are rated at 30,000 Btus per hour. Sometimes, it's referred to as a **fry-top range** (see Illustration 13-10).

You can also order griddles with a broiler below, but this is only recommended for low-volume operations (less than 100 meals in one seating) because of the obvious discomfort that kind of heat causes the operator. Most recently, several manufacturers have introduced infrared griddles, touting energy efficiency and cooking temperatures that remain constant no matter how much food is loaded on them.

STEAM GRIDDLE. In recent years, at least two manufacturers have introduced steam technology to griddles. When water is heated under pressure—in this case, in a sealed chamber beneath the griddle plate—it creates superheated steam with a temperature of up to 400°F. Because steam produces amazingly even heat, the temperature does not vary more than 2°F anywhere on the plate. This eliminates hot and cold spots and allows foods to be cooked closer together. Steam griddles also come with top lids (instead of plates). When the lids are closed, steam is injected into the cooking area for even faster results.

TEPPANYAKI GRIDDLE. Another variation of the floor-model griddle is the **teppanyaki griddle**, specifically designed for the exhibition cooking so popular at Japanese steakhouses. The oversized teppanyaki is 8 feet long and 6 feet wide, with a countertop on three sides that allows seating of up to eight guests around a 12-square-foot cooking area. The chef stands on the fourth side—slicing, dicing, and searing the customers' food to order in a fast-paced presentation that's as tasty as it is fun. The 3-inch trough is closest to the chef, with a large drain hole and grease drawer beneath. The teppanyaki griddle plate is ¾ inch thick; gas burners are rated at 30,000 Btus per hour each; electric burners have a 208- to 240-volt rating.

CLAMSHELL GRIDDLE. The **clamshell griddle** cooks food on both sides simultaneously, therefore reducing cooking time by half. Imagine a big waffle iron (without its patterned surfaces) with a 2-foot hood that is lowered over food on the griddle below. The hood automatically ignites when lowered, reaching a temperature of 165°F within seconds. The hood is spring loaded and can be raised to add or remove food without substantially affecting the cooking speed. Lift it up, and it turns off automatically. The hood of a clamshell griddle does not close all the way, allowing a 5-inch clearance for the food when it is in its lowered

ILLUSTRATION 13-10 A fry-top range is a griddle on top with an oven below.

Courtesy of Lang Manufacturing Company, Everett, Washington.

TABLE 13-3

Griddle Cooking Guide (in Minutes)

	CLAMSHELL	GRIDDLE	CHARBROILER
4-ounce boneless chicken breast	3½	7	6
Hamburger (4-inch patty)	1½	3½	3½
Steak (¾ inch thick)	3½	7	6½
Grilled sandwich	1½	4	n/a
Halibut (¾ inch thick)	3½	7	6
Snapper	1½	4	n/a
Cheese melting	½	n/a	n/a
Hash browns	4	8	n/a
Sausage	2	4	n/a

Source: Lang Manufacturing Company, Everett, Washington.

position. Its cooking speed allows a 36-inch clamshell to boast the same food output as a 60-inch griddle (see Table 13-3).

The gas-powered clamshell has a stainless-steel frame and 1-inch-thick griddle plate. The plate can be flat or *sloping*. A ***sloping plate*** is tilted slightly to allow grease to drain into a large, removable pan. The hood contains a single infrared burner; each foot of griddle surface is heated by a specially designed burner (27,000 Btus per hour) and controlled by a thermostat. Clamshells come in widths (lengths) of 3 feet and 4 feet and can be ordered with grooved griddle plates in 1-foot widths.

CONTACT GRILL. The electric answer to the gas-powered clamshell is called a *contact grill*—and there's that troublesome term *grill* popping up again. At any rate, that's what it is called. The contact grill is a square griddle plate, smaller than a clamshell but also able to envelop the food and cook it on both sides at once. Models are available on legs or casters, with shelves or small steam tables mounted beneath. This appliance requires 12 kilowatts of electric power per hour.

Two more incarnations of the two-sided cooker are worth mentioning. One is Vulcan-Hart Corporation's Duplex Cooker. Every 12-inch griddle section has a top layer (called a *platen*) that can be lowered onto the griddle. The plate height can be readjusted to accommodate different types of foods, and each griddle/platen zone has its own controls (see Illustration 13-11). The Instant Burger (made by Smokaroma of Boley, Oklahoma) is another two-sided appliance. This one is unique because it uses direct energy transfer to cook hamburgers in half a minute. No oil is necessary for the Instant Burger griddle, and it requires neither preheating nor an exhaust hood. Like a microwave, it uses energy only when there's a product being cooked. This specialty

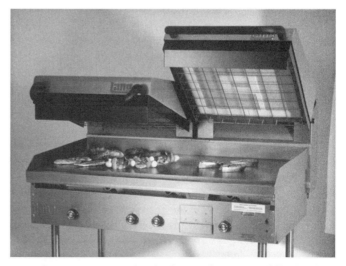

ILLUSTRATION 13-11 The duplex cooker is a griddle divided into sections. Each section has its own hood that lowers, like a clamshell, onto the cooking surface.

Courtesy of Lang Manufacturing Company, Everett, Washington.

ILLUSTRATION 13-12 The panini grill is a type of clamshell griddle that grills or toasts both sides of a sandwich at once.

Courtesy of Star Manufacturing International, Inc., St. Louis, Missouri.

griddle does well in small operations such as convenience stores and delis. Hot dogs, sausages, and chicken breasts can also be cooked in it.

PANINI GRILL. Paninis—toasted, grill-marked sandwiches stuffed with fresh ingredients and melted cheese— are popular and easy to make using the panini grill, a type of contact grill that cooks both sides of a sandwich at once. Sometimes called a *sandwich press*, it is versatile enough to use with breakfast items, flat breads, Reubens, and other sandwich types as well (see Illustration 13-12). Most panini grills are countertop models that do not require special ventilation. Multiple combinations of flat or grooved surfaces are available for top and bottom platens that can heat up to 570°F. Look for controls that allow you to adjust top and bottom cooking surfaces separately.

The biggest difference between models is the material the surfaces are made of. Aluminum is nonstick and heats quickly, but also it loses heat quickly. Cast iron retains heat well but takes longer to heat. Stainless steel heats quickly, is known for its even heat distribution, and is the most durable of the three.

MONGOLIAN BARBECUE. The *Mongolian barbecue* is a flat-surface griddle, 4 feet in diameter, which may run on either electricity or gas. At its namesake restaurant, the customer fills a bowl with thinly sliced vegetables and chicken, beef, or fish. The bowl is handed to an attendant, who tosses the ingredients on the griddle and sears them in less than one minute. The story behind the name *Mongolian barbecue* is that the Mongols, who invaded China centuries ago, placed their shields over fires and used them to cook on. The first griddles!

13-6 ■ BUYING AND MAINTAINING GRIDDLES

One of the most important considerations when shopping for a griddle is its heat recovery time. This is critical if you use foods straight from the freezer instead of from the refrigerator, because the colder the food is when you place it on the griddle, the lower the surface temperature drops. In tests conducted by the American Gas Association, it took 77 Btus of gas energy to bring 1 pound of beef from 40°F to 140°F. However, when the beef was frozen to a temperature of 0°F, it took 196 Btus of gas energy to bring it to 140°F—more than twice the energy output for frozen food as for chilled food.

So, when a manufacturer recommends your griddle burners should each have an average gas input of 30,000 Btus per hour, don't decide it's way too much. You may use that kind of energy only when the griddle is being preheated or when it is totally covered by frozen foods, but you do want to have that capability, and more Btu power helps cut your heat recovery time.

Another way to decide if the gas input is sufficient is to take the total Btu input figure you get from the manufacturer and divide it by the total number of square inches of surface cooking space. The figure you come up with should be at least 100 Btus per square inch. If it's not, the griddle might not meet your needs, especially if you cook a lot of frozen product.

To control the heat on the griddle surface, there are several options. Basic manual controls are ideal for users who require immediate control over the flame and careful monitoring of the food by the cook. For consistent heat that does not require monitoring, there are three types of thermostat:

1. *Hydraulic modulating.* The user sets the temperature level and a thermostat monitors the surface temperature, automatically adjusting the burner flames below as needed. Accuracy of this type of thermostat is plus or minus 20°F to 30°F.
2. *Snap action.* This is an electric analog thermostat. Again, the user sets the temperature and the thermostat adjusts the burner flames—but because, as its name implies, it snaps into action quickly, the temperature variation is not as wide, only 7°F to 10°F.
3. *Solid state.* This is the most sophisticated of the thermostat types; it adjusts temperature within 2°F to 5°F.

There should be one thermostat for each burner on your griddle.

The griddle plate can be ordered to specifications. Manufacturers have developed griddle plate surfacing that restricts the amount of heat radiated by the griddle, and they offer several choices of materials and thicknesses, so think about exactly what you cook. If you need to preheat quickly and make frequent temperature adjustments, a thinner (⅜–½ inch) griddle plate is recommended. If you use mostly frozen products, a thicker (¾–1 inch) plate holds heat without extreme temperature fluctuations. In *Food Equipment Facts*, Scrivner and Stevens describe these griddle materials from which to choose:

Cast-iron plates. These hold heat extremely well but are porous and do not contract and expand. Frozen foods tend to stick to cast iron, and the surface requires constant cleaning.

Polished steel plates. These eliminate the food sticking problem, but the shiny surface causes some product shrinkage.

Cold-rolled steel plates. These have excellent heat transfer properties; the product adheres well without sticking or shrinking; and they're easy to clean.

Chrome-finish plates. Again, these have excellent heat retention and transfer properties, and they're easy to clean using a blade or scraper, cold water, and a special dry chemical powder.

No matter what type of griddle plate you use, you must season it, both before first use and after each cleaning. By now you're probably familiar with **seasoning**, which means to lightly coat a cooking surface with oil. This prevents food from sticking while protecting the surface from rust. Failure to season a griddle consistently results in poorly cooked products and, eventually, permanent damage to the griddle plate.

When your griddle is first installed, preheating it to 200°F helps remove the heavy rust-preventive coating applied at the factory. Use a cloth dampened with cooking oil or a grease solvent to remove the coating when it is softened by heating.

One manufacturer suggests users then follow this five-step seasoning method:

1. Set the griddle temperature at 375°F to 400°F.
2. Pour unsalted cooking oil onto the griddle surface.
3. Allow the oil to heat until it begins to smoke.
4. Wipe the griddle surface thoroughly with a porous cloth.
5. Put on a second coat of seasoning oil and repeat steps 3 and 4.

Seasoning should be done only after cleaning—which, in this case, means removing all bits of burned and carbonized food from the surface. Usually this is done using a metal spatula or griddle scraper, with long, firm strokes. Never strike the griddle surface with the edge of the spatula, as this can easily nick the surface and eventually affect its cooking ability.

Griddles should be cleaned when they're warm but not too hot; manufacturers suggest between 150°F to 175°F. Some manufacturers sell specially made griddle cleaner that is spread across the plate, allowed to stand for 5 or 10 minutes, then scraped or wiped into the trough. Afterward, the plate must be sponged thoroughly with water to remove all cleaning residue before seasoning. Some chefs prefer to use water or ice to clean the griddle; the ice provides gentle friction to help loosen baked-on particles, although it is critical that the griddle be absolutely cold when ice is used or the temperature shock may warp the plate. Cleaning may also include lemon juice or vinegar to cut grease and leave the griddle plate looking shiny. Finally, it is always easier to maintain a clean-as-you-go policy than to scrape accumulations of burned food and carbon off the griddle.

A couple of other notes about griddle care: Electric griddles can be scraped clean with a **griddle stone,** also called a *brick* or *pumice,* which is pulled over the plate in the same direction as the grain of the metal. As tempting as it may be to use steel wool on some of the tough, baked-on burned food, don't do it. It will scratch the surface. (The other thing that warps a griddle is using it as a rangetop. Don't set full pans or kettles on it.)

The griddle's controls and connecting cables should also be wiped clean with a nonabrasive cleanser, and grease troughs should be emptied at least once a day.

13-7 TILTING BRAISING PANS

We include the tilting braising pan in this chapter because it is actually a griddle with raised sides, which gives it the appearance of a big, flat-bottomed kettle. It is often called a *tilting skillet, frypan,* or *brazier.* Whatever the name, it is perhaps the most versatile piece of equipment in a foodservice kitchen (see Illustration 13-13). When professionals have to choose between a tilting steam kettle and a tilting braising pan, they consistently choose the tilting braising pan.

Why? First, it is designed for easy operation. It consists of four parts: a large griddle surface for heat transfer, vertical sides to contain liquids and spatters, a tilting pan body to make loading and unloading food easier, and an attached cover to retain heat and control the cooking process. The braising pan is heated from the bottom (beneath the griddle area) by gas burners or electric elements. This, and its hinged cover, allows the pan to perform the duties of a rangetop, a griddle, a kettle, a steamer, a cook-and-hold oven, a bain-marie, and more. Its sturdy construction conserves energy, and its ability to handle so many cooking tasks mean less labor because it saves staff the chore of lifting and transferring foods from one pot or pan to another. Floor models can be put on casters for wheeling product out to serving areas, although they must be able to anchor to a wall to prevent them from tipping over when tilted.

A restaurant kitchen could use its tilting braising pan at breakfast as a griddle for cooking scrambled eggs, potatoes, and bacon; at lunch for making large batches of chili, soups, or stews; and at dinner for poaching salmon, steaming vegetables, slow-roasting a pot roast, or mixing a chicken casserole.

Let's walk through the typical functions of this appliance to show you exactly how this is possible. Note that the temperatures we suggest are estimated ranges; specific recipes may call for experimentation.

1. *Griddle cooking.* Its raised sides may get in the way a bit, but the griddle area of a tilting braising pan comes in sizes ranging from 9 square feet to almost 24 square feet. Keep the cover open, set the thermostat at 300°F to 400°F, and the griddle area will fry pounds of bacon, sausage, and burgers, or grill sandwiches and French toast.

2. *Pan frying.* Heavy loads of chicken or fish, or doughnuts to be deep-fried, are no problem here. Pour the oil into the pan, heat it from 350°F to 400°F, and keep the cover open for frying. A thermostat regulates the temperature to minimize oil breakdown.

3. *Kettle cooking.* Tilting braising pans can do all types of liquid batch cooking, with capacities that vary from 10 to 40 gallons. Wide temperature ranges provide for anything from simmering to rapid boiling. Having a griddle at the bottom enables the operator to brown meat and then pour everything else in and switch its use from griddle to kettle. Close the cover during kettle cooking and set the thermostat at 200°F to 225°F.

4. *Steam cooking.* Add 2 inches of water to the bottom of the pan, and it becomes a steamer. Racks can be purchased to hold as many as three full-size pans above the water. Make sure the pans are perforated to allow the steam to surround the food in them. The lightweight racks are easily removed for storage. During steaming, close the cover and set the temperature at 300°F to 400°F.

5. *Oven roasting.* The same wire racks can be loaded with meats to be roasted at very low temperatures (175°F to 275°F) to minimize shrinkage and hold in natural juices. Of course, the cover is closed during roasting.

6. *Braising.* A speedy way of preparing meat is to braise it. This means quickly browning it with an open cover, then lowering the temperature, adding a small amount of liquid to the pan, and simmering until fully cooked and tender. The recommended temperature for simmering is 225°F to 295°F.

7. *Poaching.* Add 2 inches of water to the bot-tom of the pan and set the thermostat at 200°F to 225°F, closing the cover for cooking. If poaching eggs, be sure to acidify and salt the water first.

8. *Proofing.* Add 2 inches of water to the bottom of the pan, place the pan of dough on a wire rack so it doesn't touch the water, turn the thermostat to a gentle 80°F to 100°F, and wait until the dough rises.

9. *Holding, thawing, serving.* Fill the tilting braising pan with 4 inches of water and set the thermostat at 175°F. With that, you've transformed it into a bain-marie, a steam table that keeps vegetables or sauces warm on a serving line. You could also hold rolls, bread, or pastries in it without using water and with the cover closed. If the unit will be used regularly in this way, rolling casters and a quick-disconnect feature for its gas hookup are recommended.

ILLUSTRATION 13-13 The tilting braising pan is actually an appliance, not a pan, that can fry, roast, steam, braise, poach, and proof. It can be square or round, and its large size and ability to tilt make it handy for large-volume cooking.

Courtesy of Groen, Unified Brands, Jackson, Mississippi.

Now that we've discussed the full range of capabilities, let's cover the construction details. Most tilting braising pans have ¾-inch-thick stainless-steel griddle plates at the bottom; even thicker plates are recommended if your menu requires preparing a lot of sauces. The sides of the pan are 10-gauge stainless steel, either 7 or 9 inches tall. Coved corners are recommended for easier cleaning. Several manufacturers mark lines on the sides to indicate capacities (10 gallons, 20 gallons, etc.).

The hinged cover is made from a solid sheet of 16-gauge steel, with a one-piece tubular handle welded on. The cover is supported by heavy hinges and a spring-loaded activator. At the rear of the cover, a *drip shield* catches condensate and funnels it back into the pan.

Because the pan is too heavy to lift manually, its tilting function is accomplished with a hand crank or an electric motor. Both are self-locking to ensure precise control.

The heating unit, located beneath the griddle plate, runs more than 100 small heat transfer lines around the cooking surface to ensure even heat distribution. It has two thermostats: one to control cooking temperature and the other to detect overheating and shut the unit off automatically if its temperature exceeds 450°F. When that happens, it takes about 30 minutes to allow the pan to cool down and reset the high-limit thermostat.

Tilting braising pan units stand on stainless-steel legs no more than 3 inches tall, making the total height of the appliance 20 to 23 inches. Where space is at a premium, the pan can be mounted in a cabinet to add extra storage space. Tabletop versions of the tilting braising pan require as few as 28 square inches of counter space, with capacities of 10 to 15 gallons. Even a 30-gallon model can fit into a space less than 3 feet square. They can be fueled by gas or electricity—both types can reach their cooking temperature within 10 minutes—or even hooked up to special batteries for portability. The pan is supported by a **trunnion** (a term usually used for cannons!), which means a set of sturdy steel pins that allows it to pivot.

The type of culinary operation often signals the size of pan you need. The highest-volume uses require a 30- or 40-gallon pan. The type of product you prepare most frequently also helps with the decision. If you stew foods, the formula is the same as for a steam-jacketed kettle: Multiply the number of portions by the portion size, then divide that figure by 128 ounces.

If you use it as griddle space, think about the number of portions that can be cooked on 1 square foot, then determine how many portions you have to cook at one time. This yields the total square footage needed.

Optional accessories and features may not be as optional once you see everything the tilting braising pan can do. Steamer conversion kits include wire racks and perforated pans; casters are recommended to allow easy movement of the unit for cleaning and repositioning. Fill faucets can be ordered to put water directly into the pan from a faucet mounting bracket so employees need not carry pots of water to the pan. Also, request the gallon marker feature, which allows you to deliver a measured amount of water into the pan without hand-measuring it. Sealed, water-resistant controls are preferable because liquids are so often part of this type of cooking. Coved (rounded) corners make for easier stirring and cleaning. Work trays and pan carriers can be ordered to fit most pans. Pan carriers make it easier to pour large quantities into the pan, no matter how it is tilted.

Finally, address these two considerations for installing your tilting braising pan. First, because the pans give off steam and other vapors that are natural by-products of the cooking process, most cities require that they be installed under an exhaust hood. Traditionally, they are located next to the steamers or steam-jacketed kettles because they share the characteristics of large-batch cooking equipment. They must be able to tilt in any direction without obstruction.

Second, consider installing the pan so that its pouring lip is located over a floor drain or at least over a recessed floor area. Occasional spills can't be avoided, and this makes cleanup much easier.

A cousin of the braising pan is the **skittle** or *combi-pan*, which tilts like a braising pan and can be used to accomplish most of the same cooking methods. It's used most often as a steamer. In fact, it's the only type of kitchen steamer that does not require a boiler, which results in lower maintenance costs.

■ SUMMARY

Many of the most popular types of surface cooking equipment are discussed in this chapter, including broilers, griddles, and grills. The three basic broiler types are overhead, charbroiler, and specialty broilers. Overhead broilers have their heating elements over the food; a charbroiler's heat comes from beneath the food, which is why it is sometimes referred to as an *underfired broiler*. Charbroiling uses flames, smoke, and radiant heat.

One popular specialty broiler is the food finisher, a countertop appliance that quickly reheats, browns, or crisps foods just enough to finish them before serving. Its cousins include the versatile salamander, a miniature broiler that can also be used as a plate warmer; and the cheesemelter, which does just what its name indicates and can also brown, poach, and boil small batches of food. The conveyor broiler is another great option for places that have high-volume cooking needs.

Broilers should be purchased based on the type of menu, amount of food that must be cooked at one time, and the type of heat source that works best for what you broil. One thing all broilers have in common is that they must be kept clean. Food and carbon buildup can be scrubbed away with a special broiler brush. They all also need an adequate ventilation system for smoke and grease, which are inevitable by-products of broiling.

The griddle is the flat-top appliance you have no doubt seen in quick-service eateries, used to cook hamburgers, pancakes, hash browns, and more. The cooking surface of a griddle is called its *plate*, and most griddles are divided into sections, called *heat zones*, with separate temperature controls so you can heat just as much space as you need. You can order an appliance with an oven or broiler under the griddle; a specialty griddle that heats on two sides simultaneously, known as a *clamshell*, is also available. There are several styles of clamshells and *contact grills*, so called because a heated upper plate as well as a heated lower plate comes into contact with the food.

Griddles must be cleaned regularly, but they must also be seasoned—lightly coated with oil to prevent rust as well as food sticking.

And finally, an incredibly versatile appliance—the tilting braising pan—combines the large, flat surface of a griddle with raised sides to contain liquids or spatters and a tilting body to make it easier to load and unload food in large quantities. The pan is heated from the bottom, either by gas or electricity. You can use the tilting braising pan to griddle, pan-fry, roast, braise, poach, steam, and more. Operators also like it because its tilting function minimizes the need to lift large amounts of food in heavy pans.

■ STUDY QUESTIONS

1. What is the difference between an overfired broiler and an underfired broiler? For what would you use each of them?

2. What is the grid (or grid assembly) of a broiler?

3. Many broilers come in three models or styles: drop-in, countertop, and freestanding. What is the difference, and why would you choose each of them?

4. What is a salamander, and where in the kitchen would you put it?

5. List three things you should check if your broiler does not seem to be getting hot enough.

6. How does the temperature of the food you are cooking affect a griddle?

7. How do you clean and season a griddle?

8. What is a clamshell?

9. What makes a tilting braising pan so versatile?

10. How do you decide what size tilting braising pan you need for your foodservice operation?

14

STEAM COOKING EQUIPMENT

■ INTRODUCTION AND LEARNING OBJECTIVES

It would be difficult to select a single chief advantage of cooking with steam in today's food-service businesses. Some would say it is speed; others cite the ability to cook multiple foods simultaneously without compromising their aromas or flavors. Still others cite the health benefits of steam cooking, a method that retains the natural consistency and nutrient content of many foods, even prepared in large quantities.

Institutional foodservice led the way in the acceptance and use of steam equipment, but it took new technology and health-conscious dining trends to prompt wider interest in steam cooking. Restaurant chefs sometimes turned their noses up at steam cooking because it certainly isn't the showy hands-on, open-flame type of preparation. Today, however, the programmable combination oven/steamer (commonly known as a *combi-oven*) is a staple in many commercial kitchens. As you will learn in this chapter, the **combi** offers three powerful cooking methods: hot air, steam, and a combination of the two, which amounts to the work of several appliances, saving money and space. It also holds hot food at its optimal temperature safely and without loss of moisture, flavor, or overall quality.

In this chapter, we'll cover what steam equipment can and cannot do, and how to select the right type of steam equipment right for your kitchen. Whether you want to make baked potatoes or bratwurst, to warm enough danish for an army, or cook just a single serving of fresh vegetables, you'll learn everything you need to know to choose and use steam equipment. In Chapter 6, we introduced the rudiments of steam energy. Now, we explore your equipment choices.

After reading this chapter, you will be able to:

■ Identify the important features of steam-jacketed kettles.
■ Identify the important features of pressure steamers.
■ Identify the important features of high-pressure and low-pressure steamers.
■ Identify the important features of pressureless steamers.
■ Identify the important features of combination pressure/pressureless steamers.
■ Identify the important features of specialty steamers.
■ Identify the important features of combination oven/steamers.

We also detail the sizes of equipment on the market, accessories available to use with steamers for all manner of applications, plus tips for choosing, installing, and maintaining steam equipment.

HOW STEAM COOKING WORKS

The principle behind steam cooking is simple: When water reaches its boiling point at 212°F, steam is its gaseous form. It takes only 180 Btus (British thermal units) of energy to turn a 1-pound block of ice into boiling water, but to evaporate the same water and make it into steam vapor requires 970 Btus. The energy of these Btus is contained within the steam and used to cook foods quickly but gently.

The secret to steam cooking is its efficiency as a heat transfer medium. The energy carried by the steam—about six times more energy than boiling water—transfers immediately upon contact with a cooler surface. This transfer can occur directly (e.g., in a steamer) or indirectly (through the wall of a steam-jacketed kettle). The *heat transfer rate* is the measurement in Btus of how much heat goes from one substance to another in a given time and space—actually, per square foot per hour per degree of temperature difference. How's *that* for a complex formula? Steam's heat transfer rate is 300 Btus, compared to a rate of 3 for air, 7 for circulated air in a convection oven, 38 on a griddle, and 85 when boiling in a pot of water.

To illustrate the difference, if you stick your hand into an oven cavity preheated to 400°F, you'll feel the heat but you won't burn your arm; but if you put your hand over a boiling teakettle, the 212°F steam will scald your hand immediately. The latter is simply a more effective method of heat transfer. In a closed steamer, however, food will not burn. As long as the temperature of the food is less than 212°F, the steam will continue to condense on the food surface and deliver 970 Btus of energy as it cooks.

Steam cooking uses less energy overall than conventional methods because power is needed only to keep the steam up to pressure while the equipment is operating. But in order to achieve higher temperatures than 212°F, the steam must be held under pressure. A quick physics lesson about steam: Water at its boiling point (at sea level) cannot exceed 212°F, no matter how much additional heat is applied. Instead, we must raise the pressure in the cooking chamber to raise the temperature of the steam. At 5 psi (pounds per square inch), water and steam temperatures rise to 222°F; at 10 psi, 239°F, and at 15 psi, the temperature hovers at 250°F. Conversely, lowering the pressure decreases the temperature.

This is often referred to as *superheated steam*. The name sounds impressive, but superheated steam doesn't offer any big advantages in the cooking process. Its most common use is to "dry out" the steam to cook certain types of foods that do not require as much direct contact with moisture. Steam with a small amount of superheat in it is often called *dry steam*.

Steam's reduced cooking time allows cook-to-order scheduling for many delicate foods that don't keep well under heat lamps or on steam tables. In some commercial situations, a small pressureless steamer gives the option of cooking vegetables to order, which might allow a restaurant to stock smaller quantities of more exotic fresh produce. In some steam appliances, almost any liquid can be used, including broth, a basic stock, or even wine—each of which imparts a distinct flavor.

Fresh vegetables are often steamed because this method preserves their flavor, color, texture, and nutrients. Comparison tests show that vegetables prepared in a steamer contain one-third to one-half more of their natural nutrients than the same vegetables prepared in a stockpot (see Table 14-1). Steam cooking also tends to yield more than other methods because the product doesn't shrink as much. Food handling is minimal, as most items can be prepared, cooked, and plated right out of the same pan. This results in less labor cost, with less stirring and handling of the food.

Interestingly, steam-cooking equipment actually reduces the number of pots and pans to be bought, used, and washed. The food simply does not burn on the pan, which reduces the need to scour and scrub. You can use the same unit for initial food preparation, reheating, and reconstitution. The equipment itself eliminates the lifting of heavy stockpots on and off the range.

For all its versatility, there are a few important cooking methods that steam cannot accomplish: It cannot bake, and it cannot brown. Braising is possible in a steam-jacketed kettle, but

---

TABLE 14-1

Steaming or Boiling?

This chart shows the dramatic savings of nutrients when steam is used in preference to a stockpot. The relationship among nutrients, color, flavor, and texture is close.

VEGETABLE	COOKING METHOD	LOSS OF DRY MATTER %	LOSS OF PROTEIN %	LOSS OF CALCIUM %	LOSS OF MAGNESIUM %	LOSS OF PHOSPHORUS %	LOSS OF IRON %
Asparagus	Boiled	14.0	20.0	16.5	8.8	25.8	34.4
	Steamed	7.09	13.3	15.3	1.4	10.4	20.0
Beans, string	Boiled	24.6	29.1	29.3	31.4	27.6	38.1
	Steamed	14.2	16.6	16.3	21.4	18.8	24.5
Beet greens	Boiled	29.7	22.2	15.9	41.6	44.9	43.1
	Steamed	15.7	6.9	3.8	14.1	14.0	24.5
Cabbage	Boiled	60.7	61.5	72.3	76.1	59.9	66.6
	Steamed	26.4	31.5	40.2	43.4	22.0	34.6
Cauliflower	Boiled	37.6	44.4	24.6	25.0	49.8	36.2
	Steamed	2.1	7.6	3.1	1.7	19.2	8.3
Celery	Boiled	45.4	52.6	36.1	57.1	48.7	—
	Steamed	22.3	22.3	11.6	32.4	15.7	—
Celery cabbage	Boiled	63.2	67.1	49.7	61.6	66.1	67.6
	Steamed	38.3	33.5	16.3	32.6	30.2	44.1
Spinach	Boiled	33.9	29.0	5.5	59.1	48.8	57.1
	Steamed	8.4	5.6	0.0	17.8	10.2	25.7
Beets	Boiled	30.9	22.0	18.7	30.9	33.6	—
	Steamed	21.5	5.4	1.5	29.4	20.1	—
Carrots	Boiled	20.1	26.4	8.9	22.8	19.0	34.1
	Steamed	5.1	14.5	5.1	5.6	1.0	20.7
Kohlrabi	Boiled	33.6	23.2	27.8	40.4	27.7	51.7
	Steamed	7.6	1.0	1.0	14.3	7.7	21.3
Onions	Boiled	21.3	50.2	15.6	27.8	40.2	36.1
	Steamed	11.0	30.7	7.1	15.7	31.5	15.9
Parsnips	Boiled	21.9	13.3	11.4	46.8	23.7	27.6
	Steamed	4.6	20.0	4.2	18.2	5.7	8.1
Potatoes	Boiled	9.4	—	16.8	18.8	18.3	—
	Steamed	4.0	—	9.6	14.0	11.7	—
Sweet potatoes	Boiled	29.0	71.5	38.3	45.3	44.4	31.5
	Steamed	21.1	15.0	22.1	31.5	24.3	25.1
Rutabagas	Boiled	45.8	48.6	37.1	42.7	57.2	50.0
	Steamed	13.2	15.7	13.4	3.4	24.6	14.3
Average for all vegetables	Boiled	39.4	43.0	31.9	44.7	46.4	48.0
	Steamed	14.0	16.0	10.7	18.6	16.7	21.3

Source: North American Association of Food Equipment Manufacturers, Chicago, Illinois.

meat cannot be browned. For both browning and baking, you need a combination of steam and dry heat, which is possible in the combi units discussed later in the chapter.

As handy as it is, steam equipment is effective only if it fits your kitchen layout. Be sure to look at the locations of water lines and drains, and for steam-jacketed kettles, ventilation hoods are necessary. Does it make sense to centralize steam equipment in one part of the kitchen, or can you work more efficiently with a large steamer on the hot line and a couple of smaller ones in the à la carte area?

STEAM AND WATER QUALITY

Steam equipment uses a great deal of water, and not all of it to create steam for cooking. Many gallons are used to clean the equipment and to cool the condensate to a safe enough temperature (less than 140°F) to allow for disposal into a floor drain.

Before we introduce the types of steam equipment, we cannot stress strongly enough that the steam used in cooking must be clean and produced from potable water. Most experts recommend that water used to make steam be treated to prevent many service problems, usually by installing a water softener.

It is also a good idea to protect steam equipment from possible damage by chloramines. *Chloramines* are now added to many city water systems as a replacement for chlorine because—unlike chlorine—chloramines do not dissipate in the water system. However, when exposed to temperatures above 180°F, chloramines pit and rust stainless steel. Therefore, a hollow carbon filter designed to remove chloramines is recommended for any stainless-steel equipment that produces temperatures above 180°F and comes into contact with treated water—such as steam kettles and combi-ovens.

Most of the time, municipalities monitor the level of chloramines in their water. To be safe, though, it is best to consult with local water authorities to determine if local tap water contains chloramines and if your own system should include filters to eliminate them.

The recommended water quality standards are:

- Total dissolved solids of no more than 80 parts per million (ppm)
- Water pH factor of 7.0 to 8.0
- Water hardness not to exceed 2.0 grains
- Total alkalinity not to exceed 20 ppm
- Maximum allowable silica 13 ppm
- Maximum allowable chloride 30 ppm

The amounts of iron, chlorine, and dissolved gases in the water all affect steam generation. The most frequent problems with steam equipment are caused by mineral buildup over time, on the interior sides of the units and on the parts that water comes into contact with, such as heating elements and sensors. Regular cleaning can minimize but not completely eliminate these problems. Because water supplies vary from location to location, consult the local water treatment agency (usually the municipality) before you install any steam-generating equipment.

Also, be aware that warranties and/or service agreements on steam appliances do not cover repairs for breakdowns created by water problems. And like many other appliances, steamers must be level for proper operation. Your warranty won't be honored if your steamer is installed improperly. You'll learn more about water quality and its impact on steam generation in the sections entitled "Steam Generators" and "Cleaning and Maintenance of Steam Equipment," also in this chapter.

14-2 STEAM-JACKETED KETTLE

The *steam-jacketed kettle* cooks much more quickly than a stockpot on a rangetop, and uses less energy. It can stir-fry, stew, and more. No kitchen today should be without this versatile piece of equipment, but choosing the best unit for your needs requires careful calculation, as they come in sizes from 1 quart to 160 gallons. Each type of kettle has particular qualities and operational requirements. Kettles are used to prepare soups, stews, gravies, puddings, sauces, pasta, eggs, and rice. They can partially precook foods, like fresh vegetables, for finishing later, and they can finish-cook foods that are precooked in other equipment. A kitchen that prepares whole cuts of meat, fish, and poultry can use a kettle to cook the trimmings for other purposes rather than waste them. Kettles can simmer all day, combining the leftover meats and julienne of vegetables to make stocks. We'll continue extolling the kettle's virtues after we look at some of its more technical aspects.

A steam-jacketed kettle works like a combination double boiler and bain-marie. One round, hemispheric bowl is sealed within another, with about 2 inches of space separating the bowls. Steam is introduced into this space (see Illustration 14-1). The interior bowl is welded along its top to the outside of the exterior bowl (the jacket). Specially constructed baffles within the jacket provide even distribution of heat inside the kettle. The pressure of the steam can be adjusted—for rapid cooking, it is increased; for slower cooking, it is lowered.

Steam does not actually come in contact with the food; it stays in the 2-inch space as described. As it enters and condenses on the inner surface, it transfers its heat to the stainless-steel wall, which, in turn, transfers heat to the food. The steam expands to fill the entire space between the kettle walls, with no hot spots or temperature variations. Because the kettle is almost always covered during the cooking process, it is fast and allows less heat to escape into the kitchen. And because the kettle transfers heat through its entire jacketed sides and bottom, it offers three to four times more surface area for heat transfer than the same size of stockpot. In fact, the kettle is about 65% more efficient than its rangetop counterpart.

Look back at Table 6-2 in Chapter 6 for the variety of steam pressure settings and the corresponding temperatures they generate. Remember, only a few degrees are lost as the steam condenses on the wall of the inner bowl, so most of the heat goes right to the food without scorching or burning it.

To summarize, the usefulness of the steam-jacketed kettle is based on its superior ability to transfer heat rapidly and maintain uniformity of temperature throughout the heated surface of the kettle. In a busy kitchen, it can be a real workhorse. In addition to the uses mentioned earlier, these kettles can boil large quantities of water, reheat food, slow-cook sauces, and even make coffee or tea in mass quantities with efficiency and consistency. For banquet service or other special functions, coffee can be drawn from the fixed kettle directly into large coffee pots or urns for serving. (There's another option for coffee, too, which we'll mention later.) Illustration 14-2 shows a common restaurant kitchen setup for three kettles.

ILLUSTRATION 14-1 The construction of a steam-jacketed kettle allows it to cook food slowly and evenly. As shown here, the steam does not come in contact with the food.

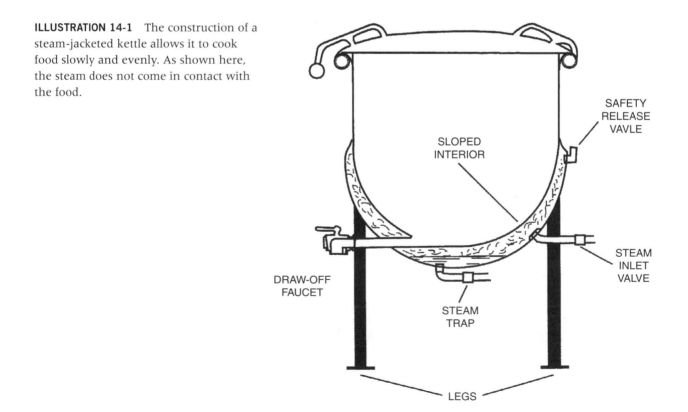

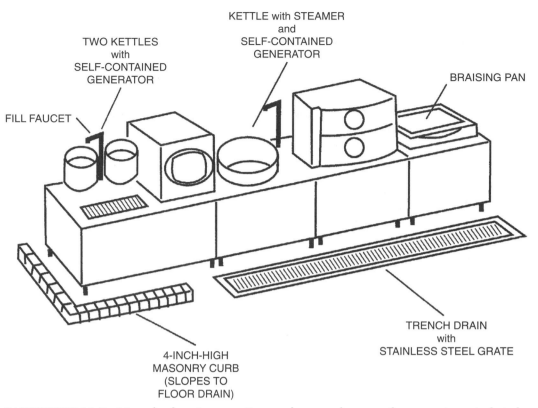

FILL FAUCET

TWO KETTLES
with
SELF-CONTAINED
GENERATOR

KETTLE with STEAMER
and
SELF-CONTAINED
GENERATOR

BRAISING PAN

4-INCH-HIGH
MASONRY CURB
(SLOPES TO
FLOOR DRAIN)

TRENCH DRAIN
with
STAINLESS STEEL GRATE

ILLUSTRATION 14-2 Many foodservice operations make room for more than one steam-jacketed kettle.

Source: Carl R. Scriven and James W. Stevens, *Manual of Equipment and Design for the Foodservice Industry* (Albany, New York: Thomas Learning, 1989).

Most steam-jacketed kettles are made of stainless steel. Choose one with an especially high grade of stainless-steel lining (called a *316-type lining*) to prevent corrosion if you'll be preparing highly acidic products, such as tomato-based soup, sauces, and pizza toppings.

During the cook-chill process, which is the topic of Chapter 15, the steam-jacketed kettle is used to both cook and refrigerate food. If yours is a direct-steam kettle, you can connect it to a source of cold water or ice and use it to cool hot foods or mix cold salads. You'll learn more about direct-steam kettles in a moment. One enterprising school foodservice worker uses the kettle to soften the burned-on crusts in baking pans before washing.

At least one manufacturer has created an insulated kettle—an important achievement, because workers can brush against it without worrying about getting burned. No matter how hot the inside is, the exterior remains cool and safe to the touch. Another plus is that the kitchen stays cooler.

VARIETIES OF STEAM-JACKETED KETTLES

To harness its versatility, manufacturers have developed a wide variety of kettle types and sizes using different mounting devices and sources of steam.

DEEP-JACKETED. This kettle is more cylindrical than round (hence the name *deep*) so it is a good alternative where floor space is limited but output must be high. It is stationary and can be mounted on the floor or wall. Floor mounting may be on tubular legs or a pedestal (see Illustration 14-3). It usually includes a one-piece hinged cover, steam inlet and outlet valves, a safety valve, a draw-off line, and a faucet. Stock or soup can be drained through a removable strainer that fits into the draw-off pipe.

ILLUSTRATION 14-3 A deep-jacketed kettle is deeper and narrower than a regular steam-jacketed kettle, and can fit where space is tight.

Courtesy of Vulcan-Hart, a division of ITW Food Equipment Group, LLC, Louisville, Kentucky.

You get more versatility with two 30-gallon kettles than with a single 60-gallon model.

SHALLOW OR FULL-JACKETED. Similar in construction to the deep-jacketed type, this kettle is wider and shorter. Its width makes it easy to empty, but it is not desirable for certain products. For instance, when cooking a large quantity of chicken, the sheer weight of the meat might mean that the bottom layers of chicken come out looking squashed instead of appetizing.

TWO-THIRDS JACKETED. A deep cylinder with a hinged cover, two-thirds of its jacketed area is surrounded by steam. Like their deep-jacketed cousins, these kettles are used for foods with high liquid content that are easy to stir and require little attention. The height of the kettle allows for vigorous stirring without spillover. Again, this kettle requires minimum floor space for the amount of work it can do.

TILTING OR TRUNNION. Designed with a pouring lip, this kettle is mounted on pivots (called *trunnions*) that allow the appliance to be easily tilted so its contents can be emptied completely and comfortably without having to lift it. A self-locking feature is included to tip the kettle at whatever angle you need. Larger tilting kettles use a hand-cranked wheel or motor-driven tilting device; smaller countertop models use a hand lever (see Illustration 14-4).

Tilting kettles are also available in smaller, tabletop sizes. Ranging in capacity from 10 to 40 quarts, tabletop models are installed individually or in groups on counter space or tables. The table can be custom-made and fitted with a small gas or electric steam-generating unit, but the trend is toward self-contained electric kettles that are simply plugged in, without retrofitting an existing countertop. They also come with their own stands.

These kettles are ideal for preparing small quantities of vegetables, sauces, and cereals. They are useful in bakeries to warm syrups and fruit or chocolate sauces or to make puddings. The self-contained electric unit is factory equipped with rust inhibitors and a water purifier to prevent mineral buildup within the kettle steam chamber. The unit also features a sight glass to check the water level in the chamber.

SIZING AND SELECTING KETTLES

Let's assume you have already considered cost, operating life, availability of repairs, and so forth. Now your task is to choose which kind, what size, and how many kettles you want to buy. Although it sounds like a simple job, you have a lot to think about. You'll go through three steps to make the decision:

1. Determine the kinds of foods you serve and estimate the volumes you need, especially peak volumes.

2. Look carefully at your available counter and floor space. How much room do you have for kettles? You may be able to fit a stationary kettle into a hot line, but a large tilting kettle must be freestanding to tilt correctly.

3. Consider your workers' ability to handle the food amounts, especially if you'll be buying a large kettle.

One calculation for determining kettle size is to estimate its output, which is roughly 8 pounds of meat or vegetables (or 4 pounds of poultry) for each gallon of kettle capacity. These numbers take into account shrinkage (which is minimal) and allow some **head space**—the term for the empty space between the top of the food and the rim of the kettle. Most manufacturers recommend that 15% of the total kettle volume be left for head space.

Many manufacturers also have charts that suggest kettle sizes based on either numbers of meals to be served or portion sizes of those meals. Table 14-2 is a sizing chart, reprinted with permission from the *Handbook of Steam Equipment* by the North American Association of Food Equipment Manufacturers.

Typical kettle sizes in today's market range from a few quarts for tabletop models to 160 gallons for floor models. Always consider the space available for installing a steam-jacketed kettle, and find out building regulations in advance of purchase. Don't be surprised to discover that kettles must be installed under exhaust hoods. Be flexible as you make trade-offs. If you're short on space, you may have to use one larger kettle instead of two smaller ones, simply because the big one takes up less room.

ILLUSTRATION 14-4 A tilting or trunnion kettle can be locked into place for filling and cooking or tilted for pouring and easier cleaning.

Courtesy of Groen, Unified Brands, Jackson, Mississippi.

TABLE 14-2

Kettle Sizing Guide

NUMBER OF MEALS SERVED DAILY	NUMBER AND SIZE OF STEAM-JACKETED KETTLES
100–250	One 20-gallon kettle*
251–350	One 30-gallon kettle
351–500	One 40-gallon kettle
501–750	Two 30-gallon kettles
	or
	One 60-gallon kettle
751–1000	Two 40-gallon kettles
1001–1250	Two 40-gallon kettles and
	One 20-gallon kettle
	or
	One 60-gallon kettle and
	One 40-gallon kettle
1251–1500	Three 40-gallon kettles
	or
	Two 60-gallon kettles

*Obviously, two 10-gallon kettles could be substituted.

Source: North American Association of Food Equipment Manufacturers, *Handbook of Steam Equipment* 2000 (Chicago: NAFEM).

Here are guidelines gathered from a combination of foodservice industry publications and personal experience:

■ Use 10- and 20-quart kettles for gravies, sauces, and the like.
■ Use a 20-quart tilting tabletop kettle for preparing small quantities of sauces.
■ Use a 20-gallon kettle for vegetables.
■ Use a 40-gallon kettle for stewing.
■ Use a 50-gallon kettle for soup stock.
■ Use a 30-gallon kettle for 600 meals.
■ Use a 40-gallon kettle for each 800 meals served during peak periods.

POWER SOURCES FOR STEAM-JACKETED KETTLES

Either kettles are *self-contained* and generate their own steam, or they are *direct-steam* models and obtain their steam from a remote source—although nowadays, few modern facilities have access to direct steam sources. Self-contained kettles recycle water condensate and operate without water or drain plumbing connections; their temperatures range from 150°F to 298°F, which offers plenty of heat for warming, simmering, boiling, and braising duties. All such kettles have a safety valve that releases jacket pressure at preset levels and provides a way to add distilled water to the system. The factory-installed water distiller, rust inhibitor, pressure gauge, temperature controls, water sight gauge, and low-water cutoff make this a desirable piece of equipment because it's almost foolproof.

ELECTRIC (SELF-CONTAINED) MODELS. Steam is generated by immersing electric heating elements in a water reservoir located below the kettle jacket. Typical electrical ratings for floor-mounted models vary from 10.8 to 16.4 kilowatts. The voltage varies from 208 to 480 volts; three-phase installation is the norm, although some manufacturers provide single-phase installation for 20- or 40-gallon kettles.

Tabletop models require as little as 30 kilowatts; they are available in a single- or three-phase configuration and use 208 to 480 volts.

GAS (SELF-CONTAINED) MODELS. In these kettles, gas or propane burners boil the water to produce steam. Most gas models are stationary, but several manufacturers offer tilting models. The energy usage varies quite a bit with these models, depending on their size. Floor models range all the way up to 129,000 Btus per hour; tabletop models require as few as 31,000 Btus per hour for a 20-quart model.

DIRECT-STEAM MODELS. The remote energy these kettles receive can come from a main boiler designed for a whole facility, a boiler designated solely for kitchen use, or a boiler located near the kettles solely to provide steam for their needs. Boiler units are built to comply with the American Society of Mechanical Engineers (ASME) Code for unfired pressure vessels. The steam is piped into the jacket and flows to the inner wall, where it releases its energy load, condenses, and drops to the bottom of the jacket. The piping is usually fitted with a valve to regulate steam flow; it may also have a pressure-reducing valve. Once the condensate (now hot water) collects, the steam trap functions as an automatic valve and discharges the water. This valve action prevents loss or variation in the steam pressure during the cooking process. The condensate can flow into a floor drain or be returned to the steam boiler, depending on the design. A safety valve is sensitive enough to correct for as little as 2 to 5 pounds past the desired pressure.

A chief advantage of the direct-steam kettle is that cold water can be pumped into the jacket instead of steam. This is useful for chilling foods or cooling them after cooking.

If your facility is located in an older building with *process steam*—that is, steam used for heat and moisture rather than for power—it can be advantageous to use direct steam. In many cases, when an older building is upgraded and systems are replaced, the process steam is eliminated. This may in turn require you to replace steam equipment that has been operating with direct steam.

MOUNTING STEAM-JACKETED KETTLES

Mounting a kettle means securing it to a surface so it is safely anchored and doesn't tip unless you want it to. Don't be confused by the fact that you can attach a floor model to a wall. In fact, steam-jacketed kettles can be mounted in four ways:

1. *Pedestal mount.* A stainless-steel base is factory-welded to the kettle; the circumference of the pedestal is flanged down vertically to seal it to the floor. Being on a pedestal provides open space, making it easy to clean under the base. Because you need a way to get water out of the chamber, a 2-inch draw-off pipe is part of the kettle. Another option is a stainless-steel yoke stand or table, with or without a tilting mechanism.

2. *Tubular leg mount.* Stainless-steel adjustable floor flanges support tubular legs, making a tripod that is welded around the kettle. Again, you can install trunnions to enable the kettle to tilt.

3. *Wall mount.* This installation provides the highest degree of sanitation and is often used in conjunction with modern buildings' power distribution systems, which are known as **raceways**. In other words, the kettle can be installed on a cooking line and attached directly to the building's major power (and water) sources. Many manufacturers produce wall-mounted units especially designed to attach to raceway systems. It is critical to coordinate this installation with the kettle manufacturer as well as the raceway system designer. With or without a raceway, the wall must be strong enough and constructed correctly to support a full, heavy kettle, with or without a tilting mechanism.

4. *Cabinet mount.* These kettles can also be tilting or stationary, but they are usually part of a cooking battery. The complete assembly is available with legs as a standalone unit, or the entire cabinet can be wall-mounted, which is often referred to as a *console unit*. Keep in mind that this mounting usually features a built-in drain to attach directly to a permanent floor drain, so placement is affected by plumbing (see Illustration 14-5).

ACCESSORIES AND SPECIAL USES

Here are the options, both standard and nontraditional, available for most kettles.

Cooking baskets. Not unlike fryer baskets, these stainless-steel inserts are used to load, cook, and remove products prepared in boiling water. They are especially useful when cooking vegetables, pasta, rice, and potatoes. Normally, single baskets are used, but floor-mounted models can accommodate triple-basket inserts.

Covers. Some stationary models come with covers, and most manufacturers offer them as an option on all their models. As mentioned earlier, covering the kettle retains more moisture and nutrients, speeds cooking time, saves energy, and keeps the kitchen cooler.

Tables and stands. A manufacturer will mount all self-contained tabletop

ILLUSTRATION 14-5 A cabinet-mounted kettle is, as its name implies, a kettle that can be fitted into a cabinet or hot line.

Courtesy of Cleveland Range, A Welbilt Company, Cleveland, Ohio.

ILLUSTRATION 14-6 Stands and tables, like this one, can be ordered to fit gas and electric kettles.

Courtesy of Vulcan-Hart, a division of ITW Food Equipment Group, LLC, Louisville, Kentucky.

kettles on tables if you request it, and the same goes for direct-steam models. Usually two or more tabletop kettles can be mounted on one table. When this happens, a pouring sink usually is included and positioned in the pouring path of both kettles. It's worth shopping around for this handy option; some manufacturers also include a hot- and cold-water faucet with a swing spout to service the two kettles. A kettle on a stand is shown in Illustration 14-6.

Agitator mixer. This is a mechanical mixing unit that virtually eliminates the need for hand-stirring. Mixers make it easy to prepare large quantities of delicate cream sauces and other food products that need constant stirring. Heat is evenly distributed by twin-shaft agitators and scrapers.

Steam-jacketed coffee urn. You can make from 20 to 150 gallons of coffee without a hot-water source. The fast-heating action of the kettle quickly brings cold water to the proper temperature for coffee brewing.

Steam-jacketed oyster cooker. You shuck 'em, and there's a special kettle to cook 'em! This is a direct-steam piece of equipment; one manufacturer sells them in 32-, 64-, and 72-ounce capacities.

CLEANING STEAM-JACKETED KETTLES

Steam-jacketed kettles are relatively easy to clean and maintain. We'll talk about cleaning in detail later in the chapter, but here are the basics: It is best to turn the steam off and let the kettle cool down before emptying it for cleaning. Scrub the walls with a plastic brush, and open the drain to let the soapy water drain out as you work. Be sure the kettle is fully rinsed, inside and out. Clearly, the kettle should be installed where it can drain into a floor drain. For tilting kettles, the drain placement should align with the pouring path. Tabletop units should have a sink to drain into. If you don't have time to clean your kettle immediately after use, at least fill it with water and turn on the steam to heat the water and start the cleaning process by loosening any food particles.

A helpful accessory for smaller tilting kettles is the drain stand, which has a built-in shelf catch pan that extends from the base to catch cleaning water and divert it to the rear of the catch pan and into a drain line. This wastewater can then be easily routed to a floor sink behind the equipment, saving you valuable time to clean and dump the kettle.

14-3 PRESSURE STEAMERS

As you consider a steamer purchase, think first about your menu. Delicate items, such as fish fillets, benefit from pressureless convection steamers because the cooking process keeps them whole and preserves their texture. Sturdier items, such as pork loins, can be started in a high-pressure steamer for a terrific tenderizing effect and finished later on a grill. Also think about the numbers of portions you prepare at a time: Whether you produce single portions, small batches, or hundreds of portions at a time, there is a pressure steamer model for you.

It is important to check with local building officials prior to purchasing your steamer to identify the requirements for a boiler-based supply system. Many jurisdictions now require local boiler inspection for these types of units due to the potential hazards pressure boilers present. Finding out exactly what is required by code aids in your decision-making process.

Pressure steamers, also known as pressure cookers or **compartment steamers**, are ideal appliances for cooking fresh, defrosted, or loosely packed foods. They cook at up to 228°F, with steam pressures that range from 2 to 15 psi. Where high volume is required and all meals are served at once, the compartment steamer is the appliance of choice. Generally, it is a floor model, available with two, three, or four compartments. Its downside is that you must be careful not to overcook. Also, if not properly cleaned between uses, the pressure steamer may transfer residual flavor—in other words, you'll taste the last dish along with the one you're now cooking. Popular applications of pressure steamers include:

- Cooking stews, pot roasts, ham, and roast beef. Shrinkage is minimal, and steamed meats are tender, moist, and flavorful.
- Cooking potatoes and other root vegetables quickly in large quantities: 100 pounds per compartment in 35 minutes. Most other vegetables are cooked in 5 to 10 minutes. Frozen vegetables take even less time.
- Boiling pasta and rice faster than in pressureless steamers.

We already know that steamers cook by transferring heat from steam onto a cold food product. In pressure steamers, however, the cold air that comes off the product is vented away from the cooking chamber as steam is introduced. As temperature inside the chamber increases and cold air is removed, a steam trap slowly closes and steam pressure builds up inside the chamber. This pressurized steam (also called *lazy steam*) works its way through the food and cooks it from the outside. Frozen block products don't cook well in this situation because the outside becomes overcooked before the inside is cooked properly.

Pressure steamers come in high-pressure (15 psi) and low-pressure (2 to 5 psi) models. Low-pressure compartment steamers provide higher productivity at lower operating cost than pressureless models when cooking single items in volume. However, high-pressure units are usually faster and more economical. They cook at about 250°F and are regularly used for such basic kitchen duties as blanching and reheating. With most steamer models, you can put the food on sheet pans or steam table pans.

Pressure steamers may run on electricity (24 to 48 kilowatts per hour) or gas (170,000 to 300,000 Btus per hour). A typical unit is 36 inches wide and 33 inches deep; four-compartment units are as tall as 67 inches. When shopping for pressure steamers, be sure they are equipped with: heavy-duty doors and gaskets and pan slide racks that are all easy to clean; compartments with seamless fabrication to prevent leaks; an adequate steam condenser and drainage system; and pressure gauges and safety devices. A separate timer for each compartment is also recommended.

LOW-PRESSURE STEAMERS

The power requirements for low-pressure steamers are the same as for pressure steamers. They differ from standard pressure steamers because their own boiler is built into a stainless-steel cabinet underneath the steamer. You can also purchase low-pressure steamers that use direct steam from a central source or have steam coil heat exchangers. The steam coil heat exchangers can take incoming steam from other building systems at 35 to 50 psi and boil it into clean steam for use by the low-pressure steamer. If you're using direct steam, the experts recommend 40 to 50 psi of incoming steam with a flow of 34.5 psi per hour per compartment. A pressure-reducing valve can reduce compartment pressure to as low as 2 psi.

The low-pressure compartment steamers feature a handy combination pan-and-shelf slide unit. Pans may be pulled out two-thirds or more of their length without tipping to check the progress of cooking food. Simply remove the center assembly or partition of the slide unit. Each compartment can hold standard-size sheet pans.

HIGH-PRESSURE STEAMERS

High-pressure steamers do not offer the high-volume cooking capacity of their compartment steamer or low-pressure counterparts. A high-pressure steamer can hold only three standard-size

TABLE 14-3

Approximate Steamer Cooking Times

FOOD	POUNDS PER 100 PORTIONS	COMPARTMENT STEAMER (5 PSI)	HIGH-PRESSURE STEAMER (15 PSI)
Apples	30	15 minutes	10 minutes
Asparagus	36	15	8
Broccoli	40	12	8
Brussels sprouts	24	15	8
Cabbage	16	20	10
Carrots	16	25	12
Lima beans	20	20	10
Peas	30	8	5
Rice	26	30	15
Spinach	36	7	4
Potatoes	30	30	20

Source: Foodservice Equipment and Supplies Specialist, a publication of Reed Business Information, Oak Brook, Illinois.

hotel pans, compared to eight for the compartment steamer. However, they are in heavy use in places where small batches are in continuous demand, such as large table-service restaurants. The high-pressure steamer operates with a total pressure of about 15 psi, which raises the cooking temperature of the food and therefore reduces the cooking time, as shown in Table 14-3.

The high-pressure steamer comes in countertop and cabinet-mounted models, both gas and electric. The gas models use 40,000 Btus per hour, and the electric ones run on 12 kilowatts per hour.

High-pressure direct-steam countertop steamers, small and quick, are recommended for small batches of frozen vegetables. Despite their size (18 × 22 × 26 inches), they can produce up to 900 (2½ ounce) servings of vegetables within an hour. Be sure to use a steam coil model, which is the only one capable of purifying the steam for culinary use.

14-4 PRESSURELESS STEAMERS

Pressureless steamers, also called *convection steamers*, cook food at 212°F. There are two main differences between pressurized and pressureless steam cooking. First, during the pressureless process, the steam comes into direct contact with the food. Second, the doors may be safely opened at any time during the cooking cycle to check, rotate, or season the food—unlike other types of steamers, where this is not recommended.

Pressureless steamers are smaller and don't cook as quickly as their pressurized counterparts, but they are so easy to use that they account for more than half of all steamers sold in the United States. They provide higher-quality results for frozen foods, fresh vegetables, and seafood. The combination of moisture and relatively low temperature provide a better taste and texture. And, like other steamers, there's no problem with food burning, sticking, scorching, or drying out. Most professionals suggest that small-quantity batch-cooking needs are best met by a pressureless steam cooker.

In the convection steamer, the steam is injected into numerous inlets arranged to create a *spray* that circulates inside the cooking chamber. The heat transfers from steam to the food by the turbulence this creates—a kind of forced convection.

Some units have a fan to help circulate the steam around the chamber, and a vent system is used so air is continually eliminated from the cooking chamber. This technique provides fast, moderate cooking at a steady temperature of about 212°F. In addition, venting prevents the buildup of gases, odors, or any other by-products of the cooking process that could compromise food quality. Some people compare pressureless steaming to microwaving, only without the microwave's drawback of unevenly cooking or drying out some foods.

The preferred pan is perforated, 12 × 20 inches in size or smaller. However, solid pans may be used for potentially messier foods, such as pasta, rice, meat loaf, casseroles, or stews. The pressureless pure steam environment is capable of 300% more efficiency than a combination of air and pressurized steam. And because the steam is not pressurized, the door of the unit can be opened any time during the cooking process—and the temperature remains at 212°F.

Here are three practical kitchen realities for which pressureless steaming can be helpful:

1. Untrained personnel are less likely to have accidents with the pressureless steamer.
2. If food must be seasoned or handled during cooking, you can open the door.
3. You can cook large quantities faster than, say, in a stockpot or oven—although it's still slower than pressure steaming.

In recent years, manufacturers have created higher-power and higher-capacity units to enable the pressureless steamer to compete with the low-pressure steamer. New models, with Btu ratings of 150,000 to 300,000, have greatly increased volume capacity and accelerated cooking times. You can also purchase a large unit with more than one boiler for more than one compartment, two powerful blowers in each compartment, and two large steam entry ports for better steam distribution.

The basic pressureless steamer is usually less expensive than pressure steamers, and it saves energy as well. The typical features of a pressureless steamer include:

■ Multiple steam generators (one for each cooking cavity) do not require pressure reheat valves or gauges because they are not pressurized, so there are fewer components to fail. Energy can be saved by using only those compartments that are needed.

■ A powerful blower is installed in each cavity to increase the velocity of steam within that cavity.

■ A simple, dependable 60-minute timer is necessary. You'll also want a warming mode that keeps the boiler simmering between uses and the cooking cavity warm and ready to power up within seconds.

■ Easy-clean features include an automatic deliming process. There's usually a removable side panel where you pour a deliming solution into the appliance as well as a warning light to tell you when to do it to prevent overheating or other malfunctions.

■ A specially hinged door has a heat-resistant gasket and an easy-open latch. Always look carefully at appliance doors because they're sure to be opened and closed thousands of times every year. Also, look for a door that is easy to change (to open from either left or right) in case you want to move the steamer. A magnetic switch should cut power to the blower whenever the door is opened.

■ Simple utility connections mean faster, inexpensive installation and easy disconnection if necessary for cleaning or servicing. A wide sink with a drain under the door of the cooking compartment reduces the hazard of wet or slippery floors near the steamer and provides a convenient landing spot for hot pans.

CONNECTIONLESS STEAMERS

The traditional steamer is sometimes called a *connection steamer* (referring to the need to hook it up to a boiler that makes the steam). So it makes sense that the newest type of convection steamer touts itself as connectionless and/or boilerless, requiring no traditional plumbing hook-ups. Water is added manually to a tank that contains one or more heating elements, so all you need is a water source and an electrical outlet to plug into. This steamer has a couple of advantages: It has no boiler to clean, and it is easy to move as needed. A fan circulates steam within the cooking cavity.

The connectionless steamer is not a substitute for higher-volume, fully plumbed steamers, but it can be useful in low- to medium-volume operations. It can hold three to six sheet pans. This type of unit uses less water—perhaps 8 gallons *a day*, compared to 30 gallons *an hour* for a traditional steamer—and no water filtration system is necessary. It is electric; some manufacturers submerge the heating elements in the water tank and others do not. Some models have low-temperature vacuum steaming capability. The boilerless, connectionless steamers are quite new; their reliability and popularity are still unproven. But they are yet another example of the evolution of foodservice equipment to meet a wide variety of needs.

STEAM AND POWER REQUIREMENTS FOR PRESSURELESS STEAMERS

Most facilities, particularly small- and medium-size restaurants, opt for a pressureless steamer with a self-contained generator that makes its own steam. However, pressureless steamers can also use direct steam from a central supply or an external boiler in buildings that already supply steam for water heating or temperature control.

Unless your unit is self-contained and creates its own steam, you must determine whether the available steam supply is sufficient to run your pressureless steamer. For example, a steam boiler rated at 2.2 BHP (boiler horsepower) can generate enough steam for most two-compartment (six-pan) steamers but *not* enough to run *two* steam appliances in the kitchen at the same time. The manufacturers are clear about each unit's steam requirements—roughly, it takes 0.75 BHP to run one steamer compartment, or 1 BHP for every 20 gallons of steam-jacketed kettle capacity. Water pressure should be no more than 60 psi and no less than 30 psi. Keep in mind the distance steam must travel—it loses pressure on its journey through pipes—as well as the sizes of piping and fittings used.

Countertop models have their own self-contained steam boilers, which might be powered by gas or electricity. Either way, they need a clean, direct water supply and appropriate drain lines. Countertop units go through a preheat phase before the cooking process begins. The ones that preheat faster use more power.

Electrically powered steamers require from 5 to 48 kilowatts; a 48-kilowatt model is capable of delivering 4.7 BHP. Gas-operated units use anywhere from 45,000 to 250,000 Btus. Under normal conditions, they take 6 minutes or less to preheat.

SIZING AND SELECTING PRESSURELESS STEAMERS

Because we've been discussing countertop models, let's begin with their typical dimensions. On the outside, a countertop model is 26 inches wide, 19 inches high, and 29 inches deep, including its utility connections. The stand on which it can rest measures 24 inches wide, 25 inches high, and 29 inches deep. These figures increase proportionally with a stacked system, where two units are placed atop each other, or the quad system, where two units are stacked on top of two more, for a total of four steamers side by side (see Illustration 14-7).

ILLUSTRATION 14-7 Pressureless steamers can be stacked, one compartment atop another, or located side by side.

Courtesy of Cleveland Range, A Welbilt Company, Cleveland, Ohio.

TABLE 14-4	
Foodservice Steamer Needs	
MEALS PER HOUR	COMPARTMENTS
0–200	1 single-compartment steamer
200–400	1 double-compartment steamer
400–600	1 triple-compartment steamer
600–800+	1 quadruple- or 2 double-compartment steamers

Floor models usually have two compartments, and their overall size is 24 inches wide, 33 inches deep, and 58 inches high, although some are 36 inches wide. This includes the base.

Generally, the cabinet base of floor models houses the boiler and is made of stainless steel with a full-perimeter angle frame and reinforced counter mountings. Most have hinged doors and adjustable stainless-steel legs. The desirable height for a steamer is 32 inches.

Although overall dimensions are important for planning your floor space, you must also think carefully about the interior capacity of the steamer. Table 14-4 provides estimates for pressureless steamers based on the number of meals served per hour. We assume each compartment can hold three or four 12 × 20 × 2½-inch pans or two or three 12 × 20 × 4-inch pans. As with steam-jacketed kettles, most manufacturers offer cooking capacity guidelines with their steamer models.

Manufacturers also provide suggested cooking times for various foods, although these are all relative. As with ovens, ranges, or any other cooking appliances, a good chef makes allowances for temperature fluctuations; food freshness, size, and shape; the depth and type of pan used; and the desired degree of doneness.

A common complaint among new steamer users is that they are so fast, it's easy to overcook items. Use the manufacturer's manual until you are comfortable deciding on cooking times yourself, and remember to set the timer. It automatically cuts off the steam at the appropriate moment and alerts you with its loud buzzer. With a little practice, you'll find steamers are great for blanching items such as chicken before breading, cooking thinly sliced meats that don't need browning, freshening stale breads, and reheating almost any type of food.

14-5 SPECIALTY STEAMERS

This category includes a few highly specialized steam appliances designed for limited but common purposes.

FLASH STEAMER. The *flash steamer*, also called a *steam food cooker*, is a countertop model used to heat sliced meat sandwiches, melt cheese, and warm rolls and other pastry products without making them soggy. There are *closed* and *open* flash steamers. In both cases, distilled water is heated and converted into steam inside a sealed cavity. In the closed steamer, the steam escapes through tiny holes and enters the food, which creates pressure and velocity in the system. In the open steamer, a steam generator sprays water over the bottom heated surface. This creates steam, which naturally rises into the food area.

Flash steamers come in two sizes: The full-size model accommodates a 4-inch-deep steam table pan, and the half-size model takes a 2½-inch-deep pan. There are two operating styles as well. You can either push a button to energize the water pump or solenoid valve, which allows water to enter the steam generator, or press down on a handle or arm lever to start the process. Flash steamers require 120-volt electric power (15 to 20 amps) and have simple controls for rapid reheating of precooked foods. The steam food cooker can also hold cooked foods until needed.

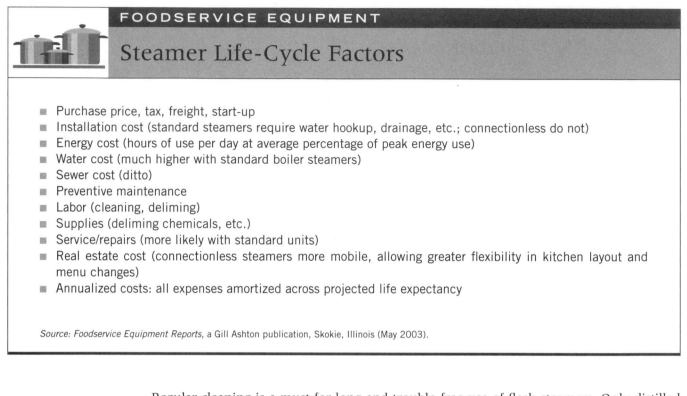

Steamer Life-Cycle Factors

- Purchase price, tax, freight, start-up
- Installation cost (standard steamers require water hookup, drainage, etc.; connectionless do not)
- Energy cost (hours of use per day at average percentage of peak energy use)
- Water cost (much higher with standard boiler steamers)
- Sewer cost (ditto)
- Preventive maintenance
- Labor (cleaning, deliming)
- Supplies (deliming chemicals, etc.)
- Service/repairs (more likely with standard units)
- Real estate cost (connectionless steamers more mobile, allowing greater flexibility in kitchen layout and menu changes)
- Annualized costs: all expenses amortized across projected life expectancy

Source: Foodservice Equipment Reports, a Gill Ashton publication, Skokie, Illinois (May 2003).

Regular cleaning is a must for long and trouble-free use of flash steamers. Only distilled water should be used because mineral deposits find their way onto the heating surfaces.

SPATULA STEAMER. The spatula steamer—so named because the insulated handle on its perforated basket makes it look like a spatula—is for reheating individual portions. It's great for things like late-night hotel room service and à la carte restaurant orders. It can be used either to cook or to reheat, shooting steam at the food as many as 14 times in a 2½-minute cycle. It contains a built-in boiler that uses plain tap water to generate steam when plugged into a 110- or 208-volt electrical source. A release valve at the back of the unit allows excess steam to be safely released.

NEEDLE INJECTION STEAMER. Finally, here is just the right steamer for a perfectly warmed Danish! The *needle injection steamer* offers a fast, simple, and consistent way to reheat or warm porous food products. In this appliance, needles are arranged in a specific pattern to heat one type of product. When food is pressed down onto the needles, steam is released directly through the needles into the food. Although this direct internal steam penetration is efficient from a heat transfer point of view, it makes sense to have this steamer only if the type of food you'll be heating with it makes enough money for you to justify the cost. Pastries do well in this environment, but dense items, such as baked potatoes or bratwursts, do poorly.

PASTA/RICE STEAMER. Other steamers can cook pasta and rice perfectly, but there are *pasta and rice steamers* especially designed to cook an 8-ounce portion of refrigerated pasta in 30 seconds or the same serving of frozen pasta in 90 seconds. This piece of equipment releases steam at high velocity into a small enclosed cavity. According to its manufacturer, it's also ideal for quick reheating of other precooked items.

STEAMER/FRYER. The public's interest in more healthful cooking led to the development of a *steamer/fryer*. For foods that already contain enough oil to brown, the steamer/fryer can cook a product with the same taste and texture as if it were fried. After the food is steamed, the steamer/fryer switches to "brown" mode to create a crisp, golden-brown exterior. Potatoes, chicken pieces, breaded cheese sticks, onion rings, and some types of seafood turn out well in the steamer/fryer.

14-6 COMBI-OVENS

Combi-oven is the term for single appliance that can cook via steam or convection (the circulation of hot air)—or both at once (see Illustration 14-8). This unit, with its variable humidity controls, consumes up to 60% less energy than traditional cooking appliances. Today's combis are technologically sophisticated and highly programmable. They can be set to turn on and off automatically and programmed to cook dozens of products with specific settings and steps. However, most of these are a combination of three distinct cooking modes:

1. Steam mode emulates the speed of a pressure steamer—without the pressure.
2. Hot air mode allows convection oven heating with the added feature of humidity.
3. Combi cooking uses a combination of superheated steam plus hot air.

The combi-oven moves air around mechanically within the moisture-filled cooking cavity. By combining convection heat with the steam's moisture, the result is a moister, fresher cooked product with a longer holding life. The moisture keeps nutrients and flavor in the product while the air movement speeds up the cooking process. Browning also takes place faster and more evenly in the presence of moisture.

The chief benefits of combi-ovens are their speed and versatility. They are handy in preparing all sorts of foods, from roast chicken to steamed vegetables, poached salmon to pot pies, soups to desserts. This is also the unit to use when preparing meats or baking rolls, breads, or pastries that require perfectly browned crusts. With a combi-oven on the hot line, waste is no longer an issue. Vegetables can be prepared just before they're served instead of sitting and getting soggy. Daily specials can be plated and cooked in advance, then steam-reheated and sauced for service. Pasta can be cooked in advance and reheated in less than a minute.

The first combi-oven was electric, introduced in the 1970s by Rational Cooking Systems, Inc., of Schaumburg, Illinois; in the mid-1980s the company added a gas-heated model. The original units were enormous and expensive, found only in high-volume foodservice sites such as hotels, casinos, and hospitals. Over the years, smaller and simpler combis have been introduced; these are popular for take-home and supermarket deli applications. In high-volume settings, food may be cooked early in the day, rolled into a blast chiller for a quick cool-down, then plated,

ILLUSTRATION 14-8 The combination oven/steamer, nicknamed *combi-oven*, combines moisture and convection heat and can take the place of several pieces of cooking equipment.

Courtesy of Rational Cooking Systems, Inc., Schaumburg, Illinois.

garnished, covered with plastic wrap, and put in a display case. When a customer selects it, the combi can retherm it right in the store.

Combi-ovens not only cook many kinds of foods; they cook them fast. Production times are up to 40% quicker than conventional ovens. For example, a typical restaurant rotisserie cooks a whole chicken in about 1 hour and 20 minutes. If you forgo the enticing appearance of the traditional rotisserie, a combi-oven has the same chicken ready to serve in 35 minutes, with the same taste, moist interior, and nicely browned exterior texture. The oven's convection airflow means it can cook at lower temperatures, increasing the yield of meat products up to 30%. Many combis use computerized temperature probes to sense when meat is fully cooked and automatically stop when the meat reaches the desired core temperature.

The combi's high-temperature settings rival a pressure steamer for cooking time. Fresh steam is directed into the cavity from a self-contained steam generator and rapidly circulated by a power blower. All frozen and most fresh vegetables do well in this mode, if you remember that load size and the type of product influence its cooking time. Shellfish, seafood, meats, and poultry also cook nicely at high-temperature settings with no liquid added. Dehydrated items, however—rice, pasta, and cereals—must be covered with water. The ratio of water to rice is 2:1; water to dry pasta, 3:1.

Low-temperature steam enhances the cooking of items like frozen chopped spinach, fresh leafy vegetables, and fresh broccoli or asparagus. The low-temperature setting simply circulates the steam more gently.

The hold mode is not intended for primary cooking but rather to keep foods warm or reheat something that's already been cooked and then refrigerated. The hold temperature is preset at 145°F with a light moisture level to preserve food quality, although it's probably not wise to hold most foods for more than an hour, as their overall quality and appearance is bound to suffer a bit. Large cuts of meat should be tempered (marinated or tenderized) before putting them in the hold mode, where they can then remain for several hours. For instance, if you know a roast will be sitting for a while, adjust the combi's hold temperature to 10°F below the desired final temperature.

Combi-ovens save time, increase yield, and are versatile. Did we mention they also save space? Most models can be stacked. They must be installed under an exhaust hood, but they are smaller than many hot line appliances, so not as much hood is needed. (When you consider that restaurant exhaust systems cost between $800 and $1,400 per linear foot of hood, every inch counts.)

The air inside a combi-oven circulates through a *fat filter*, trapping food odors and grease particles before the air reenters the cooking chamber. This filter is mounted on the baffle at the rear of the oven, where it is easily removable after the oven racks are removed. (Make sure the oven is cool before you try to remove the filter.) At least one manufacturer recommends having two fat filters so you can wash one while the other is being used. They're easy to wash, either by hand with a mild soap or by running them through the dishwasher.

Tabletop combination oven/steamers can hold from 10 to 20 steam table pans. If you prefer using sheet pans, most can accommodate a stainless-steel slide rack that fits neatly, is easy to remove for cleaning, and can hold up to 10 sheet pans. Most manufacturers also have stands on which to put tabletop units.

Some floor-model combis can be adapted to fit on rolling carts, which is handy if your operation does a lot of cook-chill preparation. The loaded cart can roll handily from oven to refrigerator. The dimensions of these rolling units vary, but they generally come in full and half sizes. When determining the correct size for your kitchen, remember to calculate another 6 inches of height for rolling legs.

WATER, STEAM, AND POWER REQUIREMENTS FOR COMBIS

In electric combi-ovens, the air is heated by a heating coil in the oven interior, which requires from 10 kilowatts for countertop models to 72 kilowatts for floor models. Gas-operated units are floor models only; their Btu requirements range from 90,000 to 225,000 per hour.

In the electric combis, you have a choice of steam production systems: a built-in steam generator or steam injection. Choosing steam injection does away with a generator and its associated

cleaning and deliming—water is simply sprayed into the hot oven cavity, where it turns immediately to steam. It works well for small quantities of product and costs less to operate than a
boiler, but the boiler is still the best choice for high-volume cooking. For gas models, a built-in
generator is the only choice. It is located either in back of or beneath the cooking cavity.

Most models have a steam generator standby mode that allows the generator to switch
pressures without a cool-down period. Steam generators empty from inside the oven, flushing
their systems completely to eliminate water buildup, scales, or lime deposits, which saves
money you may otherwise spend on costly water treatment. As mentioned earlier, however,
a water-softening system is recommended to reduce the risk of contamination. Manufacturers
recommend even lower levels of total dissolved solids (less than 60 ppm) for combi-ovens
than they do for pressureless steamers (less than 80 ppm). In other respects, however, water
requirements are similar: an alkaline content of less than 20 ppm and a pH factor greater than
7.0. Combi units are sensitive enough, and used often enough, to recommend a regular cleaning schedule.

They require a ¾- to 1¼-inch cold-water pipe with 45 to 75 psi of water pressure as well
as a separate indirect waste pipe with air gap. They also need a water supply shutoff valve and
a backflow preventer. As you might have guessed, it is best to have a professional plumber
handle installation.

All units have automatic controls to preset and regulate the oven temperature. The
amount of moisture inside the oven is relative to the temperature setting; that is, even when
cooking with steam, as the temperature increases, humidity decreases. Obtaining full saturation, or 100% relative humidity, is not possible because steam in the combination mode is
pressureless and combi cooking almost always requires hot temperatures (above 212°F).
However, you can adjust and control the oven's humidity to some extent.

Many models offer programmable, computerized controls that do more than implement
cooking combinations. An employee can choose the amount and temperature of the steam as well
as the fan speed to circulate it. A computerized combi can be programmed for 99 cooking modes
or more, each of which can contain multiple steps. A correctly programmed combi-oven allows a
cook to put a certain food into the oven
chamber and press a button. What a luxury!

Combi controls, whether solid state or
computerized, are extremely sensitive to
temperature. If the unit will be installed
next to other hot line equipment (broilers,
fryers, ovens, etc.), a *side shield* is needed to
protect the combi-oven from heat damage.

VACUUM STEAM COOKING

This variation on the combination oven/
steamer is commonly known as a **vapor
oven**. Preparing food with this low-temperature steam appliance means the food is
never exposed to high or uneven temperatures. Vapor ovens are relatively new pieces
of equipment, electrically heated (using
from 4 to 6 kilowatts per hour), and built
into a heavy-duty stainless-steel cabinet
(see Illustration 14-9).

Here's how vacuum steam cooking
works. The cooking chamber holds water,
which is preheated to the desired temperature. Inside the chamber, a vacuum pump
reduces the air pressure by pulling air out
of the cooking chamber. This lowers the

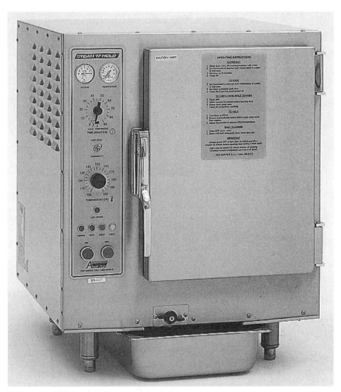

ILLUSTRATION 14-9 The vapor oven or vacuum steamer is a
combination oven and low-pressure steamer.

Courtesy of AccuTemp Products, Inc., New Haven, Indiana.

boiling temperature of water to a (relatively) cool 140°F or 150°F, creating a low-temperature steam with a superior heat transfer rate. This means food can be cooked with a minimum of temperature difference between the food and the steam. Steam transfers heat uniformly—that is, it condenses on all exposed surfaces depending on their temperature rather than their proximity. More steam condenses on the colder parts of the food. The steam continues to condense there until the food's surface temperature reaches the same temperature as the steam itself. This process naturally lowers the temperature in the cooking chamber, which causes the thermostat to turn the heater back on. The ongoing cycle means food is cooked evenly and thoroughly without having to turn it, baste it, or watch it constantly. A timer on most units shuts off the steam and puts the oven in hold mode when the food is done.

This type of cooking is perfect for delicate items like fresh green beans, broccoli, and fish fillets. Vapor oven temperatures range from about 140°F to 200°F. Timer, thermostat, and cooking cycles are all set using an electronic control panel.

Energy savings is one big advantage of vacuum steam cooking. It uses only 10% of the power of a conventional steamer because it makes only as much steam as it needs. There is no excess to be condensed and drained.

14-7 STEAM GENERATORS

There are three sources of steam for steam cooking equipment:

1. From a central, outside source, usually a building's existing steam supply, that may be tapped into with piping to run appliances
2. From a freestanding **boiler** or *steam generator* (the terms are used interchangeably in the industry) that heats water and makes steam for two or three appliances at once
3. From an appliance equipped with a built-in boiler

We've already mentioned the importance of "clean" steam, uncontaminated by chemicals in the water. Centrally supplied steam from an outside source may be economical, but the piping for such a system is expensive, which may mean grouping your steam equipment to save money. Central steam supplies are becoming a rarity, though, so most foodservice businesses have freestanding steam generators or appliances with built-in boilers.

Steam production is measured either in pounds per hour (pph) or in boiler horsepower (BHP). A good general guideline is that 1 BHP is needed for every 20 gallons of kettle capacity, and ¾ BHP is needed for each steamer compartment.

Steam generators can be purchased to run on electricity, gas, or steam coils. If you're shopping for a generator, first look at what is available in the market because models vary quite a bit. Then be sure to match the capacity of the boiler to the total demands of the steam appliances it is designed to service.

If you attach multiple pieces of equipment to the same boiler, turn them on one at a time, never all at once, to give the generator time to provide sufficient steam and pressure. A few more boiler basics follow.

ELECTRIC STEAM GENERATOR. Constructed according to standards of the American Society of Mechanical Engineers (ASME), when you turn on its heat and water switches, the unit fills with water automatically. It also drains automatically under pressure when the switches are turned off. It uses three-phase power, between 24 and 48 kilowatts, depending on its size. An electric unit can generate steam at 5 to 15 pounds psi. If two pressures are needed—for example, 5 psi for one unit and 15 psi for another—a pressure-reducing valve to provide the 5 psi is necessary.

Available dimensions range from 24 to 36 inches wide, 33 to 36 inches deep, and 28 to 33 inches high. The stainless-steel cabinet is mounted on a 24- or 36-inch-wide base, typically on four 6-inch stainless-steel legs. Its two front doors have magnetic latches, and all piping is confined within the cabinet.

GAS STEAM GENERATOR. Also constructed according to ASME guidelines, the gas-fired model is rated (sometimes called its *firing rate*) between 140,000 and 300,000 Btus; about 60% of that figure actually reaches the food being cooked.

The gas-fired boiler provides 15 psi of pressure and has an optional valve that lowers pressure to 5 psi if necessary. The dimensions and control systems of the gas model are similar to the electric model.

When writing specs for a gas generator, always specify the elevation where it will be installed. In communities higher than 2000 feet above sea level, special gas orifices must be used.

STEAM COIL STEAM GENERATOR. You can use a coil system with either electric or gas. Steam coils circulate steam to heat water, but they need another type of boiler to create the steam. To be part of a steam generation system, coils need a minimum incoming water pressure of 20 psi and maximum incoming pressure of 50 psi. They can produce 2.6 BHP and 85 pounds of steam per hour. A 120-volt single-phase electrical connection is necessary to operate the coil controls. Consider using a reducing valve to facilitate the steam-operated equipment connected to this generator.

Boilers are generally blown out, or drained, at the end of a busy day and refilled the next day. To maintain its efficiency, a boiler must be rigorously cleaned to prevent buildup of lime, scales, and other mineral deposits on the boiler's walls and around the water-heating element. This buildup also forms acidic conditions inside the boiler, accelerating spot corrosion and causing eventual leaks. The frequency of this cleaning depends on whether your area has hard or soft water, but it must be done even if the boiler has its own water filter. Some boilers are easier to clean than others; a port may allow you to add descaling agents or inspect the interior, or a warning light may comes on when cleaning is due. Metallic corrosion can be controlled by installing a *cathodic protector*, a metal device mounted inside the boiler, suspended in the water. You cannot eliminate acidity and scaling, but you can minimize it with a good water treatment system and regular maintenance. Service companies say that forgetting to delime is the most common cause of repair-related calls.

Finally, no matter which power source you use, remember to comply with the manufacturer's water quality requirements. Equipment failure caused by inadequate water quality is not covered under warranty.

14-8 STEAM EQUIPMENT CLEANING AND MAINTENANCE

Now that we've cleaned the boilers, let's turn our attention to the rest of the steam equipment, where scale buildup is also a problem. It forms on any surface of the appliance, from interior walls to water-sensing probes to heating elements. In fact, each ¼ inch of lime-scale thickness increases the energy required to produce sufficient steam for the appliance by 40%. It pays to descale and delime, and it is a lot less expensive to do so than to replace equipment. Some full-featured appliances, including combi-ovens, have automatic deliming programs that sense buildup and alert the operator to run a deliming cycle.

You'll get greater energy efficiency from steam-operated equipment if you use it at full capacity. That means a couple of smaller steamers, well loaded, are usually better than one large one that is used less or operated half full. Whenever possible, cook similar-size pieces of food together; it's faster, and the food will be more uniformly prepared.

Another energy-saving practice: Place an external steam generator as close as possible to all the steam-operated equipment to reduce heat loss in transit. Remember, cooking speed can triple with units that introduce steam directly onto food (e.g., vapor ovens).

Also explore the option of cooking food to near doneness in steamers and then finishing it on a griddle, fryer, broiler, or even in the oven. You can save energy by steaming first, then using the more conventional cooking appliances to add the color, crust, or flavor your customers crave.

Steamer maintenance means making sure steam is properly contained and safety valves work to prevent ruptures of the cooking cavity from too much interior pressure. Check all gaskets regularly for leaks, and look at the door mechanisms while cooking to ensure they fit tightly. Once or twice a week, lift and check the safety valve to make sure it is not corroded to the point of being ineffective under excessive pressure. Check pressure gauges, pilot lights, timers, and thermostats for proper calibration. Always keep the user's manual handy on the job site.

Stainless-steel surfaces, both interior and exterior, are key to prolonging the useful life of these appliances. They are usually easy to clean. You can apply a degreasing spray to the interior, then activate the steam mode to loosen any food particles. A built-in spray hose rinses away the grime. Some experts consider steam equipment a pot washer's best friend because of its ability to steam off caked-on food particles on pots and pans. Here are a few more cleaning specifics:

- Allow the unit to cool with the door open. Remove the shelves by unlatching them; scrub them in the pot washing sink with detergent and hot water; rinse and drain.

- Always scrub both interior and exterior with a soft brush. With every cleaning, check the compartment drain to clean out any clogs.

- Rinse the inside and outside of the steamer with clean, hot water; then, close the door and turn on just enough steam to heat the interior. You must wipe down the exterior with a clean, dry cloth; the stored heat should be enough to dry the interior.

- Install and latch the shelves.

For the interior oven cavity, the latest technology in combi-ovens is a self-cleaning model. Clip a device into a slot on the back of the oven, and enter a number onto the control panel that indicates the intensity of cleaning desired (from 1 for light cleaning after bread-baking to 7 for heavy, greasy jobs like roasting chicken). The cleaning device has arms that move around, dispensing correct proportions of water and detergent, then a drying agent.

■ SUMMARY

Steam cooking is an efficient way to transfer heat to food. Steaming also preserves the flavor, color, and texture of fresh foods; keeps them moist; and reduces waste and shrinkage. You can use a steamer to prepare food to the point of near doneness, then finish it on a griddle, broiler, and so on. The faster cooking time is useful when you need to prepare delicate foods quickly, especially those that don't keep well under heat lamps or on steam tables.

There are several types of steam equipment, all popular in foodservice. The steam-jacketed kettle is constructed like a bowl within a bowl. Steam is pumped into the small space that separates the two bowls and cooks the food without coming into direct contact with it. Its large area of surface heat makes the steam-jacketed kettle extremely versatile. Kettles range in size from 1-quart tabletop models to 200-gallon floor models. They must be installed under a ventilation hood. They can be mounted in cabinets, on walls, on legs, or on pedestals.

A related appliance, the pressure steamer or compartment steamer, can be used to boil rice or pasta and cook meats or root vegetables. Inside this steamer, the cold air that comes off the food product is vented away from the cooking chamber as steam is introduced. The pressurized steam cooks the food from the outside in. There are low-pressure and high-pressure steamers as well as pressureless models. The designations refer to the pounds per square inch (psi) of steam that may be generated, which determines the maximum temperature the steam can reach. Pressureless steaming is actually steaming at very low pressure, which is best for delicate foods.

During pressureless steaming, the steam comes into direct contact with the food. Unlike with other steamers, the door or lid may be opened safely at any time to check or season the food. Pressureless steamers are also called *convection steamers*.

Combination steamers can be used either with or without pressure; and combination oven/steamers can also cook with or without steam. The major benefits of combi-ovens are their speed and versatility, which allow them to take the place of several pieces of equipment. Today's combi is a technological marvel, programmable to cook dozens of foods even when the cooking processes require multiple steps.

Steam equipment requires steam-generating capacity. There are three ways to achieve this: by tapping into a building's existing steam system; by purchasing a freestanding steam

FOODSERVICE EQUIPMENT

Placement of Steam Equipment

Your choice of steam equipment often is determined by the floor space, plumbing connections, and power sources available in your kitchen. Here is a summary of the seven factors to keep in mind when sizing up a potential steam installation site:

1. Most steamers require a floor drain for removing condensate from the cooking compartment and the boiler.

2. Determine the water quality; provide for a water softener if needed.

3. Install floor drains in front of tilting pans and kettles to help contain liquids that spill on the kitchen floor.

4. Consult the manufacturer's requirements for sizes of water inlets and drains. An air gap is recommended between steam generator and floor drain.

5. Locate the floor drain outside the perimeter of the equipment to prevent the venting steam from rising and condensing on the unit's electrical components.

6. Look at your city or county ordinances for proper ventilation and fire suppression requirements for steam-operated equipment.

7. Consider the proximity of electrical power to the equipment. Even if the boiler is gas-fired, you may need electricity to power the appliance itself.

Source: National Association of Food Equipment Manufacturers, *Handbook of Steam Equipment* (Chicago: NAFEM, 2000).

generator (or boiler), which can create steam for several appliances; and by purchasing equipment with its own built-in boiler. Freestanding generators should be placed as close as possible to the equipment to reduce the loss of heat and pressure as the steam is piped to its destination. In all types of steam generation, water quality is a critical factor. Mineral solids from water build up over time and can coat the insides of appliances, heating and sensor systems, and the like. That's why filtration and regular cleaning are requirements for proper steam equipment operation.

■ STUDY QUESTIONS

1. Which qualities make the steam-jacketed kettle a superior heat transfer unit?

2. What are chloramines, and what precautions do they require from owners of steam equipment?

3. Is wall mounting an option if you buy a floor-model appliance?

4. Briefly explain the benefits and disadvantages of pressure versus pressureless steaming. Why would you choose one or the other?

5. What is superheated steam, and how or why is it used in cooking?

6. What is the desirable height for a steamer to rest above the floor?

7. How accurate are suggested food cooking guidelines? What are some considerations in determining cooking time?

8. When would you prefer a pasta and rice specialty steamer to any of the more versatile units?

9. Of all the steam-cooking units described in this chapter, which has the highest heat transfer ratio?

10. Your budget says you can afford only one type of steamer unit for your new quick-service restaurant. Which one would you choose, and why?

Jim Hungerford

OWNER, BOISE APPLIANCE AND
REFRIGERATION COMPANY, BOISE, IDAHO

Jim Hungerford is the owner of Boise Appliance and Refrigeration Company, a busy full-service equipment maintenance and repair firm with clients that include schools and military bases as well as numerous restaurants in southern Idaho. Hungerford has been in the foodservice equipment business since 1982.

Q: When a restaurant first opens, can you tell whether it will be able to stay in business based on the equipment decisions the owners make?
A: With any new business, you have to keep a close watch on your cash flow. What I have experienced with first-time restaurant operators is they try to save money by purchasing used equipment when they would probably be better off to buy new. Most dealers of used equipment only warranty it for 30 days, but new equipment carries a full 1-year parts and labor warranty. That alone keeps their maintenance budget down, at least for the first year.

It's extremely hard to say whether or not a restaurant will make it in today's marketplace because there is so much competition. In my opinion, it is very hard for the mom-and-pop operation to succeed without a lot of capital and a unique specialty item or concept to interest the customer.

Q: Most chefs seem to have strong opinions about using gas or electric appliances. From a service standpoint, what do you think?
A: The local utility rates are a factor in choosing the type of equipment, for obvious reasons, but my personal preference is for all gas equipment. It's quicker, it's easier to maintain, and, to me, gas appliances give off a more even heat.

With electrical equipment, you have a lot more components that make up the inner workings of the appliance. From a service standpoint, electrical equipment is generally more expensive to repair than gas equipment. Electrical problems can also be harder to diagnose due to the more prevalent use of solid-state controls that contain integrated circuitry.

Q: What is the most common service problem you see with ranges and ovens?
A: Most often, it is calibration of equipment and the inexperience of the people who operate it. All new restaurant owners should familiarize themselves fully with the owner's and operator's manuals. We receive a lot of service calls simply because the owner is not familiar with how to use the equipment—and because this is not a factory warranty situation, the owner is responsible for paying for the service call.

Q: Is cleanliness of the equipment a problem?
A: It's a really big problem. The fryers are a prime example; we have had customers bring fryers in that are in such bad shape—because they were never cleaned—that it is actually cheaper for them to buy a new one than to try and repair the old one. The wiring and electrical controls can become so grease-soaked that they either stop working or become a fire hazard.

Q: So how often should you wipe them down?
A: I believe all cooking equipment should be wiped down and cleaned on a daily basis. This should be done at the end of every workday, and there should be one person designated to perform these tasks. I would also suggest purchasing a fryer with a built-in filter system, or at least a portable oil filter. Oil is expensive, and filtering it will save on your oil costs.

Q: What about griddles? They look indestructible, but I know there are some maintenance tips for them. What's the most common problem you see?
A: The extreme heat of the griddle can sometimes cause the side and backsplash to separate from the cooking surface. Should this happen, you have the possibility of oil leaking

into the burner assembly [on gas equipment] and into the controls [on electrical equipment].

It is also important that the correct utensils be used on the griddle surface to prevent grooves and gouges. You can have a griddle buffed, but it's a very expensive procedure.

Q: How about steam-jacketed kettles?
A: Steam-jacketed kettles are virtually maintenance-free. The most important thing is to use distilled water only, to prevent contamination within the system.

On convection or pressure steamers, a steam generator is used to produce steam for the cooking compartments. In areas where the water has a high mineral content, it is important that the generator be opened up at least twice a year. There are service agencies that can inspect and descale the generator to remove mineral deposits. Failure to do this regularly can ultimately destroy the steam generator and might even be hazardous to employees working with it.

We suggest a water-filtering system for convection steamers. I'm also a big fan of a process called *reverse osmosis.* It's a filtering system with three or four cartridges, a pump, and a membrane filter that purifies and stores water in a tank for the steam appliance. The appliance uses water out of the storage tank instead of tap water.

Q: And how about refrigeration units?
A: Of course, not cooling is the biggest problem. It is the restaurant's responsibility to clean the condenser coils to keep them free of grease and dirt that blocks airflow. You can keep coils cleaner if you buy a roll of thin black foam that can be used as a filter. Cut a piece off to fit right over the condenser. Change it every so often.

I would also recommend checking with your local service company to see if they offer a quarterly preventive maintenance program for refrigeration equipment.

Q: If you were a restaurateur, would you designate certain people in the kitchen to take responsibility for certain pieces of equipment?
A: That is a good idea, but the biggest problem is high turnover of personnel. If you knew you were going to have the same people, that would be a great idea. I've found the cleanest places are usually schools, nursing homes, and government facilities. It's partly because they have more regulations, and also because their people have been there longer than in a restaurant. Some even have their own maintenance people.

Q: Let's talk about common problems you see with dishwashing systems.
A: Dishwashers are pretty simple to use. There are high-temp machines, which use 150°F wash water and 180°F rinse water; and low-temp machines, which use standard household-type hot water and clean and sanitize dishes with the use of chemicals.

On the high-temp machines, the most common problem is that the booster heater isn't heating the water hot enough. On low-temp machines, spotting is the most common problem. It's usually due to not adding the right amounts of wash and rinse chemicals.

Q: How about waste disposal systems?
A: The most common problem is jamming the turntable. This is usually caused by something other than food, like silverware, dropping into it. Also, very often, they'll mop the floor with those stringy mops and dump the mop water into the sinks. The mop strings get wrapped around the seals, causing the unit to leak water down into the motor housing and destroying the upper and lower bearings.

Q: Now if you're a foodservice operator and you want to get a service contract from a company like yours, what should you expect? What is reasonable to expect?
A: We try to run our business so that a service call is a same-day call. If it's an emergency—if a refrigeration unit is down or something they absolutely can't get along without—we try to get somebody out there within 2 or 3 hours. But, normally, a 24-hour period from the time you call is reasonable. Truthfully, though, the customers don't feel that way. They think you should have somebody sitting here waiting around to answer their call within 10 minutes! The fast-food chains are the worst.

Q: When someone calls you for service, what do you need to know over the phone?
A: The first thing I always try to get is the brand name, and then I ask, "What is the machine doing? Is it running at all?" We can fix a lot of things right over the phone. You'd be amazed at the calls we get where the appliance has been accidentally unplugged, or there's a reset button but nobody knows to push it. If I had any advice, it would be: Just try to keep your people as informed as possible on the use and care of the equipment.

15

COOK-CHILL TECHNOLOGY

■ INTRODUCTION AND LEARNING OBJECTIVES

Most foodservice businesses prepare some foods in advance. *Cook-chill* is an advance preparation technique used to streamline production, minimize waste, and improve food safety at the same time. Cook-chill is the process of cooking food in quantity and then rapidly chilling it. The cooked food is not frozen, but cooled so quickly that it does not stay in the danger zone (from 41°F to 135°F, or 5°F to 57°C) long enough to support harmful bacteria growth. Once it's been quickly chilled, holding the food under proper refrigeration (at 34°F to 40°F) prolongs its shelf life, allowing it to sit for at least 5 days and, in some cases, up to 21 days before serving. A product that stays fresher longer means certain menu items can be prepared twice a week instead of daily, saving time and money.

The technique was developed in Germany more than 50 years ago for the government-run hospitals there. It was an attempt to control labor costs, which is still one of its advantages today. Cook-chill technology is not a way to stockpile leftovers, although it can be used as such. It is a system of quantity food preparation designed to create a stock of safely prepared and refrigerated foods that can be used as needed. That's one reason why it is sometimes referred to as the *cook-to-inventory* method of food preparation.

After reading this chapter, you will be able to:

- Explain the uses, benefits, and drawbacks of cook-chill technology.
- Describe the cook-chill process, including recipe adaptation and food safety.
- Identify the equipment used in the cook-chill process.

Cook-chill is not to be confused with an old but popular technique called *ice shock*, when food is cooked and then immediately plunged into ice water. Instead, it is a complete cooking and cooling system that requires the purchase of several pieces of equipment. There is a certain mystique about it because the process separates meal preparation from meal service. However, the advantages are obvious. Let's take a look at them.

15-1 WHY USE COOK-CHILL?

It seems there has always been a debate in commercial food preparation between cooking from scratch, or cooking to order, and using convenience products. You might consider cook-chill a way to satisfy both sides of the argument. The food *is* made fresh, from scratch—it is just packaged and chilled quickly for later use instead of being served immediately.

Users of cook-chill agree that consistency—the uniformity of product—is its major selling point. At one time, the perception was that this system would be useful only in institutional kitchens with constant high-volume needs, such as prisons or school systems. But in recent years, cook-chill technology has received the blessing of respected culinary professionals. A restaurant can hire a gourmet chef, for instance, to establish recipes with food quality, flavor, and presentation standards that may then be carried out uniformly by a staff with far less training.

From an investment standpoint, centralized production allows restaurant operators with multiple locations to produce high-end signature items with less cash outlay for real estate, conventional kitchen equipment, and expensive ventilation systems. The food is delivered cold to multiple locations. The chilled items are removed from the refrigerator as needed and brought to serving temperature by heating (commonly referred to as **rethermalization**, or *retherm*) in smaller, more portable appliances.

When done correctly, the cook-chill process produces foods with no loss of color, flavor, texture, or nutrients. The advantages are many.

Suitable for any size operation. The advantages to high-volume feeding situations are obvious, and, as we've mentioned, chain restaurants can use it to centralize production and distribute to smaller units. But cook-chill isn't only for large or multisite operations. Even a modest-size restaurant can make use of cook-chill technology for its banquet and catering needs and/or for safe storage of extra portions. Manufacturers now offer smaller, undercounter models of their blast chillers.

It can also be used as a way to keep a stock of all menu items on hand at all times by precooking and properly storing them.

Effective time management. Kitchen staff members' time can be organized for best results, cooking high-volume items when business is slow and having them ready to use when the kitchen is busiest. Labor savings (of 10% to 40%) can be realized because highly skilled workers can produce the core menu items while relatively unskilled workers can keep up with tasks like portioning and reheating.

Other resource management. Equipment and space can be used more efficiently because a central kitchen can turn out product for several sites. Also, ingredients can be purchased in larger quantities, providing savings.

Menu flexibility and diversity. Because foods are prepared in advance, experimentation is possible—and mistakes are likely to be caught in production instead of on the guest's plate! Chefs can embellish precooked ingredients to offer a greater overall selection of dishes because each is not cooked individually. (Traditional recipes sometimes must be modified slightly, which we discuss elsewhere in this chapter.)

Ability to serve special needs. For hospitals, senior centers, and schools, nutritional requirements are easier to meet because foods can be pre-portioned and special diet restrictions can be taken into account. Food safety for these higher-risk dining populations is improved.

Increased food safety. Time-temperature abuse is one of the biggest factors in food-borne illness outbreaks—that is, preparation and storage methods that keep foods in the danger zone for bacteria growth too long. The whole cook-chill concept is a series of procedures designed to minimize this critical period, and scientific research shows it works well.

Service improvements. Most cook-chilled foods need only be reheated, so the kitchen and waitstaff have more time to garnish, improve presentation, and attend to the needs of customers. A restaurant or cafeteria can also offer a wider variety of foods no matter the time of day.

Reduced waste, improved portion control. No matter how the number of guests fluctuates, only the meals ordered must be reheated. This means no more partial batches of your uneaten daily special hit the trash can at the end of the day. It's easier to keep tighter control over food waste and portion sizes when you can package them individually, or in bulk, as needed.

A more relaxed work atmosphere. The critical time factor between cooking and serving each item is greatly reduced, relieving pressure on production staff and, quite simply, improving productivity. The kitchen isn't on a constant deadline.

Increased profitability. If food can be ordered and served promptly, your turnover rate improves and so does your profitability. Serving sites can be smaller and can keep minimal inventories on hand. Tighter control over expensive kitchen expansion or hiring extra kitchen employees is possible. In short, it can mean more efficiency, greater choices for guests, and better service.

Several manufacturers offer systems that combine a combi-oven and a blast chiller in a side-by-side configuration that allows food to be transferred instantly from oven to chiller. Such innovations mean cook-chill systems can be used to save time and money in every type of foodservice venue, from fine dining to in-flight commissaries, luxury cruise lines, and supermarket deli counters.

15-2 HOW COOK-CHILL WORKS

The theory behind cook-chill is simple: Hot foods must be cooled through the bacterial danger zone quickly (within 90 minutes) and uniformly. Once chilled, the food can be stored in standard reach-in or walk-in refrigerators. When ready to serve, it just needs to be rethermed. The cook-chill production system is simple to operate if it is well managed. Illustration 15-1 shows the basic steps; the flowchart in Illustration 15-11 explains them in greater detail.

Quick-chilling requires more sophisticated refrigeration than a walk-in, which can't do the job fast enough. This includes introducing high-velocity cold air all around (above and beneath) each container of food. The U.S. Department of Agriculture (USDA) mandates that cook-chill equipment be capable of bringing food from a cooking temperature of 180°F to its chilled storage temperature of below 40°F in less than 2 hours (see Illustration 15-2). Today's commercial equipment is, in fact, capable of doing the job in less than 1 hour, depending on the product. (Most manufacturers' estimates of chilling time are based on incoming food product that is between 140°F and 160°F.) Before we discuss the equipment used to accomplish this super-chilling, let's cover the all-important steps that lead up to that point.

SELECTION OF RAW MATERIALS

As with any cooking endeavor, your final product is only as good as the quality of the raw ingredients. Especially if you are working with fresh meats or seafood, it is vital to check your vendors' handling and distribution methods.

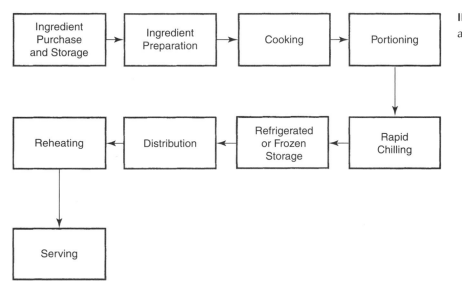

ILLUSTRATION 15-1 The basic steps in a cook-chill system.

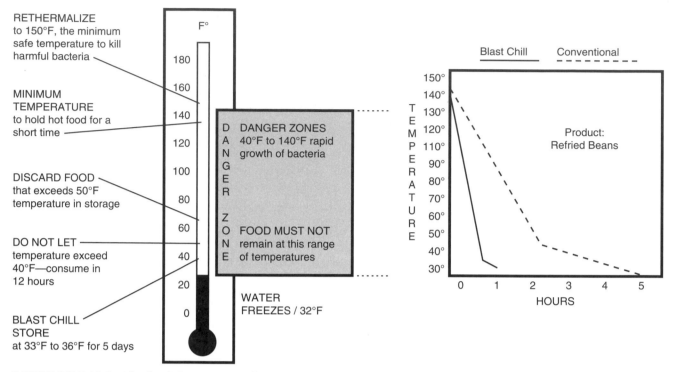

ILLUSTRATION 15-2 The food danger zones that affect time-temperature concerns. As you can see, a blast chiller does the job far more quickly than conventional refrigeration.

Courtesy of the Electric Foodservice Council, Fayetteville, Georgia.

STORAGE CONDITIONS

After purchasing the best-quality raw products, keep them in safe, sanitary storage so they are in prime condition when you cook them. This means following basic hygiene principles and using a storage area that is the appropriate temperature and humidity. Thaw frozen materials in the refrigerator, not at room temperature or in the microwave. Either of these exposes food to high enough temperatures that bacteria could grow. It can also thaw them unevenly, leaving cold spots at the core of the food that may result in uneven cooking.

PREPARATION

Prepare the food for cooking as if you were going to cook and serve it that day. Again, the use of sanitary surfaces and separate hand tools for different raw foods is a must. Ideally, this prep work takes place in an area separate from the cook-to-order and plating areas of the kitchen.

Special care must be taken to make the quick-chill process more efficient for meats. Make sure pieces of meat do not weigh more than 6 pounds or measure more than 2½ inches in thickness.

RECIPE MODIFICATION

Because the cooked foods rest before being served, minor recipe alterations may be necessary. Flavors continue to develop; spices get spicier; starches change consistency. The first step in cook-chill recipe development is to create the optimum flavor and presentation profile for a single dish, then convert it to batch production. Equipment manufacturers suggest reducing sugars and salt content by 25%, other spices and seasonings by 10%, and the amount of leafy herbs called for in a recipe by 20% to 30%—particularly sage and bay leaves, which grow stronger with time when added to foods. They also recommend not adding all the liquid called for in a recipe until the end of the cooking process, using it to correctly adjust the texture just before cooling.

It is smart to create recipes that call for standard quantities of ingredients—full cans, round numbers—to minimize errors in volume production and eliminate half-empty containers that must be stored.

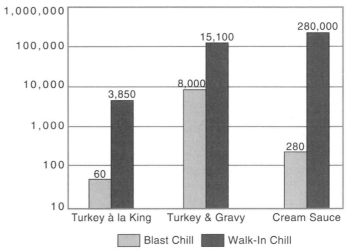

ILLUSTRATION 15-3 An example of the exponential growth of bacteria on just a few products, using two types of chilling. Courtesy of the Electric Foodservice Council, Fayetteville, Georgia.

COOKING

You cook the food in the same manner, using the same equipment you would normally use. Most high-volume operators understand the benefits of batch cooking, and cook-chill methods can increase your batch output by three or four times! Favorite batch-type items, such as casseroles, meats, sauces, and soups, all lend themselves to this preparation.

No matter what you cook, it is essential that its core temperature reach 165°F and be held at that temperature at least 2 minutes to destroy any pathogenic bacteria or microorganisms that may be present (see Illustration 15-3). A variety of probes and thermometers may be used to check temperature; remember, thermometers can be out of calibration, so check the accuracy of yours regularly—at least every 3 months.

Now we come to the major steps that make cook-chill distinct from any other type of food preparation.

PRECHILL PREPARATION

Once the food is cooked, the chilling process must begin as soon as possible—that is, within 30 minutes. Load the food into shallow stainless-steel steam table pans, 2½ inches deep, 12 × 10 inches or 12 × 20 inches. Don't load the cooked food any higher than 2 inches deep in the pan. Deeper containers may be used, but only if the chilling unit is efficient enough to lower the food temperature fast enough. Some foods may require even smaller portioning to facilitate chilling. This depends on the size, shape, and density of the food; the denser the product, the longer time required to chill.

You may wonder why we recommend covering the food (with plastic wrap or flat stainless-steel lids) when the pans would cool even faster uncovered. This is done primarily for safety reasons, but keeping pans covered also reduces surface dehydration of the food, eliminates flavor transfer when chilling several foods at the same time, and prevents excessive icing on the coils of the chiller's refrigeration system when too much moisture circulates inside the unit. The moisture content of the food, its ability to retain heat, and its temperature when it reaches the chiller all affect total chilling time.

Containers made of materials meant to insulate food also affect chilling time. Containers made of disposable materials must be stored under completely sanitary conditions before use, or you risk introducing bacteria as you fill them with hot food. As you can see, there are many variables, almost all of them health related.

Each cook-chill system user must decide whether to portion foods before or after chilling takes place. There is no clear-cut choice, so you must consider how and when you will use the food. For a banquet, you might take it from the chilling pan, arrange it in a serving pan, and chill it that way. For room service or on a cruise ship, you might make individual portions, perhaps on covered plates or in vacuum-pack bags. For a buffet, the food might be put into a chafing dish insert. Whatever the choice, portioning must be done within 30 minutes of cooking so the food can be safely chilled. Regular measurement of food temperature during this time and throughout the cook-chill process is recommended.

There are two kinds of quick-chilling. *Blast chilling* circulates cold air at high velocity around pans of food to cool it quickly. *Tumble chilling* immerses packed food products in cold liquid.

BLAST CHILLING

Blast chilling is appropriate for solid foods such as chicken parts, beef rounds, whole casseroles, burger patties, and so forth. The food containers are covered and then either placed on carts to roll inside the *blast chiller* or put directly on shelves inside the chilling unit. (Illustration 15-4 shows a roll-in chiller.)

Blast chillers range from conveyor belt systems for high-volume production all the way down to undercounter models that hold three pans. Like refrigerators, they are available in reach-in or roll-in models. Some feature rotating racks to circulate food as it chills; some are on casters to be easily rolled between production areas.

Inside the blast chiller, powerful fans blast the food with cold air at speeds of up to 1300 feet per minute to quickly draw off the heat. The unit must be capable of reducing the temperature of a 2- to 3-inch layer of food from 150°F (or so) to less than 40°F within 90 minutes, fully loaded. It should also contain an accurate temperature display and built-in food probes with digital displays as well as a timer that alerts kitchen personnel with an audible buzz or bell when it's done. Most blast chillers have a holding mode that kicks in automatically when the blast chilling is complete. This important feature allows you to chill a batch of food at the end of one day, then come back and remove it the next morning. Most have multiple modes:

- A "delicate" or "soft chill" cycle for light loads, low-density products, and items like salads
- A "hard chill," to quickly cool foods to a core temperature of about 40°F
- A "flash-freeze" or "blast-freeze" mode for freezing foods to a solid 0°F
- A "hold" mode that keeps the food at whatever temperature is preset by the operator.

As you might imagine, these units require a lot of power. The typical 10- to 12-pan chiller has a 1- to 2-horsepower motor—essentially the same as a 240-square-foot walk-in cooler. The largest blast chillers in institutional settings are capable of handling from 90 to 400 pounds of food per chilling cycle, which means a daily output of between 540 and 2400 pounds of food. The 90-pound capacity chiller can accommodate ten 2-inch-deep hotel pans (the flat, 12 × 20-inch standard pans); the 400-pound chiller can handle 44 such pans. The electrical requirements are 120/208–240 volts, 60-cycle, single-phase power for the smaller models; 208–230 volts, 60-cycle, three-phase power for the larger models. Larger condenser unit motors are between 3 and 10 horsepower.

ILLUSTRATION 15-4 A roll-in blast chiller.

Courtesy of Alto-Shaam, Inc., Menomonee Falls, Wisconsin.

The capacities of the smallest chillers, which hold only a few pans, range from 18 to 30 pounds of food per load, with compressors that run on ½- to ¾-horsepower motors. Medium-size chillers can handle 45 to 100 pounds of food per load. A bonus for the undercounter chiller user is that some feature a self-contained evaporator, making a nearby floor drain unnecessary.

In many kitchens, blast chillers are located close to the cooking areas to help meet that critical 30-minute deadline between stove and chiller. For the medium-size units, you need an area about 5 feet square and 7 feet tall. For the larger models, it's more like 9 feet tall, 9 feet long, and 5 feet deep. (Of course, the huge industrial models require more space.) The condensing unit takes up the most room and can be located with the chiller or, with proper installation, elsewhere in your kitchen.

Some chillers have optional printers that can be linked to their temperature probes. You can receive a printout that lists the product's core temperature, performance details about the chilling cycle and the refrigerated cabinet itself, and the compressor's running temperature (to check for overheating or icing up). An audible alarm feature can be set to alert you if any of these components fails. Other chillers contain optional ultraviolet lights that sanitize utensils overnight.

Pay close attention to the design of the chiller. Check for easy access to the evaporator components because, like any refrigerator, these must be cleaned regularly. Look for removable racks, shelves, and shelf slides, as they are also easier to clean. Automatic defrost and evaporation features save time and hassle in operating your blast chiller.

BUYING A BLAST CHILLER

Selecting a blast chiller is not just a matter of size and temperature. Other choices:

Direct versus indirect airflow. Direct airflow is cold air blown directly onto the pans. It is effective but can mean more spattering or drying out of foods that are not covered. The velocity of the airflow and the way it is balanced are critical considerations. A blast chiller without sufficient air velocity or balanced airflow cannot chill its contents evenly.

Size considerations. One issue is the footprint of the appliance. The other is the volume of product that can be chilled *per cycle*. The manufacturer's specifications should include two measurements of volume: the number of pounds and the number of pans the chiller can accommodate per cycle.

Reach-in versus roll-in. Reach-ins are designed to hold a certain number of pans. Their shelves may or may not be adjustable (adjustable is, of course, better). Roll-ins are designed as open cavities into which roll-in racks full of product are wheeled.

Self-contained versus remote compressor. How much available space do you have to install the chiller? Smaller units are self-contained; larger ones have the option of a separate standalone compressor that requires installation of refrigeration lines between compressor and chiller, space (for the compressor/condenser), and another electrical connection.

Chilling versus freezing. Some cook-chill systems can also be used as cook-freeze systems. You can buy a blast chiller that has a blast-freeze mode and freeze the finished product after quick-chilling. Comparing the two processes, a University of Wisconsin study determined it takes 18 times more Btus of energy to prepare, chill, store, and reheat meat loaf in a cook-freeze system than it does in a cook-chill system. But for long-term food storage, freezing is the way to go.

Compatibility with other equipment and your product flow. These are the logistical considerations. What types of equipment are used to cook the food being chilled? Make sure the pans or roll-ins fit into the chiller, or you'll have to spend time dishing hot food into different containers for chilling. Is there adequate reach-in and walk-in space for the amount of product the blast chiller produces?

ILLUSTRATION 15-5 An agitator kettle, which can mix large quantities of food.

Courtesy of Groen, Unified Brands, Jackson, Mississippi.

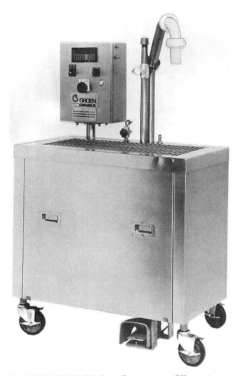

ILLUSTRATION 15-6 The pump-fill station.

Courtesy of Groen, Unified Brands, Jackson, Mississippi.

TUMBLE CHILLING

Blast chillers are generally considered more versatile than tumble chillers because they work with just about any type of solid food product and are available in a great variety of sizes. However, using them to cool liquids—soups, stews, gravies, and sauces as well as some types of pasta—presents practical challenges. Putting these foods in shallow pans gets messy, and the process holds potential for contamination. High-velocity cold air can cause them to spatter, but placing them in buckets in a walk-in doesn't chill them fast enough. For these foods, **tumble chilling** is most appropriate. And when it comes to reducing food temperatures quickly, the tumble chiller is much more efficient than the blast chiller. Some experts say tumble chilling doubles the safe shelf life of foods compared to blast chilling.

The tumble-chilling process requires several pieces of equipment. First, foods that can be pumped are cooked in a special steam-jacketed kettle (called an **agitator kettle** or *mixer kettle*), which includes an agitator or mixing arm (see Illustration 15-5). The agitator is needed for uniform suspension of solids and continuous mixing action as the food is pumped from the cooking vessel into containers. Manual mixing isn't precise enough to accomplish this. The goal of the agitator is to lift and fold the food gently so it isn't damaged in the process. Use of a kettle is also handy because, by putting cold water in the kettle jacket, the food can be precooled to a certain extent when cooking is complete.

The smallest kettles hold 60 gallons and the largest hold up to 200 gallons, but you never want to fill them any fuller than 180 gallons to facilitate mixing. Agitator kettles must have a 3-inch steel valve at the bottom to flush foods through the kettle and on to the next step, the **pump-fill station**. Here, kettle-cooked foods are transferred (pumped) directly from the kettle into flexible plastic tubes designed to withstand the temperature extremes of both hot filling and rapid chilling (see Illustration 15-6). To flush efficiently, the solids in foods (chunks of vegetables in soup, for instance) should not exceed 1 inch in diameter.

The plastic casings are disposable and come in sizes from 1 quart to 3 gallons. When full, they are sealed, weighed, and labeled in an area some kitchens refer to as their **metering station**. They can be refrigerated, frozen, or reheated.

The pump-fill station has a pump with a variable-speed motor, a sink with a drain, a rack to hold the plastic casings, a clipper to snip off the ends of full casings, and a label printer. The whole thing can be put on a rolling cart for use with multiple kettles; it takes up a space about 3 square feet and almost 7 feet tall. The 1½-horsepower motor uses 208/230-volt, 60-cycle, single-phase electricity.

By the time the casings leave the pump-fill area, they should be filled, sealed, trimmed, and labeled. From here, the casings are manually transported to the **tumble chiller** or, in very large operations, perhaps placed on a conveyor belt. The tumble chiller is a rotating drum full of cold water (34°F) that gently kneads the full casings as they roil around inside (see Illustration 15-7). This speeds the cooling process to less than 1 hour and can drop the food temperature from a steaming 180°F to 40°F. The ice water is supplied by an ice machine (called an **ice builder**) that can be installed, with correct plumbing connections, either indoors or outdoors.

Tumble chillers come in sizes for 85 to 300 gallons of product per cycle. The smallest units are 6½ × 4 × 8 feet; the largest take up 10 × 6½ × 10½ feet. Some tumble chillers employ a chute that allows loading as the unit is running. Tumble chillers require a 2-inch water supply pipe and a 3-inch drainpipe.

Many tumble-chilled foods can be stored safely for 35 to 45 days before rethermalization.

OTHER CHILLING OPTIONS

There are a few old-fashioned ways to chill food faster. They probably seem inconvenient compared to blast and tumble chillers, but they are used nonetheless, especially for small batches.

Experienced chefs know that the first step is to reduce the size or quantity of what you're trying to cool. Any food cools more quickly in small or shallow containers. After food is divided into smaller quantities, an *ice-water bath* may be used. Place the pan full of food into a larger container (another pan or a sink) filled with ice water, still loaded with plenty of ice. This is labor intensive but effective.

ILLUSTRATION 15-7 The tumble chiller is a rotating drum full of cold water.

Courtesy of Groen, Unified Brands, Jackson, Mississippi.

Large batches of food can be precooled (before blast or tumble chilling) by stirring them with a hollow, heavy-duty plastic paddle filled with water and frozen. Of course, this requires workers to fill and freeze the paddles and to do the stirring.

Steam-jacketed kettles can be used to cool food by filling them with icy-cold water instead of steam. Some recipes that require water, such as soups, can be prepared using less water; the remaining water can be added cold to help cool the finished product.

You can turn your walk-in cooler into a super-chiller with special ***conduction shelves***. A solution of water and antifreeze circulates through the shelf interiors, cooling food rapidly as the pans sit on them. Its manufacturer claims the refrigerated shelves also decrease the work the walk-in has to do without affecting other foods stored inside (see Illustration 15-8). Some shelf systems are on rolling casters, equipped to hold pans or individual plates, and come in standard sizes that can roll directly from a blast chiller to a retherm unit.

For large cuts of meat (hams, turkeys, etc.) or other foods not suited to kettle cooking, you can buy a combination ***cook-cool tank***, also known as a *chiller-cook tank* or a *cook-chill tank*. In this appliance, uncooked food is vacuum-packed into its plastic casings, then loaded onto wire racks. The full racks are stacked inside the tank. Hot water circulates inside the tank, slow-cooking meats in their own juices and producing a very high yield. Then, on a pretimed cycle, the hot water is drained away and replaced with icy-cold water to begin the rapid-chill process. Cook-cool tanks can be used to prepare most types of meats and vegetables. The refrigerated shelf life of food prepared this way is similar to that of tumble-chilled items.

Cook-cool tanks come in sizes ranging from 200- to 2500-pound capacities. The largest models contain multiple rows of wire trays on which to put food. They use a 3-horsepower recirculating water pump and need floor space that's about 8 feet long and 4 feet wide. The tank should include a meat probe to record internal cooking temperatures. Electrical requirements depend on the size of the tank, but they'll be 208-, 230-, or 460-volt three-phase.

ILLUSTRATION 15-8 Conduction shelves are refrigerated to cool food faster.

Courtesy of Thermodyne Foodservice Products, Inc., Fort Wayne, Indiana.

A final word about ice builders, which are needed to provide sufficient amount of ice water for both cook-cool tanks and tumbler chillers. Your total daily production figure determines the capacity of the ice builder. The ice builder makes ice water and pumps it to a heat exchanger located on the tank (or chiller). There the ice water chills the distilled water used to cool the food. As mentioned earlier, ice builders need not be installed directly adjacent to the appliances they serve, and they can be located either indoors or outdoors. For added efficiency, they can also be used to produce ice during off-peak hours.

15-3 STORAGE AND DISTRIBUTION

Now that you have a batch of precooked, chilled food, where do you put it until you need it? The typical commercial refrigerator in general use in your restaurant kitchen is not the answer because it is opened and closed constantly. Your goals are to keep the precooked food at a constant temperature and to prevent cross-contamination between the precooked food and other foods in the refrigerator. Your best option is to install a dedicated reach-in or walk-in unit specifically for cook-chilled foods. Most are available with monitoring systems and alarms to alert you if the temperature strays from that ideal range of 32°F to 38°F. If it is not possible to install a separate refrigeration unit for your cook-chilled products, consider installing conduction shelves, mentioned above, in an all-purpose walk-in.

Employ a first-in, first-out method of stock rotation so none of the food exceeds the expiration date on its label. Labeling is the other key to successful storage. The labels on each container must specify the type of food, date of preparation, and destination if the food will be used at an off-site location.

If for any reason the food reaches a temperature over 41°F but not more than 50°F, it should be used within 12 hours; if it is accidentally allowed to reach more than 50°F, it should be destroyed because it is potentially dangerous to eat. It is a good idea to regularly send samples of your stored foods for safety testing at a laboratory.

Despite these cautions, cook-chilled food that has been correctly prepared and stored will make you proud to serve it. It will look and taste as good and fresh as the day it was prepared. Most of the time, food travels only a few feet to its point of rethermalization, but this is not always the case. As you've discovered by now, temperature control is key to the success of this system, and it is especially important when transporting food from its preparation site to other locations. Insulated carriers, refrigerated carts, and refrigerated vehicles are all used for off-site catering and satellite distribution. In just a moment, you will also learn about carts made for reheating and serving cook-chill foods. At the least, prechilled, insulated coolers should be used for short journeys.

When the food reaches its destination, a temperature reading should always be taken to ensure it has not warmed to a temperature above 45°F. Then it must immediately be placed in the appropriate refrigerated storage until use. If the food must be transported hot because there is no retherm capability at its destination, its core temperature is critical. It must be 140°F or higher.

15-4 RETHERMALIZATION

Rethermalization, or reheating, is the conclusion of the cook-chill process and also a key control point in the prevention of contamination. The guidelines:

- If a cook-chill food is supposed to be eaten cold or at room temperature, it should be consumed within 30 minutes of leaving the refrigerator.
- If it must be reheated, this should take place promptly. This means taking the food through the temperature danger zone once again, as quickly as possible—bringing it to

an internal temperature of 165°F for at least 15 seconds within 2 hours of its removal from refrigeration. A probe thermometer inserted into the center of the food item is the best way to determine if this core temperature is reached.

■ If the rethermed food is allowed to cool to room temperature, it should never be reheated or returned to chilled storage. For safety reasons, it must be destroyed.

Specially built rethermalization units are designed to work with cook-chill systems, but you can use just about any cooking appliance: convection ovens, steamers, kettles, combi-ovens, and infrared units. If a traditional oven is used, take care the food is not dried out. To retain moisture, food should always be covered when it is rethermed. Your choice of methods and appliances depends mostly on how the product is packaged. Many experts feel the microwave oven is the best way to retherm individual portions that have already been plated. Food that has been bagged and chilled should always be heated in moist conditions—by placing the bag in a bain-marie, a steamer, or a combi-oven—or you can remove it from the bag and heat it on a rangetop or in an oven. High-volume commissaries can retherm covered dishes with conveyor impingement ovens.

The *rethermalization cart* is a recent invention to meet the needs of healthcare facilities and others that want roving capability. These specialty carts are capable of bringing preplated cold food up to proper serving temperature in less than 1 hour and holding it at optimum temperature for extended periods (see Illustration 15-9). Sometimes the carts roll; sometimes the trays of food can detach and be rolled.

Retherm carts offer a wide variety of options. Units that can be switched from moist (convection) heat to dry (radiant) heat offer the most flexibility—lasagne, for instance, requires moist heat, while dry heat retains the crispness of fried chicken. Some carts have a small refrigerated compartment in addition to heating capability; some can be connected to an electrical outlet and programmed to start heating at a certain time; others can be plugged into their own compatible freestanding chiller units. The controls for the carts can be manual or automatic, and most have secured control systems to guard against tampering. Hazard Analysis of Critical Control Points (HACCP) compliance is easy when an onboard computer monitors every push of a button, every internal temperature, and every door-opening.

ILLUSTRATION 15-9 A rethermalization cart can warm plates of food as it is wheeled from place to place.

Courtesy of Alto-Shaam, Inc., Menomonee Falls, Wisconsin.

FOODSERVICE EQUIPMENT

Cook-Chill at Work: Fresher, Safer, and More Cost Effective

The Beau Rivage Resort and Casino, a 1740-room hotel resort on the Mississippi Gulf Coast with 12 restaurants and 4 bars and lounges, turned a storage warehouse into an on-premise cook-chill food processing plant for three reasons, according to George Goldhoff, vice president of food and beverage: "Food quality, economics—saving money on labor and food costs—and food safety."

A fourth reason may be that the resort had to solve facility issues. For instance, it needed a larger main kitchen and was at maximum capacity in terms of water usage. With the new plant, the hotel was able to get another main water line from the city, which improved water pressure on the property as a whole.

The plant uses state-of-the-art cook-chill equipment, including specialized chilling tanks, to increase efficiency, food freshness, and consistency of product, and to eliminate many food contamination worries. Take, for example, the hotel's signature gumbo. In 1 week, it serves approximately 650 pounds of gumbo—which, with cook-chill technology, now can be prepared all at once, once a week, making it more consistent and raising food safety standards.

"Most food is cooked at 160°F and stored at 40°F; anything in between is dangerous," Goldhoff explains. With the chilling equipment, some foods can be dropped from cooking temperatures to below the danger zone in as little as 20 minutes. "Most kitchens take pots and dump them in an ice pack. Now we chill [our gumbo] all down in 40 minutes rather than, say, 4 hours."

The process also means vegetables are kept crisper and fresher. Nothing comes from a frozen package anymore. It also gives food a longer shelf life—14 to 21 days, according to Executive Chef Joseph Friel. "And we're not adding preservatives or anything," he says.

The hotel now produces all its own soups, stews, and salad dressings and roasts its own deli meats in large kettles that use computer technology to control temperature. A blue line records the actual temperature in the pot, while a red line indicates the actual temperature of the ingredients in the pot. Everything is recorded, labeled, and kept on file.

"All the information on how to cook a batch of food, like marinara sauce, is saved on file—at what temperature, how long to hold it, and how to store it," Goldhoff says. Such efficiencies also make transport and storage issues easier and neater, meaning fewer hygiene problems.

The full equipment costs—which included juicers, coffee-bean roasters, gelato-making equipment, and more in addition to the cook-chill operation—were $400,000, but Goldhoff says he expects to see a full return on the investment after 2½ years.

"Now that we're not buying any prepared foods," he explains, "it makes it easier to control food costs."

Source: Excerpted from *Hotels* magazine, a publication of Reed Business Information, Oak Brook, Illinois (February 2005). Courtesy of Beau Rivage Resort and Casino, Biloxi, Mississippi.

When looking at retherm carts, ask about the dishes that can be used with it. Some carts accommodate disposable paper and plasticware, while others require the purchase of compatible plates because the dishes have direct contact with a heating element. Some of the carts have mechanisms to lock ceramic dishes in place as they roll to prevent breakage; others do not. Other considerations depend on your intended use for the cart, but you should have a good idea of these issues:

■ Any size constraints, including loading docks, freight elevators, or regular elevators to maneuver through

■ How quickly the food should be heated, and how long it must be held at that temperature

■ The type of electrical power available at your remote locations

■ The flexibility of your menu

The last point is important because serving from a cart sometimes means changing the way the food is presented to customers. For space, or quicker reheating, you may have to slice baked potatoes in half, for instance, or wrap some foods in aluminum foil. It's hard to get toast to stay crisp and dry in a retherm cart, but soups, stews, and hot cereals adapt perfectly.

15-5 BUYING AND USING A COOK-CHILL SYSTEM

When selecting cook-chill equipment, the first thing to determine is whether its capacity is sufficient to meet your peak production needs. This is important for two reasons: to determine overall output and to facilitate careful timing of the process so newly cooked food can always begin rapid chilling within half an hour.

The best way to begin shopping for a system is to visit foodservice facilities that already use cook-chill. If you are looking for a retherm cart, ask a hospital employee how convenient it is to push around to patients' rooms. Some foodservice businesses form committees with goals of making sure the entire operation understands and supports the cook-chill process before they purchase the components. Many manufacturers also allow you to test the pieces of equipment on-site for a few days before finalizing your purchase.

Let's look at how cook-chill technology is used in specific foodservice situations. The first example is provided by Williams Refrigeration Limited, a British company with U.S. offices in Maywood, New Jersey. The company describes a small resort restaurant, open 6 days a week, serving approximately 600 meals per week. It seats 50, serves two meals per day, and turns each table once during those meals.

The restaurant's owner realized business was being turned away on the 1 day (Monday) the restaurant was closed but saw no other way to give his kitchen staff a day off—until he purchased a blast chiller, capable of chilling 175 meals per day. By using the cook-chill system 4 days a week, he could increase his sales by 100 meals (by staying open Mondays) and allow the kitchen staff to take 2 days off per week instead of 1. Look at the schedule in Table 15-1 of meals produced (and consumed) before and after the cook-chill system. See the difference?

In this operation, an upright blast chiller was used, capable of handling 50 pounds of food per cycle in standard hotel pans. Each chill cycle lasts from 68 to 87 minutes, and the blast chiller automatically reverts to storage mode following the completion of the cycle.

TABLE 15-1

Meal Output Comparisons—Small Resort Restaurant

MEAL OUTPUT *BEFORE* INSTALLATION OF BLAST CHILLER	M	TU	W	TH	F	SA	SU
In storage	0	0	0	0	0	0	0
Production (cook-chill)	0	100	100	100	100	100	100
Sales (meals sold)	0	100	100	100	100	100	100
Inventory (meals on hand)	0	0	0	0	0	0	0
MEAL OUTPUT *AFTER* INSTALLATION OF BLAST CHILLER							
In storage	0	75	150	50	125	200	100
Production (cook-chill)	175	175	0	175	175	0	0
Sales (meals sold)	100	100	100	100	100	100	100
Inventory (meals on hand)	75	150	50	125	200	100	0

A suburban school district in Duncanville, Texas, has used cook-chill technology for more than a decade to plan its foodservice production and deliver more than 7000 meals per day to 12 schools from a single kitchen. Six cooks staff the central kitchen, located in a 2000-student high school. The major components of the 5000-square-foot production center include:

- 1250 square feet for ingredient storage
- 532 square feet for preparation
- 345 square feet for test kitchen
- 720 square feet for cooking area

The hot-food area uses a total of three agitator kettles: one 200-gallon, one 100-gallon, and one 60-gallon. Along with the requisite pump-fill station and tumble chiller, this equipment provides about 90% of the food items served at the schools. An air pressure system is used in the preparation area and during repacking to vacuum-pack some prepared items (such as cut produce) for longer storage life.

In the bakery area, three full-time bakers turn out all the buns, cookies, doughnuts, and sheet cakes for the district. Here you'll find two 80-quart mixers, automatic dividers, round and sheet pans, a large roll-in proofer, a rotary oven, and a doughnut maker.

Finished products await distribution in an 8000-square-foot storage and distribution center, which features about 740 square feet of walk-in refrigerated space. The storage area is divided to hold about 20% prepared foods, 40% frozen foods, and 40% dry goods.

ILLUSTRATION 15-10 (a)   Beau Rivage cook Linford Clocke transfers food cooked in kettles to flexible plastic casings for quick chilling.

Courtesy of Beau Rivage Resort and Casino, Biloxi, Mississippi.

ILLUSTRATION 15-10 (b) Beau Rivage cook Linford Clocke at the agitator kettle control panel, which controls and monitors heat, water levels, and speed of agitator.

Courtesy of Beau Rivage Resort and Casino, Biloxi, Mississippi.

In each satellite school, mini-kitchens of about 450 square feet contain combination rangetop and convection ovens, plus steam table setups. Food items arrive in bags and are boiled or baked and placed in steam table pans for cafeteria service. The satellite kitchens do prepare a few of their own items: french fries, some cookies, and cornbread. For cobblers, the filling is cooked and chilled and the shell is made at the central bakery; at the satellite kitchen, the cook simply pours filling into the shell and bakes it in the convection oven. At the end of the lunch period, satellite kitchens send dirty dishes and silverware back to the central kitchen's warewashing system.

The benefits of this system to the Duncanville School District include the ability to offer students a choice of three hot entrées daily, adding variety to the menu. Waste is reduced because the cooks at the satellite kitchens open only as many bags of food as they need. The rest remain refrigerated for later use. Labor costs and energy use are minimized, and quality control is improved. A computer program stores and analyzes recipes, helps in meal planning, and tracks sales to determine which items are the biggest hits with students.

Illustration 15-10 provides a summary of the cook-chill process. As you can see, the convenience and multiple benefits of cook-chill technology make it an expense worth considering no matter your foodservice needs.

15-6 CLEANING COOK-CHILL EQUIPMENT

All the food safety precautions you are taking with cook-chill are for naught if you don't scrupulously clean the equipment between uses. Sanitize the agitator kettle between batches using a bleach-and-water solution of 5 tablespoons bleach for every 20 gallons of kettle capacity and

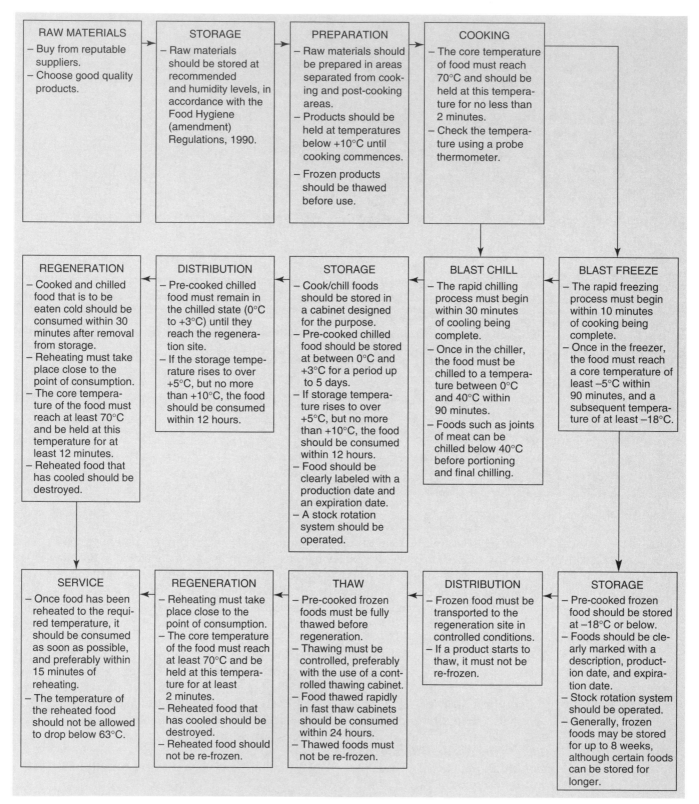

ILLUSTRATION 15-11 A flowchart summary of the cook-chill process.

Source: Williams Refrigeration, Inc., Maywood, New Jersey.

1 cup bleach for a 60-gallon kettle. Heat this solution to lukewarm (about 80°F) and then pump it from the kettle through the food product fill hoses and into a drain. Some pumping stations have constant pumping functions, which can be used with a detergent- and warm-water solution to clean between batches of product, but only bleach can provide the sanitizing function. Whenever a system is cleaned or sanitized, the process must end with a clean-water rinse to flush out detergent or bleach residue.

Some components of tumble chillers, such as the food pumps, can be removed and disinfected separately, which you should do daily. Remove the pump, brush it clean of food particles, soak it in the manufacturer's recommended solution, and lubricate both sides of the pump seal with an approved food-grade grease before reinstalling. The agitator arm, scraper blade(s), and strainer for the water circulator pump can also be removed and cleaned. Again, lubricate all seals as you reinstall parts.

Most manufacturers include complete disassembly and cleaning instructions with the equipment. Read them, learn them—and then keep them on hand for new employees to learn as well. Illustration 15-11 provides a summary of the cook-chill process.

■ SUMMARY

Cook-chill is the process of cooking food in quantity and then rapidly chilling it. The cooked food is not frozen, just quickly and thoroughly chilled to prevent harmful bacteria growth. This method creates a stock of ready-made refrigerated foods for the busy kitchen.

Government regulations specify that cook-chill equipment be capable of bringing food from a cooked temperature of 180°F to its chilled storage temperature of below 40°F in less than 2 hours; most commercial equipment can do the job in less than 1 hour.

In the cook-chill process, food is prepared just as if it were to be served immediately. The chilling process must begin within 30 minutes after the food is cooked. Some foods should be divided into smaller portions to facilitate quick chilling. There are two kinds of chillers: blast chillers, which circulate cold air at high velocity around the food; and tumble chillers, which immerse packaged food in cold liquid. Tumble chillers require an ice builder, a machine that produces ice water to chill the food.

Foods with high liquid content are packaged and chilled in a pump-fill station, where they pass through flexible plastic tubing and into plastic casings to be refrigerated, labeled, and frozen or reheated. The reheating process is known as *rethermalization*—retherm for short. Safe rethermoving means reheating the food to a temperature of 165°F for at least 15 seconds within 2 hours of taking it out of refrigeration. If cook-chill food reaches a temperature of more than 50°F before you are ready to serve it, then it should be destroyed. A variety of refrigerated and insulated containers are available for safe food storage, but it is always wise to check and recheck temperatures.

Because cook-chill foods are not going to be used right away, it is especially important to label them with their contents, production date, and often their weight. A first-in, first-out method of stock rotation is recommended for cook-chilled products, just like all others.

■ STUDY QUESTIONS

1. Describe two of the advantages of the cook-chill method, and explain why you think they are important.

2. What is the optimum temperature range at which cook-chill food should be stored until it is served?

3. Why is it especially important to check a potential food vendor's storage methods before buying ingredients to use in the cook-chill process?

4. What are the special considerations for cooking and chilling meats, and why?

5. At least six factors affect the time it takes to chill cooked foods. Name three of them.

6. What is rethermalization? Do you need to set aside space in your kitchen, or use a particular piece of equipment, to carry out this step of the cook-chill process?

7. What information must be provided when you label a cook-chill product?

8. Why should cook-chill food be stored separately from other types of food whenever possible?

9. What is an ice builder?

10. What is the difference between tumble chilling and blast chilling? What factors determine which of these two methods you should use?

16

DISHWASHING AND WASTE DISPOSAL

■ INTRODUCTION AND LEARNING OBJECTIVES

The dishwasher may well be the most expensive piece of foodservice equipment you will ever buy. A medium-size restaurant, for example, will spend as much as $20,000 for the machine itself. Outfit the entire dish room—with tables for soiled and clean dishes, a few dish racks, and a booster heater for the final rinse—and you're looking at a total investment of $40,000 or more.

In addition to "money pit," the dish room has another unfortunate designation: In most businesses, it receives the least attention and typically is run by the least trained (and perhaps the least appreciated) employees. The purpose of this chapter is to instill a new appreciation for this critical foodservice sanitation function so you'll never take your dish machines—or the people who use them—for granted.

After reading this chapter, you will be able to:

■ Describe sizes, types, and ratings of dishwashers, and how to evaluate their features.

■ Explain the purpose and function of booster heaters.

■ Explain the steps for safely cleaning glassware, plateware, and flatware.

■ Identify the functions of food disposals and waste pulpers.

■ Identify the functions and features of specialty washers.

Let's begin by defining *dishwashing* as the process by which dishes, glasses, flatware, and so on are cleaned using a combination of three things: hot water, detergent, and motion. Mechanical dishwashing has three essential parts:

1. *Sufficient amounts of water at hot enough temperatures to sanitize.* These temperatures are mandated by NSF International, and dish machine manufacturers comply with the NSFI rules. The temperatures are measured utilizing Heat Unit Equivalents (HUEs). The intention is to have enough flow of hot water to raise the surface temperature of the ware high enough to sanitize them. NSF guidelines specify the amount of water, minimum temperatures, and length of time of the wash and rinse cycles to ensure sanitation (see Table 16-1).

2. *Detergents that can soften water, penetrate food particles, and loosen them from dish surfaces.* This is known as a detergent's **wetting action**. Detergent is not inexpensive and can become a real budget drain if not used carefully. Water temperature works in tandem with the detergent to clean and sanitize.

TABLE 16-1

Specification Requirements for Hot-Water Sanitizing Machines

	MINIMUM WASH TEMPERATURE	MINIMUM SANITIZING RINSE TEMPERATURE	MAXIMUM SANITIZING RINSE TEMPERATURE	SANITIZING RINSE PRESSURE (RANGE)
Stationary rack/ single temperature	74°C (165°F)	74°C (165°F)	90°C (195°F)	138 kPa ± 34 kPa (20 psi ± 5 psi)
Stationary rack/ dual temperature	66°C (150°F)	82°C (180°F)	90°C (195°F)	138 kPa ± 34 kPa (20 psi ± 5 psi)
Single-tank conveyor	71°C (160°F)	82°C (180°F)	90°C (195°F)	138 kPa ± 34 kPa (20 psi ± 5 psi)
Multiple-tank conveyor	66°C (150°F)	82°C (180°F)	90°C (195°F)	138 kPa ± 34 kPa (20 psi ± 5 psi)

Source: NSF International, Ann Arbor, Michigan.

3. *Adequate time for water recirculation.* This includes sufficient volume and velocity of the water as well as a machine designed to spray the dishes for the correct amount of time to rinse them fully.

In the table, note that rinse pressures are measured in *kPa* as well as *psi*; kPa is the abbreviation for **kilopascal**, the metric air pressure measurement equivalent of pounds per square inch (psi). One kilopascal is equivalent to 1% of atmospheric pressure. These factors and more are covered in detail in this chapter.

16-1 CHOOSING A DISHWASHING SYSTEM

Numerous variables enter into setting up your dishwashing area. As you write your equipment specs, think about these items:

- Space available in your kitchen
- Proximity to service areas, such as waitstations, and (equally important) distance from eating area
- Type of service: how food is delivered to guests
- Power and plumbing sources and their locations
- Water quality in your area
- Energy-saving and water-saving features
- Local building codes and health ordinances
- Detergent quality and uses
- How dishes are handled (and where they are placed) both before and after washing
- Your proposed budget
- Access to reliable workers

A word about the last item: The cost of labor for the dishwashing process is more than 1 cent per dish. This doesn't sound like much until you see just how many dishes are used per guest per day. The averages can be found in Table 16-2, and these don't include pots and pans or the preparation and cooking utensils. Selecting the proper dishwashing system has a significant impact on your bottom line. Most foodservice consultants believe the first places to cut costs are the non-income-producing, unskilled labor areas of a restaurant, which includes the dish handlers. However, making the process more automated (so it uses fewer workers)

TABLE 16-2	
Dish Needs in Foodservice	
TYPE OF OPERATION	DISH COUNT PER GUEST
Fine dining	20
Casual, family	15
Cafeteria	10–12, plus tray
Counter service	6–8

costs more initially than a basic system. In short, it's a trade-off you must decide on. First, however, you must analyze the entire warewashing function.

■ *Determine the volume and types of wares and utensils you'll be washing.* How can they be cleaned quickly and at the least cost?

■ *Study individual job functions.* How many times are dishes handled before they reach the dish machine? Do employees have to lift the full, heavy bus tubs in order to put the dishes onto racks to load into the machine, or are the tubs kept on the same level for unloading and racking? Are there logical places to put the clean dishes?

■ *Consider working conditions.* Most dish rooms are small, noisy, and overheated; employees work in tight spaces and are constantly exposed to hot water and harsh chemicals. It is not unreasonable to ask manufacturers for the decibel ratings of their dish machines or for content and safety information about the cleaning chemicals.

■ *Determine your peak activity hours.* If you can wash and return dishes to service quickly, you can maintain a smaller inventory, saving money and still taking good care of your guests.

■ *Decide how your guest fits into the process.* In some foodservice settings, the guests clear their own tables. A conveyor system can be installed to send them directly from dining area to dishwashing area.

■ *Analyze the service records of the dishwashing equipment.* If your dishwashing staff is faced with frequent breakdowns, dish breakage, or general inefficiency, you should consider purchasing new machines or perhaps redesigning the system.

It helps to know a bit about how mechanical dishwashing works. If you are a dirty dish, you have only two ways to go: Either you move through a spray pattern on a conveyor, or you stand still in a dish rack and the spray pattern moves around you. The motion is necessary to cover all of the dish surfaces completely. To perform these functions properly, as well as to control costs, the foodservice manager should understand six fundamental rules:

1. Maximize the use of sanitation personnel in the kitchen.
2. Control detergent use by prescraping or prewashing.
3. Use enough water (and the energy to heat it) without wasting either.
4. Control machine use by running only fully loaded dish racks through the washer.
5. Use the proper wetting agent, which promotes faster rinsing and drying and minimizes spotting.
6. Lay out the dish room efficiently, which helps reduce breakage.

Speaking of layouts, there are as many configurations as there are dishwashing systems (see Illustration 16-1). Resist the temptation to choose whatever model seems to fit best in your square footage. Instead, design the system with productive workflow, safety, and sanitation issues in mind.

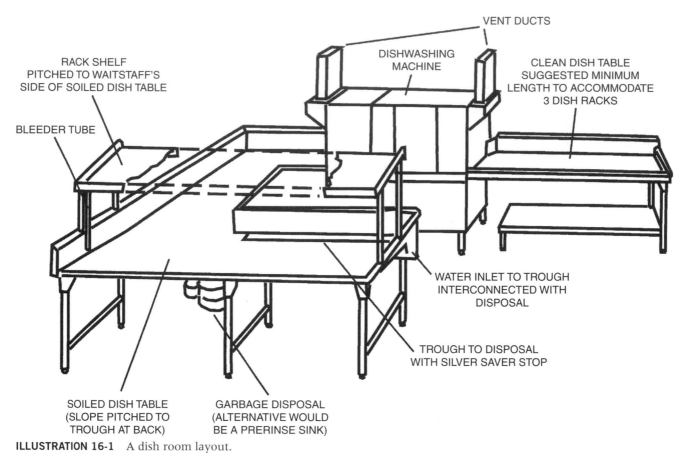

ILLUSTRATION 16-1 A dish room layout.

Source: Carl R. Scriven and James W. Stevens, *Manual of Equipment and Design for the Foodservice Industry* (Albany, New York: Thomas Learning, 1989).

HIGH-TEMP OR LOW-TEMP?

The amount of room you need overall depends in part on the type of dish machine you choose. Dish machines often perform their sanitation functions at a steamy 180°F, so you'll need space for ventilation hoods, which are normally required. The dishwasher that rinses at 180°F is known as a *high-temp* model. A high-temp dish machine requires a **booster heater**. Ordinary 140°F dishwater is piped into the booster heater to bring it up to the necessary 180°F for use during the rinse cycle. Most newer dish machines have their own built-in booster heaters, but others offer them as a separate unit, which requires more space and separate installation.

More advanced dishwashers now include heat recovery units to capture heat that would normally be vented off and reuse it to heat incoming water. This allows for more efficient operation and reduced utility costs. It also means reduced exhaust fan speeds because less treated air has to be removed from the building.

Where venting is a problem, most cities allow the use of a *low-temp* dishwasher. This model rinses at the lower, regular wash temperature of 140°F but uses chemicals instead of super-hot water to do the sanitizing during the rinse cycle; this is usually some type of chlorine, stored in a separate unit that can be wall-mounted or located under a counter. This unit can be obtained from (and usually installed by) your detergent supplier. It works by injecting a preset amount of chemical solution directly into the rinse water. Some cities also require that a litmus-paper test be used periodically to check the strength of the sanitizing solution. Table 16-3 lists a few guidelines for chemical concentration and rinse-water temperature. The abbreviation *ppm* means parts per million, and *NaOCl* is sodium hypochlorite, the chemical term for the chlorine solution.

Many dish machines are **field convertible**, which means they can be changed from hot-water sanitizing to chemical sanitizing or vice versa by following the manufacturer's instructions.

TABLE 16-3		
Data Plate Specifications for the Chemical Sanitizing Rinse		
SANITIZING SOLUTION TYPE	FINAL RINSE TEMPERATURE	CONCENTRATION
Chlorine solution	min: 49°C (120°F)*	min: 50 ppm (as NaOCl)
Iodine solution	min: 24°C (75°F)	min: 12.5 ppm, max: 25 ppm
Quaternary ammonium solution	min: 24°C (75°F)	min: 150 ppm, max: 400 ppm

*For glasswashing machines that use a chlorine sanitizing solution, the minimum final rinse temperature specified by the manufacturer shall be at least 24°C (75°F).

Source: NSF International, Ann Arbor, Michigan.

Points to consider in the high-temp versus low-temp debate: Using a chemical sanitizer can save money on energy costs, as the water doesn't have to be quite so hot. It also reduces ventilation requirements. But some utensils don't react well to the chemicals, notably aluminum, pewter, and silver-plated items. The sanitizing solutions should be a minimum of 50 parts per million (ppm), per NSF International guidelines. If the sanitizing chemicals aren't properly balanced to the amount of water, spotting and a detergent taste can linger on the dishes.

One benefit of a high-temp machine is faster drying time. Of course, it takes an adequate power supply to provide that boost of heat to the water, and the plumbing lines must be nearby to handle high volumes of water that go from booster to dish machine.

The rule about dishwashing is that dishes soiled by foods that contain protein usually require very hot water to dissolve and power off the food residue; but coffee cups and most glassware don't need scalding water to get them clean. The only glassware exception may be wineglasses: Purists feel they should be cleaned with scalding water and a minimum amount of detergent (or no detergent at all) to prevent residue on the glass that interferes with the taste and clarity of the wine. Some wineglasses are also quite delicate, and it isn't unusual for fine-dining establishments to assign someone to hand-wash them. Several manufacturers, recognizing the need for delicate washing of stemware, have added "soft wash" cycles to their machines. This setting reduces the amount and force of the water used to clean the glasses.

Both high-temp and low-temp machines are affected by your area's water quality. They have the same scaling and chemical buildup problems as any other appliance that uses tap water. In addition, food particles can plug up lines and solenoids. Water filters and, in some areas, water softeners are necessary components of successful mechanical dishwashing.

DETERGENT USE

Environmental awareness has led to a push for cleaning products that minimize pollution, and dish detergents are no exception. The typical foodservice business runs thousands of gallons of soapy water down its dish room drain every day. In the early 2000s, environmental activists in several states (Illinois, Maryland, Minnesota, Vermont, and Washington) successfully introduced legislation to remove phosphate from household dish-cleaning products. A total of 17 states passed similar bans.

High phosphate levels kill fish by depleting oxygen in lakes and streams, causing excessive algae growth. Phosphate also is found in farm and lawn fertilizers, so dish detergent was hardly the only culprit, or even the primary one. The Soap and Detergent Association (now known as the American Cleaning Institute) requested more time, asked to use phosphates in smaller amounts, and called the phosphate scare "alarmist." But by 2010, makers of dish detergents had quietly reformulated their products without the phosphate. They said it was too difficult to make different versions of their products for different states. Today, some users are convinced the phosphate-free formulas are not as effective.

It is important to consider ecological implications of the products you select. In the European Union, eco-labeling has been voluntary for home-use dish detergents since 2003, and the standards provide a good checklist when shopping for an eco-friendly product for commercial use. To name just a few, check for:

- Biodegradability of the surfactants
- Dangerous, hazardous, or toxic substances or preparations
- Washing performance
- Fragrances
- Packaging

Their environmental impact aside, cleaning products are expensive—and in foodservice, you use a lot of them. Be sure to ask the equipment manufacturers about how much dish detergent is required per load, and figure that into your long-term costs.

DISHWASHER SIZES AND RATINGS

Here's a step backward for you: How about deciding if the volume of dishes you'll be using is sufficient to require a mechanical dishwasher in the first place? Most foodservice operators decide almost immediately they couldn't live without one—not only for sanitation reasons but also to save time. Even the least sophisticated machine can complete one of its cycles in a couple of minutes. An undercounter machine may take 90 seconds to 4½ minutes. You may not need a continuous dishwashing operation; you might store the soiled dishes and wash them periodically in a flurry of activity. Or you may decide that a larger machine with a lower rate of water flow per rack saves water and energy when compared to a smaller machine that must be run more often.

Let's talk first about a small breakfast-and-lunch operation that serves no more than 50 meals per day. In this case, a single-tank, undercounter machine suffices. A typical commercial operation serving 200 meals per day should have a single-tank door-style dishwasher. An operation serving 300 to 500 meals per day would probably choose a conveyor-style dishwasher. And a large, high-volume operation (such as a hotel banquet department or airline kitchen preparing meals for passengers) needs a flight-type conveyor-style washer, which comes in lengths from 9½ feet up to 60 feet. Later in this chapter you will learn more about each of these options. Meanwhile, Table 16-4 lists the sizing suggestions from the authoritative handbook *Food Equipment Facts* by Scriven and Stevens.

Dishwashers are often sold with promises that they will clean so many dishes, or so many racks of dishes, per hour. Although these claims are not necessarily untrue, in real-life dish

---

TABLE 16-4

Dishwasher Needs in Foodservice

MEALS PER HOUR	STYLE OF DISHWASHER
Up to 50	Counter or undercounter
50–250	Single tank, door style
250–400	Single tank, conveyor
400–750	Single tank, conveyor with prewash option
750–1500	Double tank, conveyor with prewash option
1500 or more	Flight-type conveyor (or, where space is tight, a carousel)

Source: Carl R. Scriven and James W. Stevens, *Food Equipment Facts* (New York: John Wiley & Sons, Inc., 1999).

TABLE 16-5

Dimensions of a Dish Room

MEALS PER HOUR	DISH ROOM AREA SQUARE FOOTAGE
200	100
400	200–300
800	400–500
1200	600–700
1600	800–900

Source: Carl R. Scriven and James W. Stevens, *Food Equipment Facts* (New York: John Wiley & Sons, Inc., 1999).

---

rooms a machine rarely achieves its full potential. A good general rule is to calculate the capacity of the machine at 65% to 75% of the factory rating.

Scriven and Stevens also devised basic space guidelines for the dish room (see Table 16-5). You may also refer to Chapter 4 of this textbook for additional dish room space guidelines.

Most commercial dishwashers are rated in several basic categories.

MACHINE RATINGS. The number of full dish racks per hour that a machine can wash is its rating, an indication of its maximum mechanical capacity. A typical dish rack measures 20 × 20 square inches. A basic rule is to estimate the machine capacity at 70% and the average rack capacity:

- ■ 16 to 18 9-inch plates per rack (which varies, of course, with dish diameter)

- ■ 25 water glasses per rack (again, this depends on glass size)

- ■ 100 pieces of flatware per rack

Different types of racks can be purchased to accommodate different types of dishware: peg racks for cups or bowls, mesh cups to contain flatware, and so forth.

Other variables that go into figuring actual productivity are the dish room layout, the length of time the dirty dishes sit before washing, the hardness of the water, the efficiency of the machine operator, and fluctuations in the flow of soiled dishes into the dish room.

PUMPS AND MOTORS. Highly efficient motors and pumps ensure the proper volume of water at the required pressure levels to remove food from dishes. Motors range in size from ½ to 5 horse-power. If dishes or glasses are plastic, you might consider the use of *reducing collars*.

When the dishwasher fills with water, a pump circulates it. Pump efficiency is measured in gallons per minute (gpm) and ranges from 45 for small, undercounter models to 240 for the largest flight-type machines. An energy-saving option is the low-flow pump, which can cut water consumption in half for the final rinse, saving more than 100 gallons of water per load.

HEATING EQUIPMENT. The water heater keeps water hot in the tank(s) of the dishwasher. The smallest electric models use only 1.2 kilowatts—the largest, 23 kilowatts. Gas is also used to heat dish-water, resulting in new technology similar to infrared gas burners. Gas is extremely efficient and puts out less radiant heat in the kitchen. The decision to use gas or electricity depends largely on what is available in your area and how much each costs.

In some areas, it is necessary to fit the dishwasher with a water softener to keep the minerals in hard water from condensing on the heating element or clogging water lines.

RINSE WATER. The rinse cycle is critical for sanitation. Allowable water pressure for this cycle is between 15 and 25 pounds per square inch (psi). It can be checked at the pressure gauge installed at

TABLE 16-6

Energy Star Efficiency Requirements for Commercial Dishwashers

CATEGORY	HIGH-TEMP EFFICIENCY REQUIREMENTS PER RACK	LOW-TEMP EFFICIENCY REQUIREMENTS PER RACK
Undercounter	1.00 gallons	1.70 gallons
Stationary, single-tank door	0.95 gallons	1.16 gallons
Single-tank conveyor	0.70 gallons	0.62 gallons

the inlet side of the final rinse valve and read while the rinse water is flowing. The *rate of flow* is also important; it means the machine must be replenishing its storage tank even while a rinse cycle is taking place. Finally, the rinse water temperature should be 180°F in most situations. As already mentioned, it may take a booster heater to accomplish this.

CONSERVATION FEATURES. The other important consideration in today's resource-conscious environment is water and power consumption. In the past decade, manufacturers have reduced electrical, water, and chemical use by as much as 50% on their machines. However, it is still important to evaluate each machine for cycle times, conveyor speed, and the like to make certain it meets your warewashing needs.

Another feature is the use of electric eyes and other types of sensors in the machine that detect the presence of dish racks and shut down the pumps or exhaust fans when the machine is empty. This "idle time" feature can save significant energy. Also ask about the amount of insulation in dish machine construction; the better insulated it is, the more comfortable the ambient temperature in the dish room, which saves on utility costs for cooling the space and typically reduces noise as well. Table 16-6 lists requirements for the U.S. government's Energy Star rating.

16-2 TYPES OF DISHWASHERS

The dish machine is the major component of your dish room operation. Now let's look at specific types of dishwashers for different foodservice operation sizes.

UNDERCOUNTER DISHWASHERS

At peak performance, an undercounter dishwasher can wash from 18 to 40 racks per hour, one rack at a time, making it ideal for small operations. Some machines have slides on which to rest removable racks; others contain racks that remain in the machine on rollers and can be slid in and out. Removable racks are recommended because this allows the operator to rack dirty dishes in advance and have them ready to load as soon as others are washed (see Illustration 16-2).

Undercounter dishwashers use from 3 to 5 gallons of water per wash cycle and only a few kilowatts of power to maintain the water temperature in the tank. The time for each wash cycle varies from 1:30 to 2:20 (minutes:seconds). The ideal pump capacity is 45 gpm.

High-temp models may have a built-in booster heater that provides just over 1 gallon of super-hot rinse water, 40°F hotter than the wash cycle; most models feature an optional booster with a 70°F heat increase.

Low-temp undercounter machines are usually smaller, only 1½ gallons in capacity. They get their water at 120°F from a standard hot water heater and do their sanitizing with chemicals.

The best interior design seems to be dual wash arms at both top and bottom of the washing area, sending high-pressure sprays that clean dishes thoroughly. The double arms are more expensive than a single arm; you may not need them if your dishes are not heavily soiled.

The height of the dishwashing cavity is between 11 and 17 inches. The largest height seems handiest because it allows the operator to load and unload 16 × 18-inch cafeteria trays as well as bulky but often-used preparation items such as mixing bowls and electric mixer accessories.

As its name implies, the undercounter machine is usually installed under a counter with a drop-down front door or beneath a specially designed dish table with a sink. The sink setup may contain a heavy-duty sprayer for pre-rinsing dishes; a removable, perforated *scrap basket* to scrape waste into; an electric food disposal unit; a corrugated drainboard area; and an overshelf. Some models also have a switch that activates a deliming cycle. Plumbing is located beneath the drainboard to save floor space.

There are also freestanding models with stainless-steel sides and top, which measure a little less than 3 feet in height and 2 feet square in width and depth. They can be mounted on 6-inch legs or on an 18-inch stand. The latter is handy because it raises the machine to counter level, which reduces bending to load and unload.

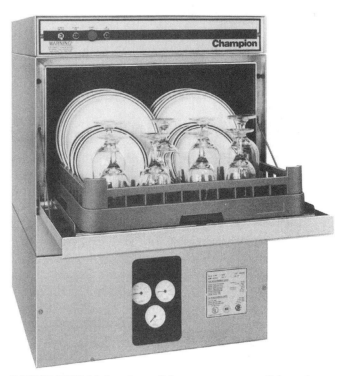

ILLUSTRATION 16-2 A small low-temperature dishwasher designed to fit beneath a counter.

Courtesy of Champion Industries, Huntington, West Virginia.

Like most mechanical dishwashers, the wash and rinse cycles are fully automatic, controlled by electric timers when the door is closed and latched. If necessary, the cycle can be interrupted by lifting the door handle; it resumes when the door is relatched. A light goes out when the cycle is complete. On the front door, gauges indicate wash and rinse temperatures as well as rinse pressure.

Undercounter models are powered by electricity only, at 120 volts. Models with built-in heaters require 208/240 volts. Some newer undercounter units are actually capable of notifying the service company if repairs are needed.

GLASSWASHERS

A variation of the undercounter dishwasher is the glasswasher, often installed in bar settings. Getting glasses clean (without breakage) is such an art form that at least one manufacturer's machine has eight cycles for different types of glassware. The normal glasswasher washes and rinses with tap water, provides a final rinse with sanitized water, and blow-dries the glasses. A heavy steam cycle ensures removal of food particles; a low-temperature cycle cleans plasticware without melting it.

Water temperatures in these units range from the 150°F wash cycle to a super-hot 212°F for quick drying. This is a benefit because the faster the glasses dry, the less likely they are to emerge with water spots on them. The full wash-rinse-dry cycle takes about 20 minutes and uses 3 gallons of water for each fill. The required water pressure is 20 psi.

Most glasswashers look like undercounter dish machines, with the pull-down door and a variety of racks designed to hold different types and sizes of glasses. However, you can also get a conveyor-type glasswasher that turns out nearly 1000 clean glasses per hour (see Illustration 16-3). This machine has three pumps to dispense precise amounts of detergent, sanitizing solution, and rinse aids, and a 3-kilowatt water heater. Glasses enter the machine on a polypropylene conveyor belt, are washed at 140°F to 160°F, and are rinsed twice. The final rinse is with cool, distilled water so the glasses come out ready to use instead of hot to the touch.

A pass-through-type fully automatic glasswasher is capable of washing 2000 glasses per hour. The conveyor belt on this unit can run in either direction at 14 inches per minute. Installed at countertop height, its width varies from 48 to 72 inches depending on the amount of space you want for loading and unloading at either end of the machine.

SINGLE-TANK DOOR-STYLE DISHWASHERS

A common name for the single-tank door-style model is the ***stationary rack dishwasher***, and it is the most widely used of all dish machines. For a foodservice operation that serves 50 to 200 people and generates 750 to 1250 dishes and related items per hour, this is the typical recommendation. The door-style dishwasher is easy to operate and can handle from 35 to 55 racks per hour. You'll usually find it installed between two dish tables, one for stacking soiled dishes, the other for stacking clean ones. (A helpful guideline is to make sure the soiled-dish table is 50% larger than the clean-dish area.)

Water and detergent are mixed in a single tank, so the machines are sometimes referred to as *dump and rinse*. Most of them store the rinse water and use it again in the next wash cycle. Soil and detergent is skimmed off the top of the water in the tank, so a constant replenishing by the final rinse keeps the machine full.

Most manufacturers offer single-tank machines with a hot-water sanitizing mode or two chemical sanitizing modes. A normal-duty chemical sanitizing machine requires a water temperature of 120°F and is able to wash 62 racks per hour; a light-duty model requires 130°F water and is able to handle up to 80 racks per hour. Many single-tank machines are ***field convertible***, which means they can be set up for either high-temp or low-temp sanitizing.

A 1-horsepower motor powers the unit and, depending on its size, needs from 100–120 volts to 400–460 volts of electricity. The water tank holds 16 gallons, kept hot by a 5-kilowatt electric heating element or, in the case of gas heat, an energy-saving burner with an automatic pilot light igniter (so the pilot light is not on continuously). The pump capacity is 160 gpm. The stationary rack machine uses about 1.2 gallons of water per rack.

The unit is made of stainless steel, including the wash arms inside. Located above and below the dish racks, the arms are removable and interchangeable. Two nozzles are mounted on the arms to spray and sanitize during the rinse cycles (see Illustration 16-4). Be sure to check the dishwasher arms once in a while to make sure they rotate easily.

The newest dishwashers feature microcomputer controls that can be programmed to time each cycle. The dishwasher begins automatically when the door is lowered, stops immediately if it's opened during the process, and resets itself. Draining and overflow prevention is also regulated, with a large bell-type automatic overflow and drain valve. Perforated strainers and removable scrap baskets submerged beneath the unit make it practically clog-proof.

The dimensions of the average stationary-rack dishwasher are 67½ inches high, 26½ inches wide, and 29¼ inches deep; in short, the machine fits into a 4-square-foot floor space. The door opening can range in width from 15 to 19 inches, which is a consideration if you'll be washing serving trays in it. Because the booster heater is almost always built into the unit, you don't have to allow additional space for it. However, your site selection may depend on accessibility to an exhaust canopy, as most local ordinances require this machine be vented at a rate of 100 cubic feet per minute (cfm). Machines with heat recovery require lower exhaust rates, which ultimately saves on treated air being vented from the building.

Handy options for your single-tank dishwasher include fabricated steel ***dish tables***. You'll want to order them with a 6-inch-high backsplash and perhaps an overshelf to store glass

ILLUSTRATION 16-3 A glasswasher is designed to wash and sanitize cups and glasses without water-spotting or chipping them.

Courtesy of Champion Industries, Huntington, West Virginia.

racks. A drain tube on one edge of the shelf allows water to drain onto the table below. Undershelves are also available; they're 6 inches off the ground and handiest below the clean-dish side of the table.

The dirty-dish side should include a sink for rinsing and scraping duties. Most are 20 inches square and 6 inches deep. The center opening at the bottom of the sink should be at least 3½ inches wide to accommodate installation of a drain or disposal unit and a removable scrap basket. If no electric disposal is installed, a sink flange with a 2-inch lever drain is recommended. Also ask for a heavy-duty sprayer to use by hand to pre-rinse dishes.

MOVING DISHWASHERS

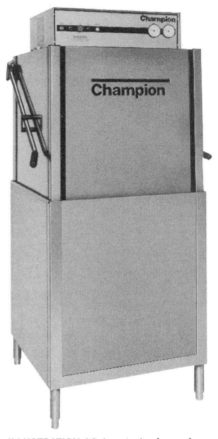

ILLUSTRATION 16-4 A single-tank door-style dishwasher.

Courtesy of Champion Industries, Huntington, West Virginia.

Large operations that routinely serve more than 200 meals in a short period require a dish machine that moves dishes through wash and rinse cycles on a conveyor belt. The most basic of these models is the *rack conveyor,* designed to transport full racks of dishes through wash and rinse. A single-tank rack conveyor can wash from 125 to 200 racks per hour; a double-tank machine can do 250 to 300 racks per hour (see Illustrations 16-5 and 16-6).

Rack conveyors generally include more sophisticated options than stationary dishwashers, such as recirculating prewash or power prewash cycles, corner scraper units, and automated activators, which run the machine only when racks are in it. Human labor requirements depend mostly on the number of racks to be handled; when a multiple-tank machine is running at full demand, two people may be needed to load and unload racks. However, you'll soon read about automatic loaders and unloaders that could cut labor needs to a single employee.

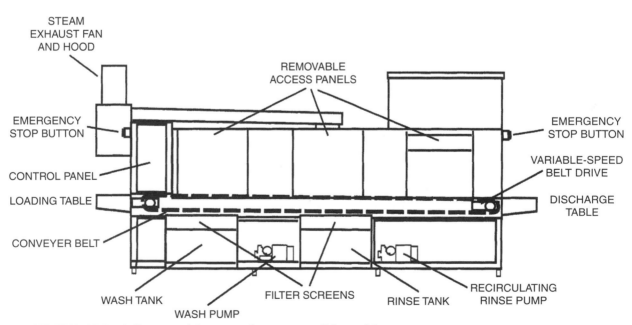

ILLUSTRATION 16-5 A diagram of the parts of a conveyor dish machine.

ILLUSTRATION 16-6 A double-tank conveyor dishwasher.
Courtesy of Champion Industries, Huntington, West Virginia.

Rack conveyors move the dishes either by chain or pawl. The *chain* works somewhat like a bicycle chain, pulling the conveyor belt along; a *pawl* is like a piston that uses a single rod to pull the racks through the machine. You can install an electric eye sensor in most machines that saves energy by stopping the conveyor or the water flow when no dishes are coming through.

The dishwasher is constructed of 16-gauge stainless steel. Flexible strip curtains made of plastic hang at both ends of the dishwashing chamber to keep water inside; curtains also are used to separate the wash and rinse functions. A large inspection door located near the front of the machine allows easy access to its interior; if the door is opened midcycle, the pump motor shuts off automatically as a safety feature. Likewise, if the movement of racks is obstructed, an adjustable overload mechanism is triggered that prevents damage to the machine or the racks. Conveyor speed is set by NSF International standards for each machine. When chemical sanitizing is used, the final rinse uses a liquid bleach solution.

The 2-horsepower motor is controlled by a single on–off switch. Its electrical specifications range from 200 to 230 volts single-phase to 400 to 460 volts three-phase. Some models use the same motor to drive the conveyor and the pump; others have a separate ¼-horsepower motor to drive the conveyor. The pump circulates water at 195 to 265 gallons per minute, and self-draining pumps drain water out of the tank between cycles.

Washing is done by two fixed spray assemblies, one at the top of the unit and one below. Both are removable (no tools needed) for cleaning. Perforated strainers cover the wash tank's water surface to prevent debris from clogging the drain. The rinse cycle is conducted by double-action rinse arms, also located above and below. The machine is plumbed so that half the rinse water goes back to the wash tank as makeup water while the other half goes down the drain. The water tank is filled manually by turning a fill valve (or, in some cases, pushing a button) in front of the machine.

A space-saving option is the ***side loader***, a cantilevered table that lets the operator load racks from the front of the machine. It is available in 24- and 30-inch lengths to accommodate different sizes of racks. An extended pawl bar connects the side loader firmly to the machine. A side loader is helpful where space is small because it allows a dishwasher to be located closer to a corner of the room.

Another popular feature is called the ***curved unloader***. This 38-inch addition helps push racks off the conveyor at a 90-degree angle to the machine. Each rack clears the machine before the next rack is discharged.

Both single-tank and double-tank conveyor-type dishwashers can be ordered with a ***prewash option***. This is a separate chamber from the regular washing area, where dirty dishes begin their odyssey. Sometimes it's referred to as a *power prewash,* and most machines have a separate 1- or 2-horsepower motor for this function. Inside the chamber, stainless-steel arms with water jets shoot a high-velocity spray to effectively scrape all kinds of tableware. The food residue is collected on perforated strainer screens (also called ***scrap screens***) and then drops into a strainer basket (or scrap basket) below.

At a length of about 7 feet, double-tank machines have four compartments for distinct duties: power prewash, regular wash, power rinse, and, finally, sanitizing rinse. Three large inspection doors allow access to the chambers. If they're opened during a cycle, the action immediately shuts down until they're relowered. Each compartment has its own set of spray arms and/or directional water jets. A benefit of these large machines is that they provide better separation of the wash and rinse cycles; also, with two rinse cycles, detergent is removed thoroughly from the dishes. If you've ever eaten at a restaurant and tasted soap on a glass or utensil, you know how important a good rinse cycle is.

Many models recirculate their final rinse water into the wash tank. The water in the tanks may be heated by electricity, gas, or steam. Inside each tank, a *thermistor* sensor continuously monitors and adjusts the water temperature, and a float device protects against the water level becoming too low.

MORE OPTIONS FOR CONVEYOR MACHINES

To eliminate the need for an exhaust system, several manufacturers include a **condenser** as an accessory. The condenser is mounted over the vent of the dishwasher, where air flowing out of the dishwasher passes over water-cooled coils inside the condenser. A ½-horsepower centrifugal blower (1000 cubic feet per minute) helps remove the moisture from the air and returns the dried air to the dish room area. Louvers on the unit can be moved to send the airflow in any direction. The condenser is activated automatically by the pump switch located on the control panel. The International Mechanical Code in the United States requires that larger dishwashers have exhaust hoods or ducts with fans, while smaller units may add treated air to the space to reduce heat and humidity in the room. Consult a mechanical engineer in your locale to determine what is required.

Because the coils are water-cooled, access to a water source is a prime installation consideration. The required incoming water pressure is 20 psi; it should flow at a rate of at least 6 gpm and at a starting temperature of no more than 55°F. As the air passes over the cool water, the water naturally heats up. It can then be piped back into the prewash tank or discharged to the drain. Another water-saving feature: The rate that cold water flows into the condenser is regulated automatically by the passage of dishes into the machine.

A condenser unit adds about 2 cubic feet to the dish machine, but because it is installed at the top, it fits anywhere there is a standard 7-foot ceiling.

A system to speed the drying of dishware is the **blower-dryer**, found in conveyor models as well as flight-type dishwashers. Blower-dryers can be all electric or steam heated. The all-electric blower-dryer is a 5-foot-long stainless-steel chamber with forced circulation of high-velocity hot air inside. It's designed to dry china, silver, and even plasticware quickly. A heavy-duty fan draws room air through an electric heater and into ducts at the top of the chamber. The heated air is forced vertically downward onto passing dish racks; baffles below the conveyor aim the air upward again. So all this hot air doesn't blast anyone who happens to be unloading the machine, large fan intakes at the outgoing end keep cooler, room-temperature air moving over the unloading area. An exhaust system to the outside of the chamber is necessary to carry off the moist air from the drying process. This is accomplished by a vent stack with a control damper, which can be connected to the dish room's main exhaust system.

Adding a blower-dryer adds 5 feet of length and almost 2 feet of height to your dishwashing area, plus additional exhaust requirements of 1200 to 1400 cubic feet per minute. You'll need more electrical capacity, too, because the unit has a 2-horsepower blower motor and a dozen 3500-watt air heaters that use 42 kilowatts of power per hour. An installer wires the blower-dryer to your dishwasher motor switch so it operates only when the dishwasher is on.

A steam-heated blower-dryer functions just like the all-electric model except for the way heat is delivered to the drying process. Instead of electric heaters, we find double rows of high-efficiency tube-type heat exchangers. Also, the condensation that gathers on each coil must be drained away. Steam requirements to dry dishes properly are 110 pounds per hour at 20 to 25 psi.

Like other types of dishwashers, conveyor machines must be ventilated. Exhaust requirements vary with the size and capacity of each machine. Overall, however, built-in hoods increase efficiency as they make for faster air drying and a more comfortable work environment.

Another option: Stainless-steel ventilating **cowls** (seen on the machine in Illustration 16-7) are hood-shaped covers that can be fastened to each end of the conveyor and then connected to an exhaust system. Cowls add 16 to 20 inches to the overall height of the unit. Dampers can be installed in the ducts of the exhaust system to ensure exhausting doesn't occur too rapidly, which could chill the final rinse water below acceptable temperatures.

Typical exhaust requirements are:

- For single-tank machines, 200 cfm at the entrance, 400 cfm at the clean-dish side
- For multiple-tank machines, 200 cfm at the entrance, 500 cfm at the clean-dish side
- For a machine with a blower-dryer unit, as high as 1400 cfm at the clean-dish side

CIRCULAR DISHWASHING SYSTEMS

The highest-volume foodservice operations might want to look at circular conveyor systems, which can handle from 8,000 to 24,000 dishes per hour. An oft-cited industry survey ranks these carousel-type conveyors in three categories based on rack capacity per hour:

Type 1: 180 racks
Type 2: 387 racks
Type 3: 480 racks

The advantage of a circular system over an equally large flight-type dishwasher is the ability to operate it with fewer personnel. Although the dishes go through the inside of the machine in a straight line, the exterior part of the conveyor belt curves to form a loop from the exit to the entrance of the dishwasher (see Illustration 16-8).

A circular dishwasher can have one, two, or three water tanks. Its rack conveyor is driven by a heavy-gauge stainless-steel chain assembly, with a motor of either ½ or ¾ horsepower. The power of the motor depends on the length and configuration of the setup as well as the size of the machine. The conveyor can move clockwise or counterclockwise, and the machine can be purchased with its inspection windows on the outside or the inside of the circle, so be sure to specify this.

A 2-horsepower motor runs the wash and rinse cycles; an additional 1-horsepower motor can be included for a power prewash cycle. Electrical requirements may be 200–230 volts or 400–460

ILLUSTRATION 16-8 A circular dishwashing system can handle the highest volume of dirty dishes with fewer employees needed to operate it.

Courtesy of Champion Industries, Huntington, West Virginia.

volts three-phase. Water temperatures are 150°F for the wash, 160°F for the rinse, and 180°F for the final sanitizing rinse. The unit uses between 130 and 348 gallons of water per hour (depending on its size); its venting requirements are 200 cfm at the loading end, 500 cfm at the unloading end. (Adding a blower-dryer increases this requirement to 1400 cfm at the unloading end.)

The curve requires a slightly different setup, often consisting of: a straight table for soiled dishes (with waste disposal and spray nozzle), the dish machine, a curved soiled-dish table, and a curved clean-dish table.

This is a big workspace, so have it designed to fit your needs. The experts generally suggest allowing no less than 6½ feet for the curved tables and about 11 feet for the straight table. At least one manufacturer, however, claims to be able to handle 130 racks per hour in a linear space of just over 9 feet. This configuration is usually not a true circle; an oval-, triangle-, or L-shaped configuration might fit your dish room better. In short, the modular nature of the components gives you a lot of options.

As long as you are designing exactly what you want, here are a few add-ons to consider, depending on your budget:

■ Heat recovery
■ Built-in electric or steam booster heaters to increase the water temperature 40°F to 70°F
■ Prewash units
■ Automatic tank fill with a manual bypass
■ Food waste disposal unit

- Blower-dryer or condenser
- Longer dish tables on either end
- Hand sinks and/or silverware soaking sinks (these require separate hot/cold water and drain connections)
- Storage shelves above or below
- Vent cowls and locking dampers
- High vertical clearance inside the machine (up to 2 feet, to accommodate large pots and pans)
- Top-mounted control panel (saves space and protects the controls from water contact)
- Single-point electrical with electrical disconnects
- Single-point water supply
- A hose reel with a 25-foot hose (for cleanup)

FLIGHT-TYPE CONVEYOR DISHWASHERS

Flight-type dishwashers are also called *rackless conveyors* or *belt conveyors*. Perhaps the name comes from the fact that the dishes run through the machine in a line as straight as an airport runway. The machine itself is the aircraft carrier of the dish room, and the length of the runway can be customized from 13 feet up, usually in 1- or 2-foot increments; the width of the machine can also be customized, up to 5 feet across (see Illustrations 16-9 and 16-10). These are real workhorses, found mostly in hospitals, prisons, dormitories, and the like. A flight-type machine is recommended if more than 8000 dishes are handled per hour, which translates to between 600 and 800 meals an hour.

The flight-type dishwasher operates on the same principle as the rack conveyor, except the dishes or cafeteria trays are hand-placed between rows of plastic-tipped pegs on a continuous conveyor. They travel through the wash-rinse-sanitize cycle in a vertical position. Dishes don't need racks, but someone's got to load them onto and off of the conveyor. Special racks for glasses, cups, and flatware are made to fit on top of the pegs; flatware may be placed first in a rack and then sorted into perforated cylinders. It usually takes two employees to keep the dishes going through smoothly, one at each end.

The heat and humidity these machines produce is formidable, and ventilation can be a challenge. You may need to vent and replace the dish room air as much as six times an hour.

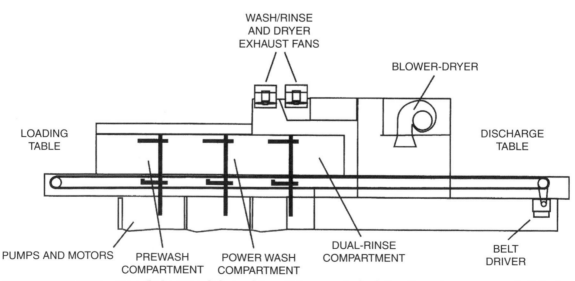

ILLUSTRATION 16-9 On a flight-type dish machine, dishes are stacked directly onto a conveyor belt that moves them through the wash and dry cycles.

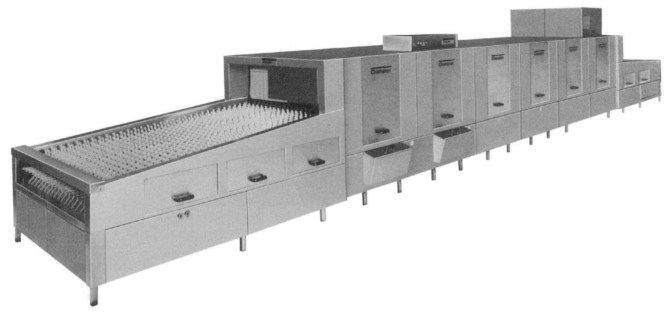

ILLUSTRATION 16-10   The flight-type dishwasher can be as long as needed to accommodate the volume of dishes to be cleaned.

Courtesy of Champion Industries, Huntington, West Virginia.

The most basic machine takes up an absolute minimum of 8 feet but requires additional space to load and unload. The typical configuration is a 3- to 5-foot loading section, a prewash section, an 8-foot power wash section, a power rinse, a fresh-water final rinse, and a drying/unloading section ranging in length from 5 to 11 feet. A total realistic length to plan on is 20 to 25 feet.

Inside the dishwasher, splash baffles and flexible plastic strip curtains separate the spray systems. The curtains are easily removable for cleaning, as are the scrap screens and baskets in the prewash section. Like other dishwashers, these can sanitize using hotter water or lower temperatures with chemicals.

Here are typical electrical, water, and steam specifications. Please note, however, that you should always make certain you have the right connections available for your final machine selection:

- Motors have different horsepower (HP) levels depending on their function: prewash (2 to 3 HP), wash (3 HP), rinse (3 HP), conveyor motor (½ HP), blower-dryer (2 HP).
- Electric heating coils, if used to make steam, need 20 kilowatts of power. A steam blower-dryer needs steam pressure of 20 psi to provide the required 75 pounds of steam pressure per hour.
- Each water tank (prewash, wash, and rinse) holds 40 gallons.
- Power requirements for water heating: 20 kilowatts for washing, another 20 for rinsing.
- Pump capacities: prewash (150 gpm), wash (240 gpm), rinse (240 gpm).
- Exhaust requirements: 500 cfm at loading end; 1000 cfm at unloading end.

Remember to match your utility needs to peak demands, not average demand. Each 3-horsepower pump requires 2.2 kilowatts each; for water-heating, an electric tank uses roughly 25 kilowatts per cycle. Water consumption ranges from 325 to 425 gallons per hour if the dish machine's output is from 8,500 to 19,000 pieces per hour.

As an energy-saving and safety feature, the pumps, final rinse cycle, and optional blower-dryer all stop whenever the conveyor stops. There's also an automatic shutoff feature whenever dishes reach the end of the unloading section, unless they are removed by the operator. Whenever inspection doors are open, the pumps and conveyor stop. A lever must be raised to open the drain. If a water tank is accidentally drained, a low-water protection device turns off the water heater until the tank is sufficiently refilled. Several manufacturers install automatic drains in their machines to minimize accidental draining of the tanks. In short, the unit almost thinks for itself!

Use this handy formula to determine the dish capacity of flight-type machines. It was created by NSF International and is reprinted here from Scriven and Stevens's *Food Equipment Facts*:

VARIABLES

C = Maximum dish capacity per hour

V = Speed of conveyor (in feet per minute)

W = Width of conveyor (in inches)

D = Distance between pegs

FORMULA

$$C = \frac{(120 \times V \times W)}{D}$$

Even with this formula, foodservice authorities suggest dishwashers be rated at only 70% of their stated capacity because not every rack or every peg is loaded perfectly by your hard-working dish room staff.

Flight-type dish machines are the largest and most expensive dishwashers on the market, so manufacturers offer quite a few options to make them worth your consideration. You usually can bargain for the length of your loading and unloading areas, extra water- or energy-saving capabilities, noise-reducing insulation packages, variable-speed conveyor controls, dryer attachments, special designs for trays, and good prices on racks and flatware troughs. For correctional facilities, there are even theft-proof and tamper-resistant features.

In addition, computer technology is making these machines more sophisticated all the time. Solid-state microcomputer controls include sensors inside the machine; visual displays or readouts of problems or service reminders; and automatic sorting, loading, and unloading units that can be programmed to gently separate plates and bowls by size. Strong magnets can pick up cutlery from the trays as they pass by. Wash and rinse arms can be preset automatically to the height of the incoming dishes to provide maximum cleaning without breakage. At the discharge end of the system, the warm, clean plates or trays can be stacked automatically into spring-loaded *lowerators* to hold their temperature for new, hot meals.

Sounds amazing? Yes, it is, and amazingly expensive, too, with the largest hospital-size system costing a cool $300,000. However, at least one facility claims its investment in such a state-of-the-art dish system has saved $35,000 in annual energy costs and cut labor costs by 12 full-time employees. Its daily dishwashing functions—some 8000 lunch trays—can all be performed in 4 hours.

16-3 BOOSTER HEATERS

The booster heater is used to raise the water temperature to 180°F for a commercial dishwasher's final rinse cycle. How hard the booster heater must work depends in part on the incoming water temperature, which is typically 110°F or 120°F but could be as high as 140°F. This means an additional boost of 40°F to 70°F. Electrical requirements vary drastically with the size of the booster and the incoming water temperature. For instance, a small unit uses 4 kilowatts of power to heat 40 gallons of 140°F water per hour or 23 gallons of 110°F water per hour. The largest booster can heat 588 gallons per hour if it comes in at 140°F but only 335 gallons if it comes in at 110°F.

Most of the new dish machines have built-in booster heaters but, if yours requires a stand-alone model, it should be located as close as possible to the dish machine so the water won't lose heat as it travels through the pipes between the two. Often it can be installed on an undershelf of the clean-dish table. The typical booster is small, less than 3 feet high and 2 feet wide. Its storage capacity is only 6 gallons, and a low-water cutoff feature eliminates the chance that the heating element will burn out if the tank is only partially full (see Illustration 16-11). Your installer should check the voltage of an electric booster heater to be sure it fits the dish machine; this prevents wasting energy.

Gas-powered booster heaters are not as popular as electric boosters, but several manufacturers offer them. They have firing inputs ranging from 50,000 to 500,000 Btus, heat quickly, and can be connected to either a natural gas or a propane gas supply. These units are vented into the same exhaust system as the dish machine. Gas booster heaters cost more initially than electric ones but, in areas of the country where gas rates are substantially cheaper than electric rates, they can be a bargain in the long run. Even the largest gas-powered boosters can pay for themselves in utility cost savings within 3 to 4 years.

The newest generation of booster heaters is infrared. They can deliver a lot of hot water efficiently without adding to the ambient heat of the dish room like the gas-powered models do.

ILLUSTRATION 16-11 A booster heater is a small tank in which hot water is boosted to higher temperatures for sanitizing dishes and flatware during the rinse cycle.

Courtesy of Hatco Corporation, Milwaukee, Wisconsin.

Steam-heated boosters often are found in large facilities, such as hotels and hospitals, where steam is generated by gas-fueled boilers and used for a variety of cooking needs, not just dishwashing. This type of booster needs steam pressure of at least 10 psi to work effectively. In some cases, it also requires electricity to heat the water to produce the steam. The steam-heated booster uses between 34 pounds of steam per hour (for an undercounter dishwasher) and 205 pounds per hour (for a flight-type dishwasher).

16-4 DISHWASHER MAINTENANCE

The cycles of dish machines often can be preprogrammed, but that doesn't mean they should be any shorter than the manufacturer's recommendations. If the water isn't hot enough, the detergent solution is weak, or the machine doesn't wash or rinse long enough, proper cleaning and sanitation will not occur.

Special care must be taken in storing and using the chemicals associated with dishwashing. Every item must have a Material Safety Data Sheet (MSDS) on hand, stored where employees can refer to it in case of emergency. This is a federal regulation, enforced by the U.S. Occupational Safety and Health Act (OSHA).

It is important to use the chemicals designed to work with your particular machine and to follow the manufacturer's instructions about correct quantities. Too much leaves residue; too little doesn't do the job properly. In both dish machines and pot sinks, you can use a test strip that checks the concentration of sanitizing solution. This should be done every time a new batch of solution is mixed with water, and perhaps up to 90 minutes later, to ensure it is not too diluted.

Assuming the installation and start-up procedures are properly performed, routine preventive maintenance keep a dish machine running correctly. Here are a few tasks grouped by how often they should be performed. They first appeared in the January 2000 issue of *Food and Service News*, a publication of the Texas Restaurant Association, and are the advice of Frank Murphy, training director of Ecolab's GCS Service Division in St. Paul, Minnesota. They are still up-to-the-minute in terms of keeping your dish machine in great shape.

TABLE 16-7

Troubleshooting Tips for Dish Machine Maintenance. Each Saves at Least a 1-Hour Repair Charge

SYMPTOM	POSSIBLE CAUSE	SELF-HELP STEPS	SAVINGS
Machine Will Not Start	Door not closed	Secure door	>$80
	Loose door switch	Tighten switch in mount	
	Main switch off	Check disconnect	
	No rack inserted	Place rack in unit	
	Overload protector tripped	Reset overload in control box	
Low or No Water	Main water supply off	Turn on water supply	>$80
	Drain overflow tub unseated	Place and seat tube	
	Machine doors not fully closed	Secure doors	
	Stuck or defective float	Check and clean float	
	Clogged "Y" strainer	Clean or replace	
Continuous Water-Filling	Stuck or defective float	Check and clean float	>$80
	Drain tube not in place	Look for drain tube in tank	
Wash-Tank Water Temperature Is Low	Incoming water temperature too low	Raise to 140°F in sanitizer models, 120°F for chemical sanitizer models	>$80
	Defective thermostat	Check the setting	
	Low steam pressure	Check and adjust if necessary	
Any Motor Not Running	Tripped due to overheating	Reset overload in control box	>$80
Insufficient Spray Pressure	Clogged pump intake	Clean intake screen	
	Clogged spray pipe	Clean	
	Scrap screen full	Must keep clean and in place	
	Low water level in tank	Check drain and overflow tube	
Insufficient or No Final Rinse	Inproper setting on pressure rinse setting	Set PSI flow to 20–22 psi on pressure-reducing valve	>$80
	Clogged rinse nozzle and/or pipe	Clean	
	Clogged "Y" strainer	Clean	
Low Final-Rinse Temperature	Low incoming water temperature	Check that booster temperature setting is at 180°F	>$80
Poor Washing Results	Wash arm clogged	Clean	>$80
	Improperly scrapped dishes	Check scraping procedures	
	Wares improperly placed in racks	Use proper racks and don't overload	

Reprinted with permission of *Food and Service News*, a publication of the Texas Restaurant Association, Austin, Texas (January 2000).

DAILY MAINTENANCE

- Leave the dish machine door open when the equipment is not in use. If there are curtains, remove and clean them and hang them up to dry at the end of each day.
- Drain, clean, and flush the water tanks.
- Brush and/or spray the scrap screens until they are clean. Do not clean them by banging them against something to loosen the scraps!
- Remove the spray pipes and flush them clean. Take care to reinstall them correctly.

- Look at the final rinse nozzles; brush off any hard water deposits.
- Check water pressure and temperatures.
- Inspect the water pump shafts for leaks.
- Refill the chemicals.

WEEKLY MAINTENANCE

- Check all the water lines and drain/overflow tubes for leaks and make sure they are tight.
- Remove mineral scales that may have accumulated on heating elements.
- Clean any detergent residue off the machine exterior.
- Remove and inspect each spray arm. You can unclog the rinse nozzles with a straightened paper clip.
- Check the pawl bars for signs of wear or restricted movement. (In a conveyor dishwasher, pawl bars are part of the assembly that keeps the conveyor moving.)
- Check the idle pump and final rinse levers for restricted movement.

Dishwashers must also be delimed periodically. Before the deliming process can begin, take the time to flush any sanitizing chemicals out of the lines by running the machine empty for six cycles. Additional troubleshooting tips are listed in Table 16-7.

16-5 CARE AND CLEANING OF DISHES

Now, let's discuss what you will be washing in your busy dish room. The investment you will make in tableware is considerable, partly because it never seems to end. Most foodservice businesses replace an average of 20% of their glasses, dishes, and utensils per year.

Experts insist that there is no such thing as unbreakable tableware. But, as one manufacturer aptly puts it, "Dishes don't break . . . they are broken!" So, a continuous training program is a must for restaurants, bars, and other foodservice businesses that use real dishes rather than disposables. Safe handling of tableware means less breakage, higher productivity, and less likelihood of accidents and injuries.

Here is a brief roundup of care and cleaning advice for glassware, plateware, and flatware.

GLASSWARE

Glass breakage occurs for two reasons: ***Mechanical impact*** results when the glass hits another object, however slightly or accidentally. This contact can cause tiny abrasions that might not be visible at first but weaken the glass and make it susceptible to breakage with the next impact. ***Thermal shock*** is a quick, intense temperature change that can cause stress on the glass to break it. A glass that comes out of the dishwasher steaming hot, for instance, should not be filled with ice and put directly into service. The thicker the glass, the more time it needs to reach room temperature.

To ensure a longer, safer life for your hardworking army of glasses, here are eight tips:

1. Keep enough inventory on hand so that you are not forced to reuse glasses that are still hot from the dish area.
2. Organize incoming soiled dishes so that bussers can place glasses, plates, and flatware in separate areas, and load glasses directly onto divided racks appropriate for the glass size.
3. Use bus tubs that have separate flatware baskets, so bussers won't be tempted to pile flatware on top of empty glasses.

4. Check dishwasher temperature before every shift. And regularly replace worn-out cleaning brushes in hand-washing areas like the bar sinks.

5. Use plastic scoops in ice bins; metal scoops can chip the rim of the glass more easily. And *never* scoop ice with the glass itself!

6. Don't pour hot water in an ice-cold glass or put ice in a hot glass. When pouring hot drinks, preheat the glass by running it briefly under warm water.

7. Never pick up multiple glasses at a time, bouquet style, in one hand. In fact, avoid glass-to-glass contact whenever possible. Unload dish racks or bus tubs one piece at a time, not in stacks.

8. *Pyramiding* is the best way to store glassware (see Illustration 16-12) if you don't have proper dish racks for the particular type of glass or cup.

Sparkling clean glasses say more about your establishment than almost any type of advertising. In most situations, it is acceptable to run glassware through the same dishwasher as plateware and flatware. However, if you need quick turnaround of your glassware supply in busy surroundings, consider a separate washing area for glasses only. Restaurants that do a high-dollar fine-wine business sometimes wash wineglasses separately, in superheated water with no detergent. This avoids the problem of soap residue and spotting on glasses. We've already mentioned glasswashers often used in bar settings.

For regular glasses (not wineglasses), mechanical or manually operated brushes can be used to scrub the insides with detergent, even before the final wash cycle. During the wash itself, water and more detergent is forced onto the glassware, then a 180°F rinse provides the sanitizing process. Most professionals suggest a wetting agent in the detergent, which allows the rinse water to run freely from the glasses and prevent spotting. Air-drying is recommended; towel-drying is inefficient and likely to leave bits of towel fuzz as well as a bit of residue from the laundry soap the towels were washed in on the clean glasses. However tiny the amount, detergent has a negative effect on carbonated beverages; soda, beer, and champagne will quickly lose their fizz when they come in contact with even a hint of detergent.

Over time, glassware may develop a cloudy look, which is not easily removed with brushing or normal detergents. This film can be the result of hard-water minerals or a combination of food protein and detergent. Beer glasses seem to be especially susceptible to this buildup. Special detergents and more brushing usually will remove it.

After the glasses are washed and dried, storing them in glass racks will save time and handling. A major cause of breakage is storing glassware in an improper rack. The only exception to rack storage is with bar glasses, which may be displayed at the back bar or in special overhead racks built onto cabinets. It is important to note, however, that some health codes prohibit public use of hanging glasses in an open-air overhead racks. You can see the potential sanitation concern if they are not used regularly.

ILLUSTRATION 16-12 Stacking trays of glasses in a pyramid shape allows more storage without damaging the glasses.

PLATEWARE

If plates are stored for more than a week between uses, protect them with a clean cloth or plastic cover, or, at the very least, store them in cabinets with doors. Plateware should always be stored by size; plates in stacks, cups in racks. Don't stack the cups inside each other, which is a common cause for chipping and breakage. Even if plates are stackable without scratching each other, never stack them more than 12 inches high. Storage shelves should be made of high-grade, nonmagnetic stainless steel to keep dishes free of dust and unsightly metal marks.

Storage is especially critical in the soiled-dish area, where the American Restaurant China Council estimates that 75% to 80% of all dish breakage occurs. China in active use should be stored at or below the food plating area; more breakage occurs when employees must reach into overhead storage to bring the plates down.

If you have servers, teach them to load trays properly; the typical tendency is to overload. When stainless-steel or silver-plated covers are used in banquet situations, the cover should fit correctly to minimize scratches and chips on plates. When clearing tables, bussers should be taught not to overload the bins, also called **bus boxes**, and make sure they are plastic or plastic-coated. Soiled plates should be placed upside down in the bus box, with cups, bowls, and glasses on top. Don't shove the plates together on their sides, like books on shelves; the unglazed footing of one plate can easily scratch the front of the next.

A frequent dish-handling mistake that ensures more chipped or broken dishes is when one piece of tableware is used to scrape food off another. Avoid this by properly training dish room employees, and providing plenty of proper scraping implements and ample space for scraping, stacking, and racking, so that all dishes can be correctly sorted and rinsed. This means never using metal scouring pads or utensils to scrub or scrape off baked-on food. These actions simply damage the *glaze*, when most commercial dishwashers can remove the food without the extra scrubbing. Before washing, rinse and place similar-size plates in stacks of no more than 12 inches high (a practice called *decoying*). It is important that soiled dishes be washed within 30 to 40 minutes after this first rinse. Coffee, tea, and some acidic foods can cause staining if allowed to sit for longer periods.

Check the dishwasher curtains regularly to make sure they are clean. Also check the water pressure and temperature, spray pattern, and length of each phase of the dishwasher cycle. Be careful especially not to pre-rinse soiled dishes with water hotter than 110°F to 120°F; higher temperatures tend to stain china and embed bake food particles onto the plates. If dishes continue to show persistent stains, you may have a water quality problem, hard water or too much iron content.

Clean, wet dishes can be air-dried in a well-ventilated area as they come out of the dishwasher. A rinse additive may be used in the rinse cycle to speed the drying process. China, which is more delicate than other types of plateware, should never be stacked right out of the dishwasher, as it is more likely to be scratched or pitted when it is hot and wet. Tilt the dish racks slightly to drain water from the feet of the dishes, and handle clean dishes only at the edges; in fact, using rubber gloves is preferable. Position dish carts or self-leveling carts at the end of the dish machine line for easy movement from dish room to storage area.

China can be overworked—that is, used too often—when inventory is inadequate. This will result in more stains and discoloration and more breakage.

FLATWARE

One of the biggest problems with utensils is that they disappear easily. Employee theft is certainly part of the reason, but carelessness is equally to blame. Several manufacturers have created flatware retrieval devices that can be placed over a trashcan. They look like a lid with a wide slot or opening, but the models we've seen contain magnets that catch falling flatware and hold it at the top of the container so it doesn't end up in the trash (by KatchAll Industries International of Cincinnati, Ohio), or has a small, battery-operated metal detector that sounds an alarm when silverware passes through the slot. The latter, called a Silver Chute by Golden West Sales of Huntington Beach, California, can be ordered with a wide rubber squeegee attachment to scrape the plates.

Both stainless-steel and silver-plated flatware will show film, stains, rust, and corrosion unless care is taken to store and clean them properly. Soiled flatware should be removed from tables as soon as possible and presoaked before going through the dishwashing cycle. There are three benefits of presoaking flatware:

1. Wetting utensils makes it easier to remove caked-on food particles. Add a small amount of detergent to the soak water to increase its effectiveness.

2. Presoaking minimizes the utensil's contact with acidic foods, eggs, and other substances that may tarnish or corrode it.

3. When using silver plate, the addition of a small sheet of aluminum foil at the bottom of the soaking tub prompts an interesting electrolyte reaction: The tarnish leaves the silver and adheres to the aluminum. This *does not work* with stainless-steel flatware, however, and might even damage it. Try it only with silver-plated utensils.

When it comes to presoaking, your flatware can get too much of a good thing. Limit soak time to 30 minutes at most, and use a mild alkaline detergent in low concentration—only 1 ounce per gallon of water—in the soaking tub. From here, the flatware should go immediately into the dishwasher. To allow it to dry would defeat the purpose of presoaking.

Some foodservice businesses buy flatware washers, special machines that can clean and sanitize utensils in 90 seconds. These combine mechanical and hydraulic action to "lift" and separate the flatware during washing and rinsing, no matter how tightly it is packed into the perforated wash baskets (see Illustration 16-13). Some models use the rinse water from the previous cycle as the wash water for the next cycle, thus saving water and energy.

In the dish room, never use steel wool or other metal-based scouring pads to remove caked-on foods from utensils. These can easily scratch and mar the mirror finish. If it is necessary to wipe them off, do so with a cloth or plastic pad. The other dish room enemy of flatware is, of course, the garbage disposer unit. Keep it out of the disposer, and both will last longer.

Pack flatware loosely in its perforated baskets and arrange it so that water can reach all parts of the utensils. Turn forks and spoons up, but keep knives pointed downward as a safety precaution. Separating them only nests them together and not get fully cleaned, so mix forks, spoons, and knives together at random. Temperature for flatware washing should be 140°F to 160°F for the wash cycle and 180°F for the final rinse.

There are several additions to the wash cycle other than the detergent. A wetting agent in rinse water will help water run off dishes more thoroughly, preventing unsightly spotting. In areas where water is hard (has a high mineral content), you may need a water softener to prevent spotting.

Improve the appearance of stainless-steel flatware by soaking it in a solution of one part vinegar to three parts water, then rinsing and air-drying it. This usually removes hard-water film and/or mineral deposits. Like silver plate, stainless steel should be presoaked promptly after use.

For years, flatware was air-dried (not towel-dried), ostensibly for maximum sanitation. All too often, however, air-drying does not remove water or detergent spots, so someone has

ILLUSTRATION 16-13 Flatware washers are specialty machines for fast washing of silverware.
Courtesy of Intedge Industries, Inc., Woodruff, South Carolina.

to hand-polish every piece. In recent years, technology has prevailed to save time and labor costs with the creation of machines that do the job automatically. Utensils can be loaded directly from the dishwasher into a *cutlery polisher*. Inside, they sit on a vibrating tray, which passes through a granular substance that absorbs the moisture from the wet utensils, drying and sanitizing them. Hot air that circulates through the machine evaporates the moisture from the granulate, leaving the machine dry and ready for the next load.

Cutlery polishers are rated based on the number of utensils they can accommodate per hour, from 3000 to 8000 pieces. They use 220/240-volt electricity, from 600 watts for the smaller machines to 1500 watts for the largest.

Silver plates must be detarnished occasionally, usually with a chemical compound rubbed onto the utensils. Rinse carefully after removing the tarnish, because the compound contains chloride and will actually corrode the silver if left on too long. Or you can make your own detarnishing solution by dissolving 1 tablespoon of salt and 1 tablespoon of baking soda into 1 gallon of boiling water. Put this solution into a soaking tub with a small sheet of aluminum foil at the bottom. Make sure all the utensils come into contact with both the solution and the foil. It should take no longer than 30 seconds to do the job.

You may also purchase a **burnisher**, a machine that uses tiny, vibrating stainless-steel balls to polish silver plates by removing some of the surface scratches and leaving them with a fine, soft sheen. Its proponents say, over time, burnishers make utensils harder and more durable. Different sizes of burnishers can accept up to 300 pieces of mixed flatware at one time, and the cycle lasts about 5 minutes. Burnishers require both water and electricity to operate. Where space is tight, they can be rolled into an area on casters.

16-6 WASTE DISPOSAL OPTIONS

Waste and garbage handling is not glamorous, but it is a daily concern in any foodservice operation. How you dispose of waste can affect the public's perception of your establishment as well as your health inspections—particularly when odor, insect, and rodent problems are present. We discussed the options in Chapter 8 of this text, and the dish room certainly is part of your waste management plan.

Scraping and pre-rinsing dishes before they enter the dishwasher is a critical part of ensuring cleanliness. It also helps maintain the dish machine by keeping food scraps and gunk out of the lines and equipment. Before we dissect the waste disposal unit itself, let's discuss the pre-rinsing process.

PRE-RINSE STATION

A pre-rinse station is as simple as a pot sink and a spray nozzle or faucet with a flexible hose and a food waste disposer to get rid of the scraps. One way to save water and still do an adequate pre-rinse is to install a low-flow spray nozzle. The idea is much like water-saving shower heads in hotels—it restricts water use but still allows a sufficient flow to blast food particles off most dishes. The water doesn't have to be any hotter than regular hot tap water, 110°F to 120°F.

A single nozzle in a dish room doesn't sound like it can amount to much in the way of conservation, but tests in California have proven otherwise. The California Urban Water Conservation Council in Los Angeles estimates that a medium-size restaurant that uses its dish room 6 hours a day saves 300 gallons of water per day, plus the money used to heat the water—total savings of close to $100 a month.

In a separate study, the Food Service Technology Center in San Ramon, California, tested low-flow pre-rinse spray nozzles and compared the costs associated with using them. The results for two popular flow rates (one regular and one low-flow) are shown in Table 16-8.

TABLE 16-8

Saving Money with Low-Flow Spray Nozzles

NOZZLES: GALLONS PER MINUTE	1.8 GPM	3.2 GPM
Daily usage (hours)	4	4
Gallons per day	430	430
Annual heat cost	$750	$1,310
Annual water cost	$420	$750
Annual sewer cost	$620	$1,120
TOTAL ANNUAL COST	$1,790	$3,180

Based on $0.60 therm, $2 per unit of water, $3 per unit of sewer.
Source: Food Service Technology Center, San Ramon, California.

A list of food products to be avoided when loading the disposer should be prominently displayed in the pre-rinse station. They include eggshells, onion skins, rice, corn husks, celery, banana peels, bones, and the outer leaves of artichokes, to name a few. Common nonfood culprits include flatware, rubber bands, and bits of plastic, wood, or cloth that sometimes find their way to the bottom of the sink in a busy kitchen.

ILLUSTRATION 16-14 The anatomy of a food waste disposer.
Courtesy of Somat Company/Red Goat Disposers, A Division of ITW Food Equipment Group LLC, Lancaster, Pennsylvania.

DISPOSER

For years, the *food waste disposer* has been a workhorse in the commercial kitchen. However, the current thinking is that, in this era of water conservation, both water use and maintenance issues are minimized by eliminating the disposer. It does, after all, use from 5 to 8 gallons per minute to flush ground solid waste into the sewer system. One alternative is to limit the disposal's water flow with a flow regulator or solenoid valve. Another is to use a waste collector instead, which is described later in this chapter.

Some large cities, primarily because of antiquated sewer systems, have banned food waste disposers altogether for commercial uses, so it is perhaps ironic that others encourage their use. After all, another way to think of the food disposer is as a convenient, sanitary, and environmentally conscious way of disposing of some types of waste that would otherwise have to be bagged and sent to landfills, attracting rodents and insects in the meantime.

The disposer is almost always installed as a component of the dishwashing system. However, in large kitchens where lots of fresh produce is prepared, a disposer is also installed in the prep sink with another in the pot sink. Most of this discussion focuses on the disposer as part of the dishwashing process (see Illustration 16-14).

The electric disposer provides a quick, tidy method of dealing with some food waste. It grinds up food scraps that are similar in chemical content to the human waste already flushed into sewers. The food waste goes into the kitchen's drainage system, through the grease trap, and into the sewer. The ground and liquefied waste ends up being a better bargain for most towns and cities because it costs less to deal with wastewater than the other popular waste disposal options: landfills and incineration.

Today's commercial-quality disposer can handle almost any of the waste food products generated by a restaurant, except perhaps in areas with water shortages or very old sewer systems. If your building uses a septic tank instead of being hooked up to a municipal sewage system, you may choose a pulper or collector instead of a disposer, both of which you'll learn more about in a moment.

Even where disposers are welcome, they are often subject to strict code requirements for restaurant use. These include rules for connecting them properly to the sewer system, separation of the disposer and grease trap pipes, types of valves used, and installation of vacuum breakers. A professional plumber can assess your situation and meet these specifics. But first, be sure you can have a disposer installed in the first place.

HOW FOOD DISPOSERS WORK

There are two types of commercial disposers: *continuous feed* and *batch feed*. Continuous feed is the most popular. When switched on, it immediately shreds whatever is in it and keeps shredding until you turn it off. Batch-feed units can handle only a certain amount of waste at one time.

When waste is put into either type of disposer, it enters a grinding chamber. Cold water is flushed through the chamber to help wash the waste through the grinder. The water flow depends on the size of the grinder motor, from 5 gallons per minute for a ½-horsepower grinder to 10 gallons per minute for a 5-horsepower grinder. Inside the chamber, small protruding bars are attached to a revolving rotor. As the rotor spins, the bars jam the pieces of food into contact with a ring of sharp steel teeth around the inside wall of the chamber, which shreds them. The bits of waste mix with the water to form a pulp and, when the pieces are small enough, they are washed into the drain. Depending on the manufacturer, the rotor may be called a *turntable, grinding table, flywheel,* or *grinding wheel;* the steel ring may be called a *shredder ring* or a *grinding sleeve.*

Disposers are available in several sizes. They are rated most often by horsepower but sometimes by the number of meals served (not ground up!) per day. Size guidelines, based on the number of meals served and the location of the disposer, are shown in Table 16-9. A 5-horsepower unit is usually more than enough to service a large restaurant, and a 2-horsepower unit is sufficient for a pot-washing or prep area, but disposers are available with up to 10-horsepower motors! You might think more horsepower means more grinding ability, but that is not the case. Yes, you'll get more power, but you also need a larger rotor or the extra power is wasted. Rotors come in diameters from 5½ to 15 inches. The opening at the top of the disposer (known as its **throat**) should also be bigger if the motor is bigger. Throat openings range from 3 to 8 inches.

TABLE 16-9				
Recommended Disposer Sizes (in horsepower)				
PERSONS PER MEAL	AT DISH TABLE	VEGETABLE PREP	SALAD PREP	POT SINK
Up to 100	¾ hp	½ hp	⅓ hp	⅓ hp
100–150	1	¾	½	½
150–175	1¼	1	¾	½
175–200	1½	1½	¾	¾
200–300	2	1½	¾	¾
300–750	3	1½	¾	¾
750–1500	5	3	1½	1½
1500–2500	5	3	1½	1½

Source: Carl R. Scriven and James W. Stevens, *Food Equipment Facts* (New York: John Wiley & Sons, Inc., 1999).

The electrical requirements of disposers are also based on motor size: 115 to 230 volts single-phase for the lower-horsepower motors, 200 to 460 volts three-phase for the higher-horsepower motors.

The problem with using an undersized disposer is that food is more likely to jam it, and disposer malfunctions never seem to happen at convenient times. One important lesson for kitchen employees is that if their own teeth won't chew it, neither will the disposer. Of course, some of the problems are plumbing related. When installing a disposer, pay particular attention to the sizes of incoming water lines and drain pipes. For example, the drainpipe for a ½- to 1¼-horsepower motor should be no less than 1½ inches in diameter; for the 5-horsepower motor, the drainpipe should be a full 3 inches in diameter.

Look for a disposer that allows automatic reversal of the direction the flywheel rotates—a feature that increases the life and efficiency of the unit and helps free it when it jams. Most disposers come with a small dejamming wrench that should be stored near the unit. The wrench is used to free the flywheel if reversing its direction doesn't do the job. Remember, always turn off the power before you start using the wrench. Most manufacturers provide detailed instructions on this process. Read them to avoid serious—and preventable—accidents. Another option to improve the life of your disposer is a control setting to flush water through the disposer even when it is not in use, to keep the lines clear.

Have your disposer installed in a sink at least 20 inches square and 6 inches deep. Make sure there is a silver saver vinyl guard for the drain (usually 3½ inches in diameter) to prevent tableware, jewelry, bones, and so on from falling into the disposer. A removable molded scrap basket with rack guides and a spray nozzle for rinsing dishes are also recommended for the sink. The electrical controls for the disposer are located below the working surface of the soiled-dish table. Most models include, at the least, an on–off switch, an automatic reversing button, and a reset button.

WASTE PULPERS AND COLLECTORS

Buildings on septic systems or with large kitchens (more than 1500 meals a day) may want to add a **waste pulper** to their disposal repertoire because it can handle paper products as well as food waste. This machine is capable of reducing much of your trash to 15% of its original volume. Paper napkins, Styrofoam containers, plastic straws, cardboard boxes, milk cartons, aluminum foil, and more are potential fodder for this system, which uses as little as 1 or 2 gallons of water per minute, as opposed to 7 or 8 gallons needed to flush waste through a disposer. Pulpers also improve sanitation and reduce rodent and insect problems by breaking down waste that might otherwise sit in plastic bags in your dumpster for days at a time.

Here's how the pulper works: Waste is dumped into a trough that feeds the pulper's input tray (see Illustration 16-15). Inside, the waste is drawn into a vortex of water and through a flatware trap (just in case any utensils get caught in the machine). Then precision cutters mince the waste into bits. The **slurry** (consisting of 95% water and 5% waste) moves into a water press, or water extraction chamber. Here an augur spins the waste against a fine-mesh screen to squeeze out most of the water. The water is returned to the pulping tank for reuse, while the clean, semi-dry slurry (now pulp) is expelled down a chute, usually directly into bags for transfer to a dumpster.

Remote extraction equipment can be located 200 feet from a pulper, and slurry can be piped to it from one or more pulpers. The extractor can actually discharge the waste through the building wall into a dumpster outside, or the waste can be directed

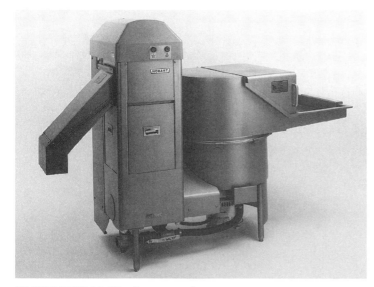

ILLUSTRATION 16-15 A waste pulper system.
Courtesy of Hobart Corporation, Troy, Ohio.

into truck dollies or other containers to be used for compost or cattle bedding. Of course, you want the waste to be totally biodegradable for composting, or paper-only waste for cattle bedding. Institutions often use the pulper/extractor for sensitive document destruction.

Very large operations have several waste pulpers at work throughout their kitchens, with custom piping to a large water press unit located adjacent to the dumpster. That way, no labor is required to haul the bags outside. This is a major consideration because, although pulpers are effective at reducing the volume of trash, the actual weight is the same—or heavier, because of the added water content.

Some manufacturers address the issue of waste odor by offering automatic dispensers of deodorant or sanitizing solution that is injected as the slurry reaches the water press. Thus, the pulp emerges odor-free as well as semidry. Advances in technology produce a drier, flaked product that accelerates the composting process. But, as with disposers, there are limits to the types of materials that can be processed by a pulper. Glass, ceramic, metal, and cloth are not suited for this type of machine.

Waste pulpers vary in size from 2 × 4 four feet (with a capacity of 350 pounds of waste per hour) to 4 × 7 feet (900-pound-per-hour capacity). They run on three-phase electrical current. Motors for the most common sizes range from 5 to 7½ horsepower, although mini-pulpers for smaller operations have motors from 1 to 3 horsepower and huge custom units have motors from 10 to 40 horsepower. The water press in each unit has its own motor. Pulpers require at least a ½-inch water line and a 3-inch drainpipe.

Perhaps the biggest drawback of waste pulping is the initial cost of the equipment. Smaller units may cost $8,000 to $15,000, but the largest ones can run as high as $125,000. Also consider that in remote systems, the cost of piping can be as expensive as the cost of the equipment depending on the length necessary (known as the *run length*) for the pipes to carry the slurry to the extractor. Manufacturers have several methods for determining optimum size. One factor is the amount of waste to be processed per peak hour of business, and the general rule is that 1 pound of waste is generated for each guest served. Variations occur, however— ½ pound cooked vegetable waste, for example, is not the same as ½ pound of waste that includes chicken bones and packaging material. Some manufacturers suggest doing a waste assessment of the operation so you know exactly what, and how much, you have to dispose of.

The other concern is noise. Grinding things into bits is not a tidy or quiet process! The best place to install the machine may be where soundproofing can be used; also, install seismic pads under the legs to help with vibration noise.

Some training is necessary to get optimum use from your pulper. It should be cleaned daily; most have an automated cleaning cycle. The augur screens should also be removed and rinsed and the internal components sprayed clean with water. When the first pulpers were introduced, it was widely believed that every day, cardboard should be the final item run through the pulper to absorb small food particles and help with odor control. It makes sense, and manufacturers say there is no harm in doing this, but it is no longer necessary.

If you do business in an extremely rural area, both pulpers and disposers may be out of the question. A third option—and one that is increasing in popularity even in areas with good sewer systems—is the **scrap collector** (or *waste collector*), a screen-perforated pot that sits in a sink that acts as a strainer as water recirculates through it. A collector can be used in a dish room, for instance, where scraps are scraped into it instead of a garbage can. The water cleans up the waste as much as necessary, dissolves part of it, and leaves the rest to be periodically dumped into the trash. Collectors are even better water-savers than pulpers because they use only 2 gallons of water per hour.

16-7 ■ WASHING POTS AND PANS

Like waste disposal, pot-washing is an unappealing but essential function in any kitchen. The goal here is to clean and sanitize all the cooking and preparation utensils and equipment; procedures are similar to those in the dishwashing area, but on a larger scale to accommodate the largest items in the kitchen. It must also suit the person responsible for the pot-washing so he or she can get it done quickly and effectively. For years, how well these items were cleaned depended entirely on the work ethic of the person doing the scrubbing. As you will see—not any more!

MECHANICAL POT/PAN WASHERS

That old standby, the three-compartment sink, is a staple in commercial kitchens and associated with the drudgery of pot-washing. Most people, even if they need a job badly, resist the idea of standing for hours over dirty, lukewarm water, scrubbing baked-on gunk off large pots and pans, and who can blame them? This, combined with labor costs and high turnover, has prompted the invention of the ***mechanical pot/pan washer***. It is better for the business owner too. Depending on the menu and volume of business, mechanizing this part of the kitchen can save up to 75% on labor costs and 50% on water and detergents.

The mechanical pot washer is similar to a dishwasher but customized for the messy, bulky work of cleaning cookware. It takes from 4 to 15 minutes to clean, sanitize, and dry just about any pot or pan, and the machine is designed to accommodate many at a time. Pots and pans can be checked periodically and taken in and out of the machine, or left in longer, as needed (see Illustration 16-16). However, NSF International standards require that the machine have a "120-second wash cycle, using at least 20 gallons of 150°F water per square foot of rack, and a minimum of 15 seconds for the final rinse cycle, using at least 6/10ths of a gallon of 180°F to 195°F water per square foot of rack."

Detergent for mechanical pot washers should be a nonfoaming type designed to penetrate baked-on food and grease. The turbulence of the water as it circulates in the machine allows it to peel away softened food residue.

At least one manufacturer offers a pellet detergent machine that reuses plastic pellets that act as little scouring pads to clean soil from dirty pots and pans.

Undercounter pot washers are designed to save space. They can wash and sanitize sheet pans, cake pans, candy molds, machine parts, or mixing bowls of up to 40 quarts in size. An undercounter machine is about 19 inches tall, 28 inches wide at the door, and operates with a 2-horsepower pump motor. A 3-kilowatt booster heater provides the additional water heat for sanitizing. The wash water is heated by an electric immersion heater that uses 6 kilowatts per hour.

The undercounter model accommodates 6 sheet pans, while counter-height machines fit 10 sheet pans. The slightly larger unit may also have a shorter wash time—as fast as 3 minutes for a full cycle. Handy pass-through models are also available, with doors on both sides. In any installation, remember to allow enough room to open doors fully for loading and unloading.

The ultimate in pot- and pan-washing is a machine that fits an entire rack of sheet pans. These are installed in a pit or used with a ramp so multitier racks can be rolled in for washing. Mechanical pot/pan washers also save labor costs because they can go virtually unattended about 75% of the time. Of course, some restaurants have an employee whose job is to load and unload the pots and pans. In most cases, this isn't the smartest use of that person's time.

As in any dish room, it is critical to have enough workspace to separate the clean items from the dirty ones. Loaded racks of pots and pans can also be quite heavy, so having adequate space for unloading and drying is important. It takes up to half an hour for a full rack to air-dry. A ***scrapping area***, where as much food and grease as possible is removed by hand, is necessary as part of the incoming soiled-pan area. If the same person does the loading and unloading, a hand sink is required for proper handwashing between tasks.

ILLUSTRATION 16-16 The mechanical pot/pan washer is also known as a pan/rack washer. It is a kitchen workhorse built to fit pots, pans, lids, and utensils.
Courtesy of Champion Industries, Huntington, West Virginia.

POWER SINKS

One creative alternative to the mechanical pot/pan washer requires more human involvement. It is a combination of high pressure and hot water in a motorized three-compartment sink known as a **power sink** (see Illustration 16-17) or *continuous motion sink*. A heater in the sink keeps the water hot so it doesn't have to be drained and refilled as often, which saves water. An industrial wash pump with an electric 1½- to 3-horsepower motor, also installed in the wash sink, creates a high-velocity water spray of about 115°F. This is enough water to blast soil from most pots and pans without hand-scrubbing them. Nozzles are installed below the water level, angled to create a rolling, bubbling hot-tub effect on the submerged pots and pans. A presoak area, a rinsing area, and a sanitizing sink make up the rest of the installation, all with swiveling gooseneck faucets. Food waste disposer, drainboards, and shelving complete this dish room setup, so you must decide whether you have the space for all of it. Power sinks come in configurations from straight lines to L-shapes and U-shapes.

The differences between competing power sinks is the way water moves around in the sink—a tornado or whirlpool action, or hot tub–like jets. The locations of the intake nozzles

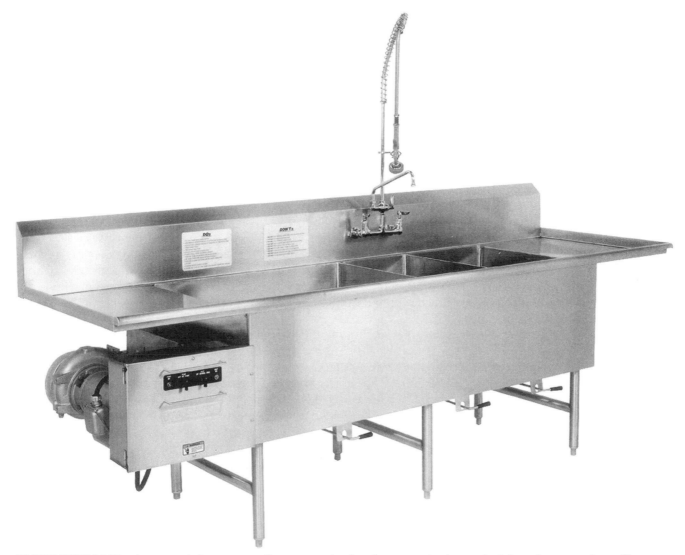

ILLUSTRATION 16-17   A power sink uses a small motor to circulate hot water in the wash sink and power grime off pots and pans.

Courtesy of Champion Industries, Huntington, West Virginia.

are different too. Be sure to view the water flow when the sink is full of pots and pans to see if it meets your needs. Tank (sink) sizes range from 61 to 112 gallons in volume and the motor requires either single- or three-phase electricity, 208 to 240 volt and/or 480 volt. Prices range from $6,500 to $8,500 for a basic 9-foot model, and up to $12,000 or $15,000 for the largest industrial models. When traditional three-compartment sinks cost between $1,600 and $2,500, it's a big expenditure—but manufacturers say with the savings in labor, water, utilities, and detergent, a power sink can easily pay for itself within a year.

Recent developments include power-wash units that can be retrofitted onto existing pot sinks. A small standalone pump mounts to the drain opening of the sink, filtering the water and sending it back churning into the sink compartment. The price of this option is about $2,000.

In sum, dishwashing is typically one of the most expensive kitchen functions in terms of labor and energy costs. As you shop for the right machines for your dish room, be aware of their hot-water requirements, energy use estimates, and energy-saving options.

■ SUMMARY

This chapter continued the discussion begun in Chapter 4 about the rigorous demands on a dish room and its crew, and the design elements that are most helpful in streamlining the warewashing process.

Commercial dishwashing requires hot water, detergent, and motion. The water must be hot enough to sanitize the dishes. Health regulations dictate the amount and temperature of the water as well as the length of the wash and rinse cycles. Motion is necessary to ensure complete coverage of all the dish surfaces.

Dish rooms are designed based on available space and utility connections, proximity to service areas, building codes and health ordinances, how food is served and how dishes are handled, and the quality of both water and detergent. Peak hours and volume of dishes are also major considerations.

Even the least sophisticated dish machine can complete a wash cycle in less than 1 minute—the smallest machines, under 5 minutes. Some dish machines require a separate tank, called a *booster heater*, that provides additional water heating for sanitizing during the rinse cycle.

Dishwashers are rated or sized by the number of full dish racks per hour they are capable of washing, but experts caution that, because most racks are not loaded perfectly, a more realistic capacity is 70% of what the machine's manufacturer claims. The rate of flow—the speed at which the machine replenishes and heats its water storage tank—is also an important consideration.

Dishwashers can be freestanding or undercounter. Specialty machines are available for washing glassware.

Commercial dishwashers have either a single tank or a double tank, depending on how much hot water you require. The largest-volume foodservice operations have conveyor or flight-type dishwashers. Other specialty machines wash pots and pans. All dish machines contain a scrap screen or basket to collect food particles. A blower-dryer can be added to dry dishes faster.

Heat and moisture make dish rooms a tough place to work unless the dish machines are installed under a vent. Local building codes often require ventilation. Some models have a built-in condenser that removes moist air from the machine, dries and cools it, and releases it into the dish room.

The food waste disposer is an integral part of the dishwashing system. Not all cities allow the use of disposers—this usually depends on the area's sewage system—but some cities encourage their use because disposers cut down on landfill. Where the latter is true, strict requirements pertain that disposers be connected properly to water and sewer systems and not interfere with grease interceptors' operation.

Commercial disposers can be continuous feed or batch feed. A continuous-feed machine shreds whatever is in it as soon as it is turned on; a batch-feed machine shreds only a certain amount of waste at a time. Their electrical requirements are based on their motor size.

Facilities with large waste output may consider installing a waste pulper, which can mash paper and Styrofoam products, aluminum, and cardboard into a watery slurry along with food. Smaller models are known as *mini-pulpers*. The advantage of a pulper is that it turns out cleaner waste that is easily composted. But the systems are expensive, and even pulpers can't accept all types of waste.

■ STUDY QUESTIONS

1. Joe's Cafeteria serves about 800 people during its daily lunch rush. According to the guidelines in this chapter, how many lunch dishes will Joe and his staff need to wash? What type of dishwashing system would you recommend for Joe's?

2. List three of the recommendations you have learned in this chapter for avoiding breakage in handling dishes and glasses, and explain briefly why they are important.

3. When discussing a dishwasher's water tank capabilities, what is the difference between the rate of flow and the pump efficiency?

4. When shopping for a glasswasher, why is it important to find out what temperature it uses to dry glassware?

5. What three things should you consider before deciding to add a blower-dryer to your dishwashing system?

6. Are there any advantages to a circular dishwashing system over a straight-line one?

7. Calculate the dish capacity of a flight-type dishwasher that moves at 5.2 feet per minute, has 2 inches between pegs, and is 45 inches wide.

8. Why is it important for a food disposer to be able to reverse its motor?

9. What is the function of the water press in a waste-pulping machine, and why do you think it is necessary?

10. Based on the information in this chapter and in Chapter 7, how would you incorporate a composting system into your dish room? Explain why you would—or would not—opt to compost food or paper waste.

17
MISCELLANEOUS KITCHEN EQUIPMENT

■ INTRODUCTION AND LEARNING OBJECTIVES

Famous chef, author, and bubbling television personality Julia Child would have turned 100 in 2012. Her great-nephew, Alex Prud'homme, has said in interviews that his great-aunt described herself as a "knife freak, frying pan freak, and gadget freak." And she knew that if she used a new appliance or utensil on one of her cooking shows, it was sure to become a hit, from blenders and stand mixers to the very first food processors, which were introduced in the 1960s.

Today, Child's kitchen is an exhibit at the Smithsonian Institution—and it seems rudimentary compared to the hundreds of handy kitchen devices on the market today. That's a good reminder that not all gadgets are worth their sometimes hefty price tags.

All are designed to streamline the seemingly endless prep work in a commercial kitchen. In this chapter, we confine the discussion to those machines that are in most frequent use across a broad scope of the industry. A few of the selections are new, and new technological developments for venerable workhorses are also presented.

After reading this chapter, you will be able to:

■ Identify the important features of food mixers and their attachments.
■ Identify the important features of food slicers, cutters, and grinders.
■ Identify the important features of blenders and juicers.
■ Identify the important features of toasters and food warmers.
■ Identify the important features of coffee brewers and espresso/cappuccino machines.

You will learn which features are important when selecting these items—and which are not—as well as tips on determining the correct sizes for your operation and caring for your shiny new appliances. As Julia Child would say, "Bon appétit!"

17-1 FOOD MIXERS

The most popular mixer in today's restaurant kitchen is the *vertical mixer,* also known as the *planetary mixer*. The latter term refers to its mixing action, which is like the rotation of a planet around the sun: The mixing arm (also called a beater or *agitator*) revolves around the inside of the mixing bowl while rotating on its own axis. The bowl itself remains stationary, except on a few large models. This motion provides thorough, effective mixing action.

Food mixers are identified by the capacity of their mixing bowls. Typical sizes range from 5 to 80 quarts, but huge 120- to 250-quart capacity mixers are available for special applications, such as bagel production. The smaller units can sit on tables or countertops; the larger ones are floor models. The standard for small- and medium-size operations is the 20-quart model; larger establishments may find a 60-quart model necessary, with a backup 20-quart mixer for occasional additional capacity. These choices should be made by first establishing:

- What foods will be mixed
- How much will be mixed in a single batch
- How often the mixer will be used in a day

As an example, if mashed potatoes are on your menu, assess how many batches you need in a day and in what periods they must be made. Your restaurant may require all its mashers in a 1-hour time window each and every day; or it may be a high-volume location that uses 8-pound batches of potatoes as many as 30 times a day. Seen this way, your answers to the questions what, how much, and how often are critical. The most frequent cause of mixer malfunction is overwork. Overloading the bowl past its rated capacity, over and over again, shortens the machine's life span and squanders your investment.

The size of the motor is a second size consideration. Don't buy a mixer based on its horsepower but rather on the kinds of products you plan to mix. For most restaurants, a 20-quart machine with a ½-horsepower (HP) motor is sufficient. But for stiff doughs, such as bagels or pizza, higher horsepower is recommended. The thicker the dough (more flour and less water) in the bowl, the greater stress is placed on the mixer motor. This is why it is helpful to understand the **absorption ratio** of the flour—that is, the weight of the liquid divided by the weight of the flour. The dough can hold only a certain amount of water to reach its desired consistency, and this amount is expressed as a percentage of the weight of the flour. (The ideal absorption ratio is 50%, which means twice as much flour as water.) Lower absorption ratios mean stiffer dough, which is harder to mix.

Be sure to allow enough room for volume increases when mixing yeast-based dough. You can't load what amounts to 20 pounds of ingredients in a 20-pound mixer and call it good! Consider the space that the product will require in the bowl after mixing, as it rises.

The most common mixer sizes in foodservice are the 12-quart and 20-quart models, but the 5-quart mixer is the standard workhorse for any small operation. It can mix dough, make mayonnaise, and mash potatoes with a ⅙-HP motor. It weighs about 44 pounds empty, so it's hefty but portable. However, caution should always be taken to set it on a firm platform. This prevents an unattended mixer from "walking," a natural consequence of the mixing action in which the vibration moves it, a little at a time, until it falls off a table or countertop. Special stainless-steel tables (sometimes called **benches**) can be purchased for tabletop appliances like mixers, with locking casters, mounting holes to hold the appliance in place, pegs to store attachments, and a shelf beneath. Because a standard 20-quart mixer weighs at least 200 pounds, these rolling tables are a smart idea.

The 30-quart mixer is a floor model recommended for facilities that make bread or other dough products in large quantity. It is the smallest mixer model with a four-wheeled **bowl dolly** (or *bowl truck*), which permits the operator to roll the mixer bowl into position rather than lifting it. This mixer also sits above the floor on four solid legs, permitting thorough floor cleaning beneath it. At this size, motors can range from ¾ to 1½ horsepower for the toughest mixing jobs. The 30-quart mixer weighs about 330 pounds.

The largest mixer models are 40 to 80 quarts; these are found in large kitchens, busy pizzerias, and retail bakeries. Mixers with a capacity of 30 quarts or more that use at least 200 volts of electricity can be ordered with a power bowl lift, a switch that raises and lowers the bowl into place. Otherwise, they are equipped with a hand-operated lift. Large mixers can also be used with smaller bowls—you can put a 30-quart bowl on a 60-quart mixer for a smaller job, for instance—as long as you fit the mixer with a **bowl adapter** to hold the smaller bowl firmly in place. These supersize mixers come with motors of up to 2½ horsepower.

ILLUSTRATION 17-1 A vertical mixer.

Courtesy of Hobart Corporation, Troy, Ohio.

HOW MIXERS WORK

All vertical mixers work basically the same way (see Illustration 17-1). The mixing mechanism is contained in a horizontal housing parallel to the table or floor. The power hub for mixing attachments is located on the front side of the housing. The motor is located in or near this upper housing; the housing is supported by a vertical column, mounted on a flat base or heavy feet, which also contains the lift mechanism that raises and lowers the bowl to the agitator. (The smallest mixers may not have a lift mechanism; you just unlock the upper housing and tilt it back to disengage the agitator and bowl.)

Attached to the base of the vertical column is the sliding yoke or arm that holds the mixing bowl. Arm and bowl are made to fit snugly together, and the bowl is stabilized by a flat wheel or lever or, in the largest mixers, by an electric drive. Some mixers allow a variety of positions for their bowls; others, only up or down. Mixing bowls have flat bottoms to help stabilize them during the mixing process.

The mixing mechanism consists of a vertical shaft of heavy steel connected to the gear drive. The planetary action allows full contact with all the contents of the bowl during every revolution of the shaft and agitators. The agitators fit onto the mixer with a push-up-and-twist motion that engages locking pins, the same as the beaters on your mixer at home. They're made of aluminum or (more expensive) stainless steel.

Electrical controls operate the motor and transmission, which drive the agitator at different speeds. Most mixers have three speeds, but some have four. Speed selection is controlled by a simple lever that shifts gears. In simple terms, mixers have transmissions that are either gear-driven or belt-driven:

- Gear-driven machines come with three or four fixed speeds, preset at the factory. They must be stopped in order to change speeds, which is a key operating point. Generally speaking, the higher the horsepower of the gear-driven motor, the more likely it is to mix heavy ingredients successfully.

- Belt-driven machines are also called *variable speed mixers* because this type of motor allows you to increase the speed while it is operating.

The gear-driven mixer usually costs more than a belt-driven mixer because it provides consistent results with longer life and fewer maintenance problems. The belt-driven machines can, over time, experience belt slippage and require lubrication, especially if they are overworked. For safety reasons, be sure the on–off controls on your mixer are large and plainly labeled. NSF International has additional safety standards for mixer construction.

Technology improvements allow the mixer to do double duty with the attachment of other tools to a hub on the motor that is called the *power take-off*, or PTO. Vegetable slicers and meat grinders are just a couple of handy assemblies that may be attached and removed as needed.

The smaller mixer sizes (5, 12, 20, and 30 quart) can operate on 110/114-volt electrical current. Models that are 40-quarts and larger require 112/120 volts three-phase; 220, 208/240, and even 230/460 volts for the largest, industrial-size models.

In an industry where change is the norm, the mixer is one machine built to last with a little simple maintenance. Wear and tear on a mixer is determined by the number of batches and types of product being mixed. Liquids, of course, are easier on the machine than solids. If dough is what you'll be mixing most, regular cleaning is essential to prevent damaging buildup of ingredients. Also remember these important maintenance tips:

- Use the attachments for the purposes they were designed. If a paddle (used to mix batter and pudding) is used for something as thick as bread dough, the motor works harder and the resulting bread is inadequately kneaded.
- Never overload the mixer. It makes the mixing task messier and more difficult and could cause the motor to burn out.
- Never bypass the safety interlocks. These are magnetic locks that prevent the machine from running if the bowl is not in its proper position. Running the mixer without these locks can seriously injure the operator.
- Never change gears while the mixer is running. Doing so strips the gears.

ATTACHMENTS AND ACCESSORIES

In addition to mixers, you need the attachments (agitators) that go with them. There's an attachment for whatever you want to do, from whipping marshmallow to cutting shortening into dough to chopping cabbage. The standard accessories include bowls and a variety of aluminum agitators, including beaters (also called *paddles*), whips (wire whisks, also called *balloon whips*), and dough hooks (also called a *dough arm*). These are seen in Illustration 17-2, and they allow you to adapt the mixer for multiple tasks. Different sizes of mixing bowls may be used on most mixers; these require attachments that work with the particular size of bowl.

Bowl guards are heavy wire shields that fit over the bowls during mixing to prevent injury to workers' hands. Once an optional accessory, the bowl guard is now a required OSHA safety feature, and mixers are designed so they cannot operate unless the guard is in place. (Older models can be retrofitted to comply with this requirement.) Most manufacturers offer a chute that can be used, even with the guard in place, to add ingredients during mixing. Lids and splash covers are also available.

To keep food inside the mixing bowl at a certain temperature, use a bowl *jacket*. The bowl fits inside it, and the jacket can be filled with ice or hot water as needed.

With its sturdy motor, the mixer is almost like a power tool. As we've mentioned, most manufacturers offer attachments that use the motor to accomplish kitchen tasks—slicing, dicing, and chopping—to increase its versatility. They also sell speed-drive attachments that more than triple the speed of the vegetable-slicing function. These extras can double the utility of the mixer, as long as you remember they also double wear and tear on the motor.

As you can now see, it's a mistake to purchase a more moderately priced mixer designed for home use. They are similar, but the commercial mixer has greater horsepower and is designed to meet the rigors of foodservice use.

Flat Beater   Wing Whip   Wire Whip   Dough Arm   Pastry Knife

ILLUSTRATION 17-2 There is a different attachment for any mixing job. Here are a few of the most popular ones.

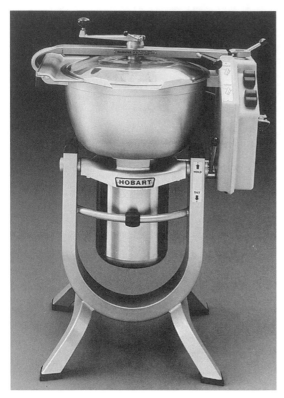

ILLUSTRATION 17-3 The vertical cutter mixer can chop, puree, and knead dough as well as mix.

Courtesy of Hobart Corporation, Troy, Ohio.

MIXER VARIATIONS

The chief advantage of the *hand-held mixer* (often referred to as a *Bermixer*) is its portability. It saves the time and mess of ladling ingredients from a pot into a mixer or food processor to be beaten. The hand-held mixer can be fitted with agitator attachments for beating, stirring, and blending. Hand-held mixer sizes reflect the capacity of the container they are able to mix, from 1 pint to 20 or 30 quarts. The agitator shaft can be from 10 to 40 inches in length. These versatile tools last indefinitely if you remember never to put them in the pot sink or dishwasher but to clean them individually by hand, keeping the motor dry.

The **spiral mixer** is designed to provide thorough kneading and smooth, evenly mixed dough in less time than the plane-tary mixer. It works gently and does not significantly increase the temperature of the dough during mixing, which helps pre-vent early fermentation of the yeast.

It has a dual timer that allows the operator to program the machine to shift automatically from one speed to the other dur-ing the kneading process. Spiral mixer capacities range from 120 to 250 quarts, but they can mix batches as small as 4 pounds if necessary. Their motors range in size from 7½ to 10 horsepower.

Originally, the **vertical cutter mixer** (VCM) came to the United States from Europe. The VCM is almost a combination mixer and blender; it has a mixing bowl with a motor mounted at the bottom and a shaft that projects up from it. Removable sleeves with cutting or mixing blades can be fitted onto the shaft (see Illustration 17-3). Its advantage is its high mixing speed. Dough or batter can be thor-oughly agitated in about 30 seconds; a bowl filled with water and a dozen heads of iceberg lettuce becomes perfectly chopped salad greens in less than 5 seconds. VCMs can be used to chop meats or vegetables, make slaw, mash potatoes, mix salad dressings, and stir cake batters. A dough-kneading sleeve can also be ordered.

The bowl is attached to the machine but has a pouring lip and tilts a full 90 degrees to pour out its contents. Clamp-down lids are transparent or have a viewing portal to allow the oper-ator to safely check on the contents of the bowl during processing. A rotating lever on top turns a baffle inside the bowl that scoops food from the edges toward the blade. The smallest VCMs stand about 2 feet tall, the largest floor models, 4 to 5 feet.

Power requirements for their two-speed motors vary depending on the size of the VCM. The 4- to 6-quart tabletop models have 1-horsepower motors; 30- to 45-quart models have 5-horsepower motors; the largest units, up to 130 quarts, are driven by 25-horsepower motors.

The last two items on the mixer variations list are not really mixers, but we're putting them here because they involve mixing action for different purposes. The **vacuum tumbler** is a motorized machine that gently kneads pieces of meat with a marinade that contains salt. The salt extracts the proteins in the meat to form a natural coating that, when the meat is cooked, seals in its juices. Because the coating is evenly distributed, the meat surface also cooks with a nice, even color. In addition to producing juicy, uniformly cooked products, vacuum tumblers can increase the yield of most meats.

Food-borne illness outbreaks have created new interest in the *salad greens washer* and *salad spinner*. Greens washers immerse produce in water and provide gentle agitation that loosens dirt and grit from hard-to-clean items such as lettuce and spinach. Some greens washers inject small amounts of ozone gas into the water, which works as a natural sanitizer. Then the greens are transferred to a salad spinner to remove the excess water. You've probably seen hand-cranked spinners for home use, but motorized spinners are required for volume drying. Look for a salad spinner that has two speeds, as more delicate greens (spinach) dry better at slower speeds while hardier items (romaine) can be dried at higher speeds. Washed-and-spun greens that are promptly bagged and sealed last up to 5 days in the refrigerator.

17-2 FOOD SLICERS

The *food slicer* is intended to replace the old chef's knife and the seemingly endless chore of slicing sandwich meat and cheese. Seen in Illustration 17-4, its use in portion control is also universally acknowledged. Because it can cut meats, cheeses, and vegetables more accurately than a knife, the results are uniform thickness and maximum output.

Choosing a slicer is a lot like choosing a mixer: You have to think about what foods you will be slicing and how much the machine will be in use. A medium-duty slicer with a 9- or 10-inch blade is sufficient for 2 or 3 hours a day, but if you'll be slicing for more than 4 hours a day, or if you'll be slicing a lot of cheese—which is notoriously tough on slicers—consider a high-volume machine with a 12-inch blade or larger.

The next decision is whether your machine should be manual (hand-operated, slicing to order), or automatic (capable of slicing large batches without hands-on supervision).

Slicers are usually identified by the diameter of the cutting knife or blade, and the size affects the machine's efficiency. The most popular are the 10- and 12-inch disk-shape blades. A typical slicer can hold products that range from 7½ to 12 inches in diameter in its carriage. The food rests in the carriage area, where the operator holds it firmly in place with a handle as the *feed grip* gently pushes the food toward the spinning blade for slicing; or the food is loaded into a chute and gravity-fed toward the blade. Automated models have a carriage drive with two speeds, low (36 strokes per second) and high (51 strokes per second). These work without an operator but should still be checked periodically because they continue the slicing motions even when there is no more food to be sliced.

Slicers can be adjusted for various thicknesses of finished product, from a paper-thin 1/32 inch to 1¾ inches thick. When not in use, setting the slicer's gauge plate (or *index knob*) at zero is an important safety precaution. Slicer accidents are common in food service—so much that some manufacturers have now made it impossible to detach the carriage from the slicer unless the gauge plate is set at zero. A *center plate interlock* prevents the slicer from being operated when the carriage is removed for cleaning. The *blade ring safety guard* is a permanent edge at the back of the blade that protects anyone handling the blade during operation and cleaning. These thin metal guard plates shield the top and back edges of the blade, and the feed grip is positioned so it cannot strike the blade or guards.

Any heavy-duty slicer should have ½-horsepower motor. Some gravity-feed slicers have standard ⅓-horsepower motors, but you'll get better results with the slightly larger motor. Automatic models can be manual, two-speed automatic, and variable-speed. A *no-volt release* prevents unintentional start-up of the slicer. After a power outage, or if it is accidentally unplugged, machines with no-volt release protection must be restarted.

Food slicers should be made of anodized aluminum or stainless steel and designed with maximum sanitation in mind. Look for seamless surfaces, rounded corners, and corrugated feed grips to prevent foods and liquids from lodging in crevices. Slicer blades are made of stainless steel or carbon steel, and debates about the merits of each and whether (like other knives) they should be cast or stamped are ongoing. The Food Service Technology Center in San Ramon, California—which we've mentioned elsewhere in this text for its state-of-the-art appliance-testing research—found that both substances and both manufacturing methods are equally well suited for the job. Some manufacturers tout chrome-plated blades, but the sharpening process soon wears the chrome off the cutting edge anyway.

In addition to the standard models, there are a few specialty slicers. Consider a specially designed bread slicer.

ILLUSTRATION 17-4   A commercial food slicer.

Courtesy of Hobart Corporation, Troy, Ohio.

A typical food slicer does the job but with too much waste. There are special slicers to core tomatoes, make french fries, and slice and stack large quantities of lunch meats. Some can even process two foods at once, then layer them alternately for commercial sandwich making.

Everyone seems to dread slicing cheese because, as it reaches room temperature, it sheds moisture and becomes gummy, making the slicing process sloppy and wasteful. Keep cheeses and meats refrigerated until just before slicing to minimize this problem.

SLICER SAFETY AND MAINTENANCE

We've already mentioned the built-in safety features of food slicers, but there are plenty of other slicer-related safety rules. The first is that no one under age 18 is legally allowed to operate a slicer. Slicer operators should use clean, well-fitting, cut-resistant gloves, especially to clean the machine; they should also avoid wearing loose-fitting clothing or jewelry that could become caught in the slicer.

Slicers should never be used to cut solidly frozen products. That's a job for a meat saw. Remove plastic or paper wrapping from foods before slicing, and never attempt to cut through a tin can or container of food using a slicer.

If the food is not being cut satisfactorily, the most common culprit is a dull blade. Built-in blade sharpeners, stored under the slicer base, are standard features. Here are guidelines for correct sharpening:

- The slicer has two types of stones, a sharpening stone and a honing stone. The latter is to smooth out burrs that develop on the blade during sharpening.
- The blade should always be completely clean and dry before it is sharpened, and it should be wiped clean of debris after the sharpening process.
- Avoid oversharpening—a few seconds is sufficient. Longer sharpening reduces the life of the blade.
- When you replace the blade, replace the sharpening and honing stones. Old stones can damage new blades.

Inexperience makes slicing dangerous, and well over half of slicer-related accidents occur when the machines are being cleaned. This is why the manufacturer's safety features and guidelines are so important. The entire slicer must be cleaned and sanitized daily, and the blade should be cleaned frequently during the day. Blades must also be lubricated with food-grade oil, never vegetable oil. The motor also must be oiled occasionally, which is a job for a service technician.

Aluminum parts should be cleaned by hand, never with bleach or in a dish machine. One innovative idea is the BladeRunner, a chunk of sponge-like material saturated with quaternary ammonia, which cleans and sanitizes the blade as it is sliced like a food product and then discarded. KatchAll Industries International makes the BladeRunner.

Finally, cross-contamination must be prevented in slicer use by cleaning or switching blades between slicing cooked and raw products; slice meats and cheeses while they are cold, and return unsliced or unused portions to refrigeration immediately.

17-3 FOOD PROCESSING

The food processor was introduced to the European market in the 1950s by the French company Robot Coupe (pronounced "ROH-boh koop") and marketed in the United States by a trio of investors who saw it at work in a New Orleans restaurant and formed Robot Coupe U.S.A. in Jackson, Mississippi, still a dominant player in the market. Cuisinart (known for creating processors for home use) and Waring (famous for its blenders) are other major manufacturers.

Food processing is the term for changing the shape, size, or consistency of food. The equipment is versatile, user-friendly, and surprisingly safe considering the inherent hazards of chopping, grating, and so on. There are two basic types of food processors—the continuous feed and the bowl style—as well as a third type of machine that can do both.

The *continuous-feed processor* has blades with openings of various shapes for cuts such as strips, shreds, coins, crinkles, and juliennes. Food is fed through a chute at the top and sliced by a blade. As many as 20 interchangeable blades (sometimes called **plates**) are available—a few are shown as Illustration 17-5—as are specialty blades for grating nuts, spices, chocolate, bread crumbs, and others. Be sure your equipment package includes a **hard-products slicer**, a blade for processing foods such as carrots, potatoes, and cheese. Also notice whether the design of the machine allows you to catch the outgoing food easily in standard-size pans. The continuous-feed processor is designed to expel finished product through a chute and keep working as new food is fed into it. It is the best choice when a consistent cut or high-volume output is needed.

The *bowl-style processor* has a removable rotating bowl and a two-blade assembly that rotates at the bottom of the bowl to do the cutting (see Illustration 17-6). Different assemblies produce the different cuts. Handy accessories include a dicer, meat chopper, and scraper that can be operated from outside the bowl as it rotates. With this type of processor, you fill the bowl, process the food, empty the bowl, and refill it. The advantage of the bowl-style processor is that it cuts quickly and without pressure, so delicate items like parsley can be minced without becoming mushy. The rotating action also helps mix the food as it's being cut. As a safety precaution, bowl-style machines cannot be turned on unless a bowl cover is locked in place, and the cover cannot be raised while the machine is turned on.

Combination food processors have a single motor base that can be used with a continuous-feed attachment or a bowl for cutting and mixing. They work well for kitchens that require maximum versatility to prep a wide variety of foods.

Like mixers, food processors require an assortment of bowls. Some are tall and narrow (to minimize spattering), and others are wide and short (for faster processing). Purchase your bowls, like the processor itself, based on the foods they will be used to process.

Food processors operate with motors of ½ to 1 horsepower and a single on–off switch. Like other small appliances, they can have belt-driven or gear-driven motors. Electrical requirements vary from 115-volt single-phase power to 460-volt three-phase power. A wise option for the bowl-style processor is a pulse function, which prevents overheating when processing large chunks of food.

How quickly do processors work? Their blades whir at more than 1700 revolutions per minute! The output for two popular sizes is shown in Table 17-1. A ½-horsepower combination-style processor is capable of handling 350 pounds of product in an hour, depending on the product and the type of cut. The best way to ensure maximum output is to outfit the machine with large feed attachments and cutting plates with large surface areas.

Cheese is the most difficult food item for any processor. The job is much easier if the cheese is refrigerated until processing and dusted lightly with cornstarch to prevent sticking. If your processor will be used primarily for grinding cheese, consider buying one with a slightly larger motor.

PROCESSOR SAFETY AND MAINTENANCE

Employees must be instructed to keep their fingers out of the hoppers and discharge chutes. Tools to push product into the processor, called *pushers* or **stompers**, should always be used instead of hands. If anything must be cleared

Slicer
(1/8 in.)

Shredder
(3/16 in.)

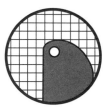

Slicer, julienne
(1/4 in. x 1/4 in.)

Cuber

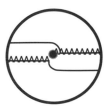

Slicer, crinkle cut
(1/4 in.)

ILLUSTRATION 17-5
Food processor plates, or blades, produce different sizes and shapes of product.

ILLUSTRATION 17-6 A bowl-style food processor.

Courtesy of Hobart Corporation, Troy, Ohio.

from inside the machine during use, it is not enough simply to turn it off and reach in. You must also unplug it so you don't accidentally trip the on–off switch with your hand in the bowl.

You'll get maximum useful life from your processor if it is cleaned after each use with a nylon brush that can reach into the crevices of the machine without damaging the seals on the motor. Stainless-steel, aluminum, and plastic parts can be wiped clean with a damp cloth and the manufacturer's suggested sanitizing solution. The removable parts—cover, discharge chute, bowls, and so on—can be washed in hot, soapy water, rinsed, and air-dried. The processor base that contains the motor must never be submerged in water.

Be sure the plates are easy to change, sharpen, and sanitize. They should never be run through a dish machine but rather hand-washed and allowed to air-dry. Drying them by hand only invites an injury. Also, never leave a blade submerged in a sink, as an unsuspecting person could reach in and get cut. Manufacturers make special plate storage racks, use them.

MANUAL FOOD PROCESSING

A final word about food processors: Nonmotorized units with simple hand cranks are also available. These are an effective and inexpensive option.

The names *food chopper* and *food grinder* are used interchangeably because most of these machines do both, depending on the attachment they are fitted with (see Illustration 17-7). The chopper/grinder is a simple machine with a single-pole toggle switch that controls a pushbutton on–off function. A rubber cap typically protects the pushbutton assembly. Larger machines may have a magnetic starter with an automatic reset button that shuts off if the unit becomes jammed, overheated, or overloaded.

The operator loads meat into the machine by pushing on a hard plastic *stomper* or *pusher*; plastic is a better choice than wood because it can be sanitized. Stainless-steel knives or plates inside the machine perform the chopping functions, and the same pressure that pushes the food into the machine forces the ground meat out through another opening. It can emerge into a pan or a sausage casing attached to the machine. The diameter of the output opening can be adjusted by hand. Electrical requirements for chopper/grinders range from 115-volts single-phase to 460-volts three-phase.

Manually operated *vegetable cutters* are designed for individual processing of items like blooming onions. Some can core, wedge, and dice; others can perform one simple but otherwise time-consuming task, such as slicing cabbage or tomatoes, coring apples, or cutting potatoes

TABLE 17-1		
Food Processing Times		
FOOD PRODUCT	**4-QUART PROCESSOR**	**6-QUART PROCESSOR**
Minced vegetables	2 pounds/20 seconds	3 pounds/20 seconds
Soft cheese	2 pounds/20 seconds	3 pounds/20 seconds
Ground meat (raw)	2 pounds/60 seconds	3 pounds/60 seconds
Ground meat (cooked)	3 pounds/60 seconds	4 pounds/60 seconds
Source: Hobart Corporation, Troy, Ohio.		

ILLUSTRATION 17-7 A food chopper/grinder is used primarily for meats.

Courtesy of Hobart Corporation, Troy, Ohio.

into french fries. Most manual cutters are tabletop units that can be solidly clamped to a prep counter or wall-mounted. They are good choices for small operations or when the budget prohibits purchase of motorized processors.

In caring for these items, remember their metal parts are susceptible to food acids; they must be handwashed frequently to keep their sharpness and always air-dried.

BLENDERS

An inventor named Fred Osius patented the first blending machine in the United States in 1933, but it was big band leader Fred Waring who made it famous. Osius knew Waring had been an engineering student who loved gadgets, so he took his blender to a concert, talked the security crew into letting him backstage, and convinced Waring to back further research for the new machine. It was first marketed as the Miracle Mixer and finally the Waring Blendor (with an "o").

High-capacity, heavy-duty blenders—the kind that can crush ice cubes in seconds and blend them into drinks—didn't come onto the commercial foodservice scene until the late 1980s. Depending on your menu and theme, you may need heavy-duty blenders or bar mixers.

With the popularity of juice bars, many companies developed blenders specific to that market segment. These blenders can be mounted to a countertop so the blender container is above the counter with the motor unit underneath. This allows for a less cluttered-looking counter and puts the blender container at an easily accessible height for the operator.

The *blender* can perform most light food-processing duties, with multiple settings for speed, power, and so on. It can chop, puree, mix sauces, liquefy, make milkshakes, and crush ice into slush for frozen drinks like margaritas. The ***bar mixer*** is designed to work primarily with ice, liquids, and solids, and is used strictly for making drinks at bars and nightclubs. It is quicker and usually quieter than a blender. Some bar mixers have their own integrated ice bins that dispense exact amounts into the container, eliminating the bartender's need to measure it by hand.

Your equipment choices in this category depend on answers to a number of questions:

■ How much blending must be done in a day's work?

■ How quickly do you need to produce a drink?

■ How much space do you have?

■ Where do you want to locate the machine?

■ How much noise are your customers and staff members willing to endure?

■ What do you blend: frozen or cooked items; especially fragile, hard, or chunky ingredients?

Both commercial blenders and bar mixers are larger than those made for home use and are built to handle about 100 blending cycles per day. They are identified by the capacity of their containers, from 1 quart to 5 gallons. They rest on a heavy, die-cast base with rubber feet for stability. Motor sizes range from ½ to 3 horsepower. The strongest ones are used for juicing vegetables.

Most commercial blenders have two speeds, but some offer variable-speed control or a pulse switch for intermittent higher-speed uses. Frankly, the blender's power is not as important as your ability to control it. If you can't select the blade speed (measured in revolutions per minute [RPMs]) and the timing for each task, consistent results are not possible. Digital technology has made the modern blender extremely user-friendly: highly programmable, with sensors that monitor and modify blade action and overall designs that are engineered for uniformly blended output. Some manufacturers even program the electronic chips in their highest-end blenders specifically for certain bars' signature drink recipes.

Inside the blender, a toggle switch activates the motor, which turns a set of blades attached to the clutch. The clutch, which you'll see when the container is removed, takes a lot of abuse—particularly with a lot of start-and-stop action—so it's constructed of steel or flexible neoprene rubber. There are four- and six-blade assemblies, easily removable for cleaning. Manufacturers offer different angles, lengths, and shapes of blades for different purposes.

The container where blending is done is most often made of stainless steel, but glass or polycarbonate materials are also used. (Polycarbonate is the material used to make aircraft windows; it is tough but translucent.) The only trouble with poly containers is that they can't be cleaned in dish machines. Stainless steel retains heat and is used for hot soups and sauces; glass is rarely used in foodservice because of the risk of breakage. Containers can be ordered with one or two handles, whatever makes lifting easier. The container is usually covered with a lid made of rubber or vinyl, often a two-piece **filler cap** so ingredients can be added during the process. Container capacities range from 1 quart to 5 gallons. The 1-quart model is sufficient for as few as a dozen frozen drinks per hour.

If the blender is not already programmable, a timer is a worthwhile option, especially for bar mixers. It frees the operator to handle other tasks and limits wear and tear on the appliance from unnecessary starts and stops.

In smaller restaurants, a **spindle blender** performs the mixing functions, from making milkshakes and sauces to whipping cream to adding ingredients to frozen yogurt. Ice cream parlors couldn't function without them! They come in countertop and hand-held models, with a single mixing head or up to five heads to make five drinks simultaneously. (A three-spindle model is shown in Illustration 17-8.) Some have three height settings to accommodate different sizes of stainless-steel mixing cups.

Even when you turn the unit on, the blender head does not start turning until an activator at its tip makes contact with the mixing cup. Cup vents, usually made of stainless steel, hold the cups in place during blending.

Spindle blenders with multiple heads have a separate small head for each motor, which means one can be serviced without affecting the others. It's important to check the motors periodically because they have transmissions that require oil. The manufacturer's manual provides recommendations.

Counter space with a water source or drain nearby is necessary for a multiple-head spindle blender. If space is tight, a single-head, hand-held model is fine. Longer rods are available for use with deeper pots and kettles.

The body or casing that houses the blender motor has sealed seams to protect it and simplify cleaning. Blenders must be kept clean and, oddly enough, protected from unnecessary moisture. This is because most units have a cooling fan in the base; water can get sucked into it and corrode internal parts. Some models have recessed cooling vents to help prevent this. Syrups and other sticky ingredients can also clog toggle switches and buttons, so it's smart to spray the machine with an alcohol-based cleaner at the end of the day.

Noise is a significant issue with all the grinding and glugging noises that busy blenders make. Most manufacturers craft some type of encasement of polycarbonate or acrylic to fit around the blender and allow access through a hinged door. The blender may or may not shut off automatically whenever

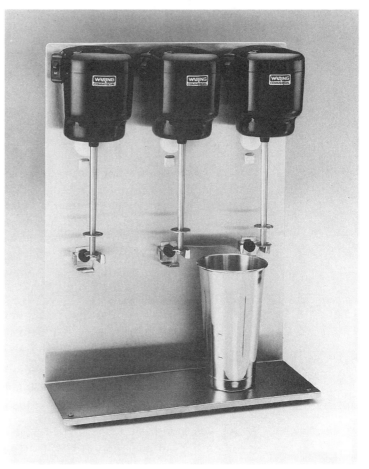

ILLUSTRATION 17-8 A spindle blender is used to make milkshakes at soda fountains.

Courtesy of Waring Products, a division of Conair Corp., East Windsor, New Jersey.

the door is opened, but it's a nice feature. While no one can eliminate blender noise altogether, the goal is to reduce the noise level to below 72 decibels, the standard loudness of a normal two-person conversation.

JUICERS

The *juicer* or *juice extractor* is a cousin to both the blender and the food processor. Nothing quite compares to the olfactory and visual appeal of a freshly squeezed glass of orange juice with breakfast, at the health club, or at the bar as part of a tropical cocktail. You can choose from manual-feed models or hoppers that can be loaded with oranges that roll automatically into an extractor. The machine, with a motor from ½ to ¾ horsepower, crushes the fruit, separates the skin and pulp from the juice, and dispenses the juice into an individual glass (smallest models) or a chilled reservoir (largest models). Of course, the fruit must always be washed before being loaded into the machine, as its external skin comes into contact with the juice during the mechanical squeezing process.

These sturdy machines are made of stainless steel and aluminum. The unit itself can't be submerged in water, but most parts are removable for cleaning—which can be a sticky job! Most juicers are appropriate for a variety of fruits and vegetables, although they are most often used for citrus fruits. Prices for commercial-model juicers range from $2,000 to about $7,000 for the largest standalone machines.

17-4 TOASTERS

The incredible variety of breads now available in foodservice requires choosing among toasters that provide versatility and ideal conditions for toasting each type of product. Do your customers expect their bagels toasted on one side or both? How much toasted bread is your kitchen expected to turn out on a busy morning? As with other appliances, the toasters you choose depend on the types and sizes of food products being toasted, the volume of business, speed of service, amount of labor available to you, and how much space is available.

The toaster is another appliance that most surely requires a commercial model instead of those designed for home use. Just one comparison says it all: The home toaster weighs up to 3 pounds; the commercial four-slot toaster weighs between 18 and 20 pounds. Technology has greatly improved commercial toasters, adding innovations like solid-state controls that allow easy time and temperature adjustments, inch-thick insulation to keep the appliance and kitchen cooler, and power-saving switches with a standby mode. Both electric and gas-powered toasters are available, but because the gas-powered models must be vented, most users specify electric ones for ease of installation.

In toasting any food, think about the process and the product. This is a three-step process:

1. You remove the moisture from the product with heat. (That is why day-old bread makes better toast. It's already drier, so it doesn't stick in slots or on conveyors, etc.). It does not take a lot of heat to evaporate moisture, but it does take time.

2. You cook the product throughout, so it is warm and crisp.

3. You add the color. In a **continuous (conveyor) toaster**, this is done by boosting the wattage at the exit end of the conveyor belt for a nice final browning.

The two basic toaster types are batch toasters and continuous toasters. *Batch toasters* are often the familiar slot variety, which lowers the bread into slots between heating coils. The finished toast pops out automatically or is manually lifted from the slots. The slot or pop-up toaster can be ordered with wider slots to accommodate Texas toast, bagels, and other thick-bread products. Its downside is that someone must select the degree of doneness, then stand there and wait for the bread to pop up, butter it, and serve it while still warm. The pop-up toaster is available in two- or four-slot models. Some four-slot models have longer, wider (1½ inch) slots to toast burger buns, with heating elements that toast only one side of the bread. These produce buns for more than 90 burgers an hour.

Commercial batch toasters can produce 125 to 190 slices of bread per hour (two-slot models) and up to 250 slices of bread an hour (four-slot models). The manual-lift models have footprints of about 9 × 14 inches. Single-slot pop-up toasters do not seem individually impressive, but banks of them are capable of large production, and they are built for heavy use.

In pop-down toasters, bread is dropped into the slot and comes out toasted, resting against a door at the bottom of the machine. They require more vertical space than pop-up models because they're taller, but they can be reloaded more quickly. Both pop-up and pop-down styles are appropriate in situations where the servers, or even the guests in self-serve settings, make the toast.

The heating elements that toast the bread are of two types: **quartz sheath** and **metal sheath**. Proponents of quartz elements claim it radiates heat better than other substances; indeed, a good one can go from cold to full toasting mode in 30 seconds. The downside: Quartz is brittle and can break if the appliance is dropped or handled roughly, or even during cleaning.

Manufacturers claim metal-sheathed elements can remove excess moisture from the bread as it browns. The metal sheath may take 15 minutes or longer to reach full toast mode. However, once it has heated up, it holds its heat longer and is much sturdier overall than the quartz sheath.

You can order toasters with two to seven heating elements. No matter what type or number you choose, an internal temperature monitor is an important feature of a good toaster.

This is because after several consecutive uses, the toaster is already hot and, therefore, starts the next batch of toast at a higher temperature. Without an internal monitor, a toaster may pop the bread up too soon, only slightly toasted.

The other popular type of batch toaster is a *drawer-style toaster*, with a drawer or platen onto which bread is placed to slide into a heated chamber for toasting. Drawer-style toasters are made with two types of heat: contact and radiant. Contact toasters work like griddles. The item to be toasted is placed on the heated surface, and a lid or weight holds it flat. This is a perfect way to toast hamburger and hot dog buns. Radiant-heat toasters look like cheesemelters or home-use toaster ovens. A drawer or tray is loaded with food and slides under a heating element that toasts the top surface of the food. This type of toaster is commonly used for heating open-faced sandwiches, a process that takes from 2½ to 5 minutes. Its footprint is about 10 × 20 inches.

The **continuous toaster** (or *conveyor toaster*) is a high-volume appliance (see Illustration 17-9) for toast output of 150 slices or more per hour. Some operators use a regular conveyor oven for this job because it can bake cookies, pizzas, and so on in addition to making toast. Food is placed on a conveyor belt at the front of the toaster. Color and doneness are controlled by varying the heat, the conveyor speed, or both. In fact, many operators prefer a single control for both because it's difficult to set separate controls to work together. Most conveyor toasters feature a switch that allows you to turn off one of the heating elements and toast food on one side instead of two. At the end of the conveyor belt, the food slides from the belt into a chute that empties into a collection tray. The biggest conveyor toasters have multiple belts running side by side, each with its own heating elements and controls.

There are both horizontal conveyors and vertical conveyors. The horizontal conveyor is the best choice for extremely high volume—more than 500 slices per hour. It has a slightly larger footprint than the vertical conveyor, but you can stack more than one horizontal conveyor toaster and run them simultaneously.

Vertical conveyor toasters feature longer cooking chambers and are known for producing high-quality product, but they cannot be stacked. The vertical designation means that bread, for example, is loaded into baskets or pockets at the top of the appliance. The conveyor revolves

ILLUSTRATION 17-9   A conveyor toaster is a type of conveyor oven, designed for high-volume output.

Courtesy of The Middleby Corporation, Elgin, Illinois.

up, over the top, and down the back of the toaster, and the finished toast is dropped onto a tray at the bottom. The available space may determine whether your operation uses a vertical or horizontal machine. The vertical models have a 22 × 15-inch footprint and stand almost 3 feet tall. Horizontal models have an 18 × 12½-inch footprint and are only 14 inches in height. Vertical toasters usually cost more than horizontal ones.

Toasters of all types require single-phase electricity ranging from 120 to 240 volts. Select yours based on the volume you need at peak production times, and remember that, no matter what the manufacturer's claims, the output of a toaster depends on several factors: the density and moisture content of the product being toasted and whether an attendant is present to keep it fully and promptly loaded. Because toasters are used to make everything from toast to frozen waffles, figures vary. Some common toaster "bragging rights"—and our own real-life estimates—follow:

- *Four-slot toaster.* Listed from 200 to 360 slices an hour, an average of 150 to 180 slices is more realistic, and only if the machine is attended by an employee.

- *Pop-down toaster.* Listed from 240 to 420 slices an hour. Again, an average of 150 to 180 slices is more realistic.

- *Batch toaster.* Fully loaded with 12 burger bun halves per batch, manufacturers suggest figures of 200 to 300 bun halves per hour, enough to make 100 to 150 burgers. In actual operation, these figures are 50% to 75% attainable.

- *Conveyor toaster.* Bun production is rated higher than toast output from the same machine, and the range is from 150 to 1200 slices per hour. Real-life figures depend on the density and moisture of the product as well as the skill of the operator in loading and unloading the machine.

Specialty toasters are used for buns and bagels. Bagel toasters are conveyor models with a higher clearance (2-inch is standard, so these are wider to allow taller bagels to pass through). They run higher wattages and have a different pattern of heating elements than standard machines. The *contact bun toaster* uses a heated platen to sear the bread quickly and evenly on both sides. The panini grill, which we introduced in an earlier chapter, is a type of contact toaster. The idea behind these quick-toast appliances is to seal the bread a bit by toasting it so it doesn't get soggy when made into a sandwich.

17-5 FOOD WARMERS

Keeping food hot while it's waiting to be served without compromising its quality is the classic foodservice challenge. Fortunately, there's a piece of warming equipment to fit almost every situation. Most fall into one of three categories: food warmer, soup warmer, or drawer warmer. Steam tables and hot-food tables are also used.

Warming or *holding* is a controlled cool-down from hot-out-of-the-oven to a safe and stable serving temperature. The type of food, the length of time it is held, and the way it is served all affect the equipment choice, but the absolute first requirement of any food warmer is that it maintain a minimum holding temperature of 135°F. This minimum temperature refers to the internal temperature of the food, not its surface temperature, and is required by law in all jurisdictions for prevention of food-borne illness.

Quick-service operators have worked with manufacturers to design and develop hot holding bin units that allow for multiple small batches of products to be held for short periods in a bin-type arrangement. These units, seen at operations like McDonald's restaurants, allow the staff to easily access a number of hot holding bins for the variety of sandwiches offered. These units occupy a relatively small footprint, making them convenient as a sandwich workstation for high-volume applications.

The ideal food-holding appliance for periods exceeding one hour is a humidified, closed cabinet with convection heat, such as the ***drawer warmer*** seen in Illustration 17-10. Why?

The closed cabinet protects the food from temperature changes, and the moisture content prevents it from drying out. A drawer warmer is a heated cabinet with one or more drawers that slide open to load or remove food. Many offer humidity control, which is a plus for some types of food, and separate controls for each drawer. Drawer warmers are available as countertop or drop-in appliances. Any holding cabinet must be equipped with sufficient power to reheat rapidly whenever the door is opened.

Where a closed cabinet is not feasible—say, for short holding periods (15 minutes to an hour) in a busy self-serve or quick-service environment—the best choice is an open warming unit (see Illustration 17-11) that provides conductive heat from the bottom and/or radiant heat from the top. Open warmers are susceptible to air drafts, which can be minimized either by raising the unit's temperature slightly or adding wattage to the radiant elements above. The disadvantage to short-term warming is that it cannot keep food moist. Some foods may be covered or wrapped for holding this way, but it's not ideal. And what to wrap it in? Tinfoil holds in moisture but reflects radiant heat rather than absorbing it, paper absorbs heat better but dries food out faster, and plastic wrap melts in the intense heat required to keep the food safely hot.

ILLUSTRATION 17-10 A drawer warmer can accommodate standard pan sizes in different drawers with separate controls.

Courtesy of Alto-Shaam, Inc., Menomonee Falls, Wisconsin.

Here are some of the food-holding choices available today: The term *overhead warmers*, of course, means the warming lights are placed above the food. These can be affixed to a counter, suspended from a wall or ceiling, or be freestanding, portable units attached to a base suitable for food placement. Because they give off so much heat, it is not advisable to mount warmers any closer than 3 inches from walls or ceilings. They may blister your paint! Most warmers turn on and off with a switch; the more expensive ones have a knob for adjustable heat control. Overhead warmers use *warmer lamps* to work effectively, but they are different from your basic heat lamp. This type of bulb uses infrared lighting to produce heat, from 250 to 275 watts. Like other types of light bulbs, there are round ones and tube-shaped ones for larger heating spaces.

Several specialty warmer models, such as the buffet warmer for banquet and cafeteria settings, are available. These have sneeze guards and, in addition to overhead lamps, insulated cables that run beneath the food pans to provide gentle warming from beneath. Thermal shelves, found in pizzerias and bakeries, are made of insulated material with lamps above that can be adjusted with a dial for different heat settings. Pass-through windows can also be outfitted with warmer capability.

You can also choose a *base-heated warmer*, either electric or gas. These operate by heating a well into which food containers are placed. Most models require that water be in the well, but some operate without water. A thermostat controls the heating element, which uses from 500 to 1200 watts. Multi-unit designs use up to 1600 watts. Base-heated warmers open at the top to accommodate food pans; adapters are available to accept various pan sizes.

Like its electric counterpart, the gas-powered food warmer is about 14 inches wide and 24 inches deep. It stands less than 1 foot in height, but adjustable legs are recommended, which add another 3 to 5 inches.

ILLUSTRATION 17-11 Food warmers can keep food hot and appetizing until it is delivered to the customer.

Courtesy of Alto-Shaam, Inc., Menomonee Falls, Wisconsin.

In buying a gas model, consider where the gas fitting will be placed so it doesn't interfere with the use of the unit. It is important that air intakes on the bottom of gas units be kept clean.

Soup warmers are cousins to base-heated warmers, and they are far more versatile than their name suggests. Use them to heat sauces, chili, melted cheese, hot fudge, and more. They come in many sizes and styles from round to rectangular and can be placed on a countertop or dropped into cooks' tables or salad bars. The countertop models may remind you of an electric crockpot, but these are made of stainless steel. The most common countertop models accommodate inserts of 4, 7, or 11 quarts. For larger volume, rectangular 12 × 20-inch pan units are available.

Water is required for most warmers, and it is highly recommended. The electric element at the base of the unit heats a reservoir of water, which, in turn, heats the contents of the immersed pan. The advantage of water is that it transfers the heat evenly to the food container, eliminating scorching and resulting in steady, even temperature controlled by thermostat.

NSF International safety standards require these warmers be equipped with a temperature gauge on the outside of the unit to be able to easily monitor interior temperatures without opening the drawers. Temperatures range from 60°F to 200°F. Drawer warmers use 120- or 208/240-volt electricity and from 450 to 1350 watts depending on the size of the unit and the number of drawers.

STEAM AND HOT-FOOD TABLES

Another option for keeping food warm between cooking and serving is the hot-food table or steam table. What's the difference?

The typical **steam table** (see Illustration 17-12) contains a shallow tank of water heated with electric or gas-fired elements to between 180°F and 212°F. Used mostly for moist foods, shallow steam-table pans full of food are placed in the water, keeping them hot until needed. An advantage of many steam tables is that they can be heated with or without water.

The *hot-food table* accomplishes the same task by heating multiple compartments of hot air. The operator fills the compartments with standard-size serving pans of food. Individual compartments are controlled with separate thermostats, so several foods can be held in the same table at different temperatures. These may be called *waterless hot-food tables* because no water is required and, therefore, no plumbing connections.

ILLUSTRATION 17-12   Steam tables are designed to keep food at warm, safe temperatures in banquet, buffet, and cafeteria settings. This one has a built-in oven below for cooking, then serving above.

Courtesy of Thermodyne Foodservice Products, Inc., Fort Wayne, Indiana.

TABLE 17-2

Common Steam Table Sizes: Gas and Electric (Depths Vary with Accessories)

GAS NUMBER OF 12 × 20-INCH OPENINGS	TYPICAL LENGTH
2	2'6"
3	3'8"
4	4'10"
5	6'10"
6	7'2"

ELECTRIC NUMBER OF 12 × 20-INCH OPENINGS	TYPICAL LENGTH
2	2'6"
3	3'8"
4	4'10"
5	6'10"
6	7'2"
7	8'4"

Source: *Foodservice Equipment and Supplies* magazine, a division of Reed Business Information, Oak Brook, Illinois (February 1999).

Both types of heated tables are available in a variety of lengths and sizes depending on how many openings are needed for different pans. The openings, or the appliances themselves, are often called *hot wells*. An individual well measures 12 × 20 inches, just enough to fit a steam table pan. Table 17-2 shows the most common table lengths.

A major concern with heated tables is that users keep the food pans covered as much as possible to conserve heat and energy. Food at the bottom of the pan may be adequately heated, but at the top of an uncovered pan the much lower temperature may be dangerous. Casseroles, sauces, and stews must also not spend too much time on the hot-food table before being disposed of.

NSF International suggests users of holding equipment play it safe by dividing large batches of food into smaller portions for holding and serving. Smaller portions ensure more thorough and even heating, minimizing food safety risks. It is also a standard rule that, no matter what it's heated in, food that has been held for more than 2 hours should be discarded.

Where sauces and soups are served, you may also have use for the **bain-marie,** a small, uncovered variation of the steam table. The bain-marie is a single container used for cooking and also for holding, with a thermostat that controls its temperature. It comes in sizes from a 2-foot square to 30 × 72 inches. The smallest electric-powered bain-maries use 3 kilowatts of power per hour; the largest ones use almost 10 kilowatts.

17-6 COFFEE MAKERS

The best-selling nonalcoholic beverage in restaurants is coffee, and the specialty coffee market has grown to more than $30 billion a year in sales, with 58% of the U.S. population telling the National Coffee Association of the USA, Inc., that they "drank coffee yesterday," and 86% of those folks saying they make coffee at home.

This represents a major cultural shift in little more than a decade, and we probably have Starbucks, the upscale Seattle-based coffee retailer, to thank for at least some of it. Starbucks reported in 2012 that it owns or operates more than 13,000 stores in 50 countries and has branched out to offer ice cream, bottled and canned coffee drinks, and related products in supermarkets as well.

The Specialty Coffee Association of America notes that Brazil is expected to displace the United States in the next few years as the world's largest coffee-consuming nation.

The best news for foodservice operators is that customers are now completely accustomed to paying $4 and up for gourmet coffees and specialty drinks, just about any time of day. Your staff had better learn to make them well, and this starts with top-notch equipment. Incidentally, there are also commercial tea-making machines, which may be useful depending on your menu and customer preferences. But we'll focus here on coffee making.

Your first considerations in purchasing coffee-brewing equipment are the amounts you need to produce and the types of coffee drinks your establishment offers. The requirements are far different for a place that needs 10 gallons of coffee per hour and a single brew versus the place that needs 1 gallon per hour of 10 brews. The market is cluttered with a wide selection of brewing options and appliances for coffee, tea, and espresso drinks. Before we introduce them, let's list the factors that influence the quality of brewed coffee. Some, but not all, are directly related to the coffee-making machine:

- The type, freshness, and quality of the coffee bean
- The way the beans are ground (finely ground coffee takes less time to brew but can be stronger in flavor than coarsely ground coffee)
- The overall water quality (you may need to filter or purify water before using it for brewing)
- The temperature of the brewing water (optimum 195°F to 205°F)
- The quality of the coffee filter
- The uniform extraction of the coffee from the ground beans (the volume and way the water makes contact with the coffee grounds during brewing)
- How well the equipment is cleaned
- The holding temperature of the finished product (optimum 185°F, held no longer than 30 minutes)
- Whether the finished coffee is stirred for consistent strength

It's almost a science, and plenty of coffee aficionados know a good cup from a bad cup and won't hesitate to tell you about it. And here's one more important variable: the ratio of coffee grounds to water. One pound of coffee can be brewed with 1¾ to 3 gallons of water, with very different results. You should learn the strength of brew your customers expect.

Speaking of water, we've mentioned in other chapters the importance of filtering it. With coffee, this is critical. What you're selling is, in fact, mostly water—so you might as well treat it like the valuable resource it is. Minimize its mineral and chemical content, and anything that might create an off taste. You'll not only brew better coffee; you'll help your brewing equipment last longer.

Coffee drinks can enhance your image as a foodservice provider, so they offer plenty of merchandising opportunity. You can purchase green coffee beans and roast them yourself or whole beans to grind as needed, both adding an aromatic component to your dining area. Gourmet blends are another way to scent the air with hazelnut, peppermint, cinnamon, and other luscious flavors that may entice customers to buy.

For all this hoopla, *coffee brewers* are surprisingly simple machines. Let's dissect the brewing process:

Water is added to the machine, either manually or by a ¼-inch water line plumbed into the unit. For the latter, proximity to a water line and a floor drain is required.

Coffee is placed in a brew basket (or filter basket) in a filter that keeps the grounds from falling into the brewed coffee. The filter may be made of disposable, porous paper or reusable metal mesh. Cone-shaped filters allow slightly more contact between water and grounds (and therefore more flavor extraction) than filters that are flat on the bottom. Because the grounds swell during brewing, it is important that the total depth of your basket be able to handle about 50% more coffee grounds than you put into it. Most brew baskets are about 3 inches deep.

The water is heated and either allowed to drip straight through the coffee grounds or sprayed uniformly over the grounds through a spray plate full of small holes. The spray plate may be flat or rounded and is usually made of stainless steel. It should be wiped down regularly to prevent buildup of lime (from the water) and the natural oils from the coffee beans, as the holes can clog easily. The pattern of holes on the spray plate should ensure the water uniformly hits and wets the whole basket of grounds, not just a spot or two in the center.

The coffee emerges from a spout or faucet into a glass or metal carafe, which sits on a warming burner, or into an **airpot**, an insulated thermal container with a pump. The airpot can hold the liquid hot and fresh for up to an hour without flavor loss and allows customers to serve themselves. The chief advantage of an airpot is that it increases the usable life of the coffee. In a traditional, open-top coffee pot on a standard burner, you can't guarantee the flavor will be satisfactory for more than about half an hour before it starts to taste stale and burned.

Self-serve coffee is a staple in hotel lobbies, banquet setups, convenience stores, and more. Large airpots are often used for this purpose, or the 1½-gallon **satellite brewer** (see Illustration 17-13). Its output exceeds the standard 10- or 12-cup pot but is limited enough that the coffee doesn't sit around for long periods. The satellite brewer can be programmed to make smaller batches as needed, is compact (18 to 25 inches in width) to fit in small spaces, and can be connected to a plug-in docking station that gently heats to maintain correct coffee temperatures and shuts off automatically when optimum temperature is reached. This is a big improvement over the single-temperature burners on most coffee brewers that are either on or off. Today's digital satellite brewing systems can be programmed so customers (e.g., in hotel-lobby service situations) can't change the controls once they're set. Other improvements include the ability to control the extraction process by adjusting the way the hot water sprays or trickles over the ground coffee to match the blend or grind of the beans.

ILLUSTRATION 17-13 The satellite system is a combination coffee maker and docking station with sensors to hold the coffee at optimum temperature until it is ready to serve.

Courtesy of Bunn-O-Matic Corporation, Springfield, Illinois.

For individual tableside coffee service, small airpots work well, and so does the **French press.** In this stylish glass carafe, ground beans and water are combined and left to steep for a few minutes. Then the server (or the guest) holds the knob of a plunger that reaches inside the pot and presses the used grounds to the bottom of the pot before pouring out the finished brew.

Coffee makers range in size from the ½-gallon decanter to 80-gallon *coffee urns* for banquets and cafeteria settings. The most common coffee urn can brew 1½ to 3 gallons within 10 minutes. Inside, a metal spray plate sprays hot water across the coffee grounds. Temperatures can be set for different strengths of brews, and most urns give "half" and "full" settings to brew a partial urn or a full one. A control panel at the front of the urn has buttons for "brew" and "standby," and some have a light or alarm to indicate when the urn is almost empty.

Coffee urns (see Illustration 17-14) generally have three serving taps, two for coffee and a third for hot water, as well as a metal drip tray beneath the taps to catch spills. Unlike the 1 square foot needed for a coffee brewer, urns are almost 3 feet tall and at least 20 inches around. They use 240-volt three-phase electricity to operate efficiently.

Remember, the natural oil from coffee beans creates a film on coffee-making equipment. The machine works best if you wipe it down daily, including spray plates, heads and faucets, and carafes.

ILLUSTRATION 17-14 Coffee urns are the largest coffee makers, sometimes with several taps and capacities of up to 80 gallons.

Courtesy of Bunn-O-Matic Corporation, Springfield, Illinois.

The newest coffee-brewing technology allows even quick-service restaurants to brew single cups of coffee or espresso to order; Burger King and McDonald's were among the first to jump on the pod-brewing bandwagon. The coffee comes premeasured and packaged in pods or discs that fit neatly into the top of the machine. No fools, these manufacturers: The little packages are created to fit a particular machine. Fine-dining establishments also are embracing pod brewing because it's easy and speedy. The cost per cup is higher, but this may be offset by less labor and greater efficiency for waitstaff and baristas, the folks who make specialty coffee drinks at coffee bars.

ESPRESSO MACHINES

The name *espresso* (and no, it's not *ex*-presso) means "to press" or "to press out" in Italian. It was coined as a name for that country's strong coffee in small, made-to-order servings. Coffee beans are ground into a fine powder just before use—much finer than regular coffee grounds—and packed into a small filter basket. Hot water is pressed through the powder to create a strong, aromatic cup of coffee.

The name *cappuccino* is a bit more difficult to translate into English. This traditional Italian morning beverage is made simply from espresso coffee topped with milk steamed into a puffy foam. The foam often has a little peak on top, so the drink got its name from the peaked hoods worn by the Capuchin monks. (Use that trivia tidbit to impress someone next time you're in a coffee bar!)

The popularity of these specialty coffees and others has created a profitable market niche in restaurants and bars as well as a demand for the espresso machines used to make them and for experienced operators, who are called *baristas*, a term adapted from the Italian word for bartender. But until recently, the complexity of espresso-making equipment kept many food-service businesses out of the loop, unwilling to learn all the tricks, train the baristas, and figure out where to put the equipment.

The basic espresso maker is actually two small electric machines on the same base: a coffee grinder and a pump that adds water to the grounds to brew the coffee (see Illustration 17-15). Machines have several heads, which determines how many cups, or shots, can be brewed at once. Be sure the cups you want to use fit under the heads, because some machines are designed only for the traditional small demitasse cups. Most also have a stainless-steel valve on one side of the machine that uses high-pressure steam to foam the milk needed for cappuccino and latte drinks.

There are four degrees of automation for espresso machines:

1. *Manual machines* do not have a motor. They operate with a lever driven by a hydraulic pump. The manual machine requires the most training to use because the barista grinds the beans, loads each individual filter basket, watches as the espresso fills each cup, and discards used grounds.

2. *Semiautomatic machines* brew with the touch of a button, but the barista must stand and wait (up to 30 seconds) for the shot to be completed, then manually stop the action by turning off the pump. There are no controls to adjust the strength of the brew, so the barista is responsible for loading the correct amount of coffee grounds into the basket, then waiting, dumping out grounds between uses, and so on.

3. *Automatic machines* start by pushbutton and stop themselves when the shot is finished. Some machines sense the cup size and stop before it overflows. The barista grinds and loads the coffee and cleans the machine.

4. *Superautomatic machines* are the only ones that grind the coffee, load it into the filter baskets, brew, and steam or froth milk, all automatically. These models also discard the used grounds and sterilize machine components to stand ready for the next brewing cycle.

Then again, how much automation can you afford? The superautomatic machines range in price from $6,000 (a small-volume unit, for up to 50 cups a day) to $20,000 for high-volume coffee bar use. This is a significant investment, but industry figures indicate an 85% gross profit return on specialty coffee drinks, and the more automated the machine, the more labor savings you realize. These machines can also be leased, which is a good idea if you're not sure how well your employees will adapt to using them.

Another major consideration is how quickly you want to be able to make the drinks. Brewing time per shot can range from 20 seconds to 1 minute. An efficient barista can make about 70 cups of espresso per hour with each head on the machine, but latte or cappuccino drinks take longer. Some machines are capable of fewer than 100 servings per hour; others, up to 350 servings per hour.

Italian-made machines are known for their all-metal brewing chambers, which use hotter water and produce a darker brew.

ILLUSTRATION 17-15   Espresso machines create the water pressure, temperature, and steam necessary to make these strong coffee drinks.

Courtesy of Franke Coffee Systems Americas, Smyrna, Tennessee.

Swiss-made machines most often use a plastic brewing chamber and produce a slightly lighter espresso. Look for both Underwriters Laboratories (UL) and NSF International listings, and quiz the seller to be sure local repair and maintenance service is available in your area. To give you an idea of the complex engineering involved in espresso making, consider that steam is produced inside the unit at 220°F, while milk (inside the same unit) must be held below 39°F to meet NSF International safety standards—all within a space 2 feet wide or less! The control panel features from 6 to 16 buttons; the manufacturer can help you program your system for custom drinks or to specific taste preferences, with one of the control panel buttons set for decaffeinated coffee. Computerization of some machines allows the operator to track numbers of drinks or sales information by server.

Espresso machines require 208- to 220-volt electricity, a water line, and permanent drain. They can take up a lot of space, too, up to 4 feet in length and at least 2 feet in depth and height. Used coffee grounds (the spent grounds are called *coffee cakes*) can be disposed of in a drawer at the bottom of the machine that must be emptied periodically, so you must have a trash can handy.

Proper cleaning is critical because the tiny holes on the spray plates become clogged easily. A water filtration system is essential, and the recommended daily cleaning includes putting a cleaning solution (sometimes in tablet form) into the boiler; the solution runs through the milk lines and into the frothing nozzle. The process takes from 2 to 10 minutes. Milk must be flushed from the system daily. The more computerized machines turn on a warning light when cleaning is necessary, and you should pay attention to it. Manufacturers say the number-one reason for service calls is lack of cleaning.

■ SUMMARY

In this chapter, you learned about the wide variety of kitchen equipment available to perform specialized preparation tasks. Perhaps the most popular appliance is the vertical mixer, with an electric motor and agitator arm(s) to take the muscle strain out of mixing and kneading. Mixer sizes are based on the capacity of their mixing bowl and the size of their motor; choose yours based on the kinds of products you'll be mixing.

Hand-held mixers are smaller, portable versions of the vertical mixer. Vertical cutter mixers (VCMs) let you cut, mix, and blend foods. Electric slicers can cut foods more thinly and accurately than a knife, improving portion control.

Food processors have interchangeable blades to slice and dice food into a variety of shapes and sizes. The most difficult food for any processor to handle is cheese, which slices best when it is still cold from the refrigerator. There are also food choppers, grinders, and blenders—their names indicate their role in the kitchen—but today's appliances are capable of performing numerous tasks depending on the attachments that can be used with their motors.

Commercial toasters are sturdier and faster than home models. They turn out more than 200 slices of toast per hour! There are pop-up models, pop-down models, and drawer models, where food is placed on a drawer that slides into a heated chamber. High-volume operations may use a continuous toaster, placing food on a conveyor belt with variable speeds for different degrees of doneness.

Food warmers are available in many types. Some are heat-producing lamps; others are wells or tables with heated compartments where food containers are placed. Heating is accomplished with heating elements beneath the food pans, using water and steam or dry heat. Strict health and safety requirements mandate minimum internal food temperatures—not just surface temperatures—for cooked foods being held before serving, and generally, food that held for more than 2 hours should be discarded.

Finally, this chapter introduces coffee-making and espresso-making equipment. What you buy depends on the amounts you need to produce and the types of coffee drinks your establishment offers. There are many variables in good coffee brewing, including the grind (coarse, medium, fine) of the beans, the type of filter, and the design of the spray plate that sprays hot water over the grounds, but the most important factor is water quality. Filtered water is always preferable to minimize off-tastes.

Espresso makers range from manual to superautomatic. Your choice depends on how skilled your baristas are and how much work you want them to have to do. The espresso maker uses an internal boiler to heat water for brewing and to make steam to froth milk.

No matter which kind of machine you use or which type of container you use to store coffee, it must be kept clean. Coffee beans produce oil that creates a film buildup on equipment.

■ STUDY QUESTIONS

1. Why would you want to know about the PTO on a commercial mixer?
2. When would you need to use a bowl adapter?
3. Why should you always set the gauge plate of a slicer to zero when it's not in use?
4. What is the best way to grate cheese in a food processor?
5. What is the difference between a blender and a spindle blender?
6. How and why would you use a honing stone?
7. Why would you use a vertical conveyor toaster instead of a horizontal one?
8. What is the minimum temperature at which a food warmer should operate, and why?
9. Why should you keep foods covered while on a steam table?
10. Technically, what is the difference between a coffee maker and an espresso maker?

Glossary

Absorption ratio The percentage of water a particular type or amount of flour can hold as it is made into a dough.

Activated water Water to which a small electrical charge is applied in order to attract and suspend dirt particles and sanitize better than ordinary tap water. Unlike *electrolyzed water*, activated water does not contain salt.

Adaptive reuse Refurbishing or rehabilitating an old or historic building.

Aggregate A substance added to flooring material to make it more slip-resistant. It might be sand, clay, tiny bits of gravel or metal, carborundum chips, or silicon carbide.

Agitator kettle A special steam-jacketed kettle that includes an agitator or mixing arm for stirring while cooking; also called *mixer kettle*.

Agitators The beaters on a mixer.

Air door A curtain of air circulated across the opening of an oven instead of a solid door. The air curtain prevents heat from escaping. Also called an *air deck*, this feature is most often found on pizza ovens.

Airpot An insulated thermal container with a pump, used to store hot beverages such as coffee.

Air shutter A device in a gas rangetop burner that can be adjusted to control the amount of air that comes into the burner.

À la carte Food items cooked individually instead of in batches; also called *to-order*.

Allowable pressure drop The amount of pressure lost as steam is piped from its source to the equipment it powers or heats.

Alternating current (AC) An electric current that reverses its flow at regular intervals as it cycles. Alternating current can be transmitted over long distances with minimal loss of strength and without overheating wires and appliances.

Ambience A feeling or mood associated with a particular place or thing. (See also *atmosphere*.)

Amp An ampere; the term for how much electric current flows through a circuit.

Antimicrobial A property or substance that naturally retards bacterial growth, used in manufacturing items such as cutting boards, slicer blades, ice machines, and protective gloves.

Array In solar power applications, a module of solar photovoltaic cells grouped together to collect sunlight.

Ash dolly A metal tray on wheels used for storing and transporting ashes from a wood- or charcoal-burning fire.

Assembly-line flow See *straight line*.

Atmosphere The overall mood of a restaurant, determined by both practical and aesthetic elements such as lighting, artwork, noise levels, aromas, and temperature. (See also *ambience*.)

Atmospheric burner An open gas burner that uses air from the surrounding atmosphere to mix with the gas and ignite the flame. Refers to gas burners on ranges and in fryers and broilers.

Atmospheric vacuum breaker A small shutoff valve on a water line that allows it to drain completely after a faucet is shut off.

Augur (1) A flexible metal rod used to clean out plugged drains; also called a *snake*. (2) In an ice machine, a corkscrew-shaped tool used to shave ice off the interior walls of the machine.

Average check The average amount each guest spends for a meal at a restaurant.

Back bar The rear structure of the bar behind the bartender, where glasses, liquor, and other equipment are stored.

Backflow Water that flows back or returns to its source, usually because of pressure differences. Because backflow can contaminate fresh incoming water, a backflow preventer may be installed to avoid this problem.

Bain-marie A piece of equipment that holds heated or cooked foods and keeps them warm using steam or hot water.

Balanced system A series of ducts, hoods, and fans designed to work together to circulate fresh air, remove stale air, and prevent drafts in a building.

Ballast A device that limits electric current coming into a fluorescent light fixture and acts as a starting mechanism for that fixture. There are magnetic and electronic ballasts.

Ballast factor The ratio of the light output of a fixture with a particular ballast to the maximum potential output of that fixture under perfect conditions.

Banquet kitchen A separate kitchen facility specifically used for banquets or on-premises catering; also called *service kitchen*.

Banquette An upholstered couch or booth fixed to a wall, with a table placed in front of it.

Bar die The vertical front of a bar; it separates the customers from the back bar area.

Barista A person who specializes in making coffee drinks using an espresso machine, adapted from the Italian word for bartender.

Bar mixer A blender.

Base The area beneath a table that accommodates the table legs. The base should allow customers a comfortable amount of leg room. Popular options include the spider base (with four legs) and the cylindrical mushroom base.

Basket rack A place to hang multiple fryer baskets.

Batch cooking Preparing numerous servings of food at the same time; cooking in quantity.

Battery A solid line formed by fitting together heavy-duty rangetops in almost any combination of grate tops, hot tops, and griddles; also called *hot line*.

Beer box A refrigerated compartment designed to hold a keg of beer as part of a draft beer system; also called *tap box*.

Belly bar A front rail on a range that keeps cooks from touching the hot surface if they lean on the range as they work.

Benches Special stainless-steel tables for tabletop appliances with a shelf beneath.

Biofilm The technical term for slime caused by buildup of bacteria, fungi, and other microbes in moist areas.

Biomass The term for organic materials (from plants or animals) used to create a fuel source such as electricity, biodiesel, or ethanol.

Blast chiller A machine that looks like a refrigerator, available in reach-in and roll-in models, used for *blast chilling*.

Blast chilling A type of quick chilling that circulates cold air at high velocity around pans of food to cool it quickly.

Blower A fan that circulates air in a steamer or oven compartment.

Blower-dryer An electric or steam-heated compartment in a dishwasher, used to speed the dish-drying process.

Boiler A small steam generator that boils water to be made into steam.

Booster heater A heated tank that can increase ("boost") the temperature of water before it enters an appliance. Used in conjunction with a regular water heater when hotter temperatures are needed, as in dishwashing.

Bowl adapter An adapter used to fit large mixers with smaller bowls.

Bowl dolly A wheeled platform that allows a mixer bowl to be rolled instead of lifted; also called *bowl truck*.

Bowl guards Heavy wire shields that fit over mixing bowls to prevent injury to workers' hands while the mixer is running.

Brass rail The metal foot rest that runs the length of the bar, about 1 foot off the floor.

Brick oven One name for a wood-fired or wood-burning oven.

British thermal unit (Btu) A standard measurement of energy use: 1 Btu is the amount of heat needed to raise the temperature of 1 pound of water 1°F.

Broiler/griddle A dual-purpose cooking appliance that uses tubular burners for the griddle and infrared bulbs for the broiler.

Bubble diagram A drawing to help organize the layout of workstations or duties. A circle ("bubble") represents each station or duty; intersecting circles indicate related areas.

Building-related illness (BRI) A medical condition in which 20% or more of a building's occupants experience similar symptoms of discomfort that disappear when they leave the building; also called *sick building syndrome*.

Bulletin A document issued to a design or construction team that identifies a change to the original construction documents.

Burner ports The holes in a gas burner from which the flames emerge.

Burner tube The section of a gas burner where gas mixes with air and ignites; also called *mixer tube*.

Burn-in A procedure used with recently purchased cooking equipment to prepare it for regular use. The burners and/or ovens are turned on and heated to burn off any protective coating.

Burning speed The speed at which a flame shoots through the mixture of air and gas in a gas burner.

Burnisher A machine that polishes silver plate by removing some of its surface scratches.

Bus boxes The plastic bins used to clear soiled dishes from tables; also called *bus tubs*.

Bypass line A cafeteria layout in which foods are grouped into sections with indentations of space between them. Customers can line up at the sections they want and bypass the others; also called *shopping center layout*.

Cable Several electrical wires bound together and insulated.

Cappuccino Espresso coffee topped with foamed milk.

Capture velocity The ability or strength of an exhaust hood to pull stale air up and away from the cooking line.

Carbon soot A black buildup that coats and clogs the parts of a gas burner when it is poorly adjusted.

Casters Wheels on racks, carts, or pieces of heavy equipment that allow them to be rolled from place to place.

Cathodic protector A metal device mounted on the inside of a steam generator, suspended in the water, which works to control corrosion.

Ceiling-type refrigeration A single fan mounted on the ceiling inside a refrigerator that circulates cold air.

CE Marking The logo affixed to equipment, toys, and medical devices sold in the nations of the European Economic Area, denoting that the product meets minimum health, safety, and environmental protection standards.

Charbroiling Cooking foods using flames, smoke, and radiant heat.

Cheesemelter A specialty type of broiler used primarily to melt cheese and heat prepared foods.

Chinese range See *Wok range*.

Chloramine A combination of chlorine and ammonia used in some municipal water systems as a disinfectant in the water treatment process.

Chlorocarbon A chemical byproduct of chlorine that is thought to be a carcinogen.

Chromaticity The warmth or coolness of light, measured in degrees Kelvin (°K).

Circuit The complete path of an electrical current, from its source through wiring and into an appliance or socket. A *closed circuit* is complete and working; an *open circuit* has been interrupted to stop the power flow.

Circuit breaker A switch that automatically trips into the "off" position when it begins to receive a load of electricity higher than its capacity.

Clamshell griddle A griddle with a heated hood that raises and lowers, cooking food on two sides at once. Electric clamshells are called *contact grills*.

Class K The type of fire extinguisher required in commercial kitchens. (The "K" stands for *kitchen*.)

Classroom-style seating A banquet table setup in which people sit on only one side of the table.

Clean-air vacuum A vacuum in which incoming dust is filtered into a dust bag before entering the motor.

Clean steam Pure steam, not contaminated by chemicals.

Cobra Multiple lines or hoses with a single dispensing head to serve soft drinks and water.

Cogeneration Using the same power source for two jobs. See also *heat recovery*.

Coil A piece of coiled copper or stainless-steel tubing found in appliances.

Cold plate A chilled metal sheet at the bottom of an ice bin; it helps chill the lines of an automatic dispensing system for liquids.

Cold zone In a fryer kettle, a space about 2 inches deep located between the heating elements and the bottom of the kettle, where crumbs and debris fall during the frying process.

Color rendering index (CRI) A numerical scale from 0 to 100 that indicates how bright a color appears based on how much light is shining on it.

Color temperatures A scientific comparison of colors based on their warmth or coolness, measured in degrees Kelvin (°K). Whether a light source appears warm or cool is its *correlated color temperature* (CCT).

Comal A broiler used primarily in restaurants for display cooking of fajitas.

Combi Short for "combination," this term refers to a multi-functional cooking unit such as an oven/steamer.

Combination steamer A steamer that can be adjusted to be either pressurized or pressureless.

Combined heat-power (CHP) A term for *heat recovery* or heat reclamation.

Combi-oven An appliance that combines three cooking modes in one oven: steam, circulation of hot air, and both steam and hot air.

Combi-wave A small oven that combines microwave and convection cooking techniques.

Combustion The burning of trash to minimize waste.

Comfort zone A combination of external temperature and humidity level that makes people comfortable.

Commercial Kitchen Ventilation (CKV) A term for the unique needs of a heating, ventilation, and air-conditioning (HVAC) system in foodservice; requires changing out the ambient air to remove heat, odors, grease, and moisture from work areas, primarily the hot line and dishwashing areas.

Compartment steamer Another term for a *pressure steamer*.

Composting The natural breakdown of organic materials over time that turns them into a soil enhancer useful in gardening and farming.

Compressor The part of an air-conditioning or refrigeration system that acts as a pump, which affects the temperature and pressure of refrigerant gas.

Concentrating Solar Power A method for increasing the efficiency of a solar power application by using mirrors to create a higher-intensity light source. Abbreviated *CSP*.

Concept The idea for a foodservice business. It encompasses menu, theme, décor, and other factors that create an image in the minds of customers.

Concept development The process of identifying, defining, and collecting ideas to create an image for a new business or product.

Condensation In a refrigeration system or steam-jacketed kettle, the steam that cools enough to become liquid again and must be drained or removed from the system.

Condenser The part of an air-conditioning or refrigeration system where the refrigerant releases its heat and becomes liquid again. A condenser may be filled with either air or water.

Conduction The simplest form of heat transfer, in which heat moves directly from one item to another.

Conduction shelves Shelves that contain a solution of water and antifreeze to cool food rapidly.

Conductor Another name for an electrical wire; a material or object that allows electricity to flow easily through it.

Conduit Wires encased in a rigid steel pipe.

Constant multiplier The number used by a water meter reader to multiply the readings from a meter for an accurate count of how much water was used.

Continuous cooker A rotisserie oven that contains a vertical, moving ladder with product on each rung.

Continuous toaster An appliance for high-volume toasting; food is fed into the heated cabinet on a conveyor belt; also called *conveyor toaster*.

Controlled evaporation cooker A type of fryer that uses both heat and steam to hold cooked food at its peak quality; also called *CVAP* or *collectramat*.

Convection Cooking with hot air that is circulated by fans in the oven cavity, providing excellent heat transfer.

Cook-and-hold oven A low-temperature roasting appliance that cooks foods slowly and then keeps them warm until serving.

Cook-chill The process of cooking food in quantity and then rapidly chilling it.

Cook-cool tank A tank used to cook large cuts of meat or other foods not suited to kettle cooking; also called *chiller-cook tank* or *cook-chill tank*.

Cooking island A freestanding combination of kitchen equipment that does not have to be placed next to a wall, allowing kitchen staff to stand on both sides of it as they work; also called *cooking suite*.

Cooking zone The area in a fryer kettle where hot oil circulates and frying takes place.

Cook-to-inventory Another term for the *cook-chill* food preparation method.

Cowl A hood-shaped cover fastened to the end of a conveyor dishwasher that is connected to an exhaust system. Makes the system more efficient by removing moisture quickly.

Cradle A device to transport beer kegs or other drum-shaped objects and hold them steady as they roll.

Critical control points (CCPs) Any place or procedure in a food-service operation where loss of control may result in a health risk.

Crumb tray Located at the bottom of a fryer kettle, this removable plate captures food particles until they can be discarded.

Curved unloader An addition to a dish room that helps push finished dish racks off a conveyor belt.

Cycling thermostat A temperature-monitoring device on fryers that regulates heat up to 400°F.

Damper A valve or plate that regulates airflow inside the ductwork of a heating, ventilation, and air-conditioning (HVAC) system.

Deck The bottom or floor of an oven cavity. In pizza ovens, called a *hearth*.

Deck oven An oven that has more than one cavity and a separate set of controls for each cavity; also called *stack oven*.

Decoying The scraping and stacking of dirty dishes by waitstaff or bus persons, to prepare them for washing.

Dehumidifier An appliance needed in some heating, ventilation, and air-conditioning (HVAC) systems to remove excess moisture from the air.

Demand charge An electric bill based on the highest amount of power used during its busiest (peak) period; also called *capacity charge*.

Demand meter A device that tracks the maximum amount of power used by a business during its busiest times, in a fixed number of minutes.

Demographics The characteristics of potential customers: age, income, education levels, and so on.

Desiccant A substance used as a drying agent in a heating and air-conditioning system to reduce the humidity of incoming air. Silica gel and titanium silicate are commonly used desiccants.

Design In a space plan, the definitions of sizes, shapes, styles, and decorations of the facility and furnishings.

Deuce A table that seats two people; also called *two-top*.

Diffuser Another term for a vent in a heating, ventilation, and air-conditioning (HVAC) system.

Digital meter A modern type of electric meter that has a digital readout instead of a series of dials to be read.

Direct current (DC) The type of electricity that has a constant flow of amps and volts. Not widely used in the United States, direct current cannot travel through power lines for long distances and sometimes can overheat appliances.

Direct lighting Light aimed at a specific place or object to accent it.

Discharge system The drainage system plumbed so that liquids from kitchen, sinks, and restrooms flow into it.

Dish table A metal table, often with shelves underneath or glass racks above. Used as landing space in a dish room.

Dishwashing The process by which glasses, dishes, and flatware are cleaned using hot water, detergent, and motion.

Display kitchen A kitchen where much of the cooking and food preparation takes place in view of the customers.

Dolly See *hand truck*.

Dome strainer A stopper for sinks, perforated to prevent bits of solids from going down the drain; also called *sediment bucket*.

Double broiler A broiler with two separate heat controls and cooking areas.

Downlighting Positioning a light fixture to provide direct lighting to the area directly beneath it.

Drawer warmer A heated cabinet with one or more drawers that slide open to load or remove food.

Drift The problem when an electronic or digital scale doesn't hold a steady readout when weighing an object—the numbers shown on the digital readout drift up or down. This may indicate a temperature problem or a source of static electricity that is interfering with the scale's balance.

Drip pan A removable tray that catches grease drippings and crumbs at the bottom of a broiler; also called *drip shield*.

Drop-in burner Modular cooking surfaces that come in configurations of one or two burners that drop in to the rangetop, where they are connected to a power source.

Dry steam Steam with a small amount of superheat in it.

Dry storage Storage for canned goods, foods, paper products, and other items that don't require refrigeration.

Dual voltage Large commercial appliances sometimes require more than one type of electric current: one for the motor and a more powerful voltage for the heating element, for example.

Duct-type refrigeration A system inside a refrigerator that circulates cold air between a forced-air unit on the ceiling of the interior cabinet and a series of small air ducts on the back wall of the cabinet.

Dump pan A pan into which food is unloaded after leaving a fryer.

Dunnage racks Racks for storage of bulky items, such as bags and cases, which would otherwise take up valuable shelf space.

Economizer An addition to a rooftop HVAC system that senses outdoor air temperature and opens dampers to allow comfortable outdoor air into an indoor space, thereby saving on mechanical cooling costs.

Efficacy The ability of a lighting system to convert electricity into light.

E-lamp A long-lasting bulb that uses a high-frequency radio signal instead of a filament to produce light.

Electric discharge lamp A fixture that generates light by passing an electric arc through a sealed space filled with a special mixture of gases. Sometimes called *gaseous discharge lamps*, these include mercury vapor, fluorescent, halide, and sodium lamps.

Electric service entrance   The physical point at which a building is connected to its source of electricity.

Electrolyzed water   A type of water used for sanitizing hands and food contact surfaces; created by applying an electrical current to a weak solution of water and salt, which produces a superacidic, sanitized water that contains powerful oxidants. It does not look, smell, or taste any different than tap water; also called *E-water* or *ozone water*.

Elliptical reflector (ER) lamp   A specially shaped bulb for recessed light fixtures.

Energy audit   An analysis of a building's utility use, with suggestions for saving energy and money, performed by a utility company or consultant. A basic audit is called a *walk-through*; a more detailed one is called an *analysis audit*.

Energy charge   The amount of money a business pays for the total amount of electricity it uses during a single billing cycle; also called *consumption charge*.

Energy-efficient rating (EER)   An estimate of how much energy a system (such as heating and air conditioning) or appliance uses.

Energy index   The total amount of energy used by a business, calculated by reducing each type of fuel to Btus (British thermal units) and adding them.

Environment   The conditions in a public business that make guests feel comfortable and welcome or that enhance a safe and productive work setting for employees.

Environmental psychology   A study of the reasons behind people's feelings about and impressions of the components of their surroundings.

Equipment specifications   Concise written statements about a piece of equipment that spell out exactly what is needed so potential sellers can supply what the customer wants. Nicknamed *specs* or *spec sheets*.

Ergonomics   The science of studying the characteristics of people and arranging their activities to be performed in the safest and most efficient manner.

Espresso   Strong coffee in small, brewed-to-order servings. The word means "to press" or "to press out" in Italian.

Evaporator   A series of coils in a refrigeration system, filled with liquid refrigerant. The refrigerant evaporates inside the coils, creating gas vapor.

Even-heat plate   Another term for a *uniform-heat top* or hot top on a commercial range.

Expediter   A person hired in busy kitchens to act as a liaison between kitchen staff and waitstaff, to facilitate ordering and delivery of finished plates of food; also called *wheel person* or *ticket person*.

Extinction pop   A type of flashback that occurs when a gas burner is turned off. It is a popping sound that sometimes accidentally extinguishes the pilot light flame.

Extraction   A method of kitchen grease removal in which exhaust air moves through a series of specially designed baffles, throwing off the grease particles by centrifugal force.

Fat filter   A filtering device inside a combi-oven that traps airborne grease and food odors. It can be removed for cleaning.

Fat turnover   In frying, the amount of fat (cooking oil) used each day compared to the total capacity of the fryer kettle.

Feasibility study   A written document, prepared in the planning stages of a new business, that compiles the research done to justify opening the business. (See also *financial feasibility study* and *market feasibility study*.)

Feed grip   The part of a slicer that holds the food steady as it is pushed toward the spinning blade.

Field convertible   Describes a dishwasher that can be adjusted to sanitize dishes using either chemicals or hot water.

Filler cap   A blender cover made of two pieces so one can be lifted for adding ingredients during blending.

Filtration system   The equipment used to clean debris out of frying oil.

Financial feasibility study   The written document that outlines income and expenses necessary to start a new business.

Finger panels   Metal panels perforated with small holes through which hot air is forced in an impinger/conveyor oven.

Finish   (1) The amount of polish or shiny surface on stainless-steel equipment. In equipment specifications, finish is measured on a scale from 1 (rough) to 7 (mirrorlike). (2) The surface of a piece of silver plate. Popular finishes, listed from most opaque to shiniest, are satin, butler, bright, and mirror.

Fins   In a refrigeration system, the metal plates that surround the copper coils of the evaporator.

Firing rate   The amount of power in a gas-fired steam generator, expressed in Btus (British thermal units).

Flaker   An ice machine that produces soft, snowlike ice beads used to keep foods or wine cold.

Flame lift   A potentially dangerous condition where some of the flames on a gas burner lift and drop without being manually adjusted.

Flame rollout   A serious gas burner fire hazard that occurs when flames shoot from the burner's combustion chamber instead of through the burner ports.

Flame spreader   The walls of a commercial gas oven that distribute heat evenly through the cavity.

Flame stability   A clear, blue-colored ring of gas flames with a firm center cone. This means the correct mixture of air and gas is being used.

Flashback   The roar or whoosh a gas burner emits as it is lit, which occurs because the burning speed is faster than the gas flow.

Flash steamer   A countertop steamer used to heat sandwiches and pastry products quickly without making them soggy.

Flat truck   A rolling cart mechanism used in storage and warehouse areas to transport cases of product.

Flatware   Utensils that guests use to eat. Often called *silverware*, even if they're not silver.

Flight-type dishwasher   A dish machine that uses a conveyor belt to move racks of dishes in a straight line through the wash, dry, and sanitize process. May also be called a *rackless conveyor* or *belt conveyor* depending on the features of the conveyor belt.

Floating flames Flames that don't assume a normal, conical shape, which indicates the need to adjust a gas burner.

Floating rod A type of broiler grate, chrome-plated and warp-proof; also called *free-floating grate*.

Floor sink A type of drain located in the floor.

Flow (or flow pattern) Like the traffic pattern of a neighborhood or intersection, this is the method and routes used to deliver food and beverages to customers.

Fluorescent lamp A long-lasting type of electric discharge lamp. The bulb is tube shaped and contains mercury vapor. The smallest bulbs are called *compact fluorescent lamps* (CFLs).

Fluoridosis Stained teeth as a result of fluoride exposure.

Food finisher A type of fast-cooking equipment used to reheat, or finish, foods that require quick browning, toasting, crisping, and so on before serving.

Food slicer An apparatus used in slicing meats and cheeses. Provides better portion control than a knife.

Food waste disposer Installed in kitchen sinks to grind and dispose of food waste; also known as *garbage disposal*.

Foot-candle A measurement of light; the light level of 1 lumen of light on 1 square foot of space.

Foot treadle A pedal or lever pressed with the foot to open a refrigerator door.

Free-floating grate See *floating rod*.

French press A type of coffee pot. A glass carafe where ground beans and water are combined and left to steep; the beans are then forced to the bottom of the pot (manually, with a plunger) when the coffee is ready to drink.

Fry powder A substance that can be added to frying oil to minimize oxidation and improve its useful life.

Fry station An area of a foodservice kitchen where fried foods (especially french fries) are prepared for serving. May include landing or storage space, fryers and filtering systems, food warmers, basket racks, and a bagging area.

Fryer basket A perforated wire mesh container that is filled with food to be lowered into the hot oil of a frying kettle.

Fryer cabinet The metal cabinet that houses a fryer kettle and, in some cases, its filtration system.

Fryer screen A perforated wire mesh panel that separates the cold zone at the bottom of a fryer kettle from the rest of the oil.

Frying kettle The receptacle in a fryer into which the oil is poured to be heated; also called *frying bin*.

Fry-top range Another name for a griddle with an oven installed below it.

Fuel adjustment charges Fluctuations in power costs that a utility is authorized to pass on to its customers; also called *energy cost adjustments*.

Fuse A load-limiting device that can automatically interrupt electricity in a circuit if an overload condition exists.

Gas-assist oven A wood-fired oven that is equipped to use natural gas to start its flames; also called *wood/gas combination*.

Gas connector The heavy-duty brass or stainless-steel tube used to attach a gas-powered appliance to a gas source.

Gauge A measurement of the thickness of metal. The lower the number, the thicker the metal.

Glare zone The angle at which most types of lighting cause reflective glare or are uncomfortable to look at.

Glassware The wide variety of containers used to serve water or beverages, such as mugs, tumblers, and stemware.

Glaze A protective coating sprayed on plateware during the manufacturing process; also called *miffle*.

Glide A self-leveling device on the leg mechanism of a banquet table.

Grade The three-digit number used to classify stainless steel by its chemical composition. Grades 304 and 420 are the ones most often used in foodservice.

Grain A measurement of water hardness; the amount of solids (calcium carbonate and other minerals commonly found in water) expressed in parts per million (ppm) and reported by the amount found in 1 gallon.

Grate top An open gas burner; also called *graduated heat top*.

Gravity-feed machine A soft-serve ice cream machine where the mix is loaded into a hopper and flows as needed into a cylinder below, where it is frozen, scraped out of the cylinder, and dispensed.

Grease guard Device installed at the base of an exhaust fan to catch grease so it does not settle on the roof of a building and create a fire hazard.

Grease interceptor An underground site where water and waste output drains. Its role is to prevent the grease and solids from entering the sewer system. It must be cleaned regularly by professionals; also called *grease trap*.

Grease tray A receptacle for catching grease runoff in the cooking process or for aiming it into a drip pan.

Grid assembly A metal grill on which food is placed to be broiled. It rolls in and out of the broiler smoothly; its proximity to the heat source can be adjusted.

Griddle A flat, heated cooking surface made of a thick steel plate.

Griddle stone A pumice stone used to scrape and clean a griddle; also called *pumice* or *brick*.

Grill A grid of metal bars on which food is placed; it is then set over a heat source for cooking.

Grille A cover on a vent that can be manually opened and closed to regulate airflow; also called *supply grille*.

Group relamping The practice of replacing all lamps in a fluorescent fixture at the same time.

Gun The dispensing head of a multiline hose (a cobra), which can dispense as many as eight beverages from a single head.

HACCP Stands for *Hazard Analysis of Critical Control Points*. A multiple-step method to identify and correct potential health and safety hazards in foodservice.

Hand truck A small sturdy platform on two wheels that is designed to support and move heavy items such as cases of canned goods; also called *dolly*.

Hard costs Costs associated with the tangible items—not the services performed—in the construction or renovation of a foodservice facility.

Hard-products slicer A blade for processing foods such as carrots, potatoes, and cheese.

Harmonics A distortion of an electric power frequency that can create interference on power circuits.

Head space The amount of room between the top rim of a steam-jacketed kettle and the food inside it.

Heat exchanger A device that takes steam from one source, circulates it through a series of coils to clean it, and pipes it to another appliance.

Heating plant The part of a heating, ventilation, and air-conditioning (HVAC) system where fuel is consumed and heat is produced.

Heating system The ductwork and vents of a heating, ventilation, and air-conditioning (HVAC) system that distribute and control heat.

Heat pipe exchanger Part of a heating, ventilation, and air-conditioning (HVAC) system that evaporates and condenses fluid in a continuous loop to take heat from supply air to return air streams, without the use of electrical power.

Heat pump water heater (HPWH) A tank that uses waste heat from an air-conditioning system to heat water.

Heat recovery The process of capturing heat vented from one appliance that would otherwise be wasted and piping or transferring it to another appliance for another use; also called *combined heat-power (CHP)*, *heat reclamation*, or *cogeneration*.

Heat recovery time The time it takes for hot frying oil to return to its optimum frying temperature after cold food is immersed in it to be fried.

Heat transfer fluids (HTF) A method of energy savings in gas-fired equipment by powering several pieces of equipment with a single gas burner, using a series of pipes and a heated fluid that runs through the pipes to the appliances.

Heat transfer rate In steam cooking, a complex formula that measures (in Btus [British thermal units]) how much energy it takes for heat to be transferred from one substance to another.

Heat zones Individually heated sections of a griddle, each with its own controls.

Heavy-duty range A range best suited for high-volume cooking.

High-density storage Storage that can increase the overall capacity of an area by 30% to 40%.

High-efficiency toilet A commode that uses less than 1.28 gallons per flush. Abbreviated *HET*.

High-limit thermostat A safety feature on fryers that detects overheating and turns off the fryer automatically.

Hob Another name for an electric induction range, or the individual burner on such a range.

Hollow-square system See *Scramble system*.

Hood (1) An exhaust canopy located directly over cooking equipment that draws smoke, heat, and moisture away from the kitchen. (2) A cover for a portable broiler to protect it when not in use.

Hot line The area in a kitchen where hot foods are produced; also called *battery*.

Hubbelite A sturdy, comfortable type of brick-red flooring made of cement, copper, limestone, magnesium, and a few other substances.

Humidifier An appliance that adds moisture to the air.

HVAC The abbreviation for a heating, ventilation, and air-conditioning system.

Hydrogenated Describes cooking oil that has been chemically combined with hydrogen molecules to make it more solid at room temperature and to increase its stability.

Ice bin An insulated storage container that holds ice produced by an ice maker.

Ice builder A refrigerated machine that supplies ice water to a tumble chiller.

Ice cuber An ice-making machine.

Ice dispenser A refrigeration unit that makes ice and dispenses it with the push of a button or lever rather than requiring someone to scoop it from an ice bin.

Ice shock When food is cooked and then immediately plunged into ice water.

Illumination The effect achieved when light strikes a surface; also called *illuminance*.

Impinger/conveyor oven An oven that cooks by blasting high-velocity hot air onto food that moves through the cooking chamber on a conveyor belt.

Incandescent lamp A filament encased in a sealed glass bulb, commonly known as a *light bulb*.

Incineration See *combustion*.

Incomplete combustion A natural gas flame that is not using the correct mixture of air and gas. Flames are unstable and tipped with yellow.

Indirect lighting Lights that wash a space with light instead of focusing directly on one object or spot. Minimizes shadows and is considered flattering.

Indirect waste An inch or more of space between a water pipe and its corresponding drain that prevents contaminated water from flowing back into the freshwater supply.

Induction Process whereby heat is created when an electric current runs from the surface of the stove into a pot or pan, heating the food and not the pan. The induction process works only with ferrous metal, such as cast iron or stainless steel.

Infinite-heat knobs Heat controls that allow for small adjustments instead of just low, medium, and high settings.

Infrared Invisible rays that have a penetrating heating effect when used in cooking.

Ingredient room An area in some kitchens where everything needed for one recipe or task is organized before it is picked up and delivered to a specific workstation.

Inlet chiller A small tank that collects water that would normally be discharged from an ice maker and recirculates it.

The cold outgoing water chills fresh incoming water, allowing it to freeze more quickly for faster ice production.

Integral ballast lamps A lamp and ballast that are a single unit.

Inventory turnover rate The number of days raw materials and supplies are in inventory, from delivery to time of use.

Ionizer The electrically charged field in an air cleaner. Air is forced through this field and into a series of metal plates that collect impurities and airborne particles.

Jet The stream of gas that flows into a gas burner.

Kelvin (K) The measurement of color temperature (see *chromaticity*) in degrees. The lower the kelvins, the "warmer" the light source.

Kickplate A protective metal plate covering the lowest edge of a door that is frequently kicked open by waiters and kitchen employees; also called *scuff plate*.

Kilopascal (kPa) The metric air pressure measurement equivalent of pounds per square inch (psi). One kilopascal is equivalent to 1% of atmospheric pressure.

Kilovolt-ampere (kVA) The power it takes to propel 1000 volts of electricity through a power line.

Kilowatt-hour (kWh) The time it takes to use 1000 watts of electrical power.

Kilowatt-hour meter A device that tracks the number of kilowatt-hours of electricity used in a building.

Lamp lumen depreciation The gradual reduction in the light output of a bulb as it ages.

Landing space Work surfaces for setting things down.

Lavatory A hand sink installed in a restroom.

Layer platform Commonly called a *flat truck*, a surface on four wheels that is strong enough to hold 10 to 15 cases of canned goods or other products.

Layout The detailed arrangement, often on paper, of a building and its floor and/or counter space.

Lazy steam Pressurized steam in a cooking cavity that works its way through food and cooks it from the outside in.

Lens A clear plastic screen or covering that shields a fluorescent bulb to diffuse glare.

Life cycle The anticipated service life of a piece of equipment or furnishing.

Life-cycle costing A method for determining the true cost of an appliance by taking into account its power and water use, durability or expected useful life, ease of use and training, and potential repair and/or service costs; also called *total cost of ownership (TCO)*.

Light fixture The base and wiring that connects to an electric power source, and a reflective surface to direct the light of a bulb; also called *luminary*.

Light-emitting diode (LED) A type of lighting technology that uses a simple semiconductor, called a *diode*, to concentrate light. It emits brighter light than incandescent bulbs while consuming less energy and generating less heat.

Lighting transition zone An area between two types of lighting that gives people's eyes a moment to adjust to the change.

Light loss factors Conditions that contribute to a light fixture putting out less light than it is capable of.

Liquefied natural gas (LNG) Gas that has been highly compressed under very cold conditions for storage.

Liquefied petroleum gas Commonly used term is *LP gas*; also called *bottled gas* or *propane*.

Load calculation An expert estimate of the size of unit needed to heat and cool a space.

Load cell A metal bar or beam inside a digital or electronic scale that bends when an object is weighed on the scale. As the load cell bends, the strain gauges attached to it also bend, measuring the weight of the object by their resistance to the change.

Load factor The amount of power consumed, versus the maximum amount of power that could be consumed, over a stated period. Expressed as a percentage, this figure allows a business to determine its overall level of energy efficiency.

Load profile A summary of how much power a building needs and what times of day its power usage is highest. The load profile is determined by tracking the daily load factor over weeks or months to get an accurate picture of power requirements.

Louver A small shade that can be adjusted to eliminate glare at certain angles.

Lowerator A spring-loaded plate holder that can be temperature controlled to provide preheated plates.

LP gas Commonly used term for *liquefied petroleum gas*; also called *propane*.

Lumen A measurement of light output; the amount of light generated when 1-foot candle of light shines from one source.

Lumen dirt depreciation A reduction in a bulb's light output because dirt or grease has accumulated on the bulb.

Maché kitchen A kitchen design with a walk-up retail counter where food is ordered and then prepared in full view of the customer.

Market feasibility study A written document that examines the competition, potential customers, and trade area where a new business plans to locate. (See also *feasibility study* and *financial feasibility study*.)

Masonry base The concrete base on which a heavy-duty appliance may be installed.

Mechanical impact Breakage that occurs when a glass hits another object.

Mechanical oven An oven used in large-volume operations where the food is in motion as it cooks inside the oven compartment. Depending on the type of motion, this is also called a *revolving* or *rotary oven*.

Mechanical pot/pan washer A special type of dish machine designed to wash pots and pans.

Merchandising doors Full-length glass door panels on refrigeration units, often seen in supermarkets and convenience stores.

Mercury vapor lamp A type of bright light used for lighting streets and parking lots; also called *high-intensity discharge lamp*.

Metal halide lamp A variation of the mercury vapor lamp, with metallic halide gases inside to improve color and efficiency.

Metal sheath   A type of heating element in a toaster.

Metering station   The area of a kitchen where food that is being cook-chilled is sealed, weighed, and labeled; often part of the pump-fill station in the tumble-chilling process.

Methane   The most commonly used form of natural gas.

Mission statement   A written statement that expresses a business concept and image; used as a guideline for creating a business.

Mixing shelf   The surface on the bartender's side of a bar where he or she sets glasses and mixes drinks.

Mongolian barbecue   A type of griddle used in restaurants that serve Mongolian-style cooking.

Monolithic floor   A one-piece floor, poured all at once and made of epoxy, polyester, urethane, or cement.

Mullion-type refrigeration   An airflow system in a refrigerator that takes in air above the top food storage shelf of the refrigerated cabinet and discharges it below the bottom shelf; also called *back wall system*.

Multistage flash distillation   A method of sanitizing seawater by boiling and then condensing it.

Musculoskeletal disorders (MSD)   Chronic health conditions, such as carpal tunnel syndrome, back pain, and tendinitis, that may be caused or exacerbated by repetitive motion or unsafe work methods.

NAFEM Data Protocol (NDP)   A computer language and system developed by the North American Food Equipment Manufacturers (NAFEM) to be a standard for linking foodservice equipment to a central computer in order to communicate data.

Needle injection steamer   A steamer that contains a series of needles that can be arranged in different patterns to heat food. When food is placed on the needles, steam is released through them directly into the food.

Negative air pressure zone   A space in which more air is removed from the area than is piped into it.

Occupancy sensor   A controller device that can turn on light fixtures when it detects motion in a room and turn them off when the room is empty.

Ohm   A measurement of electrical resistance—that is, whether a substance is a good or poor conductor of electricity.

Oil take-up   The amount of oil absorbed into food products being fried.

Opacity   The degree to which smoke blocks the flow of light in an enclosed space.

Orifice   The hole in a gas burner through which the gas flows.

Overbroiler   A broiler with its heat source located above the food; also called *overhead* or *overfired broiler* or *hotel broiler*.

Overrun   The percentage of air forced into soft-serve ice cream by a soft-serve machine to create its soft, creamy consistency.

Oyster kettle   A small steam-jacketed kettle, so named because it is often used to cook oysters.

Ozone water   See *electrolyzed water*.

Packaged air-conditioning system   A self-contained heating, ventilation, and air-conditioning (HVAC) system in which heating and cooling functions use the same ductwork, dampers, and thermostat.

Pallet   A low, raised platform on which boxes or cases can be stacked and stored to keep them off the floor.

Pan glides   Adjustable, easy-to-slide drawers or racks for placing serving pans inside refrigerators; also called *pan slides*.

Pantry   A dry storage room; the cold-food preparation area.

Parallel flow   A pattern of food preparation in which two straight lines of workers perform their tasks. For instance, they may work back to back or facing each other. Parallel flow is handy where space is tight.

Pass-through refrigerator   A type of refrigerator with two sets of doors located opposite each other. Often placed between a kitchen and service area so the waitstaff doesn't have to enter the kitchen to get items out of the refrigerator.

Pass window   An opening between the kitchen and the dining area to pass food back and forth without the waitstaff entering the kitchen.

Pasteurize   The process of (at least partially) sterilizing or sanitizing a food or liquid by heating it to temperatures that kill harmful microorganisms without chemically altering the food or liquid itself.

Peak loading   Energy used at peak demand times, usually during the daytime and early evening. Reducing peak loading can reduce your power bill.

Peak shaving   Curtailing power use or shifting to another power source to avoid peak loading and the higher costs it entails.

Photocells   Light-sensitive timers that turn on outdoor lighting when the surroundings become dark.

Plate   (1) The surface of a griddle. (2) The blade on a food processor.

Plateware   Dishes used to serve food and hot beverages; also called *dinnerware* or *china*.

Polyvinyl chloride (PVC)   The most common substance used to make heavy-duty plastic piping.

Positive air pressure zone   An area to which more air is supplied than removed.

Postmix system   A carbonated beverage dispenser that operates using two chilled tubes (*lines*), one for syrup concentrate and one for carbonated water.

Pouring station   The central part of an underbar; the area used to add juices and carbonated beverages to drinks.

Power burners   Gas burners on a rangetop that can supply twice the heat of a conventional gas burner. Power burners use more fuel.

Power grid   A building's electrical system.

Power factor (PF)   A measurement of how efficiently an individual light fixture, or a whole business, uses power. Customers with low power factors may be charged more for their electricity.

Power sink   A motorized three-compartment sink that circulates water rapidly to wash pots and pans.

Premix system   A carbonated beverage dispenser in which the syrup and carbonated water are purchased already mixed, then propelled up a single line into the dispensing head.

Pre-prep   A workstation where food is first taken from receiving or storage to be prepared for delivery to the preparation areas of a kitchen. In pre-prep, crates of produce are opened, meats are cut into smaller pieces, and so on.

Preparation area   An area of a kitchen where food is prepared.

Pressure footprint   The diagram of a heating, ventilation, and air-conditioning (HVAC) system's positive and negative air pressure and direction of airflow.

Pressure fryer   A specialty fryer that cooks food with a combination of hot oil and steam.

Prewash option   In a commercial dishwasher, a separate chamber that blasts dirty dishes with high-velocity water to remove debris; also called *power prewash*.

Primary air   The rush of air that flows into a burner tube of a gas appliance along with the first jet of gas.

Prime costs   The major foodservice expenses: food, beverage, and labor.

Product flow   The pattern of moving food and supplies from their arrival at a receiving area to preparation sites and, finally, to the guests.

Production line   The area of the kitchen where the food is cooked, plated, and garnished prior to serving; also called *hot line*.

Proofer   A heated cabinet that permits proofing of breads; also called *proofing cabinet*.

Proofing   Gently warming dough to allow it to rise before baking.

Proprietary specifications   Naming a particular brand or type of equipment in a specification or estimate so the supplier cannot offer a substitute; also called *single-source specifications*.

Protective lighting   A light fixture or lamp specially coated to prevent bits of glass or chemical from flying out in case of breakage. These have multiple uses in kitchens and food storage areas and are now required in the United States by federal law.

Psi   The abbreviation for pounds per square inch, a measurement of steam pressure.

Pump-fill station   An area that contains the equipment to transfer cooked foods directly from a kettle through tubing and into plastic storage casings. Part of the cook-chill process.

Pyramiding   A method of stacking glasses for storage.

Quartz sheath   A type of heating element in a toaster.

Quick-disconnect coupling   A shutoff device that instantly stops the flow of gas to a gas-powered appliance. A safety feature of gas connections.

Raceway   The long, horizontal section of a utility distribution system. It contains the outlets that hook the system to each piece of cooking equipment.

Rack conveyor   A type of dishwasher in which racks of soiled dishes are placed on a conveyor belt that moves them through wash and rinse cycles.

Rack oven   An oven that is tall and thin enough to accommodate whole rolling racks of food for cooking.

Radiant heat   Heat transmitted by radiation instead of conduction or convection. The sun's rays are an example of radiant heat.

Rail   The recessed area on the top surface of a bar, nearest the bartender; also called *glass rail, drip rail,* or *spill trough.*

Range   The appliance commonly known as a *stove* in home use. In foodservice, the range usually consists of the range-top for surface cooking and the range oven for baking and roasting.

Ratchet clause   A utility company may require this minimum charge for electric power, based on a customer's highest energy use period.

Reach-in refrigerator   A refrigerator with a door that pulls open to allow retrieval of items from inside.

Reactive organic gases   Cooking odors.

Receiving area   The place where food and supplies are delivered to the building, often near the back door.

Recycling   The collection and separation of specific waste materials that can be processed and reused instead of being thrown away.

Reducing collar   A device in a dishwasher that restricts the pressure of spraying water to protect dishes.

Reflector   (1) A shiny, custom-fit surface that fits between a bulb and fixture and that mirrors light. (2) A stainless-steel or cast-iron piece in a broiler that absorbs and concentrates heat from an electric or gas burner.

Refrigerant   A substance that transforms from liquid to vapor and back again in the sealed tubing of a refrigeration system. It begins as a cold liquid, releasing its chill into the system to prompt refrigeration.

Refrigerated storage   Storage for items that must be kept chilled or frozen until used.

Refrigerated ton   A measurement of the cooling capacity of an air-conditioning system; one refrigerated ton is the equivalent of how much energy it takes to melt 1 ton of ice at 32°F in a 24-hour period.

Refrigeration circuit   The machinery and system that make heat removal possible from a refrigerated space.

Refrigeration cycle   The process of removing heat from a refrigerated space.

Registers   The dials on a gas meter.

Regulation valve   A valve used to decrease water pressure.

Reheater   A part of a heating, ventilation, and air-conditioning (HVAC) system that heats and dries air that would otherwise be too moist to use in the ducts.

Relative humidity   The percentage of moisture in a refrigerated space; it can affect the appearance and rate of deterioration of many foods.

Retarder   A temperature-controlled cabinet that stops the proofing process for dough that has risen sufficiently.

Rethermalization   Reheating.

Rethermalization cart   A specialty appliance that brings pre-plated cold food to proper serving temperature in less than 1 hour and can be rolled from place to place.

Return piping   A set of pipes in a steam system through which condensation is removed.

Reverberant sound   Noise that reflects or bounces off surfaces.

Reverse osmosis   A method of distilling seawater by pumping it through a permeable membrane to remove minerals and contaminants.

Riser   (1) A folding platform used to elevate people or tables, usually in a banquet setting. (2) The tall, vertical posts on each end of a utility distribution system. Risers are either primary (containing the major electrical and mechanical connections) or secondary (containing a smaller portion of these connections than the primary riser). (3) On stairs, the height of a single step.

Roll-in refrigerator   A refrigerator large enough for carts to be rolled in and out.

Rollout   Stale or contaminated air from above the cooking line that escapes into the rest of the kitchen. An exhaust system should prevent rollout.

Rotisserie   An oven that contains rows of metal spikes (called *spits*). Meat placed on the spits rotates and roasts slowly as warm, moist air fills the oven.

Rough-ins   The location and quantity of mechanical, electrical, and plumbing connection points in a building or room before final finishes are added to walls, floors, and ceilings.

R-value   A measurement of the amount of heat resistance in a building's insulation.

Salamander   A small broiler that can be placed on a shelf or counter. Often used as a plate warmer.

Saponification   The application of an alkaline mixture to burning cooking oil or fat that creates a soapy foam to put out a fire.

Satellite brewer   A 1.5-gallon coffee maker that is easy to transport and can be connected to a small docking station that keeps the coffee warm until serving.

Satin finish   A surface for stainless-steel worktables and refrigerator doors that reduces glare (and, therefore, users' eyestrain) because it is not shiny.

Scale   (1) A person's visual perception of the size of an object. (2) A device used to weigh items.

Scaling   The buildup of lime or other chemical residue from direct contact with water; also called *scales* or *lime buildup*.

Scope creep   Adding work to a project that was not included in the original contract and that the client might expect be performed without compensation.

Scope of work   The specifics of work to be done by a service provider as outlined in contracts or other documents.

Scoring   Creating burn marks on broiled food by placing it on a grate; also called *branding*.

Scramble system   A self-service flow pattern for cafeterias that allows customers to wander between displays or stations without standing in one line; also called *free flow* or *hollow-square system*.

Scrap basket   A perforated basket in a dish room sink or dishwasher, used for scraping food waste into or collecting it during the washing process.

Scrap collector   A perforated pot that sits in a sink and captures solid waste before it goes down the drain.

Scrap screen   A screen panel located inside a dish machine that collects food residue during the washing process.

Scraping area   Part of the dish room or pot sink washing area in which dishes or pots and pans are pre-scrubbed or scraped by hand to remove food waste scraps before insertion into a dish machine or pot washer.

Screed   A sturdy strip applied to a floor, usually at the entrance to a walk-in refrigerator, to level the floor and more easily allow carts to be wheeled in and out.

Scuff plate   See *kickplate*.

Seasoning   The process of cleaning, oiling, and heating a new pan to help lubricate and preserve its cooking surface.

Seat turnover   The number of times a seat in the dining area is occupied during a mealtime; also called *seat turn*.

Secondary air   The air in the immediate atmosphere that surrounds a gas flame.

Semi-open kitchen   A kitchen design in which only part of the space is open for public viewing.

Service system   The way food is delivered to the guests—by a waitstaff, at a pickup window, or the like.

Sheet pan   A flat, rectangular pan used for baking and food storage.

Short circuit   An unsafe electrical flow that has been interrupted to follow an unplanned path; in heating, ventilation, and air-conditioning (HVAC) terminology, an HVAC system that introduces fresh air through one duct in the exhaust hood; also called *short-cycle system*.

Side loader   A table that can be fastened to a dish machine and that allows the operator to load dishes in a small space.

Single-phase current   One electrical current that flows into an appliance. Single-phase power is usually required for small appliances.

Skittle   A type of tilting braising pan that does not require a steam boiler; also called *combi-pan*.

Sloping plate   A griddle plate that is tilted slightly to allow grease to drain into a large, removable pan.

Slurry   The mixture of solid waste and water created inside a waste pulper.

Smoke point   The temperature at which a frying oil or shortening begins to smoke and take on a sharp, burned smell.

Smoker cooker   A type of oven that cooks using smoke. Its energy source can be wood, charcoal, and/or electricity.

Snap-action thermostat   A temperature-regulating device that opens fully to permit maximum heating in appliances (such as deep-fat fryers) that require quick heat recovery.

Soft costs   In construction or renovation, the costs not directly attributed to materials and equipment. Soft costs might include the design services and permits for the project.

Solar tube   An aluminum tube installed in a ceiling to bring daylight into an interior room.

Source reduction Making less waste or reducing the toxicity of the waste that is produced.

Sparkle The pleasant glittering effect of a light fixture.

Specifier A person who writes an equipment specification.

Speed rail A place on a bar to store the bottles of liquor used most often to make drinks; also called *bottle wells*.

Spindle blender A small mixer that may have multiple heads. Used to make milkshakes and other blended drinks.

Spiral mixer A mixer that works gently to knead dough.

Splash guard A 3- to 6-inch vertical piece of metal on a griddle that helps control grease spatters and prevents food from falling behind the griddle; also called *fence*.

Splash zone A surface subject to routine spills, splashes, and contamination; also called *splash contact surface*.

Split basket A half-size basket for frying or cooking smaller portions.

Stack A vertical drainpipe in a plumbing system; also called *soil pipe*.

Standard The tap or faucet of a beer dispensing system.

Stationary-rack dishwasher Another name for a single-tank door-style dishwasher. Racks of dishes are loaded into the machine and sit still as they are washed instead of moving on a conveyor belt.

Steamer An appliance with one or more enclosed compartments that cooks food using steam.

Steam injector A method of shooting pressurized steam into an appliance to create heat.

Steam-jacketed kettle A large bowl within a bowl used for making soups, sauces, and stocks. Steam or water can be pumped into the space between the two bowls to cook, heat, or cool the food inside the kettle.

Steam table A table with openings to hold containers of cooked food over steam or hot water, to keep the food warm; also called *hot-food table*.

Steam-table pans Rectangular stainless-steel pans used for food storage and serving. They can be stacked, frozen, and heated, and come in a variety of standard sizes; also called *hotel pans, counter pans,* or *service pans*.

Steam trap A valve in a steam system that helps regulate the overall pressure of the system. There are two types of steam traps, the *inverted bucket* and the *thermodynamic disk*.

Steam tube oven A specialty oven, usually used for baking, that contains hollow tubes placed above and below the cooking cavity that fill with superheated steam to cook product.

Stemware Glassware that has a thin stem separating the foot of the glass from the bowl, such as a wine goblet.

Step rate A graduated schedule of costs for electrical power, depending on how many hours are used. The cost decreases as usage increases; also called *declining-block schedule*.

Stirling cycle technology A system that uses a sealed motor developed in the 1800s; today, associated with chlorofluorocarbon (CFC)–free coolant and solar power applications. (See also *thermoacoustics*.)

Stomper A hard plastic or wooden device used to push meat into a grinder; also called *pusher*.

Stone hearth oven A wood-burning or wood-fired oven.

Straight line A type of flow pattern in which materials or people move steadily from one process to another in a straight line. Describes kitchen workstations or cafeteria customer lines; also called *assembly-line flow*.

Strain gauge A measuring device attached to the hinges of a load cell (inside a digital or electronic scale) that bend when an object is weighed on the scale. As the load cell bends, the strain gauges also bend, measuring the weight of the object by their resistance to the change.

Strip curtains Heavy plastic strips hanging directly inside a walk-in refrigerator door that help keep cold air from seeping out when the door is open.

Sump A floor drain with a large strainer to trap solids before they reach the drainage system.

Super-cooker A multitask oven designed for cooking very quickly using two cooking technologies: microwave and convection, for instance, or microwave and heat impingement.

Superheated steam Water vapor heated to a temperature above the boiling point of water.

Sustainable construction Building structures or planning businesses with conservation of natural resources in mind; also called *green building*.

Sweep system A device that can be programmed to turn lights on or off sequentially as needed.

Tablet arm The small table attached to an individual chair for casual dining. This type of chair looks like a school desk.

Tankless water heater An appliance that saves energy by heating water only on demand rather than heating and storing it.

Tariff analysis An overview report of the billing, rates, and rules of a utility company, which should be reviewed and revised as these rates and rules change. A tariff analysis can be done for individual buildings based on their history of energy use.

Temperature shock An abrupt temperature change in a fryer kettle caused by dropping a block of cold, solid shortening into hot frying oil.

Tempered Describes a makeup air system that can either heat or cool incoming air.

Teppanyaki griddle An oversize griddle used in Japanese-style exhibition cooking, with customer seating around its perimeter.

Theme The characteristics or qualities that make up the mood of a dining experience; for example, a restaurant decorated to resemble a medieval village.

Thermal head space The upper space in the interior cavity of a wood-fired oven. The more thermal head space an oven has, the more heat stays inside the oven cavity instead of being vented away.

Thermal shock A quick, intense temperature change that creates enough stress on a glass to cause breakage.

Thermoacoustics The use of sound waves to cool foods. (See also *stirling cycle technology*.)

Thermocouple A pair of electrical wires joined at one end and attached at the other end to an appliance.

Thermoelectric control A safety feature on a gas pilot light appliance that heats two metal wires and creates a very low electric voltage to hold the gas valve open.

Thermosiphon A type of solar water heater that circulates water or antifreeze without the use of a heat pump.

Thermostat An automatic device for regulating temperature. On gas appliances, this device is a knob or dial called a *throttling control* or *modulating control*.

Three-phase current Three streams of electrical current that peak in a regular, steady pattern. Three-phase power is usually required for large, heavy-duty appliances.

Throat The opening at the top of a food waste disposer into which scraps are fed to be ground up.

Tilting braising pan A versatile cooking appliance made up of a flat griddle bottom with vertical sides and a tilting body.

Time-of-day option A billing option for electricity that allows a business to be charged for power use based on the time of day it is open, giving lower rates to businesses that use power at nonpeak times (nights, weekends).

Total cost of ownership (TCO) See *life-cycle costing*.

Total harmonic distortion (THD) ratings A measurement of how likely a ballast is to distort the power line by creating electromagnetic interference.

Traffic flow The movement of guests and employees through a building. The ideal flow minimizes backtracking and crossing paths for safety and efficiency.

Trap A curved section of pipe in a discharge system that retains some water.

Trough A recessed area that surrounds a griddle top on three sides; grease and food bits are scraped into it to be discarded; also called *gutter*.

Trunnion A device that allows heavy kitchen equipment to pivot for easy loading and unloading.

Tumble chiller A rotating drum full of cold water that gently kneads and chills packages of food as part of the cook-chill process.

Tumble chilling A method of cooling cooked food quickly by immersing the packaged food in cold liquid.

Tuning The precise adjustment of lighting.

Underbar The part of the bar directly beneath the top surface, where supplies are stored out of view of customers.

Underbroiler A broiler with its heat source located under the food being cooked; also called *underfired* or *open-hearth broiler*.

Underfoot air distribution A type of heating, ventilation, and air-conditioning (HVAC) configuration in which ducts are placed so that air can enter a room at floor level rather than near the ceiling. Abbreviated *UAD*.

Uniform charge A single rate for each hour of electricity used, found most often on residential power bills.

Uniform heat top Flat surface that distributes the heat over a larger area than the burner; also called hot top or *even-heat plate*.

Untempered Describes a makeup air system that moves air without the capability to heat or cool it.

Upfeed system The natural pressure of water inside piping that allows it to move through a series of water lines, even to upper floors of a building.

Utility cart A rolling shelf unit most often used in restaurants to transport dishes and cleaning supplies.

Utility distribution system A central location for all necessary kitchen utility services, connected directly to the exhaust canopy. Contains a single connection for each utility, plus emergency shutoffs and inspection panels.

Vacuum tumbler A motorized machine that gently kneads salt-based marinades into meat to coat it and seal in juices when it is cooked.

Vapor compression A popular type of commercial air conditioning in which a refrigerant gas is compressed into liquid form and then evaporates back to its gaseous state.

Vapor oven A type of combination oven/steamer that cooks food using low-temperature steam.

Ventilation System that provides fresh or properly treated air to a room and sends a corresponding amount of stale or used air out of the room.

Venting system The part of the wastewater discharge system that prevents water from being siphoned from the grease trap. A series of vents equalize pressure in the system and circulate enough air to reduce pipe corrosion and remove odors.

Ventless fryer A fryer with an internal air filtration system that does not require venting waste air to the outside and therefore can be used without being placed under a ventilation hood.

Vertical cutter mixer (VCM) A combination mixer and blender with attachments that can be used to mix or mash foods.

Vertical mixer A motorized appliance that mixes batters, dough, and the like; also called *planetary mixer*.

Volt The force that pushes an ampere through an electrical wire.

Wallsaver legs Chair legs that extend slightly beyond the back of a chair to prevent it from hitting the wall behind it.

Warewashing The dishwashing process, including collection of the soiled dishes, scraping, rinsing, washing, sanitizing, and drying them.

Waste audit form A means of analyzing how much waste a business is creating, with suggestions for ways to reduce the amount.

Waste heat Warm air given off by appliances or air conditioners that may be harnessed for use.

Waste pulper A machine that digests food waste and paper products, mincing and mixing them with water to create more compact waste for discard.

Watt The amount of power required for 1 amp of electricity to flow through a circuit at 1 volt of pressure.

Watt density The number of watts an appliance's heating element can produce per square inch; a measurement of its ability to provide uniform heating.

Wave guide A fan located in the top of a microwave oven cavity that propels the waves into the cavity.

Wave stirrers Devices in a microwave oven that distribute the waves throughout the cavity to ensure even heating.

Welding Joining two pieces of metal by heating them.

Wetting action The ability of a dishwashing detergent to soften water and loosen food particles from dish surfaces.

Wide-area vacuum A two-motor vacuum cleaner with a nozzle width of up to 36 inches for use in large areas.

Wok range A specialty range also called a *Chinese range* or *Oriental range*. The rings of its recessed, circular burners can be adjusted to accommodate large woks for stir-frying under very high heat conditions.

Work center An area in which workers perform a specific task, such as tossing salads or garnishing plates.

Work section An area of the kitchen that encompasses several activities, such as cooking or baking.

Zoned cooking In a conveyor oven or on a griddle, separate controls for different sections of the appliance.

Index